Biopolymers for Biomedical Applications

Scrivener Publishing
100 Cummings Center, Suite 541J
Beverly, MA 01915-6106

Publishers at Scrivener
Martin Scrivener (martin@scrivenerpublishing.com)
Phillip Carmical (pcarmical@scrivenerpublishing.com)

Biopolymers for Biomedical Applications

Edited by
Annu
School of Mechanical Engineering, Yeungnam University, South Korea

Scrivener Publishing

WILEY

This edition first published 2024 by John Wiley & Sons, Inc., 111 River Street, Hoboken, NJ 07030, USA and Scrivener Publishing LLC, 100 Cummings Center, Suite 541J, Beverly, MA 01915, USA

For more information about Scrivener publications please visit www.scrivenerpublishing.com.

Wiley Global Headquarters
111 River Street, Hoboken, NJ 07030, USA

For details of our global editorial offices, customer services, and more information about Wiley products visit us at www.wiley.com.

Library of Congress Cataloging-in-Publication Data

ISBN 978-1-119-86502-5

Cover image: Pixabay.Com
Cover design by Russell Richardson

Set in size of 11pt and Minion Pro by Manila Typesetting Company, Makati, Philippines

Printed in the USA

10 9 8 7 6 5 4 3 2 1

This book is dedicated to Time, my Mommy and Papa, and my dear Ashish.

Contents

Preface

In recent decades, significant progress has been made in polymer science for biomedical applications. The use of biopolymers specifically attracted the focus on the development of therapeutic polymeric systems. The exclusive features of biopolymers, such as biodegradability and biocompatibility make them highly sought after, and major research conducted with them has resulted in various therapeutic systems. However, until now only a few showed potential to be appropriate for human use. Furthermore, the development of artificial biopolymers to make modules for biomedical and pharmacological therapeutic systems is challenging.

This book especially focuses on the naturally available biopolymers with their use and application in the field of biomedical engineering. It elaborates on the specific biomedical applications of the most commonly used natural biopolymers. Natural biopolymers are thoroughly discussed individually, with attention paid to their exquisite biomedical applications.

Herein you will find 20 chapters penned by experts in the field of biopolymers and biomedical engineering. This book is an attempt to provide a complete resource that systematically discusses the most widely used biopolymers and their biomedical applications, and presents all important research and developments that have occurred in the field.

Each chapter covers a single biopolymer, its properties, and biomedical applications. The chapters are arranged systematically, with the most common biopolymers discussed early in the book to give more insight into the field. Further, a specific chapter is dedicated to the application of biopolymers for wound healing. Later, specific chapters are dedicated to the application of bioplastics and biopolymers for the development of medical devices and biosensors, respectively. Additionally, a chapter is dedicated to the application of biopolymers in the field of dentistry, with a special focus on their risk to human health. Keeping in mind recent advanced technologies, a chapter is dedicated solely to the latest progress of biopolymers in 3D and 4D printing for biomedical applications. The final chapter comprehensively explains the future perspectives of biopolymers in the biomedical field.

The book is a reference for scientists, research scholars, academicians, scientists, chemists, biologists, biotechnologists, polymer technologists, industrialists, health experts, and policymakers. To better comprehend the fundamental ideas of biopolymers and their biomedical applications, this book will also benefit undergraduate and graduate students of material science, polymer science, chemical engineering, and biotechnology. The book's editor and contributors are academic and commercial researchers, and scientists who are leaders in their fields of study.

The contributing authors of this book deserve a huge thank you for their incredibly passionate efforts, as do the Scrivener and Wiley publishing teams for their help.

Annu
December 2023

1

Introduction to Biopolymers

Gurleen Kaur, Rajinder Kaur and Sukhminderjit Kaur*

Department of Biotechnology, Chandigarh University, Mohali, Punjab, India

Abstract

Biopolymers have engrossed imperative position substantially in every recognizable field of research, manufacturing, and production sector such as pharmaceutical or therapeutic sector, edible industry, textile department, agronomics, cosmetics, and automotive industries due to their biodegradable, biocompatible, and superior resilience demeanor. Chitosan, cellulose, collagen, keratin, pectin, carrageenan, alginate, dextran, curdlan, and starch are the commonly utilized biopolymers by researchers in various fields of interest. Biopolymers can originate from natural sources and can be chemically obtained from biological elements or absolutely synthesized from microorganisms. On the basis of type of origin, polymer backbone, and repeating monomer units, biopolymers can be categorized into three different categories. Biopolymers have been irrefutably acknowledged and accepted in the biomedical region owing to its phenomenal usefulness in wound healing, tissue engineering, drug delivery, medical implants, and gene therapy. Chitosan, cellulose, dextran, alginate, pectin, curdlan, and starch are few commonly used biopolymers. This chapter discusses an introduction to biopolymers, its classification, commonly used biopolymers in different sectors, commercially available biopolymers, and biomedical applications of distinct biopolymers.

Keywords: Biopolymers, biomedical applications, wound healing, tissue engineering, medical implants, gene therapy, drug delivery

1.1 Introduction

Any material made up of long repeating monomer units is termed a polymer. Polymers have been extensively used in human being's routine life for

**Corresponding author*: sukhminderjit.uibt@cumail.in

Annu (ed.) Biopolymers for Biomedical Applications, (1–24) © 2024 Scrivener Publishing LLC

their use in vast applications. Among the polymers, biopolymers or the natural polymers have acquired unique consideration due to inexhaustibility, adaptability, biodegradability, and reasonability. Biopolymers are the larger structures organized from monomeric units synthesized typically from living beings such as microorganisms, herbs, shrubs, and animals with a covalent bond formation in their units [1]. The synthesis of biopolymers has achieved heavier interest in comparison to synthetic polymers like polyethylene, teflon, nylon, polyester, and many more due to distressing surrounding ecological troubles such as non-degradability, global heating, more expenses, burning of nonrenewable synthetic polymers, and unsustainability. Biopolymers are a magnificent substitute for synthetic and chemically synthesized polymers that are invariably being scrutinized to alleviate issues generated by the use of synthetic polymers [2]. One of the most valuable aspects owned by biopolymers is their easy degradation, which can be achieved with the extensive sort of disposal mechanisms such as decomposition in soil, landfill disposition, thermo-mechanical recycling, chemical recycling, microbial degradation, anaerobic digestion, and carbon dioxide neutral incineration [3]. A wide range of applications are associated with the use of biopolymers in the field of medical and pharmaceutical industry, agriculture sector, edible industry, automotive department, cosmetics, and textile industry [47, 51–54] (Figure 1.1).

Over the past few years, biopolymeric substances have enlivened the researchers for its utilization in the biomedical sector including drug delivery, tissue engineering, wound healing, gene therapy, and medical implantation due to their indispensable characteristic features. Biopolymers are being used as film, powders, and hydrogels in cardiac and liver tissue engineering, wound healing, and wound dressing because of its biodegradability and non-destructive behavior. Natural biopolymers such as alginate, chitosan, agar, collagen, casein, carrageenan, and starch have been extensively utilized in injectable and inhalation structures, drug delivery, and scaffold development [4, 5]. Biopolymers demonstrate multifaceted performance in different functionalities within the individual's body like balance the resilience and hydration of skin, development of tissues through cell embracing, further governing the demeanor of cells by sending chemical signals to them, and providing flexibility to the gastrointestinal tract and joints through lubrication [1]. The chapter elucidates about the basic knowledge of biopolymers, classification of biopolymers, preparation of biopolymers, characterization of biopolymers, commercial availability of biopolymers, and their applications in various fields.

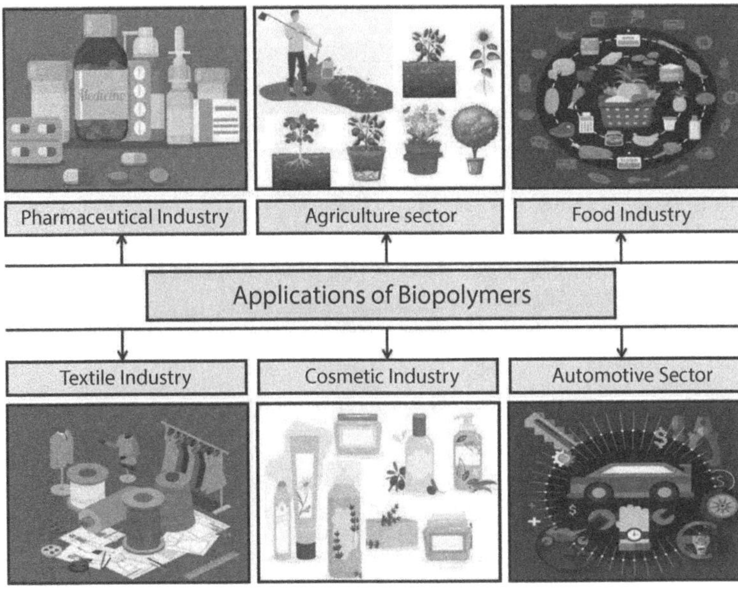

Figure 1.1 Applications of biopolymers.

1.2 Classification of Biopolymers

On the basis of type of origin, polymer backbone, and repeating monomer units, biopolymers can be categorized into three different categories as given in Figure 1.2.

1.3 Commonly Used Biopolymers

1.3.1 Chitosan

Chitosan is a derivative of chitin and one of the extensively considered polysaccharides obtained naturally. It is frequently derived from the cell wall of fungi and the firm outer covering of insects and crustaceans. Structurally, chitosan is formed by connecting glucosamine and N-acetyl-glucosamine *via* β-1,4 glycosidic linkages. Chitosan is widely employed in tissue engineering scaffold formation and wound dressing [9, 46, 49, 50].

1.3.2 Alginate

Alginate is a significant anionic unbranched linear polysaccharide also referred to as alginic acid. It is predominantly procured from the cell wall

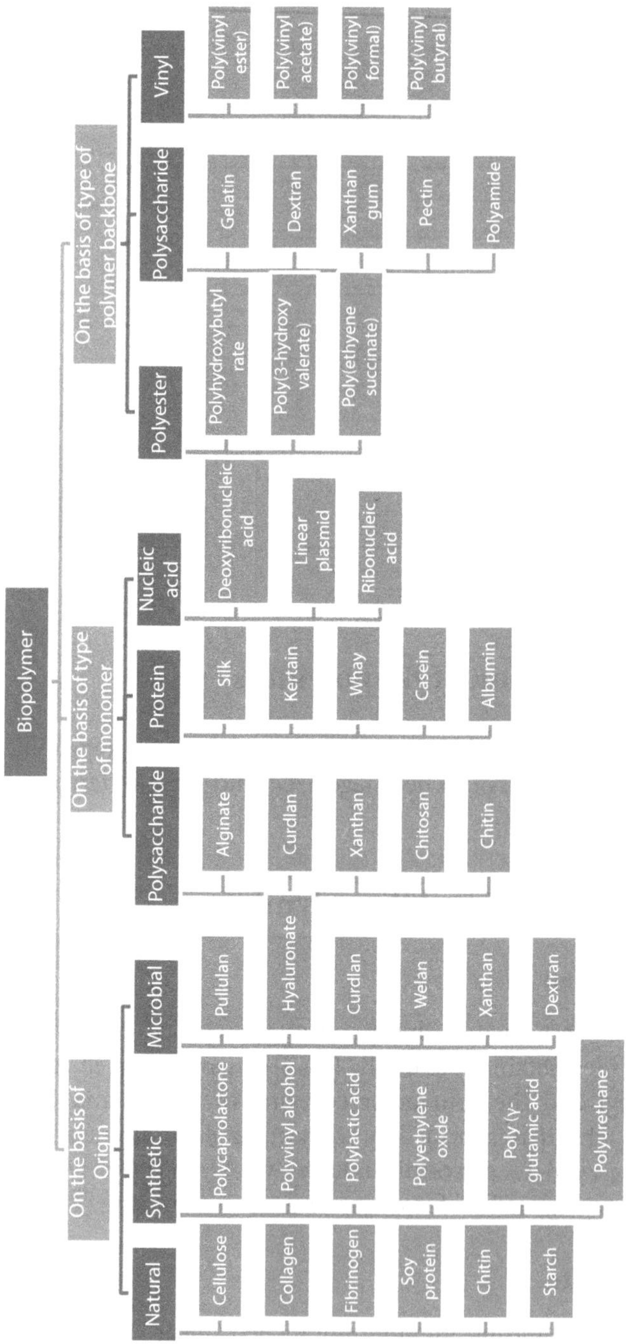

Figure 1.2 Classification of biopolymers [6–8].

of brown algae or seaweeds, especially *Laminaria* and *Ascophyllum* species. Alginate can also be extracted in extracellular form from certain bacteria. Alginate is biosynthesized through copolymerization of β-d-mannuronic acid and α-l-guluronic acid possessing diverse quantities of 1,4 linkages of both monomer acid residues and is able to bind to different molecules having compliant absorptivity, biodegradability, and a biologically functional state [9]. Alginate is principally efficacious for tissue engineering by supporting the repair of liver, cartilage, pancreas, periphery nerve, and blood vessels [10].

1.3.3 Cellulose

Cellulose is one of the extensively used existing biopolymers of glucose, allocated in plants as well as fibers like linen and cotton. *Acetobacter xylinum* is the bacterial species known to synthesize cellulose. Cellulose is made up of numerous hundreds to thousands of D-glucose monomer units linked by β(1→4) linear chains. Cellulose is an important biopolymer used in controlled drug delivery. Cellulose ether is utilized in solid tablets that permit the swelling-induced delivery of drug when anatomical fluid forms a communication with the tablet solely [11].

1.3.4 Starch

Starch is the fundamental energy reservoir plant polysaccharide which occurs in the granule form of amylopectin and amylose. Amylopectin is branched glucose polymer with high molecular weight, while amylose is a linear polymer made up of glucose monomers linked through α-D-(1-4) glycosidic bonds. The hydrogel of starch demonstrates resistance against gastric juices; therefore, it has applications in oral drug delivery and site-specific delivery systems. Starch is recognized for tissue engineering also because it assists in cartilage scaffold generation owing to its bone communicating characteristics [12].

1.3.5 Keratin

The protein keratin belongs to the class of polypeptide comprising different amino acids possessing disulphidecysteine intermolecular linkage. Keratin is highly useful in neural tissue regeneration. Keratin is classified on the basis of the presence of sulfur like soft and hard keratin [13]. Soft keratin contains lesser sulfur concentration, organized by cytoplasm filamentous clusters which are packed in unconstrained form, and offers flexibility to the epidermis, while hard keratin contains more sulfur concentration which supports epidermis stiffness comprising intermediate filaments organized in a systematic cluster entrenched in a cross-linked matrix. Two forms of keratin are known, including α-keratin and β-keratin. α-Keratin is present in hair, wool, and nails, whereas β-keratin is present in beaks and avian claws [13].

1.3.6 Carrageenan

Carrageenan is a general term assigned to the class of gelatinous and mucilaginous polysaccharide obtained from the class of *Rhodophyceaea* and red seaweed. Carrageenan is sulfated galactan having a repeated linear chain of d-galactose and 3,6-anhydro-d-galactose. Such polysaccharide possesses an antiviral aspect, by working as an interrupter of numerous enveloped viruses, antithrombotic potential through heparin co-factor II, and anticoagulant property. Kappa-carrageenan is scrutinized as a diabetic wound matrix and in regenerative medicines [9, 11].

1.3.7 Dextran

One of the complex glucans is dextran that is composed of a predominant chain of D-glucose associated through a α-(1,6) link with likely branches of D-glucose of α-(1,2), α-(1,3), and α-(1,4) links. The application of dextran includes gene transfection and nano-scale drug carrier. The generation of dextran takes place by employing a fermentation process using lactic acid bacteria in a sucrose medium. Dextran is also produced through an enzymatic pathway as it permits the generation of dextran with particular features [14].

1.3.8 Curdlan

This exopolysaccharide is constituted from monomers of glucose joined through β-(1,3) glycosidic linkages, synthesized by yeast origin like *Saccharomyces cerevisiae*, fungal species like *Poria cocos*, *Aureobasidium pullulan*, and different bacteria sources such as *Paenibacillus*, *Alcaligenes*, and *Agrobacterium*. Since many years ago, curdlan has been employed as dietary fibers for the growth of beneficial microorganisms in the gut and in pharmaceutical industries. It has also been reported that curdlan has the potential to inhibit the propagation of malaria parasite (merozoites) and therefore regarded as beneficial for the termination of malaria [15, 48].

1.3.9 Collagen

Collagen is the basic structural component of vertebrates and is an exceedingly copious mammalian protein. Fibroblasts assist in the formation of collagen and normally emerge from reticulum cells. Collagen has more affinity, no cytotoxicity, less antigenicity, and more extensile strength, biocompatibility, and degradability. Collagen exists in amino acids with a ternary helical structure including glycine, proline, and hydroxyproline [9]. Collagen is well acknowledged for tissue-based material such as vascular prosthesis and prosthetic heart valves due to elevated life expectation with a reduced risk of infection in several patients [11].

1.3.10 Pectin

This structural heteropolysaccharide is accommodated within the cell wall of terrestrial plants. Economic pectin is entirely obtained from apple pomace and citrus peel. Pectin exclusively comprises D-galactouronic acid monomers joined together through α-1,4-glycosidic bond. Pectin is greatly employed in food industry, target drug delivery, wound restoration, and tissue engineering. Pectin is endowed with biocompatibility, anti-inflammation characteristics, and biodegradability and restricts COX-2 and iNOS enzymes owing to the existence of esters in combination with galactouronic acid [16].

1.4 Preparation of Biopolymers

Biopolymers can be developed by using various techniques and methods. Majority of biopolymers formerly prevail in an environment or are synthesized by microorganisms. When biopolymers are formed with the assistance of microorganisms, particular source of nourishment and optimum environmental conditions are obligatory. Biopolymers are created either by a monomer's chemical polymerization or through fermentation. Nearly all biopolymers are compatible and consistent, with negative antagonistic influence on the biological system. The technique for the production of biopolymers from a bacterial source is understood as a consequence of their deterrence system or as a repository compound [4].

In a literature study, researchers worked on the preparation of alginate biopolymer from four *Azotobacter* strains (*Azotobacter* MB1, MB2, MB3, and MB4) [17]. Alginate is an extracellular biopolymer which exists typically in specific brown seaweeds (*Turbinaria*, *Ascophyllum*, *Laminaria*,

Lessonia, and *Sargassum*) commonly referred to as alginophyter or is secreted from natural derivatives. *Azotobacter* and *Pseudomonas* are the bacterial species known to produce alginate. Even though the vital origin for the secretion of alginate is seaweeds, bacterial sources are investigated and accepted as superior candidates for alginate production in comparison to algal alginates [18]. As *Azotobacter* is a nitrogen-fixing bacteria, alginate synthesis increases with the inclusion of a nitrogen source. Bhoir *et al.* [17] used the centrifugation method for the extraction of alginate from rhizospheric *Azotobacter* strains. *Azotobacter* was grown on Ashby's mannitol broth for 48 h. After 48 h, the culture was suspended in saline and then centrifuged to remove the supernatant. Formalin and ethanol were added to the supernatant for 30 min. The supernatant was again centrifuged. The supernatant so obtained was discarded, thus obtaining the alginate pellet (Figure 1.3). Bhoir *et al.* (2020) also carried out a preparation of alginate using molasses. Initially, distilled water and ammonium sulfate were added to sugarcane bagasse and then heated. The solution was cooled and filtered using muslin cloth and further inoculated with *Azotobacter* strain and incubated. Aliquots were removed subsequently on each alternate day;

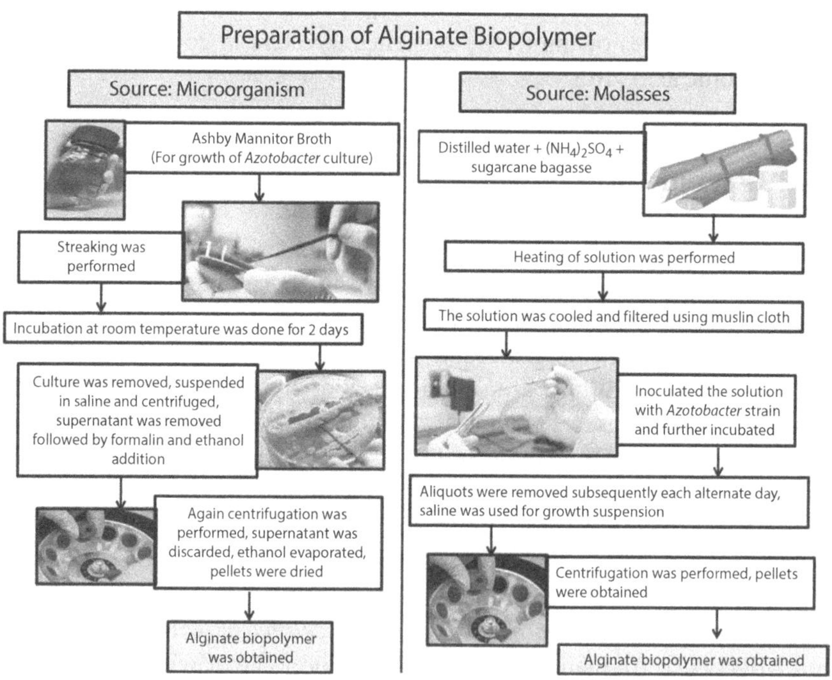

Figure 1.3 Preparation of alginate biopolymer from microorganisms and molasses.

saline was used for growth suspension, and centrifugation was performed. Finally, alginate pellets were obtained [17].

1.5 Commercially Available Biopolymers

Biopolymers find various industrial applications such as cosmetics, food packaging, pharmaceuticals, and medicine. Given below in Table 1.1 is the list of companies that provide quality-grade biopolymers to be used in various industries.

Table 1.1 Some commercially available biopolymers.

Company	Product	Biopolymer used	Application	References
Badische Anilin -und SodaFabrik (BASF)	Ecovio	Poly lactic acid	Biodegradable superior quality flexible bioplastic	[19]
Badische Anilin -und SodaFabrik (BASF)	Ecoflex	Cellulose, lignin, starch, polyhydroxyalkanoates	Durable and completely biodegradable bioplastic	[20]
Integra	NeuraGen	Collagen	Resorbable implant used for the restoration of peripheral nerve discontinuities	[21]
Integra	DuraGen	Collagen	One of the safest and highly efficacious dural grafts to repair and restore dura mater (outer membrane of brain and spinal cord)	[22]

(Continued)

Table 1.1 Some commercially available biopolymers. (*Continued*)

Company	Product	Biopolymer used	Application	References
Integra	SurgiMend	Collagen matrix	Acellular matrix developed principally for reconstructive and plastic surgery	[23]
Integra	Integra Bilayer Matrix wound dressing	Collagen, glycosaminoglycan and polysiloxane	Effective wound dresser, offer scaffold for capillary growth and cellular invasion	[24]
Integra	Integra Bovine Pericardium dural graft	Collagen	For the reconstruction of dura mater	[25]
Total Carbion	Luminy	Polylactic acid	Rigid food packaging	[26]
Anika	Orthovisc	Hyaluronic acid	Employed for the treatment of joint pain produced by osteoarthritis	[27]
Braskem	Braskem FL900PP-CF filament	Polypropylene	Used in automotive, nautical, aerospace and sports materials, provides low shrinkage, reduced dryness and resistant to chemicals and water	[28]

1.6 Biomedical Applications of Biopolymers

1.6.1 Drug Delivery

Naturally existing biopolymers, specifically the polysaccharide form, have been employed in the medicament field for the effective and target delivery of an extensive sort of analeptic agents. After cellulose, chitosan is the second most consistently existing substantial polysaccharide and is well endowed with biodegradable and biocompatible mucoadhesive property, which is why it is being considered for the formation of nanoparticles and microparticles. Such developed particles have been utilized as a promising carrier in the field of pharmaceutics for numerous therapeutic agents such as in medication, vaccines, and stimulants for parental and non-parental applications [29]. As in drug delivery, chitosan biopolymer has been examined as the most stable biologically active and pervious polymer.

The administration and formulation of a therapeutically functional compound for the objective of supporting an adept drug plasma proportion and to deliver the drug to the target site of action is referred to as drug delivery. Nanotechnology has been employed as one of the eminent and proficient techniques in the establishment of unique drug delivery mechanisms *via* encapsulating the particular drug in a nano- or micro-particle arrangement [29]. The inclusion of a pharmacological agent within the polysaccharide polymeric matrix may strengthen the stability of a bioactive component from deterioration, enhance absorption, target drug delivery, and improve the pharmacological response. Alginate, chitosan, and carrageenan are the predominantly consumed polysaccharides in numerous medicinal applications [30].

In a literature study, researchers worked on the formation and interpretation of biopolymer-based nanoparticles to enhance the disintegration of curcumin within a gastrointestinal environment. Curcumin is a polyphenol having pharmaceutical applications such as anti-tumor, antiviral, anti-inflammatory, and antioxidant characteristics. Ionotrophic gelation was used for the development of numerous hydrogel matrices based on sodium carrageenan as well as alginate comprising curcumin-loaded chitosan particles. The entrapment of such polyphenol in chitosan indicated a substantial increment of disintegration of curcumin in aqueous media. After incubation, approximately 95% of curcumin release was observed *in vitro* [31]. Carrageenan principally assisted in increased curcumin delivery from hydrogel matrices. Hence, chitosan, alginate, and carrageenan act as promising biopolymers for curcumin's improved aqueous-phase solubility.

In a recent study, Dutta *et al.* [32] reviewed the transdermal drug delivery aspect of xyloglycan to overcome acne remedy failure [32]. Xyloglycan is a biodegradable, mucoadhesive, and safe polysaccharide which is obtained from vascular plants' cell wall and is generously found in tamarind seeds. Xyloglycan is usually utilized as a stabilizer, thickener, and gelling agent. Xyloglycan was used as a gelling agent for the formation of clindamycin-loaded transdermal patch. The administration of the drug was in the form of patches *via* the transdermal system to prevent it from intestinal degradation. Xyloglycan demonstrated more bioavailability and greater drug release [32]. Biopolymers demonstrate excellent potential and effectiveness to be used in the field of drug delivery for the proficient and maximum release of a drug to the target site.

1.6.2 Tissue Engineering

Since the midst of the 19th century, scientists are working for the formation of biopolymeric materials to be used in field of pharmaceutics, especially as a restorative device in tissue engineering for the provisional insert of scaffolds. Tissue engineering requires different scaffolds such as in skeletal muscle, bone, neural tissues, cartilage, and vascular tissues. Currently, numerous leading technologies and approaches are being selected to make a porous scaffold for rejuvenating tissues or organs for a convenient and acceptable application in tissue engineering [10]. The progress of tissue engineering is based upon the correct choice of biocompatible polymeric nanomaterials or micro-materials for the matrices' build-up that would further demonstrate histocompatibility, superlative anatomical structure, cogent operational aspect, and mechanical features.

Biopolymers, due to their compatible features, are recognized as a substitute material to synthetic materials in the formation of scaffold. Biswal (2020) reviewed that collagen-based NeuraGen is greatly powerful for the restoration of peripheral nerve in approximately 43% of patients [10]. In a literature study, Perez-Guzman *et al.* [33] reviewed about zein as a promising biopolymer for tissue engineering by providing resistance to microbes, antibacterial properties, biodegradability, and compatibility [33]. Zein also worked as an eminent polymer in cardiac tissue engineering application with the presence of poly-glycerol sebacate [33]. Through electrospinning procedure, zein was used in research to form porous scaffold to repair serious skin damage. In a research, the combination of zein, gum arabic, and polycaprolactone was used to imitate a scaffold similar to the skin's naturally occurring cells and tissues [34]. For the development of a skin scaffold, gum arabic and zein acted as the polysaccharide and protein component, respectively, while polycaprolactone provided firmness and resilience to the scaffold. The results demonstrated the profitable fabrication of a biopolymeric nanofibrous scaffold with interrelated fibrous organization and showed antibacterial aspect against *Escherichia coli*. The hydrophobicity of the scaffold was improved with the presence of gum arabic. Successful proliferation and adherence to the formulated scaffold was illustrated by fibroblast cells. Greater hydrophobicity, suitable porosity, surface for attachment, growth, and proliferation of cells on the fabricated scaffold supported that biopolymers are the potential candidates to be used in tissue engineering application [34].

1.6.3 Wound Healing

The disorganization or disrupt of biological and cellular continuance of tissues is inflicted by wounds causing abnormal physiological functioning. The healing of wound is thereupon a convoluted and eminent anatomical response which aids in the refurbishment of tissue integrity for accurate body functioning. The wound healing mechanism comprises a profoundly incorporated cascade of uninterrupted and coinciding biological occurrence. The presence of inchoate cells and cellular disruption likewise exposes susceptible zones for microbiological infection. Synchronized accomplishment of an array of biological episodes and prevention against infecting microbes are therefore essential for hemostasis, tissue restoration, and maturation. The fundamental target in wound administration is to cure and repair the wound in a minimum possible time period and lessen pain with no uneasiness to the patient, whereas microbial infection and nutritive inefficiency will be the aftereffects of a slow healing procedure [35].

Presently, wound healing requires materials with inherent compatible and antimicrobial aspects such as polysaccharides. In a recent study, researchers worked on the cynoflan biopolymer as a useful wound healer which is produced by *Crocosphaera chwakensis* CCY0110, a unicellular marine cyanobacterium [36]. The researchers concluded that such naturally synthesized extracellular cynoflan biopolymer elucidated the ability to be employed for skin dressing because it manifests a compatible behavior and permits the proliferation and migration of fibroblast and endothelial cells. Furthermore, maximum (100%) viability of the cells was observed, without causing tissue death and apoptosis, thus promoting optimum conditions for cell adherence and proliferation that are important for wound healing. Moreover, *in vivo* studies illustrated that, for skin wound restoration, cynoflan absolutely adjust to the wound platform without causing a local or intrinsic inflammatory or oxidative reaction [36].

Another polysaccharide—chitosan—has the potential to communicate with epithelial cells in conjugation with mucus, ultimately assisting in the rigid intracellular junction's opening and therefore improving the permissibility of the epithelium. Considering diabetic wounds, chitosan is regarded to heal the wounds by reconstructing destructed tissue, eminent to lessen the chances of limb amputation. Alginate is the anionic polysaccharide which is known to be useful in wound healing [9]. Dressing with alginate improves the formation of collagen in skin, therefore administering the major extensible component and aiding in the regulation of skin growth. Alginate has the potential to absorb water and body fluids to accordingly develop a hydrophilic gel, therefore creating a moist condition for the restoration of wound. It has been reported that alginate dressing triggers monocytes to synthesize inflated TNF- and IL-6 cytokine levels [9]. Such biopolymers have proficiency and prove to be beneficial for the wound healing process (Figure 1.4).

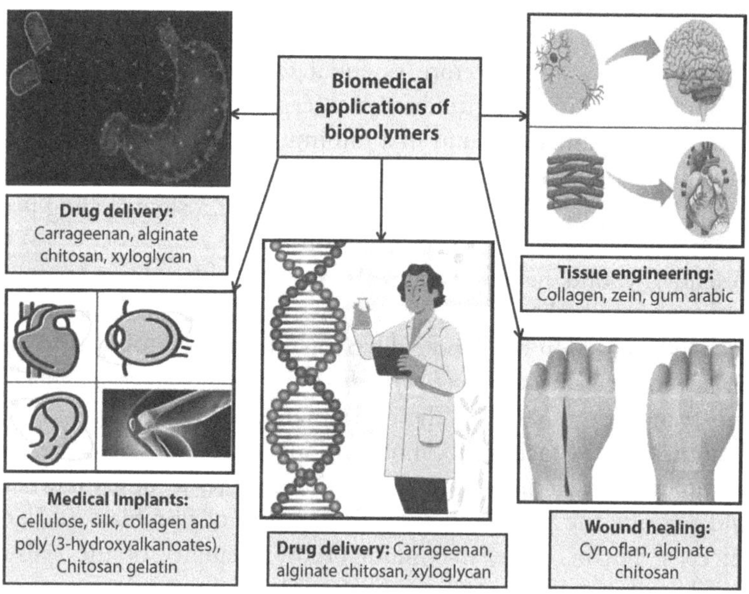

Figure 1.4 Biomedical applications of biopolymers.

1.6.4 Medical Implants

The principal purpose of an implantable device is to change the impaired structure or organ to maintain normal functioning of the body as it imitates the specific body part. The usually employed therapeutic implants comprises cardiovascular, knee, breast, cartilage, ear, and eye implants. Implants own adeptness to support an injured organ, inspect, restore, rejuvenate, and administer drugs and treatment to the target site. Ceramics, synthetic polymers, and metals were among the classic or conventional components employed as implants, but they exhibited immunological denial by the body as a serious disadvantage of traditional implants. Synthetic polymers demonstrate a serious issue of non-degradability within the body, and they follow hydrolysis for their breakdown in releasing carbon dioxide that further reduces the pH, thus causing tissue and cell mortality [37]. Therefore, biopolymers have gained much attention owing to their lesser density, biodegradability, inexpensive characteristic, and good mechanical, chemical, and thermal stability. Some of the biopolymers used in the development of implants are cellulose, silk, collagen, and poly(3-hydroxyalkanoates). Due to the biopolymer's similarity with the extracellular matrix, biocompatibility is demonstrated with no evidence of activation of a negative immune response [38]. Poly(3-hydroxyalkanoates) (PHAs) are employed for the formation of adherence obstructer, nerve and bone grafts, cardiac patches, and sutures. Actively maturing cardiovascular valve grafts are recreated with heart valve implantation as poly-lactic acid, decellularized extracellular matrix, poly-glycolic acid, and polycaprolactone support the differentiation, development, and cell growth. Poly(3-hydroxyoctanoate), a type of PHA, was used for the synthesis of trileaflet cardiac valve implant embedded with autologous cells. The valves illustrated appropriate functioning and minimum regurgitation [38]. Massive research has been conducted for the use of biopolymers to resolve the concerns associated with the field of ophthalmology, including the diverse parts of the eye such as corneal stroma retinal pigment epithelium, corneal endothelium, and limbal and corneal epithelium. Gelatin biopolymer is well known to restore ocular segments. Damage of the limbal epithelial stem cell causes suffering with pain, impaired vision, redness, and swelling, which are the major problems connected with limbal and corneal epithelium [38]. Limbal epithelial stem cell transplantation is the one and only method to regain normal vision. Chitosan and gelatin were employed for the construction of limbal epithelial stem cell substrata carrier. The activity was assessed by culturing corneal epithelial cells over distinct chitosan–gelatin membranes. As a result, biopolymers were reported with non-toxicity and greater cell proliferation features [39].

1.6.5 Gene Therapy

Cystic fibrosis is remarkably an existing and often fatal genetic disease in North American and European regions. Aberrant ion transit in the epithelium of numerous tissues occurs due to mutations in the *cystic fibrosis transmembrane conductance regulator* (CFTR) gene. As a consequence, unnaturally adhesive and dense mucus is generated, which obstructs and clogs the organs—predominantly the lung—subjecting it to cystic fibrosis pathology [40]. The dominant root of lethality in cystic fibrosis is pulmonary disorder as repetitive bacterial infection and constant inflammation happens; those further result in a dynamic damage to the lung tissues. For such a specific gene disease, the defect of the *cystic fibrosis transmembrane conductance regulator* gene requires an immediate and effective correction solution. Efficacious gene transfer structures require two segments consisting of a therapeutic nucleic acid and a transporter molecule which carries or bind to the nucleic acid. As of elevated immune response to vectors, the clinical techniques for cystic fibrosis genetic treatment have generally remained unsuccessful. However, due to a reported greater transfection proficiency, gene therapy aimed attention at viral vectors in cystic fibrosis therapy, although the application of viral vectors accelerates several issues with respect to biosafety, probable immune responses, and serious inflammation [41]. Hence, non-viral vectors stand out as an apparently secured alternative. The application of biocompatible and biologically degradable biopolymers like polylactic-co-glycolic acid and chitosan is evolving into more acceptable polymers for utilization in nanotechnology-based therapeutics for different diseases. In the literature studies, researchers worked on chitosan and polylactic-co-glycolic acid-based nanoparticles for cystic fibrosis lung gene therapy. The researchers generated the biopolymer-based CFTR-specific locked nucleic acid delivery system. The fabricated delivery method was developed in water, which was further nebulized such that it expedited a greater-performance aerosol dose transfer to the lungs, therefore maintaining biophysical characteristics with respect to particle size and indicating that such formulated biopolymer-based locked nucleic acid nanoparticles are advisable for cellular uptake and doubtlessly beneficial for clinical applications [40]. In another study, researchers evaluated and created an alendronate-formulated gelatin-based osteotropic nanocarrier for gene therapy studies. Bone-related disorders differ in their cause and behavior; however, a bone disorder's general aspect of characterization includes hydroxyapatite exposure and inflammation. Four component systems were developed, which consisted of DNA, gemini surfactant, gelatin biopolymer, and alendronate sodium trihydrate [42]. Alendronate sodium trihydrate-formulated gelatin assisted in the formation of the core of nanoparticles and targeted the bone cells (Table 1.2). The formulated gelatin was examined for its efficiency to target osteoblastic gene cells. The formulated gelatin biopolymer was optimized with various ratios of gemini surfactant and DNA so as to develop a perfect system for *in vitro* transfection of kidney embryonic cells (HEK-293) and osteoblastic cells (MG-63). The results reported that the alendronate-formulated gelatin biopolymer is regarded as an efficient and favorable bone-targeting biopolymer. The alendronate sodium trihydrate-formulated gelatin acted as a potential vector for the development of cationic nanocomplexes for increased gene delivery to bone tissues [42].

Table 1.2 Biomedical applications of biopolymers in different aspects.

Application	Biopolymer used	Mode of action	References
Drug delivery	Sodium carrageenan, alginate, and chitosan	Improved curcumin disintegration in gastrointestinal system	[31]
	Xyloglycan	Increment in bioavailability and more drug release to overcome acne treatment failure	[32]
	Ethylcellulose	Assisted in development of vaginal mucoadhesive bilayer film by working as precursor for antiviral drug controlled release	[43]
Tissue engineering	Collagen	Peripheral nerve restoration	[10]
	Zein	Important biopolymer for cardiovascular tissue engineering, provides resistance against microbes	[33]

(*Continued*)

Table 1.2 Biomedical applications of biopolymers in different aspects. (*Continued*)

Application	Biopolymer used	Mode of action	References
	Zein, gum arabic, and polycaprolactone	Mimics skin scaffold, imparts resilience and firmness to scaffold, successful adherence, and proliferation to scaffold	[34]
	Dextran	Possess water solubility, biodegradability, and biocompatible properties; does not induce immune response indicating its appropriateness for tissue scaffold formation	[44]
	Hyaluronic acid	Demonstrates anti-inflammation and non-immunogenicity features	[44]
	Chitosan	Imparts mechanical structure that imitates extra cellular matrix	[44]
Wound healing	Cynoflan	Assists in skin dressing owing to its characteristics of migration and proliferation of endothelial cell and fibroblast	[36]
	Chitosan	Useful for diabetic wound healing, reduces the limb amputation chances	[9]
	Alginate	Enhance collagen formation in skin, assist in skin growth regulation, and triggers monocytes	[9]

(*Continued*)

Table 1.2 Biomedical applications of biopolymers in different aspects. (*Continued*)

Application	Biopolymer used	Mode of action	References
	Keratin and fibrin	Reconstruction of tissues for skin wound dressing consequently indicated cell adhesion, proliferation, and cell viable properties	[45]
Medical implants	Poly(3-hydroxyoctanoate)	Development of cardiovascular valve implant	[38]
	Gelatin	Replacement of stroma	[38]
	Chitosan, heparin, and collagen	Development of bioartificial liver, showed adequate hepatocyte and blood compatibility	[38]
	Gelatin and chitosan	Aids in limbal epithelial stem cell implant, more cell proliferation	[39]
	Hyaluronic acid	Useful in ear and throat implant, formation of specialized tissues like synovial fluid and vocal folds	[38]
Gene therapy	Chitosan and polylactic-co-glycolic acid	Cystic fibrosis lung gene therapy	[40]
	Gelatin	Alendronate formulated gelatin-based osteotropic nanocarrier to cure bone disorder	[42]

1.7 Conclusion

The quest to work with biopolymers in different areas has expanded in order to diminish the utilization of non-biodegradable and non-compatible materials so as to decrease the ecological pollution and imbalance caused by the employment of synthetic materials. Biopolymers support multifarious benefits in the therapeutic and biomedical fields. The resources that can be utilized in wound repair, engineering of tissues, medical implants, target site delivery of drugs, and gene therapy ought to have a non-cytotoxic behavior, biodegradability, biocompatibility, should not induce immune response, more functional attribute, potent to assist cell proliferation and adhesion, adequate mechanical aspects, and sustain mechanical strength. Biopolymers are endowed with such characteristics and have the potential to successfully accomplish the requirements. Biopolymers have acquired abundant attention as a sustainable candidate in research studies as they have the efficiency of addressing ecological necessity and are suitable than synthetic materials. Naturally obtained biopolymers are copiously available, like alginate, chitosan, xanthan, collagen, and pectin, while certain biopolymers are synthesized using chemical processes. The preference for biopolymer relies upon their source, characteristics, and formation mechanisms to evenly exist without any complications in a specific field of application. However, the higher expense for the formation of biopolymers is obstruction if they are designed to take over traditional polymers. Therefore, in the future, additional studies and research must be conducted to generate inexpensive biopolymers that would accommodate the rigorous requirement of human well-being and ecological adverse effect studies. Future studies can also be planned to explore extraction approaches and development mechanisms that will balance the drawback of expensive production and assist in the commercialization of biopolymers with improved characteristics.

References

1. Baranwal, J., Barse, B., Fais, A., Delogu, G.L., Kumar, A., Biopolymer: A sustainable material for food and medical applications. *Polymers*, 14, 5, 983, 2022.
2. Thomas, S., Gopi, S., Amalraj, A. (Eds.), *Biopolymers and Their Industrial Applications: From Plant, Animal, and Marine Sources to Functional Products*, Elsevier, Amsterdam, Netherlands, 2020.

3. Hans, J.E. and Andrea, S.R., *Engineering biopolymers: Markets, manufacturing, properties and applications*, Hanser Publishers, Munich, 2011.
4. Mohan, S., Oluwafemi, O.S., Kalarikkal, N., Thomas, S., Songca, S.P., Biopolymers application in nanoscience and nanotechnology, in: *Recent Advances in Biopolymers*, vol. 1, pp. 47–66, 2016.
5. Biswas, M.C., Jony, B., Nandy, P.K., Chowdhury, R.A., Halder, S., Kumar, D., Ramakrishna, S., Hassan, M., Ahsan, M.A., Hoque, M.E., Imam, M.A., Recent advancement of biopolymers and their potential biomedical applications. *J. Polym. Environ.*, 30, 1–24, 2021.
6. Ibrahim, S., Riahi, O., Said, S.M. *et al.*, Biopolymers from crop plants, in: *Reference Module in Materials Science and Materials Engineering*, Elsevier, Amsterdam, Netherlands, 2019.
7. Moradali, M.F. and Rehm, B.H., Bacterial biopolymers: From pathogenesis to advanced materials. *Nat. Rev. Microbiol.*, 18, 195–210, 2020.
8. Reddy, M., Ponnamma, D., Choudhary, R., Sadasivuni, K.K., A comparative review of natural and synthetic biopolymer composite scaffolds. *Polymers*, 13, 1105, 2021.
9. Raina, N., Rani, R., Pahwa, R., Gupta, M., Biopolymers and treatment strategies for wound healing: An insight view. *Int. J. Polym. Mater. Polym. Biomater.*, 75, 1, 1–17, 2020.
10. Biswal, T., Biopolymers for tissue engineering applications: A review. *Mater. Today: Proc.*, 41, 2, 397–402, 2020.
11. Wang, Y., Wang, Z., Dong, Y., Collagen-based biomaterials for tissue engineering. *ACS Biomater. Sci. Eng.*, 9, 3, 1132–1150, 2023.
12. Song, R., Murphy, M., Li, C., Ting, K., Soo, C., Zheng, Z., Current development of biodegradable polymeric materials for biomedical applications. *Drug Des. Devel. Ther.*, 12, 3117, 2018, 2018.
13. Feroz, S., Muhammad, N., Ratnayake, J., Dias, G., Keratin-based materials for biomedical applications. *Bioact. Mater.*, 5, 496–509, 2020.
14. Diaz-Montes, E., Dextran: Sources, structures, and properties. *Polysaccharides*, 2, 554–565, 2021.
15. Chaudhari, V., Buttar, H.S., Bagwe-Parab, S., Tuli, H.S., Vora, A., Kaur, G., Therapeutic and industrial applications of curdlan with overview on its recent patents. *Front. Nutr.*, 8, 1–14, 2021.
16. Mogosanu, G.D. and Grumezescu, A.M., Natural and synthetic polymers for wounds and burns dressing. *Int. J. Pharm.*, 463, 127–136, 2014.
17. Bhoir, M., Ratnakaran, P., Durve Gupta, A., Extraction of alginate biopolymer from isolated *Azotobacter* species and its applications. *World J. Pharm. Res.*, 9, 5, 1468–1480, 2020.
18. Bonartseva, G.A., Akulinaa, E.A., Myshkinaa, V.L., Voinovaa, V.V., Makhinaa, T.K., Bonartsev, A.P., Alginate biosynthesis by azotobacter bacteria. *Appl. Biochem. Microbiol.*, 53, 52–59, 2017.
19. https://plastics-rubber.basf.com/global/en/performance_polymers/products/ecovio.html. Accessed on 12 September 2022.

20. https://plastics-rubber.basf.com/global/en/performance_polymers/products/ecoflex.html. Accessed on 12 September 2022.
21. https://www.integralife.com/neuragen-nerve-guide/product/nerve-tendon-neuragen-nerve-guide. Accessed on 12 September 2022.
22. https://www.integralife.com/duragen-classic/product/dural-repair-grafts-duragen-classic. Accessed on 12 September 2022.
23. https://www.integralife.com/surgimend-prs-collagen-matrix/product/surgical-reconstruction-plastic-reconstructive-surgery-hospital-or-surgimend-prs-collagen-matrix. Accessed on 12 September 2022.
24. https://www.integralife.com/integra-bilayer-matrix-wound-dressing/product/wound-reconstruction-care-inpatient-acute-or-integra-bilayer-matrix-wound-dressing. Accessed on 12 September 2022.
25. https://www.integralife.com/integra-bovine-pericardium-suturable-graft/product/dural-repair-grafts-integra-bovine-pericardium-suturable-graft. Accessed on 12 September 2022.
26. https://www.total-corbion.com/applications-solutions/rigid-food-packaging/. Accessed on 12 September 2022.
27. https://www.anikatherapeutics.com/products/pain_management/orthovisc/. Accessed on 12 September 2022.
28. https://www.braskem.com/usa/3d-printing-products. Accessed on 12 September 2022.
29. Ahmed, T.A. and Aljaeid, B.M., Preparation, characterization, and potential application of chitosan, chitosan derivatives, and chitosan metal nanoparticles in pharmaceutical drug delivery. *Drug Des. Devel. Ther.*, 10, 483, 2016.
30. Grenha, A., Gomes, M.E., Rodrigues, M., Santo, V.E., Mano, J.F., Neves, N.M., Reis, R.L., Development of new chitosan/carrageenan nanoparticles for drug delivery applications. *J. Biomed. Mater. Res. A*, 92, 1265–1272, 2010.
31. Guzman-Villanueva, D., El-Sherbiny, I.M., Herrera-Ruiz, D., Smyth, H.D., Design and *in vitro* evaluation of a new nano-microparticulate system for enhanced aqueous-phase solubility of curcumin. *Biomed. Res. Int.*, 2013, 724763, 2013.
32. Dutta, P., Giri, S., Giri, T.K., Xyloglucan as green renewable biopolymer used in drug delivery and tissue engineering. *Int. J. Biol. Macromol.*, 160, 55–68, 2020.
33. Perez-Guzman, C.J. and Castro-Munoz, R., A review of zein as a potential biopolymer for tissue engineering and nanotechnological applications. *Processes*, 8, 1376, 2020.
34. Rad, Z.P., Mokhtari, J., Abbasi, M., Fabrication and characterization of PCL/zein/gum arabic electrospun nanocomposite scaffold for skin tissue engineering. *Mater. Sci. Eng. C*, 93, 356–366, 2018.
35. Ghosh Auddy, R., Abdullah, M.F., Das, S., Roy, P., Datta, S., Mukherjee, A., New guar biopolymer silver nanocomposites for wound healing applications. *Biomed. Res. Int.*, 2013, 912458, 2013.

36. Costa, R., Costa, L., Rodrigues, I., Meireles, C., Soares, R., Tamagnini, P., Mota, R., Biocompatibility of the biopolymer cyanoflan for applications in skin wound healing. *Mar. Drugs*, 19, 147, 2021.
37. Rebelo, R., Fernandes, M., Fangueiro, R., Biopolymers in medical implants: A brief review. *Proc. Eng.*, 200, 236–243, 2017.
38. Volatier, T., Schumacher, B., Meshko, B., Hadrian, K., Cursiefen, C., Notara, M., Short-term UVB irradiation leads to persistent DNA damage in limbal epithelial stem cells, partially reversed by DNA repairing enzymes. *Biology*, 12, 2, 265, 2023.
39. Tiwari, N., Kumar, D., Priyadarshani, A., Jain, G.K., Mittal, G., Kesharwani, P., Aggarwal, G., Recent progress in polymeric biomaterials and their potential applications in skin regeneration and wound care management. *J. Drug Deliv. Sci. Technol.*, 82, 104319, 2023.
40. Fernandez Fernandez, E., Santos-Carballal, B., De Santi, C., Ramsey, J.M., MacLoughlin, R., Cryan, S.A., Greene, C.M., Biopolymer-based nanoparticles for cystic fibrosis lung gene therapy studies. *Materials*, 11, 122, 2018.
41. Griesenbach, U. and Alton, E.W.F.W., Progress in gene and cell therapy for cystic fibrosis lung disease. *Curr. Pharm. Des.*, 18, 642–662, 2012.
42. Mekhail, G.M., Kamel, A.O., Awad, G.A., Mortada, N.D., Rodrigo, R.L., Spagnuolo, P.A., Wettig, S.D., Synthesis and evaluation of alendronate-modified gelatin biopolymer as a novel osteotropic nanocarrier for gene therapy. *Nanomedicine*, 11, 2251–2273, 2016.
43. Jummaat, F., Yahya, E.B., Khalil HPS, A., Adnan, A.S., Alqadhi, A.M., Abdullah, C.K., AK, A.S., Olaiya, N.G., Abdat, M., The role of biopolymer-based materials in obstetrics and gynecology applications: A review. *Polymers*, 13, 633, 2021.
44. Lynch, C.R., Kondiah, P.P., Choonara, Y.E., Advanced strategies for tissue engineering in regenerative medicine: A biofabrication and biopolymer perspective. *Molecules*, 26, 2518, 2021.
45. Sellappan, L.K., Anandhavelu, S., Doble, M., Perumal, G., Jeon, J.H., Vikraman, D., Kim, H.S., Biopolymer film fabrication for skin mimetic tissue regenerative wound dressing applications. *Int. J. Polym. Mater. Polym. Biomater.*, 1–12, 2020.
46. Annu, Ahmed, S., Nirala, R.K., Kumar, R., Ikram, S., Green synthesis of chitosan/nanosilver hybrid bionanocomposites with promising antimicrobial, antioxidant and anticervical cancer activity. *Polym. Polym. Compos.*, **29**, 9_suppl, S199–S210, 2021.
47. Annu, and Ahmed, S., 1 - Bionanocomposites: An overview, in: *Bionanocomposites in Tissue Engineering and Regenerative Medicine*, S. Ahmed, and Annu (Eds.), pp. 1–6, Woodhead Publishing, Sawston, United Kingdom, 2021.
48. Aisverya, S., Annu, Ali, A., Sudha, P.N., 28 - Pullulan-based bionanocomposites in tissue engineering and regenerative medicine, in: *Bionanocomposites*

in Tissue Engineering and Regenerative Medicine, S. Ahmed, and Annu (Eds.), pp. 533–547, Woodhead Publishing, Sawston, United Kingdom ,2021.
49. Annu, Ali, A., Ahmed, S., Eco-friendly natural extract loaded antioxidative chitosan/polyvinyl alcohol based active films for food packaging. *Heliyon*, 7, 3, e06550, 2021.
50. Annu, Ahmed, S., Ahmed, S., Ikram, S., Chitin and chitosan: History, composition and properties, in: *Chitosan*, pp. 1–24, John Wiley & Sons, Ltd, Hoboken, NJ, 2017.
51. Shekh, M.I., Annu, Ahmed, S., Chapter 1 - Biopolymers: An overview, in: *Advanced Applications of Biobased Materials*, S. Ahmed, and Annu (Eds.), pp. 3–21, Elsevier, 2023.
52. Annu, Ahmed, S., Ikram, S., Perspectives of chitosan and alginate membranes for biomedical applications, in: *Natural Polymers: Derivatives, Blends and Composites*, vol. 2, pp. 157–182, Nova Science Publishers, Incorporated, Hauppauge, New York, 2017.
53. Annu, Pandit, P., Maity, S., Bhattacharya, T., Shekh, M.I., Ahmed, S., Chapter 30 - Chitosan biobased materials in textile industry, in: *Advanced Applications of Biobased Materials*, S. Ahmed, and Annu (Eds.), pp. 717–735, Elsevier, Amsterdam, Netherlands, 2023.
54. Annu, Manzoor, K., Ahmad, S., Soundarajan, A., Ikram, S., Ahmed, S., Chapter 30 - Chitosan based nanomaterials for biomedical applications, in: *Handbook of Nanomaterials for Industrial Applications*, C. Mustansar Hussain, (Ed.), pp. 543–562, Elsevier, Amsterdam, Netherlands, 2018.

2

Biomedical Applications of Chitosan and Its Derivatives

Hitesh Malhotra[1*] **and Rupesh K. Gautam**[2†]

[1]*Guru Gobind Singh College of Pharmacy, Yamuna Nagar, Haryana, India*
[2]*Indore Institute of Pharmacy, IIST Campus, Rau, Indore, Madhya Pradesh, India*

Abstract
Chitosan is a naturally occurring polymer obtained from chitin after alkaline deacetylation. Chitosan gains importance nowadays in the food and pharmaceutical industry due to its non-toxic, biocompatible, and biodegradable nature. For the past few years, scientists focus on the potential uses of chitosan as well as other possible sources. Moreover, chitosan derivatives were also developed to broaden the therapeutic potential and biocompatibility. This chapter includes the application of chitosan and its derivatives in the medical field. Chitosan is widely employed for the preparation of hydrogels, sponges, films, and transdermal patches which can be used for biomedical purposes. Chitosan possesses a wide array of pharmaceutical applications such as anti-bacterial, anti-tumor, hemostatic, anti-hyperlipidemic, and wound healing. Chitosan is also used as a carrier for drugs and a vector for gene delivery and thus helps in developing a targeted system. Furthermore, due to the unique properties and characteristics, chitosan biopolymer can be used in green chemistry. Thus, the aim of the chapter is to highlight the biomedical application of chitosan and its derivatives with the possible mechanism.

Keywords: Biocompatible, biodegradable, bioimaging, chitin, chitosan, drug delivery, green chemistry, wound healing

2.1 Introduction

Chitosan is a naturally occurring polysaccharide biopolymer linked by β-(1–4)D-glucosamine and N-acetyl-D-glucosamine mainly obtained

**Corresponding author*: hiteshmalhotra03.hm@gmail.com
†*Corresponding author*: drrupeshgautam@gmail.com

Annu (ed.) Biopolymers for Biomedical Applications, (25–52) © 2024 Scrivener Publishing LLC

from shells of crustaceans as depicted in Figure 2.1. Chitosan is a biocompatible, non-toxic, and biodegradable molecule and thus has been widely used in pharmaceutical industries. Commercially available chitosan has an approximate molecular weight in between 3.8 and 20 kDa and contains 6.89% nitrogen; it is widely utilized as a complexing agent. Chitosan is commercially acceptable compared to chitin due to the less biocompatibility and solubility of chitin. Chitosan is preferred over cellulose, pectin, agar as well as dextran due to its basic nature. Moreover, some studies reveal that chitosan markedly reduces the plasma concentration of cholesterol as well as triglycerides and thus overcome dyslipidemia [1, 100]. Furthermore, studies also reveal that chitosan combats cardiac disorders by enhancing the excretion of bile acid and dietary lipids. As compared to cellulose, chitosan significantly complexes with lipid molecules in the body, and this property contributes to the molecular mass and the rate of deacetylation [2, 98]. The synthesis of chitin and chitosan is depicted in Figure 2.1.

Chitosan and its derivatives are used widely for their anti-microbial activity against many fungi, bacteria, and other pathogens as well as preservatives in the food industry. Chitosan is also used as an anti-microbial agent for the treatment of various ailments in plant biology [3]. The plant itself produces anti-microbial chitosan via a rate-limiting enzyme chitosanases. Chitosan shows more promising anti-microbial activity than chitin due to the presence of the amine group [100, 101]. Furthermore, it has been observed that, by enhancing the degree of polymerization,

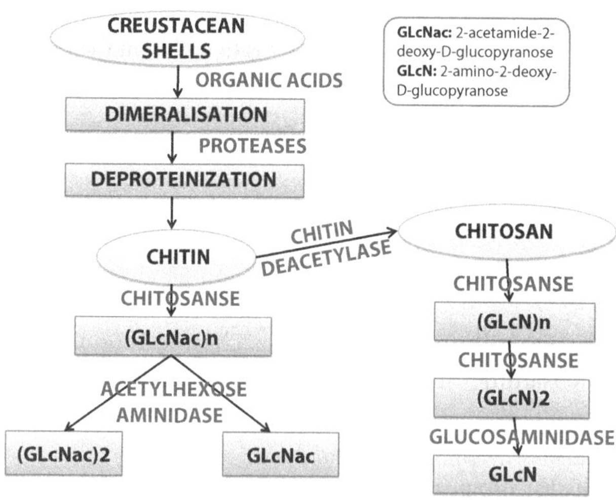

Figure 2.1 Synthesis and fate of chitin and chitosan.

the anti-microbial activity is also enhanced [4]. In addition to chitosan production, higher plants also produce phytoalexins, glycoprotein-rich hydroxyproline, reactive oxygen molecules, and beta-glucanase as a defensive mechanism. Pre-clinical and clinical studies also declare that chitosan and its derivatives can be used in humans as an anti-microbial therapy against pathogenic microbes [5, 107, 108]. Chitosan and its oligomer derivative produce non-specific immune stimulation against pathogens [6]. Certain studies also demonstrate that oligomers of chitosan modulate tumor growth by enhancing immune activity. Thus, by the enhanced production of antibodies and immune regulation, chitosan and its derivatives also possess anti-cancer activity.

Moreover, chitosan is also acknowledged as a drug carrier as it can increase the rate of drug absorption as well as targeting. Chitosan can be used as a gene carrier and also enhances the duration of gene expression. Therefore, chitosan is widely utilized for its carrier capabilities. Nowadays, chitosan is used as a carrier agent for anti-cancer as well as anti-tumor drugs [7, 90].

2.2 Characteristics of Chitosan

2.2.1 Biodegradability

Chitosan is a hydrophilic polymer with a molecular weight in the range that makes it acceptable for the carrier system. In addition, polymeric chitosan in the body degrades to form acetyl-glucosamine and glucosamine that are ultimately metabolized by the enzyme to the excretable form. The two important determinants of the rate of degradation are the rate of deacetylation and molecular weight [100]. Under *in vivo* conditions, chitosan and its derivatives are catalyzed either enzymatically or chemically—for instance, acid catalyzation, oxidation–reduction, depolymerization, and free radical reaction [8]. Chitosan in vertebrates is mainly hydrolyzed by lysozymes and colonel bacterial enzymes, and many targets for enzymes include glucosamine-N-acetyl-glucosamine, glucosamine-glucosamine, and N-acetyl-glucosamine-N-acetyl-glucosamine linkages [9, 10]. To date, around 18 isoforms of chitinase have been identified. In microorganisms, chitinases mainly target N-acetyl-β-1,4-glucosaminide linkages rather than other linkages. Therefore, the rate and degree of biodegradation of chitosan or chitin depend upon the amount of deacetylation [11]. For commercial purposes, chitosan is isolated from various bacteria like

Saccharomyces cerevisiae and *Oreochromis niloticus*, which have optimum pH of 2.5 and 9.0, respectively [12].

2.2.2 Toxicity of Chitosan

Studies demonstrated that chitosan is a non-toxic molecule that is a suitable agent for biological polymers. Chitosan is recommended for its nutritional value by FDA and various European countries [13, 14], but only certain derivatives of chitosan can be recommended after a crude analysis concerning its toxicity, residual activity, and pharmacokinetic and pharmacodynamic profile—for instance, in chitosan-based DNA nanoparticles, the pharmacokinetics will be altered due to modulation in shape and charge. Moreover, cellular entry can also be changed due to the presence of charge such as in the case of DNA complexes, where the appearance of a positive charge will lead to a reduction in cellular uptake due to alteration in cell–cell communication and thus reduced toxicity. Similarly, in covalent drug conjugation, the physicochemical properties like hydrophilicity and shape of the polymer influence the distribution as well as cellular uptake [15]. It is reported that, by altering the molecular weight and rate of de-acetylation, the Caco-2 cells, HT29-H, produce a different activity [16]. The toxicity directly links to the molecular mass as well as concentration in the case where chitosan is highly deacetylated, while if the degree of deacetylation is less, toxicity mildly links with molecular weight. However, chitosan also failed to increase the dehydrogenase activity at 1–500 µg/ml concentration on Caco-2 cells. Furthermore, studies on rat jejunum depicted no increase in LDH activity at a concentration of chitosan [17]. A clinical study conducted for 65 days demonstrated that no significant alteration had been seen in the body index profile [18], but the administration of 50 mg/kg of chitosan intravenously causes death due to platelet aggregation.

2.2.3 Biocompatibility of Chitosan

The term biocompatible means the well-matched property of chitosan with living tissue so that no toxicity, physiological reaction, or immune reaction occurs. Conventionally, studies reported that an undesirable reaction happens when some biomaterial is used for the carrier system. Thus, presently, to avoid a bio-material–tissue interaction, bioassays will be carried out either *in vivo* or *in vitro*. Biocompatibility studies classified the biomaterial into two categories, i.e., reactive and inert. The reactive substances induce severe undesirable immune reactions that may cause damage to our genetic material. Presently, several studies on humans elaborate the role of chitosan, but very few studies depict its biocompatibility [19, 101].

Chitosan is polycationic and interacts with various negatively charged molecules like anionic carbohydrates, amino acids, fatty acids, and bile acids. Thus, it is crucial to check the compatibility of derivatives before their use in therapeutics due to alteration in half-life and degradation in the body. The gradual degradation may lead to life-threatening complications, such as hemolysis, immune reaction, tissue toxicity, and cell agglutination—for instance, the biocompatible evaluation of chitosan modified with polyvinylpyrrolidone, dextran, PEG, arabinogalactan, and poly-L-glutamic acid [20, 91]. Moreover, chitosan film formed with lactic and acetic acid was evaluated for wound healing activity as compared to standard [21]. Microspheres of chitosan cross-linked with glutaraldehyde show lesser biodegradability and are thus suited for the delivery of several drugs, but such cross-linking may induce cytotoxicity and thus affect the biocompatibility. Therefore, researchers synthesize such cross-linking reagents which are non-toxic as well as biocompatible, such as genipin, which has a higher compatibility and a lesser degradation rate [22].

For biocompatibility evaluation, *in vitro* studies are to be conducted as an initial step, which includes biosafety evaluation and bio-functionality analysis. Screening studies involve hemocompatibility and cytocompatibility as well as genotoxicity studies. The tissue culture strategies can be used for such studies where *in vitro* culture involves the harvesting of cells or tissue from organisms and their analysis for growth and differentiation, such as the culture of chondrocytes when allowed to divide and placed on chitosan microspheres, suggesting its biocompatibility [23]. Two evaluation stages will be included in the *in vitro* cytocompatibility assay. Initially, cytotoxicity assays are conducted to check the test materials for external use or implantation. In such assays, morphological evaluations will be done, which will indicate the cells that are mortal or undergo any transformation [24]. Cytocompatibility assays are very crucial in certain cases like the utilization of chitosan or its derivatives in wound dressing and cultured cells, i.e., keratinocytes or dermal fibroblasts. However, for testing the cancerous cell lines, cytocompatibility assays are not employed because such cell lines are resistant to chitosan-induced toxicity. Cell cultures were evaluated for compatibility based on toxic effects such as alteration in metabolic pathways, organ functioning, cell signaling, or structural changes.

The polycationic property of chitosan molecules accounts for its toxicity, mainly thrombosis, due to interaction with surface proteins, and thus chitosan, as well as its derivatives, are not used for blood-contacting products. The mechanism behind such toxicity involves the adsorption of blood proteins on the surface of chitosan and leads to the initiation of platelet activation. In contrast, a sulfonated derivative of chitosan shows lesser activation of platelets at a pH of more than 7 [25]. The hemo-compatibility assay is mainly done for chitosan derivatives where they are utilized for intra-vascular activity, and the evaluation parameter examined includes hemolysis, fibrinolysis, red cell aggregation, platelet activation and adhesion, proteolysis, oxidative burst, and thrombosis. Currently, the chitosan derivatives used for intravascular activity include N-acyl chitosan, like hexanoyl chitosan [26]. Furthermore, genocompatibility will be carried out to evaluate the mutational capability of chitosan and its derivatives. In a study conducted on mice, it has been reported that chitosan oligosaccharides diminished the mercury-induced chromosomal abnormalities and thus concluded that chitosan suppresses genotoxicity [27, 28].

2.3 Application of Chitosan in Biomedicine

Chitosan nanoparticles have a wide range of applications in various fields such as tissue engineering, bioimaging, and cancer treatment. Bone tissue engineering, drug delivery, etc., are some of its applications in the medical field as mentioned in Figure 2.2.

2.3.1 Antimicrobial Activity

Despite the enormous progress in antimicrobial progress, there are still some infections that pose a threat to our health and for which there is no cure due to factors including the development and spread of resistant strains, the inadequacy of antimicrobial strategies, and the diminished therapeutic properties of current therapy that cause difficulty in reaching active concentration at targeted areas. The cationic nature of chitosan is responsible for antibacterial potency. Chitosan having a low molecular weight diffuses and binds with DNA and interferes with DNA and RNA functioning as compared to high-molecular-weight chitosan. It generates an impenetrable coating around the cell, alters its permeability, and prevents transit into it [29, 30].

The antibacterial activity is affected by the rate of polymerization of chitosan and the type used. It was shown that chitosan copper-loaded nanoparticles have good antibacterial activity against a variety of gram-negative and gram-positive bacteria [27]. They discovered that chitosan nanoparticles outperform copper-loaded nanoparticles in terms of antibacterial activity. Because of the negative charge and hydrophilicity

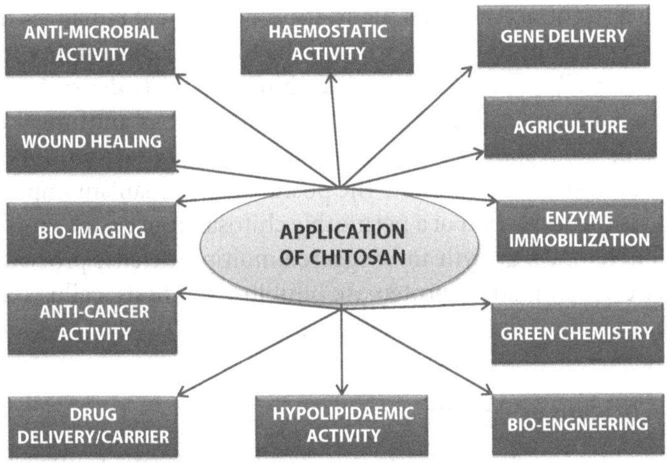

Figure 2.2 Application of chitosan in different segments.

of gram-negative bacteria's cell wall, chitosan nanoparticles have a better interaction with them than gram-positive bacteria, resulting in higher antibacterial activity. The amount of chitosan attaching to the bacterial cell wall was also said to be dependent on pH, molecular mass, and rate of

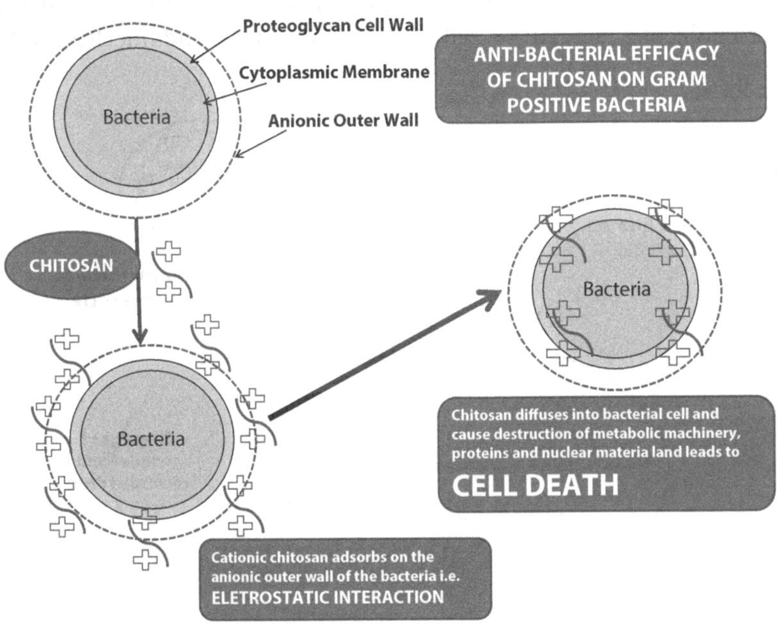

Figure 2.3 Mechanism of chitosan as an anti-bacterial agent.

chitosan acetylation. An acidic pH in the environment enhances the positive charge on chitosan polymer that leads to better diffusibility from the bacterial cell wall [92, 93]. Chitosan's antifungal activity depends upon the molecular mass and rate of acetylation. Chitosan has antifungal efficacy against *Penicillium* sp., *Botrytis cinerea*, *Phytophthora infestans*, *Alternaria solani*, and *Fusarium oxysporum*. The postulated mechanism implicated in the action is the formation of a permeable chitosan film on the crop surface that inhibited fungal growth and triggered multiple defense processes such as chitinase accumulation, proteinase inhibitor synthesis, callus synthesis, and lignifications as depicted in Figure 2.3 [94].

2.3.2 Anticancer Agent

Despite significant advancements, neoplasm remains the leading cause of mortality around the globe. Chitosan and its derivatives have been found to have an anticancer action in recent investigations [102, 103, 111]. The nanoparticles have the ability to interact directly with tumor cells' metabolism, restrict cell growth, and trigger cell apoptosis [4–7]. It was discovered that the anticancer action of chitosan derivatives is related to enhanced IL-1 and IL-2 productions, which promoted cytolytic T-lymphocyte maturation and infiltration. It also acts as an anticancer agent by boosting the body's immune system. With an increase in permeability, long-circulating macromolecules and nano-sized particles (liposomes and micelles) collect

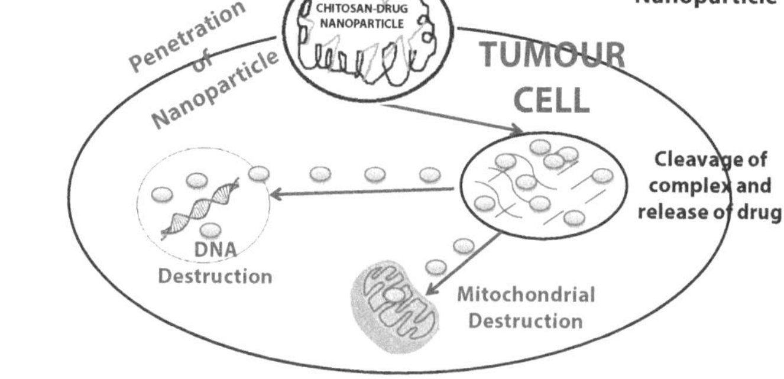

Figure 2.4 Role of chitosan in enhancing the anti-cancer potential of drugs.

at the tumor site. According to certain research, chitosan is engaged in the destruction of cancerous cells by causing cell death. The nanoparticles will aggregate on the leaky vasculature. The development of self-assembled nanoparticles for doxorubicin using arachidic acid–oligosaccharide complex has been prepared. The biodegradable polymer–drug compound of doxorubicin conjugated with a stearic acid-grafted chitosan-nanosized oligosaccharide demonstrated great cellular absorption efficiency and was effective in tumor growth suppression [31, 95] as mentioned in Figure 2.4.

2.3.3 Tissue Engineering

2.3.3.1 Bone Tissue Engineering

Bone tissue engineering (BTE) is a branch of study that focuses on the development of new three-dimensional scaffolds with highly interconnected porous structures. To reinstate damaged bone functioning, the characteristics of the replaced tissue must be matched, such as biocompatibility, osteoinductive, osteoconductive, and mechanically well-suited. The incorporation of lactate in chitosan complexed with chondroitin sulfate enhances the release of drug [32]. These strategies modulate the swelling and mechanical efficacy for adhesion and maintenance for BTE. Because of its *in situ* setup ability and bioactivity, hybrids based on chitosan and calcium phosphate have been employed as transplant alternatives. There has been an injectable cement created based on chitosan polyglactin fiber incorporating as nanospheres for bone regeneration [33, 34]. They discovered that these strategies promote the development of bone re-structuring and the breakdown of cement in bone faults. Recently, biopolymeric scaffolds based on chitosan have been fabricated by using 3D-printing technologies for tissue grafting in bone tissue engineering [99].

2.3.3.2 Cartilage Tissue Engineering

Recently, various chitosan-derived molecules have been produced for cartilage tissue engineering due to their bio-compatibility, flexibility, porous structure, non-toxicity, and elasticity. Researchers [35] used chitosan hydrogel to stabilize transforming growth factor (TGF-1) signaling for cartilage restoration. In a rodent chondral damage model, they discovered that TGF-1-coupled chitosan stimulates cartilage repair. Scientists created 3D nanostructures of chitosan and chondroitin sulfate to aid in the adhesion and multiplication of chondrocyte and found that cells were adhered, multiplied, and bio-transformed actively [36].

2.3.3.3 *Anti-Inflammatory Effects and Repair of Arthritic Tissue*

Because of the alkaline nature of the free amino group contained in chitosan, it can neutralize gastric acids. Chitosan might be employed to treat indigestion and gastric ulcers because it generates a protective membrane in the stomach. The hydrolysis of chitosan by acid leads to the production of glucosamine through salt conversion accounts for the anti-inflammatory mechanism of chitosan. In connective tissues, the structural units are mainly monosaccharide, and when these tissues are damaged or inflamed, they can be restored *via* monosaccharides' activity [37]. Thus, it is considered as an efficient therapy for preventing and treating rheumatoid arthritis as well as bone hyperplasia. In clinical practice, they are also utilized as anti-arthritic medications. Furthermore, these monosaccharides possess no negative effects and thus can be taken for an extended time.

2.3.4 Wound Healing

Wound curing is a series of biological processes that are linked to growth and regeneration. It goes *via* a variety of steps involving cellular and matrix components that work collectively to repair injured tissues and replace lost ones [38]. Many studies employing *in vitro*, animal, and clinical investigations have been published on the therapeutic benefits of chitin and chitosan on wound healing. The effect of varying degrees of deacetylation (DDA) but equivalent-molecular-weight chitin and chitosan nanoparticles on the multiplication of fibroblasts and keratin-secreting cells was examined. Their findings demonstrate that high DDA chitosan enhances fibroblast activity. The wound healing characteristics of chitosan acetate composite films have been discovered.

Wound dressing is considered to be ideal if it will be able to protect the wound from bacterial infection. In vertebrates, chronic wounds due to burns or major injuries like surgery are considered to be the most widespread treatable conditions. Wound healing involves a cascade mechanism that includes inflammation, cellular proliferation, matrix deposition, and remodeling. Moreover, the repair mechanism also involves clotting, granulation, re-epithelialization, neo-vascularization, collagen synthesis, and wound contraction. Chitosan and its derivatives accelerate the wound healing process by inhibiting fibroplasia and promoting tissue growth and differentiation. In addition, the role of topical epidermal growth factors was observed, and it was revealed that healing in burn cases accelerates while improving the quality of healing [39]. Pre-clinical and clinical studies depict that chitosan enhances the regeneration of tissue elements in skin wounds and thus affects wound healing. Chitosan, as a topical agent, also has many merits like better flow properties, non-irritancy, antibacterial activity, and a sustained release rate. Furthermore, epidermal growth factor–chitosan gel reduced the wound healing time, increased epidermal regeneration, and promoted granulation as well as tissue formation. Studies reveal that the wound healing activity of chitosan depends upon its chemical structure as well as its physical forms [40–42]. Chitosan has also been utilized in the fabrication of medical textiles such as sutures and scaffolds for wound healing applications [109, 110]. A study conducted depicts that the catalytic degradation of chitosan is caused by lysozymes present at the wound sites and forms the active molecule *N*-acetyl-*d*-glucosamine dimer. Lysozyme activity is responsible for the sustained action of chitosan and continuous source of *N*-acetyl-*d*-glucosamine dimers because inflammatory cells release lysozyme slowly and continuously in the micro-environment. Thus, the more the *N*-acetylated monomer concentration, the greater will be the healing rate, as depicted in rats treated with high molecular weight chitosan. Therefore, it was confirmed from the above discussion that high molecular weight and a high degree of deacetylation of chitosan show superior wound healing activity. Membranes composed of chitosan and alginates act as a polyelectrolyte complex that displayed greater stability to pH changes and are more efficient as controlled release membranes than when use alone. The membranes are of great therapeutic significance on highly exuding wounds with bacterial infections [43, 110]. Due to biocompatibility, capacity to absorb exudates, and film-forming properties, chitosan products are good candidates for burn and wound treatment.

2.3.5 Hemostatic Activity

Hemostatic materials have different surface properties than non-thrombogenic blood-contacting materials. Anti-thrombogenicity is a fundamental factor in the development of artificial biomaterials that come into contact with blood (e.g., prosthetic vascular grafts). Polyurethane (PU), for example, is a well-known biomaterial for biomedical applications with outstanding biocompatibility and mechanical qualities. However, due to inadequate cell attachment/growth or the stimulation of thrombogenesis on the surface, the application of PU is still limited [44, 45]. Anticoagulant modifications are commonly used when chitosan is used as a blood contact material [46, 47]. The impact of chitosan chemical structural changes and physical shape on hemostasis was recently evaluated [48]. Despite its biocompatibility, chitosan is not acceptable for use as a blood contact material.

In addition to its biocompatible and antibacterial properties, chitosan has been a desirable substance as a hemostatic agent due to its blood coagulating tendency. Chitosan's hemostatic activity is crucial for understanding how it works as a coagulant and creating medical bandages to prevent bleeding during surgery. The hemostatic mechanism caused by chitosan, according to several research, is independent of the traditional blood coagulation cascade [49–51]. Chitosan-coated sutures, N-acyl-chitosan, chitosan tropo-collagen fibers, and their N-modified fibers were inserted into the jugular and femoral veins of dogs [52]. The blood clot that developed around the fibers was examined under a microscope. When a chitosan solution was applied to a lacerated rabbit tongue, researchers examined lingual bleeding time, systemic bleeding time, and systematic coagulation time [53]. The effect of a chitosan acetate salt dressing was evaluated by monitoring blood loss, fluid use, hemostasis time, and survival in swine with severe liver injury caused by lacerated main arteries. In addition, the hemostasis performance of poly-N-acetyl Glucosamine (p-GlcNac) fiber in the swine spleen was studied until bleeding was stopped by continuous compression at the wound site. HemCon is the leading hemostatic dressing used in the prehospital treatment of injuries, and it is covered with chitosan. The ideal hemostatic drug would be able to overcome high-risk situations and not derived from animal origin. Agents formed of chitosan are cheap, absorbable, do not contain any animal derived components, and pose no risk of viral transmission and have shown tremendous promise in animal models. Chitosan-based bandages were developed with increased hemostatic characteristics in response to the issues of bleeding in war wounds. They theorized that mixing polyphosphate polymers with chitosan would result in a more effective hemostat. Polyphosphate, an inorganic phosphate linear polymer, has been demonstrated to speed hemostasis and retards fibrinolysis [54]. Polyphosphates can ionically interact with chitosan to generate polyelectrolyte complexes because in water it forms a polyanion [55]. Chitosan–polyphosphate dressings were evaluated for hemostatic potency. Silver nanoparticles were added to the dressings for antibacterial action, and thus effective hemostatic formulation was identified. Both chitosan and polyphosphate have been found to activate coagulation as amine groups in chitosan attracted on negatively charged RBC membranes, resulting in significant hemagglutination [50, 51]. Chitosan also adsorbs fibrinogen and plasma proteins, which help platelets clump together. Polyphosphate activates the contact pathway, reducing both the time gap for beginning and peak thrombosis. Simultaneously, chitosan would attract RBCs, which would increase and solidify the thrombus, resulting in a stable clot that would stop the bleeding. As a result, it was indicated that a new chitosan–polyphosphate dressing was employed to speed up blood clotting, platelet adhesion, and thrombin production.

2.3.6 Fat Binder

Because it is a complex carbohydrate that cannot be degraded by human gastric enzymes, chitosan can be classified as a dietary fiber both chemically and physiologically [56]. In rats and humans, it was shown to have significant hypocholesterolemic action. Chitosan's degree of deacetylation (DD) and molecular weight are two crucial features that set it apart from other dietary fibers [57]. Higher DD indicates that the chitosan molecule has unbound amino groups and the solution has a higher positive charge. It was shown that, when DD was high, chitosan's hypocholesterolemic action was improved, which could be due to the interaction between chitosan and fatty acid/bile acid. Another essential property of chitosan is its molecular weight, which corresponds to its viscosity. Chitosan is soluble in the stomach's acidic environment and forms a gel at high molecular weights. When fat and chitosan are consumed together in the stomach, the viscous chitosan entraps the fat droplet. Chitosan creates a precipitate in the small intestine, which has a neutral pH and hinders fat digestion [58]. Chitosan with a larger molecular weight dramatically reduced body weight growth in rats, reduced the food intake, and decreased the fat levels in plasma. The physicochemical features of chitosan will affect their hypocholesterolemic actions, he claimed in his study. The effect is better with fine particles and a higher degree of deacetylation as well as molecular weight. Chitosan's hypocholesterolemic mechanisms are likely to be electrostatic action, adsorption, and entrapment [59]. Chitosan is known to precipitate in the small intestine after forming a viscous solution in the stomach. Chitosan has a different hypocholesterolemic mechanism than dietary fiber, which increases intestinal viscosity. Interference in the lymphatic absorption of cholesterol and fat [60, 61] and increased fecal excretion of neutral steroids are all effects of chitosan [62], with fat and liver function improvement [63]. Diets high in chitosan lowered the plasma and liver lipid levels while increasing the fat and cholesterol excretion in feces. Chitosan also lowered the weight of liver and epididymal fat as well as the activities of liver hepatic and lipoprotein lipase. In comparison to rats administered with cholestyramine, rats fed chitosan had higher fecal cholesterol excretion and liver hepatic lipase activity as well as lower plasma TG levels. Because cholestyramine is inefficient at decreasing the plasma TG concentrations and has certain negative effects, chitosan is indicated as a hypolipidemic drug [64].

2.3.7 Drug Delivery

The delivery of a medicinal substance to a targeted site of the body was a significant issue in the treatment of numerous ailments [64, 96]. Adopting such strategies of drug delivery improves performance in terms of effectiveness, safety, and minimal adverse effects. Polymers utilized for targeted delivery can improve pharmacokinetics, improve therapeutic index, reduce adverse effects, and therefore increase the overall system's resourcefulness [65]. When considering such a targeted system, the drug is delivered to the location where action is required and the medication concentrates in the area, requiring a lower dose amount of therapy the next time the drug is administered [66, 97]. Chitosan nanoparticles and chitosan-immobilized nanoparticles have grown in importance due to enhanced therapeutic efficacy and release to the intended region, including good bio-dispersion and increased specificity and sensitivity. Because of its cationic functionality and aqueous media solubility, chitosan is the most prominent naturally occurring polymer with a wide range of applications as a drug delivery agent. It also has an advantage because it is non-toxic and biodegradable in nature, with no negative consequences [102–106]. Chitosan nanoparticles have attracted attention as a therapeutic delivery entity for a variety of applications, including sorting protein therapies, genetics, and adverse tumor chemical therapeutics through a variety of routes, including oral, nasal, and intravenous. The positive charge present on nanoparticles attracts to the negatively charged cell membrane and thus is responsible for the delivery of the drug to the targeted area. Chitosan's hydrophobic nature affects the efficient encapsulation of hydrophilic medicines within chitosan nanoparticles.

It was used in a topical conveyance system because of its improved bio-adhesive activity and capacity to control the release of active ingredients. Chitosan, due to the presence of an amino group, possesses a cationic charge, which describes the qualities used in medicinal conveyance systems, such as controlled drug discharge, gelation transfection, permeability enlargement, and efflux pump inhibitory properties. Chitosan nanoparticles can bind with DNA containing a negative charge and thus form a polyelectrolyte complex in gene delivery system. When DNA was in these complexes, nuclease breakdown was ineffective, resulting in a higher transfection efficiency. Using chitosan nanoparticles to store large amounts of anti-cancer therapeutics is critical for excellent cancer therapy results. Medicines encapsulated in chitosan nanoparticles can have enhanced absorption and bioavailability, thus utilized to deliver protein drugs and gene while also protecting them against enzyme destruction *in vivo*. The distribution of nanoparticles depends on the surface charge, size, and hydrophobicity of the administrated particles, and blood capillaries were capable of administering chitosan intravenously. Chitosan is an inactive component that is added to a tablet or capsule to control the drug's release. Chitosan was easily eliminated after delivery due to renal clearance; however, this only applies to chitosan with a suitable molecular weight. For very large molecular weight chitosan, enzyme breakdown is necessary. Researchers produced a chitosan/lysine copolymer and investigated the efficacy for DNA adhesion and gene transfection. The effect of particle size on the release characteristics of chitosan–montmorillonite nanoparticles stored with isoniazid therapy was examined. They discovered that a reduction in the pH and size of the nano-particles caused an increase in isoniazid release [67].

2.3.8 Bioimaging and Gene Transfer

The core of gene therapy is the cellular delivery of nucleic acids, which has the potential to heal many currently incurable diseases. Chitosan nanoparticles have several characteristics that make them an effective gene delivery technology—for instance, positively charged amino groups of chitosan interacts with negatively charged phosphate of DNA, resulting in enhanced stability and improved gene transfer capabilities. Because of its biocompatible qualities, the use of nano-chitosan in bioimaging is gaining popularity. The impact of chitosan nanoparticles was investigated on bioimaging. They investigated the use of quantum dots alone or along with chitosan in biosensing and imaging [68]. Molecular imaging allows for the noninvasive evaluation of biological and chemical processes in living cells. As a result, such strategies have the capability to improve knowledge regarding illness and medication activity at all stages of drug development, from pre-symptomatic through clinical.

Chitosan is a hydrophilic, non-antigenic biopolymer with minimal toxicity towards class cells. When chitosan composites are combined with other materials, they provide a new category of biomaterials with advanced features that could be exploited in improved imaging applications. Deacetylation of polyose derived from crustacean shells produces chitosan [poly(1,4-D-glucopyranosamine)], a polycationic biopolymer containing a hydroxyl radical and a hexosamine residue. Because of its biocompatibility, chitosan has a lot of potential as a biomaterial. It is hydrophilic, is non-antigenic, and has minimal toxicity to class cells. Chitosan is also suggested to help in drug distribution through cellular barriers and to temporarily release tight connections between animal tissue cells.

The purpose of molecular imaging is to perform non-invasive checks on biological processes. Molecular imaging is critical in overcoming the difficulties of identifying biological activities in living beings [69]. For organic processes, most molecular imaging techniques need a distinguishing agent. Endogenous and exogenous agents are the two basic types of imaging agents. Endogenous agents are created *in situ* after genetic manipulation or are naturally present within the tissue of interest. Exogenous distinction agents are administered into biological tissues, usually intravenously, and have no intrinsic properties. Because of its biocompatible qualities, chitosan is an excellent biopolymer for biological applications. It is a polycationic natural chemical molecule made up of acetyl-glucosamine and glucosamine connected by β-glycosidic linkages, with one and/or two free radical teams in repeating hexosamine residue. This complex could be replaced with lipophilic segments to improve self-assembly by escalating the unit lipophilic contacts between segments. When chitosan composites are combined with other materials, they provide a new category of biomaterials with mechanical, chemical, and useful qualities that neither native chitosan nor the integrated material can provide. When brittle phosphate or hydroxyapatite is mixed with chitosan, it produces a bio-resorbable composite with good strengthening qualities for BTE. Furthermore, including agents such as Fe_3O_4 nanoparticles could improve the hepatic tissue-targeted imaging, and the particle could potentially be used as a MR molecular imaging agent. Polyacrylic acid, an anionic molecule, is frequently combined with chitosan to create polymer composites with longer circulation durations, resulting in increased medicinal drug bioavailability. Differentiation agents for imaging purposes can also be included in the latter composites. Sonication increases oxygen delivery, and people's exchange dynamics are altered by frequency and period [68]. Scientists have created unique nanoparticles composed of chitosan–linolate conjugates as a hepatocyte-specific differentiation agent [70]. The WSC-LA conjugates self-assembled into core–shell structures in solution. The image relaxivity of the SPIO clusters formed within the hydrophobic core was extremely high. A qualitative analytical technique was used to load chitosan nanoparticles with the antineoplastic medication docetaxel, and the resulting DTX-HGC nanoparticles had a mean diameter of 345–355 nm in solution. The nanoparticles were well-scattered and durable *in vitro* for 2 weeks under physiological settings with a sustained drug profile, according to the investigators. The nanoparticles were fairly stable in the co-existence of increased albumin, indicating that they might be stable in the bloodstream as well. The authors also used a non-invasive live animal imaging technology to assess the time-dependent excretion profile, *in vivo* biodistribution, circulation time, and tumor targeting capabilities of DTX-HGC nanoparticles. In A549 carcinoma cell-bearing mice, the DTX-HGC nanoparticles showed higher anticancer effectuality by reducing the tumor size and improving the survival rate as well as significantly reducing the antineoplastic drug toxicity when compared to free DTX. Gd-DTPA has the potential to be utilized as imaging agent. Similarly, chitosan-Gd nanoparticles are prepared by adding Gd-DTPA through emulsion droplet union approach and used as an injectable device for metal neutron capture medical treatment because of their unharnessed qualities and capacity to retain Gd-DTPA within the tumor for a long time (Gd-NCT). The amount of Gd loaded in the various types of chitosan used in the production varied greatly [71].

2.3.9 Enzyme Immobilizations for Biosensing

Immobilized enzymes are preferable to free enzymes for biotechnological applications because they allow for easier enzyme recycling, improved stability and robustness, constant production, and the lack of biocatalysts [72]. Chitosan will be used for enzyme immobilization due to certain properties like non-toxic nature, biocompatibility, and large surface area [73]. Enzyme immobilization technique is used for making biosensors [74] and diagnostic strips in analytical applications. Biosensors are biological materials that can recognize specific chemical moieties and use chemical methods to signal their presence, concentration, or even activity. A transducer turns the reaction into a quantifiable response by converting the chemical messages obtained from living systems with the analyte. Biosensors have several advantages, including mobility, simplicity, adaptability, sensitivity, and selectivity and for usage in complicated contexts due to their rapid reaction [75]. Biosensors that are cheap and implantable have been the subject of research for many years, and the main hurdle in its research includes their unique catalytic, electrical, and optical properties [76]. Chitosan is widely used for this purpose due to ease of availability, non-toxic nature, compatibility, permeability, and gel formation capabilities. Furthermore, the presence of amino and hydroxyl groups can make chitosan be utilized as a modifying agent due to high reactivity [77]. Chitosan is the most widely used polymer for the construction of nanocomplexes for biosensing utilities due to its unique features. Catechol-chitosan films are used as bacterial metabolite detectors [78]. For the detection of phenolic chemicals, a tyrosinase biosensor based on (QDs/CS) (quantum dots/chitosan) nano-complexes was developed [79]. For detecting the phenolics, QDs/CS nanocomplex, i.e., enzyme electrode-based immobilized tyrosinase was created and studied. The advantages include better permeability and compatible surroundings for the immobilized enzyme with a high affinity entrapment for substrates. Scientists have described the utilization of a chitosan–carbon system for dehydrogenase enzyme-based sensing [78]. Disposable biosensors were developed with flow injection amperometric analysis based on enzymes immobilized on chitosan–gold-modified indium tinoxide electrodes [80]. Chitosan and its GAG complexes have also been shown to influence the activities of vascular endothelial and muscle cells. In addition, chitosan/heparin was created for artificial implantation of liver, and it demonstrated better compatibility, among other benefits of liver implantation [81].

2.3.10 Green Chemistry

Due to advancement in green chemistry, ionic liquids (IL) become popular green solvents. Furthermore, it is an essential requirement for environment-friendly protection. The IL have exceptional properties that can be quite valuable when typical solvents are not as active as they should be or are not suitable [82]. IL can be solubilized with a broad range of inorganic and organic compounds. The major innovative and striking characteristic of ILs as solvents will be the possibility of developing one with unique qualities required for a certain application, known as "designer solvents" [83]. The ionic liquids are now known to be used as substitute solvents in a variety of catalytic organic reactions [84, 85]. They usually dissolve in macromolecules like carbohydrates, keratin, and fibroin that contain intermolecular hydrogen bonds. The IL is also used for cosmetics, wastewater treatment, heavy metal, and rock chelation. When you look at the repeating hexosamine residue, you can see that chitosan is made up of a hydroxyl and amino group. However, the chitin and chitosan chains' strong intermolecular and intramolecular hydrogen bonding inhibit miscibility in organic solvent. Thus, to dissolve chitosan, dilute mineral and organic acids are used. These solvents are corrosive, necessitating an alkaline treatment to eradicate the acid. Furthermore, polyelectrolyte solutions have limited application in specific applications such as drug carriers and transition metal sorbents [86], while bioactive compounds may be impacted by acetic acid intake for chitosan dissolution [87]. 1-Butyl-3-methyl-imidazolium chloride can be used to capture and discharge carbon dioxide instead of amino-functionalized artificial polymers. While considering carbon dioxide's recovery process from industrial effluent gaseous mixture, it is critical to consider carbon dioxide use as a carbon source as well as environmental-related problems like global warming. Traditionally,

CO_2 fixation will be done with amines to produce ammonium carbamates and gas is emitted afterwards when they are heated. Because chitosan is a regular polyamine, it could be a good substitute for amino-functionalized polymers when it comes to CO_2 fixation. This environmentally beneficial process is capable not only of extracting the gas from industries but also helps to check the level of CO_2 [88]. Furthermore, a composite/complexing system might be used as an immobilization matrix to entrap enzymes and proteins such as hemoglobin (Hb). A thermo-gravimetric investigation shows that the Hb-chitosan–BMIM-BF4 complex has superior stability over Hb-chitosan alone. These one-of-a-kind composite materials could provide an excellent electro-chemical sensing strategy for enzymes and proteins. The gold nanoparticles, when deposited over an electrode surface through an electro-chemical process, was checked by submerging the electrode in a fluid containing benzoquinone, glucose oxidase, chitosan, and IL for enzymatic electro-deposition. The biosensor has a high amperometric response (5 s) to glucose and high sensitivity, which is 3.0 times better than a biosensor made by electro-depositing chitosan-IL-GOD biocomposite on a flat gold electrode. With the biosensor made from a gold electrode, the limit was 20 times better for glucose. The biosensor has better reproducibility, storage space stability over long periods, and acceptable anti-interference capabilities [89].

2.4 Conclusion

It is concluded that terrific scientific research has been carried out for depicting the biomedical application of chitosan. Studies demonstrated that chitosan possesses unique properties and structure which permit a wide range of applications. Chitosan and its derivatives show greater compatibility with an entire compartment of the body which reveals its therapeutic significance. Chitosan and its derivative are widely used for pharmaceutical applications such as hemostasis, anti-microbial, anti-cancer, wound accelerator, and drug delivery agent. Chitosan enhances the formation of epithelium tissue and thus promotes the healing process. Chitosan is used for developing a targeted delivery system for drugs as well as genes. Moreover, chitosan will also be utilized for treating dyslipidemia and thus lowers the cholesterol and triglyceride concentration in the blood. The key determinants of the therapeutic properties of chitosan are molecular weight and degree of deacetylation. Thus, chitosan, due to its non-toxic and biocompatible nature, is used widely in biomedicine to overcome various ailments. Therefore, it is considered that, in the future, chitosan-based products will be available on a commercial basis.

References

1. Nasti, A., Zaki, N.M., Leonardis, P.D., Ungphaiboon, S., Sansongsak, P., Rimoli, G.M., Tirelli, N., Chitosan/TPP and chtiosan/TPP-hyaluronic acid nanoparticles: Systematic optimization of the preparative process and preliminary biological evaluation. *Pharm. Res., 26,* 8, 1918–1930, 2009.
2. Thanou, M., Verhoef, J.C., Junginger, H.E., Chitosan and its derivatives as intestinal absorption enhancers. *Adv. Drug Deliv. Rev.*, 50, Suppl 1, S91–S101, 2001.
3. Agnihotri, S.A. and Aminabhavi, T.M., Controlled release of clozapine through chitosan microparticles prepared by a novel method. *J. Control. Release, 96,* 2, 245–259, 2004.
4. Kim, I.Y., Seo, S.J., Moon, H.S., Yoo, M.K., Park, I.Y., Kim, B.C., Cho, C.S., Chitosan and its derivatives for tissue engineering applications. *Biotechnol. Adv., 26,* 1, 1–21, 2008.
5. Stavroula, G.N., Ioannis, A.K., Ioanna, K., Evangelos, K., Dimitrios, B., Miscibility study of chitosan/2-hydroxyethyl starch blends and evaluation of their effectiveness as drug sustained release hydrogels. *Carbohydr. Polym., 87,* 2, 1286–1294, 2012.
6. Tiyaboonchai, W., Chitosan nanoparticles: A promising system for drug delivery. *Naresuan Univ. J., 11,* 3, 51–66, 2003.
7. Bamrungsap, S., Zhao, Z., Chen, T., Wang, L., Li, C., Fu, T., Tan, W., Nanotechnology in therapeutics: A focus on nanoparticles as a drug delivery system. *Nanomedicine, 7,* 8, 1253–1271, 2012.
8. Patel, M.P., Patel, R.R., Patel, J.K., Chitosan mediated targeted drug delivery system: *A review. J. Pharm. Pharm. Sci., 13,* 4, 536–557, 2010.
9. Wang, J.J., Zeng, Z.W., Xiao, R.Z., Xie, T., Zhou, G.L., Zhan, X.R., Wang, S.L., Recent advances of chitosan nanoparticles as drug carriers. *Int. J. Nanomed.,* 6, 765–774, 2011.
10. Termsarasab, U., Cho, H.J., Kim, D.H., Chong, S., Chung, S.J., Shim, C.K., Moon, H.T., Kim, D.D., Chitosan oligosaccharide-arachidic acid-based nanoparticles for anti-cancer drug delivery. *Int. J. Pharm., 441,* 1-2, 373–380, 2013.
11. Tripathy, S., Das, S., Chakraborty, S.P., Sahu, S.K., Pramanik, P., Roy, S., Synthesis, characterization of chitosan-tripolyphosphate conjugated chloroquine nanoparticle and its *in vivo* anti-malarial efficacy againstrodent parasite: A dose and duration dependent approach. *Int. J. Pharm., 434,* 1-2, 292–305, 2012.

12. Shi, S.F., Jia, J.F., Guo, X.K., Zhao, Y.P., Chen, D.S., Guo, Y.Y., Cheng, T., Zhang, X.L., Biocompatibilityof chitosan-coated iron oxide nanoparticles with osteoblast cells. *Int. J. Nanomed.*, *7*, 5593–5602, 2012.
13. Huet, J., Rucktooa, P., Clantin, B., Azarkan, M., Looze, Y., Villeret, V., Wintjens, R., X-ray structure of papaya chitinase reveals the substrate binding mode ofglycosyl hydrolase family 19 chitinases. *Biochemistry*, *47*, 8283–8291, 2008.
14. Collinge, D.B., Kragh, K.M., Mikkelsen, J.D., Nielsen, K.K., Rasmussen, U., Vad, K., Plant chitinases. *Plant J.*, *3*, 1, 31–40, 1993.
15. Kasprzewska, A., Plant chitinases-regulation and function. *Cell Mol. Biol. Lett.*, *8*, 3, 809–824, 2003.
16. Schipper, G., Varum, K.M., Artursson, P., Chitosans as absorption enhancers for poorly absorbable drugs 1: Influence of molecular weightand degree of acetylation on drug transport across human intestinal epithelial (caco-2) cells. *Pharm. Res.*, *13*, 11, 1686–1692, 1996.
17. Yang, Y.M., Hu, W., Wang, X.D., Gu, X.S., The controlling biodegradation of chitosan fibers by n-acetylation *in vitro* and *in vivo*. *J. Mater. Sci. Mater. Med.*, *18*, 11, 2117–2121, 2007.
18. Molinari, L.M., Pedroso, R.B., Scoaris D. de Oliveira, O., Nakamura, T.U., Nakamura, C.V., Filho, B.P.D., Identification and partial characterisation of a chitinase from Nile tilapia, Oreochromis niloticus. *Comp. Biochem. Physiol. B Biochem. Mol. Biol.*, *146*, 81–87, 2007.
19. Hartl, D., He, C.H., Koller, B., Da Silva, C.A., Homer, R., Lee, C.G., Elias, J.A., Acidicmammalian chitinase is secreted via an adam17/epidermal growth-factor receptor-dependent pathway and stimulates chemokineproduction by pulmonary epithelial cells. *J. Biol. Chem.*, *283*, 48, 33472–33482, 2008.
20. Zhu, Z., Zheng, T., Homer, R.J., Kim, Y.K., Chen, N.Y., Cohn, L., Hamid, Q., Elias, J.A., Acidic mammalian chitinase in asthmatic Th2 inflammation and IL-13 pathway activation. *Science*, *304*, 5677, 1678–1682, 2004.
21. Stirling, J.L., Nn′-diacetylchitobiase activity in Tay-Sachs disease and Sandhoff's disease. *Biochem. J.*, *141*, 2, 597–599, 1974.
22. Kuranda, M.J. and Aronson, N.N., A di-n-acetylchitobiase activity is involved in the lysosomal catabolism of asparagine-linked glycoproteins in rat liver. *J. Biol. Chem.*, *261*, 5803–5809, 1986.
23. Hollak, C.E., Weely, S.V., Van Oers, M.H., Aerts, J.M., Marked elevation of plasma chitotriosidase activity. A novel hallmark of Gaucher disease. *J. Clin. Invest.*, *93*, 3, 1288–1292, 1994.
24. Renkema, G.H., Boot, R.G., Muijsers, A.O., Donker-Koopman, W.E., Aerts, J.M., Purification and characterization of human chitotriosidase, a novel member of the chitinase family of proteins. *J. Biol. Chem.*, *270*, 5, 2198–2202, 1995.
25. Malaguarnera, L., Chitotriosidase: The yin and yang. *Cell Mol. Life Sci.*, *63*, 24, 3018–3029, 2006.

26. Thanou, M., Verhoef, J.C., Junginger, H.E., Oral drug absorption enhancement by chitosan and its derivatives. *Adv. Drug Deliv. Rev.*, *52*, 2, 117–126, 2001.
27. Illum, L., Chitosan and its use as a pharmaceutical excipient. *Pharm. Res.*, *15*, 9, 1326–1331, 1998.
28. Wedmore, McManus, J.G., Pusateri, A.E., Holcomb, J.B., A special report on the chitosan-based hemostatic dressing: Experience in current combat operations. *J. Trauma*, *60*, 3, 655–658, 2006.
29. Zheng, L.-Y. and Zhu, J.F., Study on antimicrobial activity of chitosan with different molecular weights. *Carbohydr. Polym.*, *54*, 4, 527–530, 2003.
30. Prashanth Harish, K.V. and Tharanathan, R.N., Chitin/chitosan: Modifications and their unlimited application potential-An overview. *Trends Food Sci. Technol.*, *18*, 3, 117–131, 2007.
31. Ruoslahti, E., Bhatia, S.N., Sailor, M.J., Targeting of drugs and nanoparticles to tumors. *J. Cell Biol.*, 188, 6, 759–768, 2010.
32. Santo, V.E., Duarte, A.R.C., Gomes, M.E., Mano, J.F., Reis, R.L., Hybrid 3Dstructure of poly (D,L-lactic acid) loaded with chitosan/chondroitin sulfate nanoparticles to be used as carriers for bio macromolecules in tissue engineering. *J. Supercrit. Fluids*, *54*, 2010, 320–327, 2010.
33. Meng, D., Dong, L., Wen, Y., Xie, Q., Effects of adding restorable chitosan microspheres to calcium phosphate cements for bone regeneration. *Mater. Sci. Eng. C Mater. Biol. Appl.*, *47*, 266–272, 2015.
34. Li, Z., Hassna, R.R., Hauch, K.D., Xiao, D., Zhang, M., Chitosan-alginate hybrid scaffolds for bone tissue engineering. *Biomaterials*, *26*, 18, 3919–3928, 2005.
35. Kim, K.J., Li, B., Winer, J., Armanini, M., Gillett, N., Phillips, H.S., Ferrara, N., Inhibition ofvascular endothelial growth factor-induced angiogenesis suppresses tumour growth *in vivo*. *Nature*, *362*, 6423, 841–844, 1993.
36. Silva, J.M., Georgi, N., Costa, R., Sher, P., Reis, R.L., Blitterswijk, C.A., Karperien, M., Mano, J.F., Nanostructured 3D constructs based on chitosan and chondroitin sulphate multilayers for cartilage tissue engineering. *PLoS One*, *8*, 2, e55451, 2013.
37. Boateng, J.S., Matthews, K.H., Stevens, H.N., Eccleston, G.M., Wound healing dressings and drug delivery systems: A review. *J. Pharm. Sci.*, *97*, 8, 2892–2923, 2008.
38. Xu, Z.P., Stevenson, G.S., Lu, C.Q., Lu, G.Q., Bartlett, F.P., Gray, P.P., Stable suspension of layered double hydroxide nanoparticles in aqueous solution. *J. Am. Chem. Soc.*, *128*, 1, 36–37, 2006.
39. Brown, G., Curtsinger, L., White, M., Mitchell, R., Pietsch, J., Nordquist, R., Von Fraunhofer, A., Schultz, G., Acceleration of tensile strength of incisions treated with EGF and TGF-beta. *Ann. Surg.*, 208, 788, 1988.
40. Chung, L., Schmidt, R., Hamlyn, P., Sagar, B., Andrew, A., Turner, T., Biocompatibility of potential wound management products: Fungal mycelia

as a source of chitin/chitosan and their effect on the proliferation of human F1000 fibroblasts in culture. *J. Biomed. Mater. Res.*, 28, 463, 1994.
41. Sezer, A., Cevher, E., Hatipoğlu, F., Oğurtan, Z., Akbua, J., Preparation of fucoidan chitosan hydrogel and its application as burn healing accelerator on rabbits. *Biol. Pharm. Bull.*, 31, 2326, 2008.
42. Ueno, H., Murakami, M., Okumura, M., Kadosawa, T., Uede, T., Fujinaga, T., Chitosan accelerates the production of osteopontin from polymorphonuclear leukocytes. *Biomaterials*, 22, 1667, 2001.
43. Rodrigues, A., Sanchez, E.S., da Costa, A., Moraes, Â., The influence of preparation conditions on the characteristics of chitosan alginate dressings for skin lesions. *J. Appl. Polym. Sci.*, 109, 2703, 2008.
44. Hsu, S. and Chen, W., Improved cell adhesion by plasma-induced grafting *L*-lactide onto polyurethane surface. *Biomaterials*, 21, 359, 2000.
45. Takahashi, A., Kita, R., Kaibara, M., Effects of thermal annealing of segmented-polyurethane on surface properties, structure and antithrombogenicity. *J. Mater. Sci.: Mater. Med.*, 13, 259, 2002.
46. Beena, M., Chandy, T., Sharma, C., Heparin immobilized chitosan-poly ethylene glycol interpenetrating network: Antithrombogenicity. *Artif. Cells Blood Substit. Biotechnol.*, 23, 175, 1995.
47. Chandy, T. and Sharma, C., Prostaglandin *E*1-immobilized poly(vinyl alcohol)-blended chitosan membranes: Blood compability and permeability properties. *J. Appl. Polym. Sci.*, 44, 2145, 1992.
48. Whang, H., Kirsch, W., Zhu, Y., Yang, C., Hudson, S., Hemostatic agents derived from chitin and chitosan. *Polym. Rev.*, 45, 309, 2005.
49. Hemker, H., Al Dieri, R., Béguin, S., Thrombin generation assays: Accruing clinical relevance. *Curr. Opin. Hematol.*, 11, 170, 2004.
50. Klokkevold, P., Fukayama, H., Sung, E., Bertolami, C., The effect of chitosan (poly-*N* acetyl glucosamine) on lingual hemostasis in heparinized rabbits. *J. Oral. Maxillofac. Surg.*, 57, 49, 1999.
51. Rao, S. and Sharma, C., Use of chitosan as a biomaterial: Studies on its safety and hemostatic potential. *J. Biomed. Mater. Res. Part A*, 34, 21, 1997.
52. Hirano, S. and Noishiki, Y., The blood compatibility of chitosan and *N* acylchitosans. *J. Biomed. Mater. Res.*, 19, 413, 1985.
53. Mi, F., Shyu, S., Wu, Y., Lee, S., Shyong, J., Huang, R., Fabrication and characterization of a sponge-like asymmetric chitosan membrane as a wound dressing. *Biomaterials*, 22, 165, 2001.
54. Smith, S., Mutch, N., Baskar, D., Rohloff, P., Docampo, R., Morrissey, J., Polyphosphate modulates blood coagulation and fibrinolysis. *Proc. Natl. Acad. Sci. U.S.A.*, 103, 903, 2006.
55. Mi, F., Shyu, S., Wong, T., Jang, S., Lee, S., Lu, K., Chitosan polyelectrolyte complexation for the preparation of gel beads and controlled release of anticancer drug. II. Effect of pH-dependent ionic crosslinking or interpolymer complex using tripolyphosphate or polyphosphate as reagent. *J. Appl. Polym. Sci.*, 74, 1093, 1999.

56. Razdan, A. and Pettersson, D., Effect of chitin and chitosan on nutrient digestibility and plasma lipid concentrations in broiler chickens. *Br. J. Nutr.*, 72, 277, 1994.
57. Zacour, A., Silva, M., Cecon, P., Bambirra, E., Vieira, E., Effect of dietary chitin on cholesterol absorption and metabolism in rats. *J. Nutr. Sci. Vitaminol.*, 38, 609, 1992.
58. Zhou, K., Xia, W., Zhang, C., Yu, L., *In vitro* binding of bile acids and triglycerides by selected chitosan preparations and their physico-chemical properties. *LWT-Food Sci. Technol.*, 39, 1087, 2006.
59. Liu, J., Zhang, J., Xia, W., Hypocholesterolaemic effects of different chitosan samples *in vitro* and *in vivo*. *Food Chem.*, 107, 419, 2008.
60. Ikeda, I., Sugano, M., Yoshida, K., Sasaki, E., Iwamoto, Y., Hatano, K., Effects of chitosan hydrolyzates on lipid absorption and on serum and liver lipid concentration in rats. *J. Agric. Food Chem.*, 41, 431, 1993.
61. Vahouny, G., Satchithanandam, S., Cassidy, M., Lightfoot, F., Furda, I., Comparative effects of chitosan and cholestyramine on lymphatic absorption of lipids in the rat. *Am. J. Clin. Nutr.*, 38, 278, 1983.
62. Sugano, M., Fujikawa, T., Hiratsuji, Y., Nakashima, K., Fukuda, N., Hasegawa, Y., A novel use of chitosan as a hypocholesterolemic agent in rats. *Am. J. Clin. Nutr.*, 33, 787, 1980.
63. Kanauchi, O., Deuchi, K., Imasato, Y., Kobayashi, E., Increasing effect of a chitosan and ascorbic acid mixture on fecal dietary fat excretion. *Biosci. Biotechnol. Biochem.*, 58, 1617, 1994.
64. Sgouras, D., The evaluation of biocompatibility of soluble polymersand assessment of their potential as specific drug delivery systems, in: *Pharmacy*, Keele, University of Keele, UK, 1990.
65. Wan, K.W., Malgesini, B., Verpilio, I., Ferruti, P., Griffiths, P.C., Paul, A., Hann, A.C., Duncan, R., Poly(amidoamine) salt form: Effect on pH-dependent membrane activity and polymer conformation in solution. *Biomacromolecules*, 5, 1102–1109, 2004.
66. Schipper, N.G., Varum, K.M., Stenberg, P., Ocklind, G., Lennernas, H., Artursson, P., Chitosans as absorption enhancers of poorly absorbable drugs. 3: Influence of mucus on absorption enhancement. *Eur. J. Pharm. Sci.*, *8*, 4, 335–343, 1999.
67. Yu, H., Chen, X., Lu, T., Sun, J., Tian, H., Hu, J., Wang, Y., Zhang, P., Jing, X., Poly(L-lysine)-graftchitosan copolymers: Synthesis, characterization, and gene transfection effect. *Biomacromolecules*, *8*, 5, 1425–1435, 2007.
68. Ge, Y., Zhang, Y., He, S., Nie, F., Teng, G., Gu, N., Fluorescence modified chitosan-coated magnetic nanoparticles for high-efficient cellular imaging. *Nanoscale Res. Lett.*, *4*, 4, 287–295, 2009.
69. Sahu, S., Behera, B., Maiti, T.K., Mohapatra, S., Simple one-step synthesis of highly luminescent carbon dots from orange juice: Application as excellent bio-imaging agents. *Chem. Commun.*, *48*, 70, 8835–8837, 2012.

70. Wanlop, W., Praneet, O., Ngawhirunpat, T., Apirakaramwong, A., Rojanarata, T., Ruktanonchai, U., Lee, R.J., Evaluation of chitosan salts as non-viral gene vectors in CHO-K1 cells. *Int. J. Pharm.*, 348, 1-2, 161–168, 2008.
71. Brenner, D.J. and Hall, E.J., Computed tomography-an increasing source of radiation exposure. *N. Engl. J. Med., 357*, 22, 2277–2284, 2007.
72. Skoronski, E., Fernandes, M., Magalhães, M.D.L.B., Da Silva, G.F., João, J.J., Soares, C.H.L., Júnior, A.F., Substrate specificity and enzyme recycling using chitosan immobilized laccase. *Molecules, 19*, 16794–16809, 2014.
73. Tan, Y.-L., Liu, C.-G., Yu, L.-J., Chen, X.-G., Effect of linoleic-acid modified carboxymethyl chitosan on bromelain immobilization onto self-assembled nanoparticles. *Front. Mater. Sci. China, 2*, 209–213, 2008.
74. Wilson, G.S. and Hu, Y., Enzyme-based biosensors for *in vivo* measurements. *Chem. Rev., 100*, 2693–2704, 2000.
75. Zou, Y., Xiang, C., Sun, L.-X., Xu, F., Glucose biosensor based on electrodeposition of platinum nanoparticles onto carbon nanotubes and immobilizing enzyme with chitosan-SiO2 sol–gel. *Biosens. Bioelectron., 23*, 1010–1016, 2008.
76. He, L. and Toh, C.-S., Recent advances in analytical chemistry—A material approach. *Anal. Chim. Acta*, 556, 1–15, 2006.
77. Susanto, H., Samsudin, A., Rokhati, N., Widiasa, I., Immobilization of glucose oxidase on chitosan-based porous composite membranes and their potential use in biosensors. *Enzyme Microb. Technol., 52*, 386–392, 2013.
78. Zhang, M., Smith, A., Gorski, W., Carbon nanotube-chitosan system for electrochemical sensing based on dehydrogenase enzymes. *Anal. Chem., 76*, 5045–5050, 2004.
79. Han, E., Yang, Y., He, Z., Cai, J., Zhang, X., Dong, X., Development of tyrosinase biosensor based on quantum dots/chitosan nanocomposite for detection of phenolic compounds. *Anal. Biochem., 486*, 102–106, 2015.
80. Lin, J., Qu, W., Zhang, S., Disposable biosensor based on enzyme immobilized on Au–chitosan-modified indium tin oxide electrode with flow injection amperometric analysis. *Anal. Biochem., 360*, 288–293, 2007.
81. Wang, X., Yan, Y., Lin, F., Xiong, Z., Wu, R., Zhang, R., Lu, Q., Preparation and characterization of a collagen/chitosan/heparin matrix for an implantable bioartificial liver. *J. Biomater. Sci. Polym. Ed., 16*, 1063–1080, 2005.
82. Thanou, M., Verhoef, J.C., Romeijn, S.G., Nagelkerke, J.F., Merkus, F.W., Junginger, H.E., Effects of n-trimethyl chitosan chloride, a novel absorption enhancer, on caco-2 intestinal epithelia and the ciliary beat frequency ofchicken embryo trachea. *Int. J. Pharm., 185*, 1, 73–82, 1999.
83. Cárdenas, G., Anaya, P., von Plessing, C., Rojas, C., Sepúlveda, J., Chitosan composite films. Biomedical applications. *J. Mater. Sci. Mater. Med., 19*, 6, 2397–2405, 2008.
84. Landegren, U., Measurement of cell numbers by means of the endogenous enzyme hexosaminidase: Applications to detection of lymphokines and cell surface antigens. *J. Immunol. Methods, 67*, 2, 379–388, 1984.

85. Mao, S., Shuai, X., Unger, F., Wittmar, M., Xie, X., Kissel, T., Synthesis, characterization and cytotoxicity of poly(ethylene glycol)-grafttrimethylchitosan block copolymers. *Biomaterials*, *26*, 32, 6343–6356, 2005.
86. Rinki, K., Dutta, P.K., Hunt, A.J., Macquarrie, D.J., Clark, J.H., Chitosan aerogels exhibiting high surface area for biomedical application: Preparation, characterization, and antibacterial study. *Int. J. Polym. Mater. Polym. Biomater.*, *60*, 12, 988–999, 2011.
87. Hsu, S.C., Don, T.M., Chiu, W.Y., Free radical degradation of chitosan with potassium persulfate. *Polym. Degrad. Stab.*, *75*, 1, 73–83, 2002.
88. KEGG Pathway, *Amino sugar and nucleotide sugar metabolism*, Kanehisa Laboratories, Japan, 2009.
89. Kean, T. and Thanou, M., 2017 Chitin and chitosan-sources, production and medical applications, in: *Natural Products Desk Reference*, P.A. Williams and R. Arshady (Eds.), CRC Press, India, 2017.
90. Periayah, M.H., Halim, A.S., Saad, A.Z.M., Chitosan: A promising marine polysaccharide for biomedical research. *Pharmacogn. Rev.*, 10, 19, 39, 2016.
91. Gallo, M., Naviglio, D., Caruso, A.A., Ferrara, L., Applications of chitosan as a functional food, in: *Novel Approaches of Nanotechnology in Food*, pp. 425–464, 2016.
92. Varun, T.K., Senani, S., Jayapal, N., Chikkerur, J., Extraction of chitosan and its oligomers from shrimp shell waste, their characterization and antimicrobial effect. *Vet. World*, 10, 2, 170, 2017.
93. Yadav, M., Goswami, P., Paritosh, K., Kumar, M., Pareek, N., Vivekanand, V., Seafood waste: A source for preparation of commercially employable chitin/chitosan materials. *Bioresour. Bioprocess.*, 6, 1, 8, 2019.
94. Sahariah, P. and Masson, M., Antimicrobial chitosan and chitosan derivatives: A review of the structure–activity relationship. *Biomacromolecules*, 18, 11, 3846–68, 2017.
95. Goy, R.C., Morais, S.T., Assis, O.B., Evaluation of the antimicrobial activity of chitosan and its quaternized derivative on E. coli and S. aureus growth. *Rev. Bras. Farmacogn.*, 26, 1, 122–7, 2016.
96. Castro, F., Pinto, M.L., Almeida, R., Pereira, F., Chitosan/poly (γ-glutamic acid) nanoparticles incorporating IFN-γ for immune response modulation in the context of colorectal cancer. *Biomater. Sci.*, 7, 8, 3386–403, 2019.
97. Garg, U., Chauhan, S., Nagaich, U., Jain, N., Current advances in chitosan nanoparticles based drug delivery and targeting. *Adv. Pharm. Bull.*, 9, 2, 195, 2019.
98. Gupta, M., Sharma, A., Beniwal, C.K., Tyagi, P., Curcumin coated 3D biocomposite scaffolds based on chitosan and cellulose for diabetic wound healing. *Heliyon*, 8, 11, e11442, 2022.
99. Singh, A.K. and Pramanik, K., Fabrication and investigation of physicochemical and biological properties of 3D printed sodium alginate-chitosan blend polyelectrolyte complex scaffold for bone tissue engineering application. *J. Appl. Polym. Sci.*, 140, 12, e53642, 2023.

100. Annu, Ahmed, S., Ahmed, S., Ikram, S., Chitin and chitosan: History, composition and properties, in: *Chitosan*, pp. 1–24, John Wiley & Sons, Ltd, Hoboken, New Jersey, 2017.
101. Annu, Manzoor, K., Ahmad, S., Soundarajan, A., Ikram, S., Ahmed, S., Chapter 30 - Chitosan based nanomaterials for biomedical applications, in: *Handbook of Nanomaterials for Industrial Applications*, C. Mustansar Hussain, (Ed.), pp. 543–562, Elsevier, Amsterdam, The Netherlands, 2018.
102. Annu, Ahmed, S., Nirala, R.K., Kumar, R., Ikram, S., Green synthesis of chitosan/nanosilver hybrid bionanocomposites with promising antimicrobial, antioxidant and anticervical cancer activity. *Polym. Polym. Compos.*, 29, 9_suppl, S199–S210, 2021.
103. Annu, Ahmed, S., Kaur, G., Sharma, P., Singh, S., Ikram, S., Evaluation of the antioxidant, antibacterial and anticancer (lung cancer cell line A549) activity of: Punica granatum mediated silver nanoparticles. *Toxicol. Res. (Camb)*, 7, 5, 923–930, 2018.
104. Annu, Ahmed, S., Kaur, G., Sharma, P., Singh, S., Ikram, S., Fruit waste (peel) as bio-reductant to synthesize silver nanoparticles with antimicrobial, antioxidant and cytotoxic activities. *J. Appl. Biomed.*, 16, 3, 221–231, 2018.
105. Annu, Ali, A., Ahmed, S., Green synthesis of metal, metal oxide nanoparticles, and their various applications, in: *Handbook of Ecomaterials*, L.M.T. Martínez, O.V. Kharissova, B.I. Kharisov, (Eds.), pp. 1–45, Springer International Publishing, Cham, 2018.
106. Annu, Ali, A., Gadkari, R., Sheikh, J.N., Ahmed, S., Phytomediated synthesis of cerium oxide nanoparticles and their applications, in: *Nanomaterials and Plant Potential*, A. Husen, and M. Iqbal, (Eds.), pp. 261–284, Springer International Publishing, Cham, 2019.
107. Ahmed, S., Annu, Ali, A., Sheikh, J., A review on chitosan centred scaffolds and their applications in tissue engineering. *Int. J. Biol. Macromol.*, 116, 849–862, 2018.
108. Annu, Pandit, P., Maity, S., Bhattacharya, T., Shekh, M.I., Ahmed, S., Chapter 30 - Chitosan biobased materials in textile industry, in: *Advanced Applications of Biobased Materials*, S. Ahmed, and Annu (Eds.), pp. 717–735, Elsevier, Amsterdam, The Netherlands, 2023.
109. Gupta, M., Sheikh, J., Annu, Singh, A., An eco-friendly route to develop cellulose-based multifunctional finished linen fabric using ZnO NPs and CS network. *J. Ind. Eng. Chem.*, 97, 383–389, 2021.
110. Annu, Ahmed, S., Ikram, S., Perspectives of chitosan and alginate membranes for biomedical applications, in: *Natural Polymers: Derivatives, Blends and Composites*, vol. 2, pp. 157–182, Nova Science Publishers, Hauppauge, New York, 2017.
111. Annu, Bhat, Z.I., Imtiyaz, K., Rizvi, M.M.A., Ikram, S., Shin, D.K., Comparative study of ZnO-and-TiO2-nanoparticles functionalized polyvinyl alcohol/chitosan bionanocomposites for multifunctional biomedical applications. *Polymers*, 15, 3477, 2023.

3.2 Structure and Characterization

Structurally, alginates are found to be a linear polysaccharide that consists of two uronic acid residues: α-L-guluronate (G), βD-mannuronate (M) (Figure 3.1), and its epimer, which is linked covalently at (1–4) position and has a pKa value of 1.5–3.5 [20]. Alginates with the heteropolymeric structure are commonly available in nature; therefore, a combination of both M and G residues is used frequently, whereas a monopolymeric structure is produced at the initial step of alginate polymerization through bacterial (polyM) synthesis, where *P. Aeruginosa* acts as an inactivating catalyst for alginate epimerase [21, 22]. Residues of M and G vary substantially among alginate in polymeric structures, depending on their differentiating lengths and quantities of G, M, and MG blocks. The molecular mass of alginates varies substantially according to the source of production and its growing ability. Nevertheless, it was found that algal alginates generally exhibit a high G block content, while *P. aeruginosa* has no G-block content [23]. Another major change observed in alginates is spontaneous acetylation at O-2 and/or O-3 position. In addition, acetylating alginates by chemical treatments is found only in algal alginates [24]. The polymer composition determines its physical–chemical characteristics. The inherent visco-elasticity of alginates varies and depends upon how frequently these blocks become flexible, with changes in the sequence such as GG block < MM block <MG block. One of the prominent characteristics of alginates is their capacity to bind effectively with divalent cations and, ultimately, to create interlink hydrogel scaffolding polymers [25]. The alginate affinity towards the divalent ions is observed to be increased in the sequence of $Mg^{2+} << Mn^{2+} < Ca^{2+} < Sr^{2+} < Ba^{2+} < Cu^{2+} < Pb^{2+}$ [26], depending on the type of cations engaged with the G blocks of a polymer and the hydrogels' strength, stability, and mechanical property [27]. The alginates' oligomers have a well-defined structure, showing a cytokine-inducing action that is structurally dependent.

The O-acetyl group present in the alginate structure significantly modifies the propensities of the alginates by reflecting the conformation of polymer, the growth of chain, solubility, water capacity, molecular weight, and visco-elasticity. The stronger interaction between the chains and the water molecule results in the expansion of a chain with greater solubility, which ensures more water in acetylated alginate [28]. Alginate is in great demand in different industries such as agriculture, food, textiles, cosmetics as well as the pharmaceutical industry due to its unique composition and characteristics [163]. The guluronic acid monomer present in alginate maintains

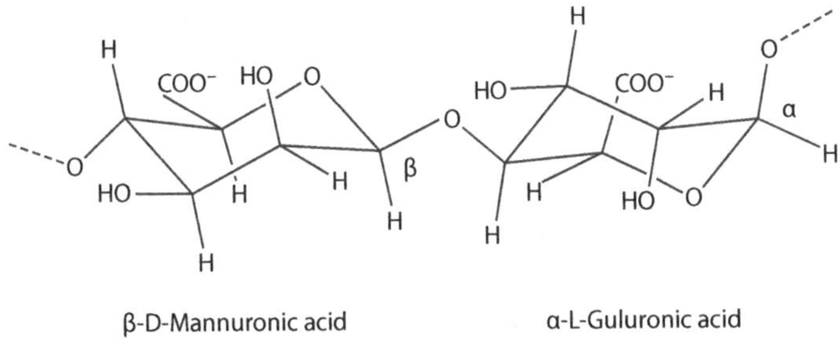

Figure 3.1 Structure of alginates [162].

mechanical rigidity and flexibility. An increase in its content increases the strength of the gel and its decreased content improves its flexible texture [29]. The pH significantly affects gel formation as a variation in its range changes the ionic form of uronic acid residues [30, 31].

3.3 General Properties of Alginates

3.3.1 Alginate Synthesis

Alginates are the most abundant polysaccharide which is commercially accessible and is generally derived from brown algae by treating it with 0.1 ± 0.2 M mineral acid and aqueous NaOH [16]. After filtering the extract, the obtained solution was precipitated either with sodium or calcium chloride to obtain the alginate salt, which was reacted with weak hydrochloric acid to change it into alginic acid. Water-soluble sodium alginate powder was synthesized once the purification and conversion step was completed [17]. The alginate content was calculated on a dry weight basis and was found to be 25%–44% for *Laminaria digitata* and 22%–30% for *Ascophyllum nodosum* [18]. Alginates can also be obtained from seaweed-derived alginate through bacterial biosynthesis (Azotobacter and Pseudomonas) by undergoing a biological reaction with extra described chemical systems. The produced alginates can retain water and form a gel with high viscosity [19].

3.3.2 Molecular Weight of Alginates

The functionality of alginate influences their molecular weight, which further determines the viscosity and gelling property, like the stiffness of the gel [32–34]. By modifying the culture conditions throughout the fermentation process, it is feasible to produce alginates with the required molecular weights [35–37]. Several studies were reported in the literature which show that the dissolved oxygen tension and oxygen transfer rate (OTR) can affect the average molecular weight and acetylation degree of the alginate [38]. The viscosity-enhancing capacity of alginate has shown its effectiveness in culture, grown-up at lower OTRmax value (2.2 mmol L^{-1} h^{-1}) compared to that obtained from the culture with higher OTRmax value (5.5 mmol L^{-1} h^{-1}). This response has been connected to a high polymer molecular weight of alginates (2,240 kDa) [39].

3.3.3 Solubility

Table 3.1 shows the description about solubility of alginates in different solvents.

3.3.4 Derivatives of Alginates

3.3.4.1 Amphiphilic Alginate

Several alginate derivatives are exploited in a broad range of therapeutic applications. One of the most commonly used derivatives is hydrophobic derivatives (e.g., alkyl chains, hydrophobic polymers) which have been synthesized into the backbone of algin [40].

The sodium alginate amphiphilic derivatives have been developed by ester bond-forming conjugation with the alginate backbone of longer alkyl chains (such as octadecyl or dodecyl). The characteristic rheological features of substantial interlinkage of gel-like networking in a half-dilution

Table 3.1 Solubility of alginates in different solvents.

Types of alginates	Solubility		
	Cold and hot water	**Fats and oils**	**Organic solvents**
Alginic acid	Insoluble	Insoluble	Insoluble
Ammonium alginate	Soluble	Insoluble	Insoluble
Sodium alginate	Soluble	Insoluble	Insoluble
Calcium alginate	Insoluble	Insoluble	Insoluble
Potassium alginate	Soluble	Insoluble	Insoluble
Propylene glycol alginate	Soluble	Insoluble	Insoluble

system, which might have been beneficial to cartilage repairs and regeneration, are shown in aqueous solutions for these derivatives [41]. The method for generating microparticles from such derivatives was conducted by dispersing them into the solution of sodium chloride, which resulted in protein encapsulation and the eventual release of the drug, by either introducing surfactants that interrupt intermolecular hydrophobic interaction or by hydrolyzing the ester linkage between the alginate backbone and the alkyl chains [42]. Doxylamine was coupled with the alginate backbone using 2-chloro-1-methylpyridine iodide as a coupled reagent *via* the formation of an amide connection. In comparison to the hydrolysis-labile products made from alginate with dodecyl ester, the hydrogels were found to be stable in an aqueous medium for a longer duration of time [43]. Using N,N′-dicyclohexycarbodiimide as a coupling negotiator and 4-(N,N′-dimethylamine) at room temperature, pyridine was indeed produced as a catalyst with water-soluble, cholesteryl-grafted amphiphilic derivatives of alginate. The derivatives were found to develop dispersed aggregates having 136 nm as mean diameter in an aqueous solution of NaCl [44]. Unlike water derivatives, the associative behavior of poly(μl-caprolactone) (PCL)-grafted alginate was highly based on the duration of the PCL chain [45]. Poly-butyl methacrylate, which has led to the extended liberation of the drug in contrast to the unmodified alginate gels, has also changed the sodium alginate hydrophobically [46].

3.3.4.2 Cell-Interactive Alginate

The cell-adhesive peptide form of the alginate derivatives has recently achieved substantial significance in the pharmaceutical industry. These derivatives are often produced utilizing carbodiimide to combine through carboxylic groups of residues by adding peptides as side chains. Alginate is naturally non-adhesive towards mammalian cells, where ligands are required to stimulate and control cellular connections, particularly in cell culture and tissue engineering applications. The widespread presence of the integrated receptors of this ligand in different cell types (e.g., $\alpha v\beta 3$, $\alpha 5\beta 1$), [47, 48] has led to the broad usage of peptides with the sequence of arginine–glycine–aspartic acid (RGD). RGD containing peptide has a water-soluble carbodiimide which may be chemically linked to the backbone of alginate [49]. A minimal amount of RGD peptide in alginate gels is required for cellular adhesion and development, which was likely cell type-specific. Minimum dosages of alginate gel, required for adherence to C2C12 and MC3T3-E1 cells in *in vitro* experiments, are found to be 10.0 and 12.5 g/mg [50–52]. It was discovered that cycle RGD peptides are stronger than linear RGD peptides at low levels and play an important role in RGD peptide affinity [53]. Diverse DGEA (Asp–Gly–Glu–Ala) and YIGSR (Tyr–Ile–Gly–Ser–Arg) peptides can amend the alginate gels and boost the adhesive attraction with different types of cells. By using water-soluble carbodiimide, alginate was modified to enhance neuronal cell addiction using YIGSR peptides [54].

3.4 Alginate and Alginate Composites as a Drug Delivery System

3.4.1 Fiber Form

The utilization of alginates as biomedical material in fiber form shows great potential due to enhanced characteristics such as ease of handling, enhanced surface area, and tendency to preserve the mechanical integrity under most circumstances. Moreover, alginates possess the absorbent nature that is essentially required for heavy oozing of wound dressing [55]—for example, in a hydro-entangled non-weaving network of calcium alginate fiber (3MTM Tegaderm™), alginate shows high integrity and gelling ability in wound dressings by easily forming a gel, maintaining a healthy environment around it, and retaining its integral form upon contact with the exudates [56]. However, the lower-molecular-weight alginate forms a fibrous network which might collapse due to being underweight when allowed to swell in water because of weak bonding among the particles present in it. By combining alginate fibers with saline-free fibers, the washing at the wound site becomes effective, which increases the overall strength of the fabric and reduces the total dressing cost. The incorporation or reinforcing of another biopolymer into the alginate fiber itself results in the formation of a composite fiber that is much stronger and is capable of providing additional cure [57].

3.4.2 Bead Form

Alginate beads are produced from a crosslinking solution which is squeezed out in the form of alginate drops from a syringe or microfluidic device. The alginate bead is commonly used in biomedical devices as a cell and drug immobilization medium. The bead can potentially be utilized as a tissue healing scaffold. However, if a substantial quantity of sodium ions present in body fluid diffuses into the bead construct, it can impair its structural integrity by exchanging with Ca^{2+}, a typical ionotropic crosslinking moiety employed in bead production. Sarker *et al.* (2021) utilized partially oxidized alginate to covalently bind with gelatine *via* the formation of Schiff's base. This was carried out by the interaction between free amino groups present in gelatine's lysine or hydroxylysine amino acid residues that reacted with the accessible aldehyde group of oxidized alginates [58]. The mechanical assets of the oxidized gelatine or alginate beads may be altered by varying their composition, resulting in a stiffer structure with a large number of oxidized moieties [59]. In addition to that, these composite beads demonstrate a high rate of deterioration than that of pure alginate beads due to the low molecular weight of oxidized alginate. By adjusting their composition, oxidation, and crosslinking level, the rate of deterioration of composite beads may be further controlled. The introduction of inorganic materials such as nano-hydroxyapatite (nHAP) and bioglass into alginates shows its potential to activate osteoblast cells that provide a nanocomposite assembly of bone, producing the collagen fibril that forms the newer apatite crystals [60]. In addition to cellular and inorganic materials, the integration of drugs and other materials, such as chitosan with alginates, also helps to improve the structure of alginate beads [61–63].

3.4.3 Hydrogel Form

An alginate-based hydrogel has an extensive range of applications both in the non-injectable and injectable types of formulations. Before *in vivo* implantation, a non-injectable type of alginate hydrogel is pre-designed in the form of liquid before reaching the appropriate site and forming gel under certain stimulatory conditions [64]. These two types of hydrogels are frequently mixed in particle or fiber form of foreign material to make alginate composite hydrogel with enhanced structural integrity and desired characteristics. When alginate is used as scaffolding material in the pure form, cell clusters are observed frequently; this is because of the absence of adhesion moieties in alginate that help in cell attachment, which results in small aggregates [163]. Therefore, gelatine or other bioactive substances are attached to the alginates to stimulate cell adhesion and imitation of the natural cellular microenvironment, which helps to increase cell growth and proliferation [65, 66]. Natural ECM has a composite design that is fiber-reinforced and compatible with the aquatic, fiber-based protein-reinforced matrix. This biomimetic approach can result in a mechanically strong composite that is adapted for the applications of tissue engineering using fibrous collagen of varied arrangements inside the alginate hydrogel [67]. When electrospun gelatine nanofiber was in line to strengthen alginate hydrogel, the tensile module and tensile strength have increased in

Table 3.2 Biomedical applications of alginates. (*Continued*)

S. no.	Material used	Substances present in formulation	Dosage form	Biomedical application	References
17.	Alginate/ gelatin/ cellulose	-	Hydrogel scaffold	Biomaterial for tissue engineering	[159]
18.	Alginate–gelatin polypyrrole	-	Hydrogel	Promising for cellular therapies, *in vitro* cell culture, and electrical stimulation-assisted tissue engineering	[160]
19.	Alginate	-	Microgel	Cell delivery in tissue engineering and regenerative medicine	[161]

3.5.1 Wound Dressing

Wound healing is a complex course of action that necessitates the treatment of pathogenic microorganisms. As a result, wound dressing has received mindfulness and is extensively used to speed wound healing and control infection. Wound dressings used traditionally consist of a sterile pad and gauze, which provide the advantage of limiting germ invasion and keeping the wound dry [69], but certain drawbacks become apparent, such as poor vapor transfer, susceptibility to bacterial infection, and its easy adherence to the site, which makes it difficult to remove, all of which significantly impede wound healing. To speed up the healing process and control the bacterial infections at the wound site, several biomaterials like proteins, proteoglycans, and polysaccharides have been studied, which provide high water content, biodegradability, and biocompatibility, thus producing a contemporary dressing [70]. The contemporaneous dressing replicates the characteristic of a natural tissue by providing a humid environment, improved permeability, and autolytic debridement which promotes re-epithelialization and accelerates wound healing [71]. Recent studies have suggested that the wound healing process is accelerated through wound dressing by providing a physical and mechanical barrier that prevents the loss of blood and resists the entry of bacteria. Wound management is a strenuous task despite the progression made in the material used for wound dressing. The cost of wound management is very high, where a growing population with an enhanced lifestyle demands an effective and affordable wound dressing [72]. Wound dressing prepared by different biopolymers, synthetic polymer, and biomaterial has resulted in an effective strategy fulfilling the gap of affordability among the population [73]. Its exceptional features like biocompatibility, affordability, non-toxicity, non-immunogenicity, and higher absorption capability can speed up chronic wound healing [74]. Alginate dressings are currently available in a variety of forms, including hydrogel, films, membranes, and sponges [75, 76]. It enhances physicochemical stability by the absorption of wound fluid, maintaining a moist environment physiologically, maintaining optimum pH, and reducing bacterial contaminations [77, 78]. The complex is formed by using different polymers which form an ionic network and helps in improving the swelling capacity, erythrocyte adhesion, cytocompatibility, antimicrobial, and drug/growth factor release. However, one drawback coincides with the use of alginate as a biomaterial, i.e., its mechanical property and the degradation pattern. The use of other polymers in combination with alginates in a different ratio helps in increasing the therapeutic efficacy by improving the mechanical strength of the formulation [163]. The crosslinkers are used for combining the different blends of alginate and polymer whose nature, time of crosslinking, and excipient define the fate of the formulation [79]. A very common combination of chitosan and alginate is used to form a bioactive inter-polymer complex in the presence of ions which form wound dressing material in the case of nanoparticles, beads, gauze, films, and hydrogels [80].

Kaygusuz H *et al.* in 2017 prepared a combination of alginate and chitosan wound dressing material crosslinked with cerium (III) solution. The elastic modulus of the sample was found to be higher due to the presence of chitosan and alginate. The film's contact angle was found to be 52.6° ± 8.5°, which forms a moderately wettable surface that helps in the linear growth of the cells. The film showed good wound dressing features with reduced growth of microorganisms during 3 h of exposure period [81]. Bagher Z *et al.* in 2020 produced alginate and chitosan (Alg/Chit) hydrogels for wound healing, which consist of different ratios of hesperidin. The SEM images revealed that the produced hydrogel was very porous, with a porosity of around 91.2 5.33 percent and linked pores. The size of the pore was found to be in the range of 63 to 144 m, which aided cell adhesion and motility. The Alg/Chit chains engage with the water molecules, causing the polymer to inflate, resulting in a 3D shape of dressing material that replicates the natural microenvironment of cells and promotes cellular proliferation, survival, and migration. The alginate provides a hemostatic effect, thus helping in plugging up the damaged site. Chitosan provides biodegradability, stability, biocompatibility, and

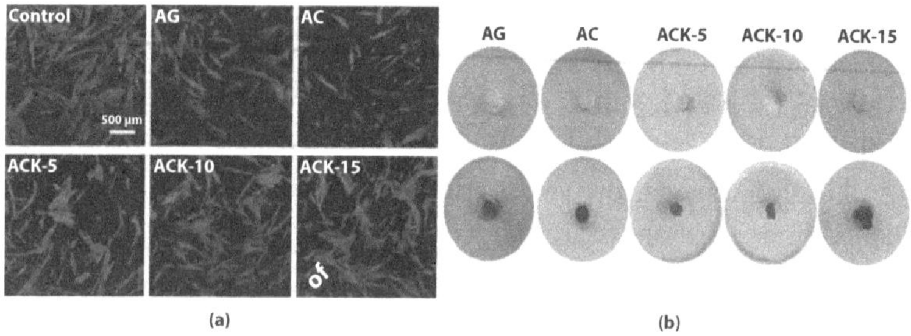

Figure 3.2 (a) *In vitro* blood coagulation test on composite sponges. (b) Cytocompatibility analysis of the sponges [84].

non-toxicity in a formulation used for the treatment of wounds. Thus, each polymer has excellent gel-forming properties [82]. In a study conducted by He *et al.* [78], where a sponge was developed utilizing alginate (AG)/carboxymethyl chitosan (CMC) containing Kangfuxin (KFX) a traditional chinses medicine. A multifunctional wound dressing was tested on rat to determine the accelerated blood coagulation, bacterial infection, and wound recovery. The result indicated the faster coagulation effect due to the presence of amine residue and a highly porous structure. The CMC binding with Ca^{2+} initiated faster coagulation of blood *via* activation of clotting factors VII, IX, and X and platelets [83]. The *in vitro* blood coagulation test and cytocompatibility of alginate-based sponges are depicted in Figure 3.2.

3.5.2 Heart and Cardiovascular Disease

Heart failure and cardiovascular diseases are the rising indigenous problems faced by the aging western population, which accounts for more than 20 million individuals globally for mortality and morbidity. The existing therapies used to treat heart failure are unable to predict its fundamental cause for the loss of cardiomyocytes. The novel strategies are being focused more, which help in the development of the therapeutic regeneration of the myocardium and the heart valve, and thus improve tissue salvage and heart function, which helps to compensate for the inefficiency of the existing strategies. New approaches are extensively studied, which embrace multicellular techniques and new material to accomplish the functional integrated vasculogenic and myogenesis in the heart/tissue [85]. Alginate, a multifaceted and flexible biomaterial that is used for the first time for cardiovascular diseases, is involved in cellular preservation and delivery and shows great potential for cardiac tissue [86]. Alginate acts as an ECM substitute in heart tissue, a delivery vehicle for stem cells of the cardiac system, and provides sustained/controlled delivery of growth factors [87]. It is found to be most effective in cardiac regeneration and valve replacement techniques, treatment of congestive heart failure and myocardial infarction, and bioengineered blood vessel fabrication using alginate-based methods [88].

3.5.3 Drug Delivery

Traditional drug delivery systems are currently not so effective in providing any acceptable medicinal profiles for the therapy of numerous illnesses. Thus, an advanced drug delivery system with a modified property that helps in the effective delivery of drugs to the specified site is required. This type of delivery system helps to deliver weakly water-soluble drug compounds, which gains the ability to transport and release a drug molecule to a target site [89]. A wide variety of polymers has been identified and tested for drug delivery into the body. Alginate is very effective in this list because of its high biocompatibility and gelation properties and has attracted interest in a variety of drug delivery applications. It has a high capacity to interact with drug molecules and forms a gel in a specified environment, which helps in modifying the drug release. An alginate gel's pore size is roughly 5 to 6 nm, which aids in drug molecule diffusion and release and thus increases the permeability of a drug [90]. Alginate can be employed to deliver microcapsules for long-term drug release, which has perked interest due to its several benefits such as biodegradability, higher loading rate of drugs, non-immunogenicity, and non-toxic nature [91]. A detailed investigation of polymers to investigate their purity and their improvement helps to increase the biocompatibility of alginate, which shows a wide application in drug delivery for multiple diseases, including tumour, diabetes mellitus, and neurological disorders. The ability to entrap various cell types, as well as the prospect of manipulating cells to create a variety of therapeutic biomolecules, has spawned a slew of new use of alginate [92]. It also responds to stimuli that show a promising drug release profile. Furthermore, glucocorticoid contained in alginate microcapsules is a new approach for treating osteoarthritis. A hollow silver alginate microcapsule successfully supplies the water-insoluble glucocorticoid betamethasone to their respective site. This has proved to improve loading efficiency and sustained release, lowering intra-articular glucocorticoid administration injury rates. The effectiveness of a pH-sensitive alginate carrier was explored, specifically for localized drug delivery systems that controlled the drug distribution based on pH sensitivity. AlgPD-BTZ hydrogel was produced by utilizing the alginate-coupled polydopamine as a structural unit [93]. The catechol group forms a covalent connection with the boron acidic group of the drug BTZ, which demonstrated that BTZ is released selectively to cancerous cells using a pH-dependent approach. In another study, hydroxyapatite was combined with sodium alginate as a bilayer coating in an iron oxide nanoparticle to obtain the delivery of therapeutic molecules in a pH-sensitive manner. Curcumin had a greater encapsulation efficiency of 93.03%, whereas 6-gingerol had a higher encapsulation efficiency of 98.78 % [94]. The electrostatic interactions of molecules might function as a pH-dependent adjustable gate, holding and releasing therapeutic molecules. Due to the leaky vasculature, the pH-triggered mechanism of drug release plays a vital role in the release of cancer drugs [95].

A drug delivery system was developed using magnetic nanoparticles. Magnetic nanoparticles can modulate diffusion and aggregation at the desired areas during the external magnetic field procedure. As a result, the system was able to achieve a high local concentration at the target locations. Because of the superparamagnetic OA-Fe_3O_4 nanoparticles, magnetic reduction-responsive alginate-based microcapsule systems have been shown to have strong magnetic targeting ability [96]. Alginate and cellulose nanocrystals were used to create a double-membrane hydrogel recently. The findings revealed that complexing antibiotic medicines had good release characteristics. The outside tidy alginate hydrogel completed the fast drug release, whereas the inner hydrogel was responsible for the extended release [97]. Zhong *et al.* in 2013 created a dual-drug delivery method using poly-(D,L-lactic) microspheres embedded into the hydrogel beads of calcium alginate. The liberation of glycyrrhetinic acid from microspheres was prolonged, while the release of BSA from a hydrogel was fast [98]. Furthermore, alginate hydrogels are a good option for protein drug delivery and have been investigated. Protein embedded in hydrogels is preserved, reducing denaturation and preventing breakdown. Alginate hydrogels containing neovascularization-adjusting agents, such as vascular endothelial growth factor (VEGF), showed a sustainable liberation of around 60% of total VEGF after 7 days [99]. The rate of protein release may be regulated by adjusting the alginate hydrogel's breakdown rate. Furthermore, sodium alginate–bacterial cellulose hydrogels have shown promise in transporting protein-based medicines such as lysozyme *via* electrostatic adsorption [100].

3.5.4 Tissue Engineering

According to the National Science Foundation, tissue engineering is a treasured methodology that may be utilized to substitute, retain, and augment tissues and organs [101]. It encompasses the acquisition of seed cells, tissue and organ manufacturing, and therapeutic applications. Cell development was encased and supported by the biological scaffold materials. Then, they were transplanted into the body and generated the required tissues gradually. Among the key research areas are bone [102], cartilage [103], vascular [104], ocular tissues [105], skin [106], and several other tissues [107].

Alginate is a well-known biomaterial with exceptional scaffold formation characteristics that are broadly investigated and used in tissue engineering to repair organ loss or failure. In most cases, physical or chemical methods have been employed to combine it with other components to develop new alginate derivative products. The numerous features and functionalities result in improved tissue engineering properties, including cellular affinity, mechanical strength, and gel-forming capability. As a result, alginate and its composites have sparked considerable interest in the field of tissue engineering [108].

3.5.4.1 Bone Tissue Engineering

Bone tissue engineering has attained a lot of consideration as a feasible alternative therapy for repairing bone abnormalities [109]. The ultimate objective of this methodology is to create bone tissue that is structurally, physically, and chemically identical to natural bone tissue [110]. Alginate is reported as a synthetic scaffolding agent for bone repair and as a source of bone tissue engineering transfer cells [111]. The mechanical asset of the scaffold is necessary for bone tissue engineering to allow the regeneration of bones. The scaffolds of alginate composite have been developed by integrating the inorganic compounds or polymers, including collagen, chitosan, and hydroxyapatite, to offer adequate mechanical properties [112]. Chitosan is used to make alginate–chitosan composites because it is the most prevalent cationic polysaccharide. For bone tissue engineering, chitosan/alginate composite scaffolds are being studied intensively. It is possible to achieve stiff strength and structural stability. Florczyk *et al.* in 2013 developed a framework of chitosan–alginate with an improved compressive strength of 0.79–1.41 MPa. It has a homogenous pore structural form, and the size of the pore is determined by the concentrations of acetic acid and alginate. When the viscosity was below 300 Pas, the cell proliferation potential was also enhanced. At 16 weeks, the highest defect closure (71.56%–19.74%) was recorded [113]. Chitosan/alginate scaffolds have been shown to promote osteogenesis and vascularization [114]. Because of its unique characteristics, such as promoting cell attachment and breakdown, collagen is commonly used to make scaffolds. Collagen I, as an important constituent of bone tissue's ECM, aids in osteoblast migration and vascular dispersion [115, 116]. As a consequence, alginate/collagen hybrid scaffolds affect the osteoblasts' osteogenic potential. It may endorse cellular proliferation, osteogenic differentiation, and cell spreading. Sotome *et al.* 2004 showed that hydroxyapatite/collagen–alginate can transport the recombinant human bone morphogenetic protein-2 efficiently. Bone development happens within 5 weeks following implantation without visible deformity [117]. Alginate-mixed HA (hydroxyapatite) composites have a structure similar to bone's natural extracellular matrix. The porous structures connected, and the high porosity promotes cell survival, proliferation, adhesion, osteoblastic phenotypic maintenance, and bone repair [118]. Lin *et al.* in 2004 discovered that the improved composite porous scaffold is alginate/HA, having an average of 150-m pore size and porosity of more than 82% [119]. The alginate/HA composite scaffolds had 84% porosity, according to another study [120]. The consistent pore architecture is beneficial to improve the elastic modulus and compressive strength, both of which are related to the amount of HA present. Mechanical strength, cell proliferation, and morphology should all be good in an acceptable scaffold to support bone tissue development [121]. N. Firouzi *et al.* in 2020 discovered that adding HA to an alginate-based hydrogel reduced the degradation rate by 41.5% and increased the compressive modulus by 2,942.5 kPa [122]. Alginate-based formulation is very attractive to use in case the host cell is unable to respond to morphogens or if the process of healing needs to be accelerated. The alginate accelerates the transplant cell population that fasten the bone formation and recovery process. One such study was conducted on mice where the transplantation of calvarial osteoblasts of the primary rat was conducted successfully in the mice in an incidence of RGD–alginate gels which enhanced the bone formation process in comparison to controlled alginate gels. It also enhanced the formation of growth-plate-like structures, which helps to replace dysfunctional epiphyses [123]. The improvement in bone formation in mice is depicted in Figure 3.3.

Silva-Barroso *et al.* in 2023 presents a study on the development of 3D scaffolds for bone tissue regeneration using a combination of tricalcium phosphate, sodium alginate, and lignin. The formulation was fabricated by the scaffolds using a freeze-drying technique and characterized their properties, including their morphology, porosity, and mechanical strength.

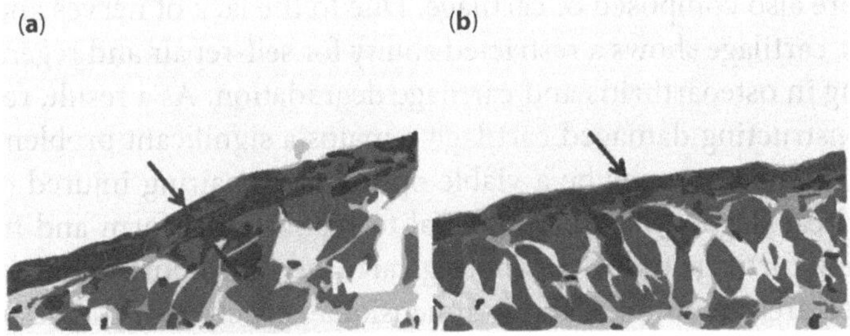

Figure 3.3 Improvement in bone formation in mice using alginate-based gel. (a). Damages skin, (b). Treated skin [6].

The results showed that the addition of lignin improved the compressive strength of the scaffolds while maintaining their porosity and biocompatibility. It was evaluated for the ability of the scaffolds to support the growth and differentiation of human bone marrow mesenchymal stem cells *in vitro*. Overall, the study demonstrated the potential of lignin-enriched tricalcium phosphate/sodium alginate scaffolds for bone tissue regeneration. The use of lignin could provide a cost-effective and sustainable way to enhance the mechanical properties of the scaffolds, while the combination of tricalcium phosphate and sodium alginate could promote bone growth and support cell growth and differentiation. Further research is needed to evaluate the effectiveness of these scaffolds *in vivo* and to optimize their design for specific bone regeneration applications [124].

Hakimi *et al.* in 2023 presents the development and characterization of a nanofiber hydrogel scaffold composed of chitosan–polyethylene oxide/clay–alginate for bone tissue engineering applications. The scaffold was prepared using an electrospinning technique, and its physical properties were evaluated, including its morphology, porosity, and mechanical strength. The scaffold was also tested for its ability to support biomimetic mineralization, a process that mimics the natural mineralization of bone tissue. The results showed that the scaffold had a highly porous structure, good mechanical properties, and the ability to support mineralization. The authors conclude that the chitosan–polyethylene oxide/clay–alginate nanofiber hydrogel scaffold has potential as a biomimetic scaffold for bone tissue engineering applications [125].

3.5.4.2 Cartilage

Articular cartilage is a highly specialized and structured tissue that is extremely hydrated (around 80%), aneural, and devoid of lymphatic vessels or blood [126]. Other bodily parts, such as the ear, nose, and bronchial tubes, are also composed of cartilage. Due to the lack of nerves and blood arteries, cartilage shows a restricted ability for self-repair and regeneration resulting in osteoarthritis and cartilage degradation. As a result, repairing or reconstructing damaged cartilage remains a significant problem [127]. Tissue engineering may be a viable option for repairing injured or deteriorated cartilage. It has the potential to replicate the form and function of human cartilage tissue, promoting cartilage development and repair in damaged areas [128]. The three-dimensional structure of the tissue engineering scaffold allows for cellular adhesion and development. Alginate scaffolds are shown to be effective in cartilage tissue regeneration. It stimulates the 2D culture-expanded redifferentiation of dedifferentiated chondrocytes, implying that it can enhance chondrocyte cell proliferation, repair injured cartilage, and maintain chondrocyte characteristics [129]. The culture of osteogenic and chondrogenic cells is aided by a chitosan–alginate scaffold. Alginate–chitosan and HA scaffolds showed a distinct and consistent porous networking structure, according to A.E. [130]. Erickson *et al.* in 2019, polymer content improved the compressive modulus and rigidity. In addition, hyaluronate was investigated to strengthen the alginate and enhance the proliferation of chondrocytes. The alginate–hyaluronic acid hydrogel has been shown to have stable physicochemical characteristics [131]. After 3 weeks of culture, these gels successfully preserved the chondrogenic characteristics. The findings showed that chondrogenic genes are efficiently expressed and cartilage ECMs are formed. Another alginate hybrid gel created by mixing polyacrylamide with alginate was shown to have exceptional mechanical strength, including high toughness (up to 9,000 J/m^2) and stretch ratio [132, 133]. Alginate–polyacrylamide also showed good tribological behavior, ultra-low coefficient, and great wear resistance [134].

Chondrocytes exhibited an ordered distribution and better integration with surrounding tissue after transplantation [135]. These hybrid gels can be utilized as cartilage implants because alginate and its composites promote chondrogenic differentiation and cell proliferation.

Krishna *et al.* in 2023 state the use of extrusion-based bioprinting for developing alginate-based hydrogels for tissue regeneration applications. The properties of alginate that make it an attractive material for tissue engineering, such as its biocompatibility and ability to form hydrogels, were studied. Additionally, the challenges associated with using alginate in bioprinting, such as controlling the gelation process and optimizing the printing parameters, was also focused on. Different approaches to overcome these challenges, including using multicomponent hydrogels that incorporate other materials, such as gelatin or fibrin, to improve the mechanical properties and cell adhesion of the printed structures were discussed. Extrusion-based bioprinting to develop alginate-based hydrogels for various tissue regeneration applications, such as cartilage and bone regeneration, was studied. The article concludes with a discussion of the future directions in the field, such as the use of more complex printing strategies and the development of new alginate-based formulations that can better mimic the native extracellular matrix of different tissues. Overall, the article provides a comprehensive overview of the current state of extrusion-based bioprinting of alginate-based hydrogels for tissue regeneration applications [136].

3.5.5 Liver Tissue Engineering

Liver illness is a danger to human well-being and is considered one of the leading causes of death. Every year, around 2 million people die throughout the world [137]. The most effective treatment for this condition is liver transplantation. However, because of a scarcity of appropriate liver donors, this is an extremely challenging process. Liver tissue engineering is thought to be the most effective means of providing liver tissue to fulfil the liver's high demand. It has the potential to increase the liver's function and build a whole organ. Hepatocyte growth can be aided by alginate composites in liver tissue engineering. Rajalekshmi *et al.* in 2021 found that a fibrin-infused injectable type of alginate dialdehyde–gelatin hydrogel efficiently promotes hepatic cell growth, proliferation, and functioning [138]. The reason for this is that fibrin has many cells adhesion pockets where cells can adhere. ECM, a liver tissue engineering scaffold, was made using oxidized alginate and glycosylated chitosan in a Schiff base reaction. The hepatocytes took on a spheroidal shape after being grown in the scaffolds. Also seen were multicellular aggregates and complete integration [139]. Another method for liver tissue engineering is the formation of microcapsules. Microcapsules may encapsulate cells and turn them into microbeads. The semipermeable membrane was then used to encase the microbeads, forming microcapsules [140]. Microcapsules offer a secure milieu for cells while also shielding them from outside influence. They multiply and are visible for at least 12 days after HepG2 cells were encased inside alginate-based microbeads. According to these studies, alginate might be a viable choice for LTE techniques [141]. Chen *et al.* in 2012 discussed the preparation and characterization of a scaffold made of oxidized alginate and covalently cross-linked galactosylated chitosan, which has potential applications in liver tissue engineering. The scaffold was created using a combination of chemical reactions and freeze-drying techniques and was evaluated for its physicochemical properties, such as porosity and swelling behavior. The results showed that the scaffold had suitable properties for use in liver tissue engineering, including good porosity and swelling behavior, and the ability to support the growth and differentiation of liver cells. The authors suggest that this scaffold has potential as a tool for the development of therapies for liver diseases or for drug screening purposes [142].

3.6 Conclusion and Future Prospective of Alginates

For various biomedical applications, alginate has demonstrated a significant useful and possible biomaterial. The most striking characteristics of alginate for these functionalities include mild gelation circumstances, biocompatibility, and simpler modifications to formulate alginate derivatives with newer properties. Alginate has been reported to be safe in clinical applications like dressing material in the healing of wounds and securely implanted in numerous therapeutic applications. The use of chemically modified alginate as a carrier to stimulate periodontal regeneration is also common. Alginate gels, like other hydrogels, have low mechanical rigidity and a variety of other physical characteristics. The use of molecules with varying molecular weights, chemical structures, and cross-linking activity will frequently result in gels that are suited for each application. Although cell encapsulation techniques are not the subject of this paper, many of the principles discussed are immediately applicable. However, the covalent cross-linking reactions may be detrimental to the cells being encapsulated. Thus, in those cases, a vigilant selection of cell-compatible chemical reagents (like initiator) and exclusion of unreacted reagents and their by-products would be required. Furthermore, our understanding of alginate's fundamental characteristics and development of new forms of cell and tissue-interactive alginate gels might pave the way for future biomedical research and engineering breakthroughs. Using genetic engineering approaches to regulate bacterial production, it may be possible to build and make novel alginate polymers with improved or unique characteristics. This method has already been used to produce several polypeptides and proteins with enhanced structural characteristics and new functionalities for medicinal applications. In contrast to the limited matter accessible from natural resources, it might reform the usage of these materials by creating newer derivatives of alginates with carefully regulated physicochemical properties suited for a particular use. For future perspective, alginates can be used for three-dimensional ALG cell culture systems, the biological activity of ALG oligosaccharides, etc.

References

1. Reddy, M., Ponnamma, D., Choudhary, R., Sadasivuni, K.K., A comparative review of natural and synthetic biopolymer composite scaffolds. *Polymers*, 13, 7, 1105, 2021.
2. Sheikh, Z., Najeeb, S., Khurshid, Z., Verma, V., Rashid, H., Glogauer, M., Biodegradable materials for bone repair and tissue engineering applications. *Materials*, 8, 9, 5744–5794, 2015.
3. Draget, K.I., Alginates, in: *Handbook of Hydrocolloids*, pp. 807–828, Woodhead Publishing, Sawston, Cambridge, 2009.
4. GheorghitaPuscaselu, R., Lobiuc, A., Dimian, M., Covasa, M., Alginate: From food industry to biomedical applications and management of metabolic disorders. *Polymers*, 12, 10, 2417, 2020.
5. Szabo, L., Gerber-Lemaire, S., Wandrey, C., Strategies to functionalize the anionic biopolymer Na-alginate without restricting its polyelectrolyte properties. *Polymers*, 12, 4, 919, 2020.
6. Lee, K.Y. and Mooney, D.J., Alginate: Properties and biomedical applications. *Prog. Polym. Sci.*, 37, 1, 106–126, 2012.
7. Kim, H.S., Lee, C.G., Lee, E.Y., Alginate lyase: Structure, property, and application. *Biotechnol. Bioprocess Eng.*, 16, 843–51, 2011 Oct.
8. Abasalizadeh, F., Moghaddam, S.V., Alizadeh, E., Akbari, E., Kashani, E., Fazljou, S.M., Torbati, M., Akbarzadeh, A., Alginate-based hydrogels as drug delivery vehicles in cancer treatment and their applications in wound dressing and 3D bioprinting. *J. Biol. Eng.*, 14, 1–22, 2020 Dec.
9. Szekalska, M., Puciłowska, A., Szymańska, E., Ciosek, P., Winnicka, K., Alginate: Current use and future perspectives in pharmaceutical and biomedical applications. *Int. J. Polym. Sci.*, 2016, 7697031, 1–17, 2016.
10. Yang, W., Xu, H., Lan, Y., Zhu, Q., Liu, Y., Huang, S., Shi, S., Hancharou, A., Tang, B., Guo, R., Preparation and characterisation of a novel silk fibroin/hyaluronic acid/sodium alginate scaffold for skin repair. *Int. J. Biol. Macromol.*, 130, 58–67, 2019.
11. Aderibigbe, B.A. and Buyana, B., Alginate in wound dressings. *Pharmaceutics*, 10, 2, 42, 2018 Apr 2.
12. Raguavaran, R., Mondal, D.B., Sharma, D.K., Jithin, M.V., Yadav, N., Biomedical applications of alginate nanoparticles, in: *Polysaccharide Nanoparticles*, pp. 311–328, Elsevier, Amsterdam, Netherlands, 2022 Jan 1.
13. Sun, J. and Tan, H., Alginate-based biomaterials for regenerative medicine applications. *Materials*, 6, 4, 1285–1309, 2013.

14. El-Sherbiny, I.M. and Yacoub, M.H., Hydrogel scaffolds for tissue engineering: Progress and challenges. *Glob. Cardiol. Sci. Pract.*, 2013, 3, 38, 2013.
15. Raus, R.A., Nawawi, W.M., Nasaruddin, R.R., Alginate and alginate composites for biomedical applications. *Asian J. Pharm. Sci.*, 16, 3, 280–306, 2021 May 1.
16. Chanda, S.K., Hirst, E.L., Percival, E.G., Ross, A.G., 338. The structure of alginic acid- Part II. *J. Chem. Soc.*, 338, 1833–1837, 1952.
17. Rinaudo, M., Main properties and current applications of some polysaccharides as biomaterials. *Polym. Int.*, 57, 3, 397–430, 2008.
18. Qin, Y., Alginate fibres: An overview of the production processes and applications in wound management. *Polym. Int.*, 7, 2, 171–180, 2008.
19. Hay, I.D., Rehman, Z.U., Moradali, M.F., Wang, Y., Rehm, B.H., Microbial alginate production, modification and its applications. *Microb. Biotechnol.*, 6, 6, 637–650, 2013.
20. Park, D., Jagtap, S., Nair, S.K., Structure of a PL17 family alginate lyase demonstrates functional similarities among exotype depolymerases. *J. Biol. Chem.*, 289, 12, 8645–8655, 2014.
21. Moradali, M.F., Ghods, S., Rehm, B.H., Alginate biosynthesis and biotechnological production, in: *Alginates and Their Biomedical Applications*, pp. 1–25, Springer, Singapore, 2018.
22. Draget, K.I. and Taylor, C., Chemical, physical and biological properties of alginates and their biomedical implications. *Food Hydrocoll.*, 25, 2, 251–256, 2011.
23. Jönsson, M., Allahgholi, L., Sardari, R.R., Hreggviðsson, G.O., Nordberg Karlsson, E., Extraction and modification of macroalgal polysaccharides for current and next-generation applications. *Molecules*, 25, 4, 930, 2020.
24. Sutirman, Z.A., Sanagi, M.M., Ibrahim, W.A., Alginate-based adsorbents for removal of metal ions and radionuclides from aqueous solutions: A review. *Int. J. Biol. Macromol.*, 174, 216–228, 2021.
25. Axpe, E. and Oyen, M.L., Applications of alginate-based bioinks in 3D bioprinting. *Int. J. Mol. Sci.*, 17, 12, 1976, 2016.
26. Brus, J., Urbanova, M., Czernek, J., Pavelkova, M., Kubova, K., Vyslouzil, J., Abbrent, S., Konefal, R., Horský, J., Vetchy, D., Vysloužil, J., Structure and dynamics of alginate gels cross-linked by polyvalent ions probed via solid state NMR spectroscopy. *Biomacromolecules*, 18, 8, 2478–2488, 2017.
27. Bhattacharyya, D., MirihanaArachchilage, G., Basu, S., Metal cations in G-quadruplex folding and stability. *Front. Chem.*, 4, 38, 2016.
28. Cunha, L. and Grenha, A., Sulfated seaweed polysaccharides as multifunctional materials in drug delivery applications. *Mar. Drugs*, 14, 3, 42, 2016.
29. Kim, H.S., Lee, C.G., Lee, E.Y., Alginate lyase: Structure, property, and application. *Biotechnol. Bioprocess Eng.*, 16, 5, 843–851, 2011.
30. Gawkowska, D., Cybulska, J., Zdunek, A., Structure-related gelling of pectins and linking with other natural compounds: A review. *Polymers*, 10, 7, 762, 2018.

31. Nair, L.S. and Laurencin, C.T., Biodegradable polymers as biomaterials. *Prog. Polym. Sci.*, 32, 8–9, 762–98, 2007.
32. Ramos, P.E., Silva, P., Alario, M.M., Pastrana, L.M., Teixeira, J.A., Cerqueira, M.A., Vicente, A.A., Effect of alginate molecular weight and M/G ratio in beads properties foreseeing the protection of probiotics. *Food Hydrocoll.*, 77, 8–16, 2018.
33. Flores, C., Díaz-Barrera, A., Martínez, F., Galindo, E., Peña, C., Role of oxygen in the polymerization and de-polymerization of alginate produced by Azotobacter vinelandii. *J. Chem. Technol. Biotechnol.*, 90, 3, 356–365, 2015.
34. Peña, C., Galindo, E., Büchs, J., The viscosifying power, degree of acetylation and molecular mass of the alginate produced by Azotobacter vinelandii in shake flasks are determined by the oxygen transfer rate. *Process Biochem.*, 46, 1, 290–297, 2011.
35. McHugh, D.J., Production, properties and uses of alginates, in: *Production and Utilization of Products from Commercial Seaweeds*. FAO. Fish. Tech. Pap., 288, pp. 58–115, 1987.
36. Moral, Ç.K., Doğan, Ö., Sanin, F.D., Effect of oxygen tension and medium components on monomer distribution of alginate. *Appl. Biochem. Biotechnol.*, 176, 3, 875–891, 2015.
37. Díaz-Barrera, A., Maturana, N., Pacheco-Leyva, I., Martínez, I., Altamirano, C., Different responses in the expression of alginases, alginate polymerase and acetylation genes during alginate production by Azotobacter vinelandii under oxygen-controlled conditions. *J. Ind. Microbiol. Biotechnol.*, 44, 7, 1041–1051, 2017.
38. Garcia, A., Castillo, T., Ramos, D., Ahumada-Manuel, C.L., Núñez, C., Galindo, E., Büchs, J., Peña, C., Molecular weight and viscosifying power of alginates produced by mutant strains of Azotobacter vinelandii under microaerophilic conditions. *Biotechnol. Rep.*, 26, e00436, 2020.
39. Gómez-Pazarín, K., Flores, C., Castillo, T., Büchs, J., Galindo, E., Peña, C., Molecular weight and viscosifying power of alginates produced in Azotobacter vinelandii cultures in shake flasks under low power input. *J. Chem. Technol. Biotechnol.*, 91, 5, 1485–1492, 2016.
40. Chen, X., Zhu, Q., Liu, C., Li, D., Yan, H., Lin, Q., Esterification of alginate with alkyl bromides of different carbon chain lengths via the bimolecular nucleophilic substitution reaction: Synthesis, characterization, and controlled release performance. *Polymers*, 13, 19, 3351, 2021.
41. Pelletier, S., Hubert, P., Payan, E., Marchal, P., Choplin, L., Dellacherie, E., Amphiphilic derivatives of sodium alginate and hyaluronate for cartilage repair: Rheological properties. *J. Biomed. Mater. Res.*, 54, 1, 102–108, 2001.
42. Leonard, M., De Boisseson, M.R., Hubert, P., Dalencon, F., Dellacherie, E., Hydrophobically modified alginate hydrogels as protein carriers with specific controlled release properties. *J. Control. Release*, 98, 3, 395–405, 2004.
43. Vallée, F., Müller, C., Durand, A., Schimchowitsch, S., Dellacherie, E., Kelche, C., Cassel, J.C., Leonard, M., Synthesis and rheological properties of

hydrogels based on amphiphilic alginate-amide derivatives. *Carbohydr. Res.*, 344, 2, 223–228, 2009.
44. Yang, L., Zhang, B., Wen, L., Liang, Q., Zhang, L.M., Amphiphilic cholesteryl grafted sodium alginate derivative: Synthesis and self-assembly in aqueous solution. *Carbohydr. Polym.*, 68, 2, 218–225, 2007.
45. Colinet, I., Dulong, V., Hamaide, T., Le Cerf, D., Picton, L., New amphiphilic modified polysaccharides with original solution behaviour in salt media. *Carbohydr. Polym.*, 75, 3, 454–462, 2009.
46. Yao, B., Ni, C., Xiong, C., Zhu, C., Huang, B., Hydrophobic modification of sodium alginate and its application in drug-controlled release. *Bioprocess. Biosyst. Eng.*, 33, 4, 457–463, 2010.
47. Lehenkari, P.P. and Horton, M.A., Single integrin molecule adhesion forces in intact cells measured by atomic force microscopy. *Biochem. Biophys. Res. Commun.*, 259, 3, 645–650, 1999.
48. Koo, L.Y., Irvine, D.J., Mayes, A.M., Lauffenburger, D.A., Griffith, L.G., Co-regulation of cell adhesion by nanoscale RGD organization and mechanical stimulus. *J. Cell Sci.*, 115, 7, 1423–1433, 2002.
49. Lee, K.Y., Kong, H.J., Mooney, D.J., Quantifying interactions between cell receptors and adhesion ligand-modified polymers in solution. *Macromol. Biosci.*, 8, 2, 140–145, 2008.
50. Rowley, J.A., Madlambayan, G., Mooney, D.J., Alginate hydrogels as synthetic extracellular matrix materials. *Biomaterials*, 20, 1, 45–53, 1999.
51. Alsberg, E., Anderson, K.W., Albeiruti, A., Franceschi, R.T., Mooney, D.J., Cell-interactive alginate hydrogels for bone tissue engineering. *J. Dent. Res.*, 80, 11, 2025–2029, 2001.
52. Serafin, A., Murphy, C., Rubio, M.C., Collins, M.N., Printable alginate/gelatin hydrogel reinforced with carbon nanofibers as electrically conductive scaffolds for tissue engineering. *Mater. Sci. Eng.: C*, 122, 111927, 2021 Mar 1.
53. Lee, K.Y., Alsberg, E., Hsiong, S., Comisar, W., Linderman, J., Ziff, R., Mooney, D., Nanoscale adhesion ligand organization regulates osteoblast proliferation and differentiation. *Nano Lett.*, 4, 8, 1501–1506, 2004.
54. Bellis, S.L., Advantages of RGD peptides for directing cell association with biomaterials. *Biomaterials*, 32, 18, 4205–4210, 2011.
55. Dhoot, N.O., Tobias, C.A., Fischer, I., Wheatley, M.A., Peptide-modified alginate surfaces as a growth permissive substrate for neurite outgrowth. *J. Biomed. Mater. Res.*, 71, 2, 191–200, 2004.
56. Karadeniz, F., Sung, H.K., Kim, H.S., Natural origin polymers: Applications as wound care materials. *J. Life Sci.*, 29, 3, 382–93, 2019.
57. S.M. Cole, D.L. Nelson, inventors; Minnesota Mining, Manufacturing Co, assignee, Alginate wound dressing of good integrity. United States Patent. US 5,197,945, 1993.
58. Zhang, H., Cheng, J., Ao, Q., Preparation of alginate-based biomaterials and their applications in biomedicine. *Mar. Drugs*, 19, 5, 264, 2021.

59. Sarker, B., Papageorgiou, D.G., Silva, R., Zehnder, T., Gul-E-Noor, F., Bertmer, M., Kaschta, J., Chrissafis, K., Detsch, R., Boccaccini, A.R., Fabrication of alginate–gelatin crosslinked hydrogel microcapsules and evaluation of the microstructure and physico-chemical properties. *J. Mater. Chem. B*, 2, 11, 1470–1482, 2014.
60. Sarker, B., Zehnder, T., Rath, S.N., Horch, R.E., Kneser, U., Detsch, R., Boccaccini, A.R., Oxidized alginate-gelatin hydrogel: A favorable matrix for growth and osteogenic differentiation of adipose-derived stem cells in 3D. *ACS Biomater. Sci. Eng.*, 3, 8, 1730–1737, 2017.
61. Boonrungsiman, S., Gentleman, E., Carzaniga, R., Evans, N.D., McComb, D.W., Porter, A.E., Stevens, M.M., The role of intracellular calcium phosphate in osteoblast-mediated bone apatite formation. *Proc. Natl. Acad. Sci.*, 109, 35, 14170–14175, 2012.
62. Sharma, C., Dinda, A.K., Potdar, P.D., Chou, C.F., Mishra, N.C., Fabrication and characterization of novel nano-biocomposite scaffold of chitosan–gelatin–alginate–hydroxyapatite for bone tissue engineering. *Mater. Sci. Eng. C*, 64, 416–427, 2016.
63. Li, H., Jiang, F., Ye, S., Wu, Y., Zhu, K., Wang, D., Bioactive apatite incorporated alginate microspheres with sustained drug-delivery for bone regeneration application. *Mater. Sci. Eng. C*, 62, 779–786, 2016.
64. Badruddoza, A.Z., Godfrin, P.D., Myerson, A.S., Trout, B.L., Doyle, P.S., Core–shell composite hydrogels for controlled nanocrystal formation and release of hydrophobic active pharmaceutical ingredients. *Adv. Healthc. Mater.*, 5, 15, 1960–1968, 2016.
65. Bidarra, S.J., Barrias, C.C., Granja, P.L., Injectable alginate hydrogels for cell delivery in tissue engineering. *Acta Biomater.*, 10, 4, 1646–1662, 2010.
66. Shachar, M., Tsur-Gang, O., Dvir, T., Leor, J., Cohen, S., The effect of immobilized RGD peptide in alginate scaffolds on cardiac tissue engineering. *Acta Biomater.*, 7, 1, 152–162, 2011.
67. Tonsomboon, K., Butcher, A.L., Oyen, M.L., Strong and tough nanofibrous hydrogel composites based on biomimetic principles. *Mater. Sci. Eng. C*, 72, 220–227, 2017.
68. Raus, R.A., Nawawi, W.M., Nasaruddin, R.R., Alginate and alginate composites for biomedical applications. *Asian J. Pharm. Sci.*, 16, 280–306, 2020.
69. Boateng, J.S., Matthews, K.H., Stevens, H.N., Eccleston, G.M., Wound healing dressings and drug delivery systems: A review. *J. Pharm. Sci.*, 97, 8, 2892–2923, 2008.
70. Jayakumar, R., Prabaharan, M., Kumar, P.S., Nair, S.V., Tamura, H., Biomaterials based on chitin and chitosan in wound dressing applications. *Biotechnol. Adv.*, 29, 3, 322–337, 2011.
71. Queen, D., Orsted, H., Sanada, H., Sussman, G., A dressing history. *Int. Wound J.*, 1, 1, 59–77, 2004.
72. Tavakoli, S. and Klar, A.S., Advanced hydrogels as wound dressings. *Biomolecules*, 10, 8, 1169, 2020.

73. Gobi, R., Ravichandiran, P., Babu, R.S., Yoo, D.J., Biopolymer and synthetic polymer-based nanocomposites in wound dressing applications: A review. *Polymers*, 13, 12, 1962, 2021.
74. Thomas, A., Harding, K.G., Moore, K., Alginates from wound dressings activate human macrophages to secrete tumour necrosis factor-α. *Biomaterials*, 21, 17, 1797–1802, 2000.
75. Saghazadeh, S., Rinoldi, C., Schot, M., Kashaf, S.S., Sharifi, F., Jalilian, E., Nuutila, K., Giatsidis, G., Mostafalu, P., Derakhshandeh, H., Yue, K., Drug delivery systems and materials for wound healing applications. *Adv. Drug Deliv. Rev.*, 127, 138–166, 2018.
76. Wang, T., Zhang, F., Zhao, R., Wang, C., Hu, K., Sun, Y., Politis, C., Shavandi, A., Nie, L., Polyvinyl alcohol/sodium alginate hydrogels incorporated with silver nanoclusters via green tea extract for antibacterial applications. *Des. Monomers Polym.*, 23, 1, 118–133, 2020.
77. Karri, V.V., Kuppusamy, G., Talluri, S.V., Mannemala, S.S., Kollipara, R., Wadhwani, A.D., Mulukutla, S., Raju, K.R., Malayandi, R., Curcumin loaded chitosan nanoparticles impregnated into collagen-alginate scaffolds for diabetic wound healing. *Int. J. Biol. Macromol.*, 93, 1519–1529, 2016.
78. Orlowska, J., Kurczewska, U., Derwinska, K., Orlowski, W., Orszulak-Michalak, D., *In vitro* evaluation of immunogenic properties of active dressings. *Curr. Issues Pharm. Med. Sci.*, 27, 1, 55–60, 2014.
79. Cattelan, G., Guerrero Gerbolés, A., Foresti, R., Pramstaller, P.P., Rossini, A., Miragoli, M., Caffarra Malvezzi, C., Alginate formulations: Current developments in the race for hydrogel-based cardiac regeneration. *Front. Bioeng. Biotechnol.*, 8, 414, 2020.
80. Varaprasad, K., Jayaramudu, T., Kanikireddy, V., Toro, C., Sadiku, E.R., Alginate-based composite materials for wound dressing application: A mini review. *Carbohydr. Polym.*, 236, 116025, 2020.
81. Kaygusuz, H., Torlak, E., Akın-Evingür, G., Özen, İ., Von Klitzing, R., Erim, F.B., Antimicrobial cerium ion-chitosan crosslinked alginate biopolymer films: A novel and potential wound dressing. *Int. J. Biol. Macromol.*, 105, 1161–1165, 2017.
82. Bagher, Z., Ehterami, A., Safdel, M.H., Khastar, H., Semiari, H., Asefnejad, A., Davachi, S.M., Mirzaii, M., Salehi, M., Wound healing with alginate/chitosan hydrogel containing hesperidin in rat model. *J. Drug Deliv. Sci. Technol.*, 55, 101379, 2020.
83. He, Y., Zhao, W., Dong, Z., Ji, Y., Li, M., Hao, Y., Zhang, D., Yuan, C., Deng, J., Zhao, P., Zhou, Q., A biodegradable antibacterial alginate/carboxymethyl chitosan/Kangfuxin sponges for promoting blood coagulation and full-thickness wound healing. *Int. J. Biol. Macromol.*, 167, 182–192, 2021.
84. Zahid, M., Lodhi, M., Afzal, A., Rehan, Z.A., Mehmood, M., Javed, T., Shabbir, R., Siuta, D., Althobaiti, F., Dessok, E.S., Development of hydrogels with the incorporation of Raphanus sativus L. seed extract in sodium alginate for wound-healing application. *Gels*, 7, 3, 107, 2021.

85. Chiu, L.L., Iyer, R.K., Reis, L.A., Nunes., S.S., Radisic, M., Cardiac tissue engineering: Current state and perspectives. *Front. Biosci.*, 17, 1533–1550, 2012.
86. Majid, Q.A., Fricker, A.T., Gregory, D.A., Davidenko, N., Hernandez Cruz, O., Jabbour, R.J., Owen, T.J., Basnett, P., Lukasiewicz, B., Stevens, M., Best, S., Natural biomaterials for cardiac tissue engineering: A highly biocompatible solution. *Front. Cardiovasc. Med.*, 7, 192, 2020.
87. Lam, M.T. and Wu, J.C., Biomaterial applications in cardiovascular tissue repair and regeneration. *Expert Rev. Cardiovasc. Ther.*, 10, 8, 1039–1049, 2012.
88. Xu, Z. and Lam, M.T., Alginate application for heart and cardiovascular diseases, in: *Alginates and Their Biomedical Applications*, pp. 185–212, Springer, Singapore, 2018.
89. Wen, H., Jung, H., Li, X., Drug delivery approaches in addressing clinical pharmacology-related issues: Opportunities and challenges. *AAPS J.*, 17, 6, 1327–1340, 2012.
90. Boontheekul, T., Kong, H.J., Mooney, D.J., Controlling alginate gel degradation utilizing partial oxidation and bimodal molecular weight distribution. *Biomaterials*, 26, 15, 2455–2465, 2005.
91. Gonzalez-Pujana, A., Orive, G., Pedraz, J.L., Santos-Vizcaino, E., Hernandez, R.M., Alginate microcapsules for drug delivery, in: *Alginates and Their Biomedical Applications*, pp. 67–100, Springer, Singapore, 2018.
92. Patra, J.K., Das, G., Fraceto, L.F., Campos, E.V., del Pilar Rodriguez-Torres, M., Acosta-Torres, L.S., Diaz-Torres, L.A., Grillo, R., Swamy, M.K., Sharma, S., Habtemariam, S., Nano based drug delivery systems: Recent developments and future prospects. *J. Nanobiotechnology*, 16, 1, 1–33, 2018.
93. Deng, Y., Yang, N., Okoro, O.V., Shavandi, A., Nie, L., Alginate-based composite and its biomedical applications, in: *Properties and Applications of Alginates*, 2021.
94. Manatunga, D.C., de Silva, R.M., de Silva, K.N., de Silva, N., Bhandari, S., Yap, Y.K., Costha, N.P., pH responsive controlled release of anti-cancer hydrophobic drugs from sodium alginate and hydroxyapatite bi-coated iron oxide nanoparticles. *Eur. J. Pharm. Biopharm.*, 117, 29–38, 2017.
95. Anitha, A., Deepagan, V.G., Rani, V.D., Menon, D., Nair, S.V., Jayakumar, R., Preparation, characterization, *in vitro* drug release and biological studies of curcumin loaded dextran sulphate–chitosan nanoparticles. *Carbohydr. Polym.*, 84, 3, 1158–1164, 2011.
96. He, S., Zhong, S., Xu, L., Dou, Y., Li, Z., Qiao, F., Gao, Y., Cui, X., Sonochemical fabrication of magnetic reduction-responsive alginate-based microcapsules for drug delivery. *Int. J. Biol. Macromol.*, 155, 42–49, 2020.
97. Lin, N., Gèze, A., Wouessidjewe, D., Huang, J., Dufresne, A., Biocompatible double-membrane hydrogels from cationic cellulose nanocrystals and anionic alginate as complexing drugs codelivery. *ACS Appl. Mater. Interfaces*, 8, 11, 6880–6889, 2016.

98. Zhong, D., Liu, Z., Xie, S., Zhang, W., Zhang, Y., Xue, W., Study on poly (D, L-lactic) microspheres embedded in calcium alginate hydrogel beads as dual drug delivery systems. *J. Appl. Polym. Sci.*, 129, 2, 767–772, 2013.
99. Silva, E.A. and Mooney, D.J., Effects of VEGF temporal and spatial presentation on angiogenesis. *Biomaterials*, 31, 6, 1235–1241, 2010.
100. Ji, L., Zhang, F., Zhu, L., Jiang, J., An *in-situ* fabrication of bamboo bacterial cellulose/sodium alginate nanocomposite hydrogels as carrier materials for controlled protein drug delivery. *Int. J. Biol. Macromol.*, 170, 459–468, 2021.
101. Griffith, L.G. and Naughton, G., Tissue engineering–Current challenges and expanding opportunities. *Science*, 295, 5557, 1009–1014, 2002.
102. Li, Z.J., Srebnik, S., Rojas, O.J., Revisiting cation complexation and hydrogen bonding of single-chain polyguluronate alginate. *Biomacromolecules*, 22, 4027–4036, 2021.
103. Nie, X., Chuah, Y.J., Zhu, W., He, P., Peck, Y., Wang, D.A., Decellularized tissue engineered hyaline cartilage graft for articular cartilage repair. *Biomaterials*, 235, 119821, 2020.
104. Zhu, J., Chen, D., Du, J., Chen, X., Wang, J., Zhang, H., Chen, S., Wu, J., Zhu, T., Mo, X., Mechanical matching nanofibrous vascular scaffold with effective anticoagulation for vascular tissue engineering. *Compos. B. Eng.*, 186, 107788, 2020.
105. Sharifi, S., Islam, M.M., Sharifi, H., Islam, R., Koza, D., Reyes-Ortega, F., Alba-Molina, D., Nilsson, P.H., Dohlman, C.H., Mollnes, T.E., Chodosh, J., Tuning gelatin-based hydrogel towards bioadhesive ocular tissue engineering applications. *Bioact. Mater.*, 6, 11, 3947–3961, 2021.
106. Yang, W., Xu, H., Lan, Y., Zhu, Q., Liu, Y., Huang, S., Shi, S., Hancharou, A., Tang, B., Guo, R., Preparation and characterisation of a novel silk fibroin/hyaluronic acid/sodium alginate scaffold for skin repair. *Int. J. Biol. Macromol.*, 130, 58–67, 2019.
107. Della Rocca, D.G., Willenberg, B.J., Qi, Y., Simmons, C.S., Rubiano, A., Ferreira, L.F., Huo, T., Petersen, J.W., Ruchaya, P.J., Wate, P.S., Wise, E.A., An injectable capillary-like microstructured alginate hydrogel improves left ventricular function after myocardial infarction in rats. *Int. J. Cardiol.*, 220, 149–154, 2016.
108. Sahoo, D.R. and Biswal, T., Alginate and its application to tissue engineering. *SN Appl. Sci.*, 3, 1, 1–9, 2021.
109. Venkatesan, J., Bhatnagar, I., Kim, S.K., Chitosan-alginate biocomposite containing fucoidan for bone tissue engineering. *Mar. Drugs*, 12, 1, 300–316, 2014.
110. Amini, A.R., Laurencin, C.T., Nukavarapu, S.P., Bone tissue engineering: Recent advances and challenges. *Crit. Rev. Biomed. Eng.*, 40, 5, 363–408, 2012.
111. Kumar, G.T., Thakur, D., Alexander, A., Badwaik, H., Krishna Tripathi, D., Alginate based hydrogel as a potential biopolymeric carrier for drug delivery and cell delivery systems: Present status and applications. *Curr. Drug Deliv.*, 9, 6, 539–555, 2012.

112. O'brien, F.J., Biomaterials & scaffolds for tissue engineering. *Mater. Today*, 14, 3, 88–95, 2011.
113. Florczyk, S.J., Leung, M., Li, Z., Huang, J.I., Hopper, R.A., Zhang, M., Evaluation of three-dimensional porous chitosan–alginate scaffolds in rat calvarial defects for bone regeneration applications. *J. Biomed. Mater. Res. A*, 101, 10, 2974–2983, 2013.
114. Li, Z., Ramay, H.R., Hauch, K.D., Xiao, D., Zhang, M., Chitosan–alginate hybrid scaffolds for bone tissue engineering. *Biomaterials*, 26, 18, 3919–3928, 2005.
115. Grinnell, F., Fibroblast biology in three-dimensional collagen matrices. *Trends Cell Biol.*, 13, 5, 264–269, 2003.
116. García, A.J., Get a grip: Integrins in cell–biomaterial interactions. *Biomaterials*, 26, 36, 7525–7529, 2005.
117. Sotome, S., Uemura, T., Kikuchi, M., Chen, J., Itoh, S., Tanaka, J., Tateishi, T., Shinomiya, K., Synthesis and *in vivo* evaluation of a novel hydroxyapatite/collagen–alginate as a bone filler and a drug delivery carrier of bone morphogenetic protein. *Mater. Sci. Eng. C*, 24, 3, 341–347, 2004.
118. Polo-Corrales, L., Latorre-Esteves, M., Ramirez-Vick, J.E., Scaffold design for bone regeneration. *J. Nanosci. Nanotechnol.*, 14, 1, 15–56, 2014.
119. Lin, H.R. and Yeh, Y.J., Porous alginate/hydroxyapatite composite scaffolds for bone tissue engineering: Preparation, characterization, and *in vitro* studies. *J. Biomed. Mater. Res. B. Appl. Biomater.*, 71, 1, 52–65, 2004.
120. Kim, H.L., Jung, G.Y., Yoon, J.H., Han, J.S., Park, Y.J., Kim, D.G., Zhang, M., Kim, D.J., Preparation and characterization of nano-sized hydroxyapatite/alginate/chitosan composite scaffolds for bone tissue engineering. *Mater. Sci. Eng. C*, 54, 20–25, 2015.
121. Petrović, M., Čolović, B., Jokanović, V., Marković, D., Self assembly of biomimetic hydroxyapatite on the surface of different polymer thin films. *J. Ceram. Process. Res.*, 13, 4, 398–404, 2012.
122. Firouzi, N., BaradarKhoshfetrat, A., Kazemi, D., Enzymatically gellable gelatin improves nano-hydroxyapatite-alginate microcapsule characteristics for modular bone tissue formation. *J. Biomed. Mater. Res. A*, 108, 2, 340–350, 2020.
123. Ho, S.S., Keown, A.T., Addison, B., Leach, J.K., Cell migration and bone formation from mesenchymal stem cell spheroids in alginate hydrogels are regulated by adhesive ligand density. *Biomacromolecules*, 18, 12, 4331–40, 2017 Dec 11.
124. Silva-Barroso, A.S., Cabral, C.S., Ferreira, P., Moreira, A.F., Correia, I.J., Lignin-enriched tricalcium phosphate/sodium alginate 3D scaffolds for application in bone tissue regeneration. *Int. J. Biol. Macromol.*, 31, 124258, 2023 Mar.
125. Hakimi, F., Jafari, H., Hashemikia, S., Shabani, S., Ramazani, A., Chitosan-polyethylene oxide/clay-alginate nanofiber hydrogel scaffold for bone tissue

engineering: Preparation, physical characterization, and biomimetic mineralization. *Int. J. Biol. Macromol.*, 233, 123453, 2023 Apr 1.
126. Armiento, A.R., Stoddart, M.J., Alini, M., Eglin, D., Biomaterials for articular cartilage tissue engineering: Learning frombiology. *Acta Biomater.*, 65, 1–20, 2018.
127. Oldershaw, R.A., Cell sources for the regeneration of articular cartilage: The past, the horizon and the future. *Int. J. Exp. Pathol.*, 93, 6, 389–400, 2012.
128. Zhang, L., Hu, J., Athanasiou, K.A., The role of tissue engineering in articular cartilage repair and regeneration. *Crit. Rev. Biomed. Eng.*, 37, 1-2, 1–57, 2009.
129. Bouhadir, K.H., Lee, K.Y., Alsberg, E., Damm, K.L., Anderson, K.W., Mooney, D.J., Degradation of partially oxidized alginate and its potential application for tissue engineering. *Biotechnol. Prog.*, 17, 5, 945–950, 2001.
130. Mahapatra, C., Jin, G.Z., Kim, H.W., Alginate-hyaluronic acid-collagen composite hydrogel favorable for the culture of chondrocytes and their phenotype maintenance. *Tissue Eng. Regen. Med.*, 13, 5, 538–546, 2016.
131. Erickson, A.E., Sun, J., Levengood, S.K., Swanson, S., Chang, F.C., Tsao, C.T., Zhang, M., Chitosan-based composite bilayer scaffold as an *in vitro* osteochondral defect regeneration model. *Biomed. Microdevices*, 21, 2, 1–6, 2019.
132. Darnell, M.C., Sun, J.Y., Mehta, M., Johnson, C., Arany, P.R., Suo, Z., Mooney, D.J., Performance and biocompatibility of extremely tough alginate/polyacrylamide hydrogels. *Biomaterials*, 34, 33, 8042–8048, 2013.
133. Liao, I.C., Moutos, F.T., Estes, B.T., Zhao, X., Guilak, F., Composite three-dimensional woven scaffolds with interpenetrating network hydrogels to create functional synthetic articular cartilage. *Adv. Funct. Mater.*, 23, 47, 5833–5839, 2013.
134. Arjmandi, M. and Ramezani, M., Mechanical and tribological assessment of silica nanoparticle-alginate-polyacrylamide nanocomposite hydrogels as a cartilage replacement. *J. Mech. Behav. Biomed. Mater.*, 95, 196–204, 2019.
135. Saygili, E., Kaya, E., Ilhan-Ayisigi, E., Saglam-Metiner, P., Alarcin, E., Kazan, A., Girgic, E., Kim, Y.W., Gunes, K., Eren-Ozcan, G.G., Akakin, D., An alginate-poly (acrylamide) hydrogel with TGF-β3 loaded nanoparticles for cartilage repair: Biodegradability, biocompatibility and protein adsorption. *Int. J. Biol. Macromol.*, 172, 381–393, 2021.
136. Krishna, D.V. and Sankar, M.R., Extrusion based bioprinting of alginate based multicomponent hydrogels for tissue regeneration applications: State of the art. *Mater. Today Commun.*, 24, 105696, 2023 Feb.
137. Asrani, S.K., Devarbhavi, H., Eaton, J., Kamath, P.S., Burden of liver diseases in the world. *J. Hepatol.*, 70, 1, 151–171, 2019.
138. Rajalekshmi, R., Shaji, A.K., Joseph, R., Bhatt, A., Scaffold for liver tissue engineering: Exploring the potential of fibrin incorporated alginate dialdehyde–gelatin hydrogel. *Int. J. Biol. Macromol.*, 166, 999–1008, 2021.
139. Chen, F., Tian, M., Zhang, D., Wang, J., Wang, Q., Yu, X., Zhang, X., Wan, C., Preparation and characterization of oxidized alginate covalently cross-linked galactosylated chitosan scaffold for liver tissue engineering. *Mater. Sci. Eng. C*, 32, 2, 310–320, 2012.

140. Majewski, R.L., Zhang, W., Ma, X., Cui, Z., Ren, W., Markel, D.C., Bioencapsulation technologies in tissue engineering. *J. Appl. Biomater. Funct.*, 14, 4, 395–403, 2016.
141. Haque, T., Chen, H., Ouyang, W., Martoni, C., Lawuyi, B., Urbanska, A.M., Prakash, S., *In vitro* study of alginate–chitosan microcapsules: An alternative to liver cell transplants for the treatment of liver failure. *Biotechnol. Lett.*, 27, 5, 317–322, 2005.
142. Chen, F., Tian, M., Zhang, D., Wang, J., Wang, Q., Yu, X., Zhang, X., Wan, C., Preparation and characterization of oxidized alginate covalently cross-linked galactosylated chitosan scaffold for liver tissue engineering. *Mater. Sci. Eng.: C*, 32, 2, 310–20, 2012 Mar 1.
143. Rajalekshmy, G.P. and Rekha, M.R., Strontium ion cross-linked alginate-g-poly (PEGMA) xerogels for wound healing applications: *In vitro* studies. *Carbohydr. Polym.*, 251, 117119, 2021.
144. Jing, X., Sun, Y., Liu, Y., Ma, X., Hu, H., Alginate/chitosan-based hydrogel loaded with gene vectors to deliver polydeoxyribonucleotide for effective wound healing. *Biomater. Sci.*, 9, 16, 5533–5541, 2021.
145. Prasathkumar, M. and Sadhasivam, S., Chitosan/hyaluronic acid/alginate and an assorted polymers loaded with honey, plant, and marine compounds for progressive wound healing—Know-how. *Int. J. Biol. Macromol.*, 186, 656–685, 2021.
146. Zahid, M., Lodhi, M., Afzal, A., Rehan, Z.A., Mehmood, M., Javed, T., Shabbir, R., Siuta, D., Althobaiti, F., Dessok, E.S., Development of hydrogels with the incorporation of Raphanus sativus L. seed extract in sodium alginate for wound-healing application. *Gels*, 7, 3, 107, 2021.
147. Zhang, M., Wang, G., Wang, D., Zheng, Y., Li, Y., Meng, W., Zhang, X., Du, F., Lee, S., Ag@ MOF-loaded chitosan nanoparticle and polyvinyl alcohol/sodium alginate/chitosan bilayer dressing for wound healing applications. *Int. J. Biol. Macromol.*, 175, 481–494, 2021.
148. Shen, Y., Xu, G., Huang, H., Wang, K., Wang, H., Lang, M., Gao, H., Zhao, S., Sequential release of small extracellular vesicles from bilayered thiolated alginate/polyethylene glycol diacrylate hydrogels for scarless wound healing. *ACS Nano*, 15, 4, 6352–6368, 2021.
149. Zhang, X., Li, Y., Ma, Z., He, D., Li, H., Modulating degradation of sodium alginate/bioglass hydrogel for improving tissue infiltration and promoting wound healing. *Bioact. Mater.*, 6, 11, 3692–3704, 2021.
150. Rezvanian, M., Ng, S.F., Alavi, T., Ahmad, W., *In-vivo* evaluation of Alginate-Pectin hydrogel film loaded with Simvastatin for diabetic wound healing in Streptozotocin-induced diabetic rats. *Int. J. Biol. Macromol.*, 171, 308–319, 2021.
151. Azad, A.K., Al-Mahmood, S.M., Kennedy, J.F., Chatterjee, B., Bera, H., Electro-hydrodynamic assisted synthesis of lecithin-stabilized peppermint oil-loaded alginate microbeads for intestinal drug delivery. *Int. J. Biol. Macromol.*, 185, 861–875, 2021.

152. Reddy, O.S., Subha, M.C., Jithendra, T., Madhavi, C., Rao, K.C., Curcumin encapsulated dual cross linked sodium alginate/montmorillonite polymeric composite beads for controlled drug delivery. *J. Pharm. Anal.*, 11, 2, 191–199, 2021.
153. Bezerra, I.C., Moma, M.S., Freitas, E.D., Silva, M.G., Vieira, M.G., Development of a modified drug delivery system through the incorporation of furosemide into sericin and alginate matrix using the experimental design approach. *J. Chem. Technol. Biotechnol.*, 96, 3, 650–661, 2021.
154. Sarheed, O., Dibi, M., Ramesh, K.V., Drechsler, M., Fabrication of alginate-based O/W nanoemulsions for transdermal drug delivery of lidocaine: Influence of the oil phase and surfactant. *Molecules*, 26, 9, 2556, 2021.
155. Ji, L., Zhang, F., Zhu, L., Jiang, J., An *in-situ* fabrication of bamboo bacterial cellulose/sodium alginate nanocomposite hydrogels as carrier materials for controlled protein drug delivery. *Int. J. Biol. Macromol.*, 170, 459–468, 2021.
156. Sun, C.K., Ke, C.J., Lin, Y.W., Lin, F.H., Tsai, T.H., Sun, J.S., Transglutaminase cross-linked gelatin-alginate-antibacterial hydrogel as the drug delivery-coatings for implant-related infections. *Polymers*, 13, 3, 414, 2021.
157. Serafin, A., Murphy, C., Rubio, M.C., Collins, M.N., Printable alginate/gelatin hydrogel reinforced with carbon nanofibers as electrically conductive scaffolds for tissue engineering. *Mater. Sci. Eng. C*, 122, 111927, 2021.
158. Distler, T., Polley, C., Shi, F., Schneidereit, D., Ashton, M.D., Friedrich, O., Kolb, J.F., Hardy, J.G., Detsch, R., Seitz, H., Boccaccini, A.R., Electrically conductive and 3D-printable oxidized alginate-gelatin polypyrrole: PSS hydrogels for tissue engineering. *Adv. Healthc. Mater.*, 10, 9, 2001876, 2021.
159. Dutta, S.D., Hexiu, J., Patel, D.K., Ganguly, K., Lim, K.T., 3D-printed bioactive and biodegradable hydrogel scaffolds of alginate/gelatin/cellulose nanocrystals for tissue engineering. *Int. J. Biol. Macromol.*, 167, 644–658, 2021.
160. Distler, T., Polley, C., Shi, F., Schneidereit, D., Ashton, M.D., Friedrich, O., Kolb, J.F., Hardy, J.G., Detsch, R., Seitz, H., Boccaccini, A.R., Electrically conductive and 3D-printable oxidized alginate-gelatin polypyrrole: PSS hydrogels for tissue engineering. *Adv. Healthc. Mater.*, 10, 9, 2001876, 2021.
161. Xu, M., Qin, M., Cheng, Y., Niu, X., Kong, J., Zhang, X., Huang, D., Wang, H., Alginate microgels as delivery vehicles for cell-based therapies in tissue engineering and regenerative medicine. *Carbohydr. Polym.*, 266, 118128, 2021.
162. Rezende, R.A., Bártolo, P.J., Mendes, A., Filho, R.M., Rheological behavior of alginate solutions for biomanufacturing. *J. Appl. Polym. Sci.*, 113, 3866–3871, 2009.
163. Annu, Ahmed, S., Ikram, S., Perspectives of chitosan and alginate membranes for biomedical applications, in: *Natural Polymers: Derivatives, Blends and Composites*, vol. 2, pp. 157–182, Nova Science Publishers, Hauppauge, New York, 2017.
164. Annu, Ahmed, S., Nirala, R.K., Kumar, R., Ikram, S., Green synthesis of chitosan/nanosilver hybrid bionanocomposites with promising antimicrobial, antioxidant and anticervical cancer activity. *Polym. Polym. Compos.*, 29, 9_suppl, S199–S210, 2021.

4

Biomedical Applications of Cellulose

Abhishek Kanugo[1], Pallavi Chaudhari[2] and Rupesh K. Gautam[3*]

[1]*Department of Pharmaceutical Quality Assurance, SVKM's Institute of Pharmacy, Dhule, Maharashtra, India*
[2]*Department of Pharmaceutics, Dr. D.Y. Patil College of Pharmacy, Akurdi, Pune, India*
[3]*Department of Pharmacology, Indore Institute of Pharmacy, IIST Campus, Rau, Indore (M.P.), India*

Abstract
Cellulose is the most extensively available biopolymer owing to its several sources (wood, plants, bacteria, pulp, and cotton). The utilization of nanotechnology enabled to achieve nanocellulose and its different types of cellulose nanocrystals, bacterial nanocellulose, and cellulose nanofibers. Nanocellulose possesses diverse characteristics such as biocompatibility, biodegradability, high surface-to-volume ratio, marked mechanical strength, cheaper price, non-toxicity, and ease of fabrication, which enable its incorporation in the biomedical field for several purposes. Nanocellulose is a versatile biopolymer and finds extensive applications in the biomedical field such as in drug delivery systems, wound dressing, tissue engineering, antimicrobial, implants, etc. Most of these applications were fulfilled by nanocellulose in the form of hydrogels and nanocomposites. The ultimate aim of this book chapter is to provide the extensive use of nanocellulose in the biomedical field and is the most promising biopolymer satisfying the requirements in several fields.

Keywords: Cellulose, biomedical applications, drug delivery, wound dressing, tissue engineering

**Corresponding author*: drrupeshgautam@gmail.com

Annu (ed.) Biopolymers for Biomedical Applications, (87–128) © 2024 Scrivener Publishing LLC

4.1 Introduction

Among the several biopolymers existing on Earth, cellulose is the most abundant biopolymer. Cellulose is mainly derived from plants or manufactured by a microorganism such as bacteria [1]. Recently, cellulose has gained tremendous significance as a biopolymer for offering diverse applications in biomedical fields. Moreover, cellulose possesses unique characteristics such as being readily available, economic, biocompatible, biodegradable, and non-toxic in nature. Cellulose is mostly insoluble in water, but various derivatives such as carboxymethyl cellulose (CMC) and hydroxypropyl methylcellulose are soluble [2].

4.2 Structure and Properties of Cellulose

Cellulose is consists of several molecules of glucose unit and is attached with β-(1–4) glycosidic bonds. Cellulose also plays a crucial role in plants by fortifying the cell wall in the form of hemicellulose along with lignin [3]. The effectiveness of cellulose is due to the presence of microfibrils situated in the cell wall of wood fibers. These fibrils are available in two forms, namely: amorphous and crystalline. The crystalline shape is identified with a well-arranged structure and randomly occurred in the amorphous form. These fibrils are categorized based on the structural and crystalline arrangements. These fibrils are recognized by X-ray diffraction, scanning and transmission electron microscopy, atomic force microscopy, nuclear magnetic resonance imaging, etc. [4].

Cellulose possesses marked tensile strength which empowers its applicability in several fields. The typical properties possessed by cellulose are chirality, hydrophilicity, and variability owing to its molecular structure. The availability of hydrogen bonds in cellulose to a greater extent renders the fibers in the crystalline form. The cellulose molecules are capable of rotating at an axis of 180° due to the presence of glycosidic bonds. In the interior of the cellulose chain, there is the existence of anhydroglucose with a 3-hydroxyl group possessing marked reactivity. Moreover, there is an existence of a hydroxyl group at the C6 position and two at C2 and C3 in the symmetry of the ring. The cellulose ring is comprised of reducing and non-reducing moieties located at the opposite ends. The consistency in the cellulose rings presents higher mechanical characteristics and is found to be increased after the elimination of hemicellulose. The existence of a hydroxyl group attached with hydrogen atoms resulted in the development of a crystalline structure [5].

Table 4.1 Types of cellulose with their characteristics.

Type of NC	Sources	Manufacturing	Dimensions	Characteristics	References
CNC	Cotton, wood, bamboo	Hydrolysis of acid	Diameter 3–50 nm	Rod-shaped crystals, possess high surface area, high mechanical strength, and excellent stability	[4, 7]
CNF	Wood, cotton, potato, hemp	High-pressure homogenization, microfluidization, grinding	Diameter 5–60 nm	Mostly derived from plants, contains a uniform crystalline and amorphous structure	[8]
BNC	*Acetobacter, Glucobacterxylinum*	Biosynthesis of glucose and alcohol	Diameter 10–100 nm	Achieved by fermentation of bacteria, extremely high mechanical strength, and thermal stability. Moreover, pure form of cellulose and flexible in nature and high degree of polymerization with biomedical diverse application	[9, 10]

4.3 Types of Cellulose

When cellulose particles existed in the nanoscale range, then they are known as nanocellulose (NC). Based on availability, structure, and morphology, nanocellulose is listed into three altered types, namely: cellulose nanocrystals (CNCs) or nano-crystalline cellulose (NCCs), cellulose nanofibrils (CNFs), and bacterial nanocellulose (BNC) [6]. The details of the several types of NC are depicted in Table 4.1.

4.4 Applications of Cellulose

The applications of cellulose in their different types (CNC, CNF, and BNC) and the form of the hydrogel, as well as its modified derivatives, gain tremendous significance in the biomedical field. The diverse applications are in the field of drug delivery, wound dressing, tissue engineering, antimicrobial properties, healthcare system, and cardiovascular, ocular, skeletal system, respectively. These several applications are depicted in Figure 4.1 [11].

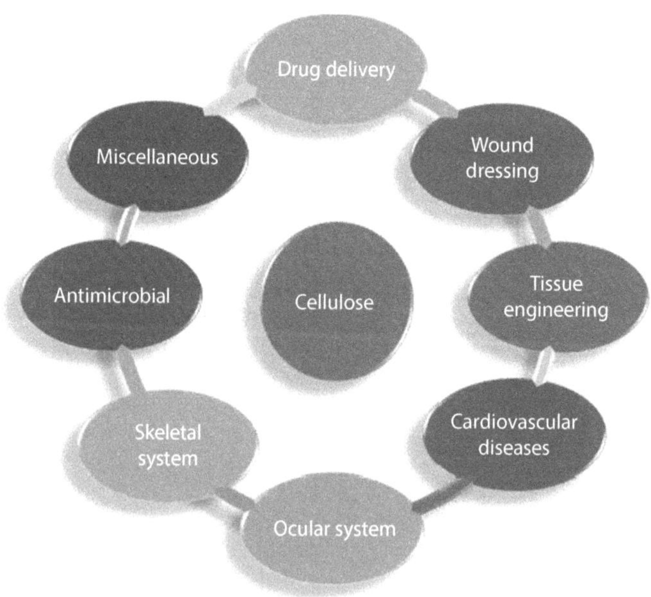

Figure 4.1 Biomedical applications of cellulose.

4.4.1 Drug Delivery

Cellulose and its various forms possess unique characteristics (nano-size, biodegradable, biocompatible, non-toxic, and tunable surface chemistry) and were found to be beneficial in several kinds of drug delivery as a versatile carrier [12, 13]. The rationale of controlled drug delivery is to transport the active ingredient at a preset rate for a definite period [14]. The drug is delivered to the body through several systems such as lipid vesicles, nanoparticles, nanocapsules, hydrogel, etc., which are dependent on the incorporation of a single or a combination of polymers or carriers [15]. The effective delivery system is capable of retaining the drug concentration within the therapeutic range for better efficacy. The release of active ingredients proceeds through dissolution, diffusion, and swelling mechanisms [16].

Controlled drug delivery is preferred for drugs with a short half-life and thereby minimizes the frequent administration of drugs and lowers the variations in plasma concentration with greater patient comfort and compliance [17]. In the form of hydrogels (CNC and CNF), cellulose avails greater bioavailability and maximum drug entrapment due to its large surface area and open pore structure as well as great water holding capacity [12, 18]. Moreover, hydrogels containing drugs are capable of delivering at the desired site in the body in the presence of external stimuli such as temperature, pH, light, and electric and magnetic field [19]. Moreover, hydrogel prepared with cellulose has a high potential of recognizing and treating various diseases. The several types and different forms (hydrogels) of cellulose applications are depicted in Table 4.2 [12].

The treatment of various types of cancer is highly challenging because of their metastasis. Conventional therapy such as chemotherapy, radiation, and surgery are invasive and generate several kinds of toxicity on the host cell. Hence, several forms of cellulose are widely utilized in the therapy of several kinds of cancer in the form of nanocomposites [41]. The CNC possesses a large number of hydroxyl groups which enabled for conjugation/modification with several moieties and ligands such as folate. In most tumors, overexpression of folic acid was observed, and thereby conjugation with antitumor agents helps in treating multiple kinds of cancer [42]. The NCC–folate conjugate that was prepared by Dong *et al.* comprised of antitumor agents for targeted drug delivery. Fluorescein-5-isocyanate was linked with NCC for further interacting with folate for a higher cellular uptake. A significant antitumor action was reported using this conjugate [43].

Table 4.2 Applications of cellulose in a drug delivery system.

S. no.	Type/ form of cellulose	Applications	References
1	CNC/NCC	CNC along with lignin nanoparticles (NPs) utilized for the targeted delivery of breast cancer and possesses theranostic applications. They are capable of causing obstacles in cell viability assay and the tendency for scattering the emitted light, thereby absorbing the assay reagents.	[20]
2	CNF	Improvement in solubility of poorly soluble agents cloxacillin and cefuroxime sodium in drug delivery.	[21]
3	CNC	CNC at 5% concentration was incorporated with collagen for achieving sustained release action. The marked swelling property was observed, which released the active ingredient at a much slower rate from microspheres.	[22]
4	CNC-hydrogel	Utilized for the sustained delivery of vitamin B12 in the form of micro-hydrogel in combination with starch.	[23]
5	CNC	Developed a double-layer hydrogel with cationic CNC and anionic alginate for the sustained delivery of ceftazidime for about 72 h.	[24]
6	CNC	Prepared CNC- and alginate-based hydrogel beads, which improves the performance of alginate beads for prolonging the action.	[25]
7	CNC	The CNC-based hydrogel was developed along with chitosan-poly (acrylic acid) for extended ocular drug delivery. The controlled action was due to the presence of CNC swelling characteristics.	[26]
8	CNF	Utilized for controlled drug delivery in the form of CNFs-hydrogel form in technetium-99m.	[27]

(Continued)

Table 4.2 Applications of cellulose in a drug delivery system. (*Continued*)

S. no.	Type/ form of cellulose	Applications	References
9	Cellulose-disc	Useful in the oral mucoadhesive formulation for its swelling, mucoadhesiveness, and sustained action; along with medicaments in dissolved form, it was capable of releasing the drug for about 4 h.	[28]
10	CNC	Utilized for stimulus-responsive drug delivery based on pH. The prepared gel exhibited greater stability in acidic conditions and thereby protect the probiotics. Hence, intestinal targeted drug delivery was achieved using CNC.	[29]
11	CNC	Modification of CNC with cholesterol resulted in a rise in hydrophobicity, which improves smoothness and retards the release of folic acid.	[30, 31]
12	CNC	The nanocomposite was developed with chitosan, and for enhancing the action, CNC was incorporated for its swelling behavior. These nanocomposites were capable of releasing the drug at a much slower rate at different pH ranges.	[32]
13	CNC	The CNC was incorporated with pluronic for enhancing the properties of the hydrogel. The conjugated hydrogel showed high mechanical characteristics and thermal stability and was useful in several drug delivery systems.	[33]
14	CMC	The hydrogel that was formulated with CMC and starch by gamma-irradiation showed greater water holding capacity, and thereby extended action was achieved in colon-targeted delivery.	[34]
15	BNC	The BNC-based nanocomposite of tetracycline was developed for controlled drug delivery.	[35]

(*Continued*)

modification, they can be utilized. Cellulose-based hydrogels possess unique characteristics such as hydrophobicity, greater porosity, biocompatibility, non-toxicity, high oxygen permeability, and diversity to bind with several functional groups as well as delivery in a controlled manner, hence gaining tremendous significance for antimicrobial properties [55]. Cellulose (CNC/BNC/hydrogel) possesses antibacterial, antiviral, antioxidant, and anti-inflammatory characteristics after their surface modification with several agents.

The antibacterial characteristics of modified cellulose gain applications in biomedicine and food. The antimicrobial action of cellulose derivatives was achieved through the surface modification of the hydroxyl group at their C2, C3, and C6 positions by oxidation, esterification, and etherification process. These agents bind covalently or non-covalently with the glucose unit, e.g., aldehydes (2,2,6,6-tetramethyl piperidinyloxy) (TEMPO), quaternary ammonium compounds (cationic porphyrin), nanoparticles (silver and gold), and metal oxide NPs (titanium, zinc, magnesium, and copper oxide, respectively) [54–56]. Several agents loaded with cellulose that show antimicrobial action are depicted in Table 4.3.

4.4.3 Tissue Engineering

Table 4.3 Cellulose-grafted materials with antimicrobial properties.

S. no.	Materials	Properties	References
1	CNF–dialdehyde–silver NP	Antibacterial activity was reported for *E. coli* and *S. aureus* in a sustained release manner.	[57]
2	Cellulose acetate-3-chloro-2-hydroxypropyl-trimethyl ammonium chloride	Cellulose-based membrane exerted a bactericidal action on *E. coli* and *S. aureus.*	[58]

(*Continued*)

Table 4.3 Cellulose-grafted materials with antimicrobial properties. (*Continued*)

S. no.	Materials	Properties	References
3	CNC–Ag–silicon NPs	Silica NP was loaded in the matrix of CNC-silver membrane and indicated bactericidal action on *E. coli* and *S. aureus*.	[59]
4	BNC–CuO nanocomposite	The BNC-based hydrogel nanocomposite membrane was fabricated with copper oxide and showed outstanding antimicrobial property against *E. coli*, *S. aureus*, and *Salmonella* species.	[60]
5	BNC–mulberry leaves	The BNC extracted from mulberry leaves showed good antibacterial property on *E. coli*.	[61]
6	CNF	Ginger cellulose film was formulated, which showed antibacterial action.	[62]
7	BNC	The BNC was prepared from *Enterobacter* sp. *FY-07* which showed antibacterial property.	[63]
8	CNC	Silver NP loaded on CNC to formulate a paper which exhibits antibacterial action.	[64]
9	CNC	Cellulose beads were developed from enzymes which showed antibacterial characteristics.	[65]
10	Cellulose	Novel soluble cellulose was prepared with chlorine termini by hyper-grafted method, and this polymer showed high antibacterial action on several bacteria.	[66]

(*Continued*)

Table 4.3 Cellulose-grafted materials with antimicrobial properties. (*Continued*)

S. no.	Materials	Properties	References
11	BNC	The cryogel was developed with hydroxyethyl cellulose and BNC which was further crosslinked with silver–titanium dioxide NP. This cryogel indicated a significant antimicrobial action of tebuconazole in a controlled release pattern.	[67]
12	BNC	Zinc oxide was loaded on BNC and formed nanocomposites which showed high antibacterial property.	[68]
13	Cellulose acetate	The nanocapsules containing essential oils such as lemongrass, peppermint, and cinnamon was loaded on cellulose acetate. These nanocapsules showed a marked antimicrobial action against *Pseudomonas aeruginosa*, *E. coli*, *S. aureus*, and the yeast *Candida albicans*.	[69]

Currently, tissue engineering has gained tremendous significance in the biomedical field for facilitating tissue repair, protection, and treatment using the basic concepts of engineering. The cellulose derivatives (CNC, CNF, and BNC) are most widely utilized for tissue engineering due to their biocompatibility, being biodegradable, possessing high mechanical strength, sustainability, water-holding capacity, 3-D porous structure, and improved cell growth as well as adhesion [70, 71]. Moreover, tissue engineering is a rapidly expanding stream in biomedical sciences and involves the linking of cells, biomaterials, and growth factors, which enabled the generation of biocompatible organs and tissues required during operative procedures, either in part or in full. The BNC is also capable of creating a 3-D cell culture by creating an extracellular matrix culture. These hydrogels can allow the movement of gases and thereby facilitate vascularization [72, 73].

The most widely used techniques for the fabrication of cellulose in tissue engineering is electrospinning, cross-linking, solvent casting, freeze-drying, the novel method 3-D or 4-D bioprinting, etc. Cellulose is a highly versatile material in the TE and covers almost all body tissues and bones in repairing and replacing such as cardiovascular, ocular, dental, vascular, etc. [74]. The cellulosic material is mostly compatible with other polymeric agents for tissue engineering such as poly(lactic acid), poly(glycolic acid), PVA, chitosan, collagen, hyaluronic acid, alginate, gelatin, etc. [70, 75].

Currently, the 3-D bioprinting technique attracts manufacturers for tissue engineering. The high rate of accuracy and precision enabled 3-D bioprinting to be utilized in tissue engineering in the biomedical field. During this technique, cells are entrapped inside the printing material gel and retain their uniformity [72]. Moreover, cellulose also finds very useful applications in skin tissue engineering for the reconstruction of skin layers, namely, epidermis and dermis. BNC expression is highly similar with that of skin layers and has a high potential of retaining moisture, with marked mechanical characteristics such as elasticity, Young's modulus, and comfortability [75, 76]. The applications of several forms of cellulose utilized for tissue engineering are depicted in Table 4.4.

Table 4.4 Applications of cellulose in tissue engineering.

S. no.	Cellulose type	Description	References
1	Plant-derived cellulose	Cellulose was derived from the plant species *Borassus flabellifer* (Linn.) and further combined with chitosan. This cellulose scaffold possesses higher mechanical and biodegradable properties and is applicable for tissue repair.	[77]
2	HEC	Hydroxyethyl cellulose (HEC) structure was altered with citric acid to create a modified scaffold for cartilage tissue engineering. This HEC-CA scaffold exhibited similar mechanical characteristics in terms of Young's modulus (10 mPa) and tensile strength (0.43 mPa).	[78]

(Continued)

Table 4.4 Applications of cellulose in tissue engineering. (*Continued*)

S. no.	Cellulose type	Description	References
3	Cellulose	α-cellulose was attached with epoxidized soybean oil, and these scaffolds are utilized for tissue engineering.	[79]
4	CNF	The CNF-based double-layered film was developed and further crosslinked with poly(globalide) for skin tissue engineering applications.	[80]
5	CNF	The CNF was synthesized from cellulose acetate and combined with hydroxyapatite and silver NPs for hard and soft tissue engineering with benefits of antimicrobial action.	[81]
6	CMC	The CMC was loaded with glycol chitosan to develop a hydrogel of lactoferrin by 3-D printing method for tissue engineering applications. This hydrogel showed biocompatibility and sustained release action.	[82]
7	BNC	The BNC showed exact similarity with ECM in their 3-D pattern and hence was used for the creation of cell culture and tissue engineering.	[83]
8	CNF	The emulsion of clove oil was incorporated in CNF-based hydrogel using high-intensity ultra-sound technology for tissue engineering.	[84]
9	BNC	The aerogel was prepared with BNC and further loaded with silver NPs and polyaniline. These nanocomposites possess high mechanical strength and are utilized for soft skin tissue engineering.	[85]

(*Continued*)

Table 4.4 Applications of cellulose in tissue engineering. (*Continued*)

S. no.	Cellulose type	Description	References
10	CMC	The CMC was used to develop nanofibrous scaffold with silk fibroin. These nanofibers were developed by electrospinning method and showed marked mechanical and swelling characteristics, therefore implemented for bone tissue engineering.	[86]
11	Cellulose acetate	The cellulose acetate was fabricated with poly-lactic acid by wet-electrospin method which resulted in a scaffold. This scaffold was coated with gelatin to behave as a nanocarrier and finds utility in neural tissuc engineering.	[87]
12	BNC	The 3-D nanofibrillar structure of BNC was developed and loaded on chitosan. This resulted in the development of nanocomposites.	[88]
13	BNC	The BNC–polyvinyl alcohol-based scaffold was generated using 3-D printing technology for bone tissue engineering.	[89]
14	CNC	3-D printed technology was utilized for the development of hydrogel scaffold using CNC, sodium alginate, and gelatin. This scaffold resembled similar characteristics for cell proliferation and exchange of nutrients and is useful in bone tissue engineering.	[90]
15	Cellulose	The aerogel prepared with cellulose by 3-D printing method showed similarity in shape and characteristics and therefore was used in conductive additive engineering.	[91]

4.4.4 Biosensing and Diagnostic

The early detection of a disease is the most favorable condition for the patients, and biosensors play a significant role in their recognition. Biosensors gain attention in the biomedical field, particularly in TE and sense, through the availability of antibodies, enzymes, receptors, and viral or microbial agents. These agents have the potential of monitoring the signals in real time such as the presence of proteins after cell injury, any kind of inflammatory response, attack of pathogens, liberation of antibodies, etc. While utilizing these biosensors, they should be patient-friendly, economic, precise, and sensitive in nature [72, 75].

Cellulose does not possess any biosensing characteristics, but in conjugation with several nanomaterials it showed diagnostic applications. Previously, these agents showed a powerful impact on the identification of blood type, HIV, and dengue. The cellulose-based paper and hydrogel showed diagnostic approaches after their structural modifications or attachments with nanomaterial [92]. Schyrr *et al.* utilized the CNC because of its versatile characteristics and conjugated it with PVA. The formed nanocomposite film was developed with a dimension of 25–70-nm thickness. This approach was based on fluorescence and showed the ability to detect protease activity [93]. Yee *et al.* used NCC to develop glucose-oxidase CNC biosensor for the detection of glucose in the samples [94]. Similarly, Esmailli *et al.* also developed a method for the detection of glucose using polypyrrole–cellulose nanocrystals as a new technique, which was further confirmed by field emission scanning electron microscopy [95].

Edwards *et al.* introduced a novel method using NCC–peptide conjugates as biosensor for sensing human neutrophil elastase enzyme (HNE). HNE is liberated in a large quantity by neutrophils during injury, which resulted in the lysis of growth factor and delay in the wound healing process. Hence, the NCC–peptide conjugate works as a prognostic marker in the identification of wound [96].

Another interesting example of a BNC-based nanopaper found its application in the formulation of an artificial tongue. The BNC paper was fabricated using multiple layers using laser printing technique and useful in optical sensor array technology [97]. The BNC was conjugated with gold NPs and utilized as a biosensor for the recognition of glucose [98]. The BNC was modified magnetically with perchloric acid and further implemented for the carrying of ligands, enzymes, and cells for prognosis in the biomedical field [99].

4.4.5 Enzyme Immobilization

The enzymes are protein in nature and contributed extensively as biocatalysts. Among the several compounds, the versatile properties possessed by NCC (high surface-to-volume ratio, mechanical strength, and hardness) attract several researchers for it to be incorporated for enzyme immobilization. The NCC-loaded enzymes showed significant catalytic reactions, improved stability, and enantio-selectivity [42]. Ariaeenejad *et al.* used CNC loaded on magnetic NPs and dopamine and thereby formulated a versatile carrier to link to enzyme immobilization. The formed nanobiocatalyst showed a high efficiency for biomass generation. This novel approach was found successful in improving the yield of fermented sugars compared with a single enzyme [100].

Li *et al.* used cellulose filter paper as a carrier for the immobilization of α-glucosidase with screening action in a rapid way and at a low cost. The cellulose structure was modified by polydopamine or polyethyleneimine using co-deposition technique and further loaded on the enzyme for its fixation. The results indicated that this process enhances enzyme reusability and is a powerful method to screen α-glucosidase suppressor [101]. Shin *et al.* utilized NCC as a carrier for the generation of polyhydroxybutyrate from syngas. For the effective transfer of carbon to produce carbon dioxide, NCC was used for the fixation. The incorporation of NCC enhances the stability and activity of an enzyme [102].

4.4.6 Wound Dressing/Healing

A disruption in the integrity of skin is an indication of the presence of a wound. The major culprits for the cause of wounds can be numerous, namely: chemical, mechanical, and thermal distresses. Thus, the prime goal of wound management is rapid closure of the wound and scar treatment [97]. The two types of wounds are, namely, acute as well as chronic, involving remedy for wound which is the active and multifarious progression of tissue regeneration that includes four phases: coagulation and homeostasis phase, inflammatory phase, proliferation phase, and maturation phase, as shown in Figure 4.2 [98].

The process of disruption of skin or mucus epithelial lining is known as wound. It is classified into acute or chronic, depending on the process for healing and its duration. The four phases involved in wound healing are the coagulation phase, inflammatory phase, proliferation phase, and maturation phase [99–106]. Cellulose is therefore used in the medical textile industry for the fabrication of dressing materials, sutures, etc. [189, 190].

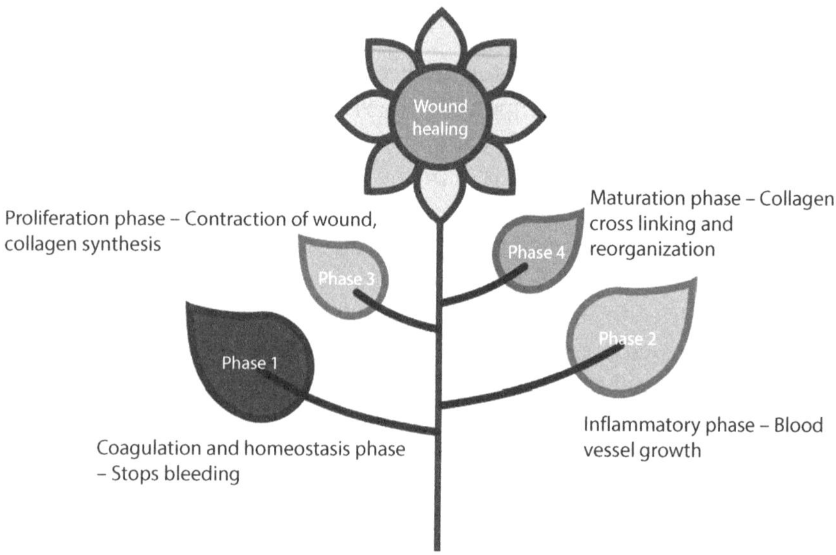

Figure 4.2 Progression of tissue regeneration.

The topical preparations proposed for wound healing that are commercially available are Bioprocess®, XCell®, and Biofill. Researchers Fontana *et al.* explained the potential of BC for skin burn [107].

The dressing BioFill®, a commercial product, had the requisite properties in terms of good adherence properties, durability, with good elastic properties that act as a barrier for bacteria, portable, etc. It has also been

Table 4.5 Application of cellulose in wound healing.

S. no.	Type/form of cellulose	Application	References
1	Cellulose/gelatin sponges	The macro- and micro-structural design, promoted wound healing	[112]
2	Nanofibers of cellulose acetate/gelatin	Utilized mat made of berberin for the treatment of ulcer of diabetic foot in the form of wound dressing. The animal studies showed good wound healing scores.	[113]

(*Continued*)

Table 4.5 Application of cellulose in wound healing. (*Continued*)

S. no.	Type/form of cellulose	Application	References
3	Cellulose-based aerogels	Prepared from organic and inorganic materials, in pure forms or composites and utilized for treatment, biosensing, and tissue scaffolds.	[114]
4	Natural extracellular matrix scaffolds	As ECM possessed the properties of tissue, it was used as a biomimetic approach that had control on cellular behavior.	[115]
5	Novel chitosan–cellulose nanofibers	Reported to possess self-healing and injectable properties and tissue-regenerating potential.	[116]
6	Nanofibrillar cellulose (NFC)	Utilized as scaffold for hASCs for cell transplantation.	[117]
7	Bacterial cellulose/ phytotherapic hydrogels	Hydrogel enhanced the healing of wound and promoted tissue repair.	[118]
8	Bacterial cellulose-based hydrogels	This hydrogel with the antimicrobial agent curcumin cyclodextrin was reported to have good antimicrobial activity.	[119]
9	Nanofibers of cellulose/ quaternized chitin/organic rectorite composites sponges (BACNF/QCR)	This showed synthesis of collagen and helped in hastening of wound healing when compared with gauze.	[120]
10	(CA/Gel/nHA) nanocomposite mats	Reported dressings that had higher wound closure, with collagen synthesis and good properties for wound healing.	[121]

proven to relieve pain in more than 300 cases [108]. Microbial cellulose finds vast biomedical application in the future [103, 109].

In adults, the succession of events for an inflammation response is of prime importance that is driven by the release of mediators [110]. As wound healing is a complicated progression, stern deformation may not result in normality, but the arrangement of unsettled tissue that have a fibrous look results in a decrease in mechanical strength and the formation of scars. Thus, biomaterials can lend a helping hand to restore this condition and prevent scar formation [111]. The applications of cellulose in wound healing are depicted in Table 4.5.

4.4.7 Applications in the Cardiovascular System

Worldwide, the major culprit for death, that affects the quality of life are the cardiovascular diseases (CVDs) and cause about one-third of deaths [122]. Development has been made for prevention and treatment of CVDS, but the use inclination towards smoking, technology-driven culture, that emphasize long working hours, less devotion to recreational activities, have elevated the chances of CVDs. Lack of physical activity, intake of sugars, metabolic disturbances are also the other reasons for CVD [123]. Due to the limitations, of current treatments for CVD, nowadays, artificial blood vessels have been used as a substitute [124]. For this bacterial cellulose has been explored for imitated vessels and impose their scientific application.

To restore the physiological functions of the patients, the use of artificial blood vessels has shaped a new era in the bioengineering field. Generally, the extensively artificial graft material used are polyurethanes, expanded-polytetrafluoroethylene (ePTFE) and (Dacron®) for synthetic blood vessels, though due to their good advantages of mechanical properties, highly biocompatibility, they pose certain disadvantages like low-resistance circulation property, poor patency rate for stenosis and thrombosis and the grafts less than 6mm, are not effective clinically. Thus, to deal with these problems, bacterial cellulose has engrossed as biomedical materials and devices.

Cellulose, a biopolymer, is water-insoluble polysaccharide, obtained from plants and microorganisms genera namely *Agrobacterium, Rhodobacter, Rhizobium, Gluconacetobacter* and *Sarcina* [125, 126]. It is considered to be the most crucial molecule of plant and major component of cell wall. Cellulose also known as anhydroglucose, in linear form and that is linked by β-1, 4, in means of the glucose residues.

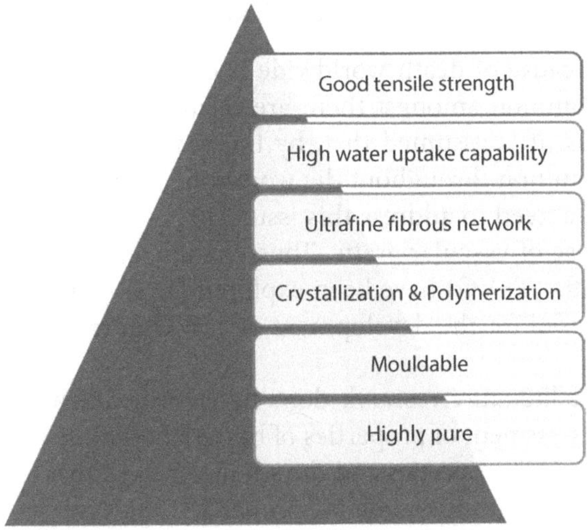

Figure 4.3 Bacterial cellulose properties.

The prime cellulose produced by bacterial genera *Gluconacetobacter* (previously *Acetobacter*) is known as bacterial cellulose (BC). BC possess distinct features of wound healing [104, 127]. The properties possess by cellulosic forms were depicted in Figure 4.3.

Bacterial cellulose revealed the potential as for atherosclerotic coronaries as artificial vessels [128]. Bacterial synthesized cellulose (abbreviated as BASYC®) have effectively used for process of implantation and ensure preservation of carotid arteries in animals [129]. The nanocomposites of bacterial cellulose of paraffin wax, starch, graphene oxide and fibrin are widely used in artificial blood vessels.

4.4.8 Cardiac Prosthetics

These are the heart valves, made of synthetic materials that are thrombogenic prospective and are implanted in patients, are maintained for life long. As the first implantation of valve in human circulatory system, varieties of prosthetic devices are used for cardiovascular diseases. Due to the advancement in technology, new prostheses have been developed.

The reports by Mohammadi *et al.* suggested a polymeric aortic valve made form PVA and BC showed the combination possess good mechanical properties and were similar to the porcine heart valve and can find application in cardiac prosthetics [130].

4.4.9 Replacement of Vascular Grafts

The leading cause of death worldwide is due to the cardiovascular diseases, the common amongst them are related to stenosis, occlusion of blood vessels. It is assumed that, by the year 2030, the mortality may rise to 23.3 million throughout the world due to cardiovascular diseases and there is a need to address this issue [131]. The treatment for this is surgery by use of vascular grafts. Thus, the upcoming tissue engineered vascular grafts (TEVG) have been explored [132]. Thus, bacterial cellulose can be used for the development of TEVG to replace small to large size arteries.

As per the literature research done by H. Zahedmanesh *et al* a comprehensive assessment of properties of bacterial cellulose tubes were done for its mechanical properties and the studies showed that tubes improved tensile strength with suture holding property [133].

Another literature reported the application of bacterial cellulose (BC) for biosynthetic blood vessels, where implantation was done in 3 different groups and was analyzed for different parameters for 2 weeks, and the results depicted that BC provides scope as biomaterial for TEVG [134].

4.5 Applications in Ocular Systems

4.5.1 Artificial Cornea

The second most common cause of blindness is the corneal disease. Though corneal transplantation has shown success, but it possesses some limitations like graft failure, shortage of corneal tissue, immune rejection etc. To overcome these limitations, a scaffold has been utilized, in form of bacterial cellulose, polymer produced by bacteria, due to its numerous properties, that possess high light transmittance, mechanical properties, that are withstand surgical sutures and intraocular pressure. Different technique namely keratoprosthesis is utilized to implant synthetic cornea, that offer ray of hope for blind patients [135].

One report as given by Brazilian research group Messaddeq, Ribeiro *et al.* devised and also got patent for confirming the use of BC to correct angle and develop contact lens for regeneration of cornea. In this mechanism, the BC membrane that are round initially are cut and compressed into stick, that will represent the semi-spherical end on the base and can lead to contact lens shape by molding it externally [136].

4.5.2 Contact Lens

The use of contact lens is wide, and the global market for it is increasing at higher rate [137]. These are used for protection of eyes, improve clarity of images and correctness of vision.

In work published by Cavicchioli *et al.* they utilized ciprofloxacin (CPX) in association with & without γ CD in BC membrane that had shape of contact lens. The results showed that the impregnated membranes did not possess toxicity, and found that BC-CPX can be good biomaterial for contact lens [138].

4.5.3 Retinal Pigments

The retinal pigment is selective barrier that regulate the photoreceptor layer, with main role in maintenance. The study of BC for retinal pigment was studied, with surface modification by using carboxymethyl cellulose and chitosan. The adhesion and proliferation of RP epithelium cells, was assessed. The surface modified BC enhanced union and propagation of RP epithelium cells. The BC along with acetylated BC had high elastic modulus, but second had small tensile strength and broke when elongated. The results showed that the acetylated form had good adhesion [139]. The application of cellulose forms in ocular system was depicted in Table 4.6.

Table 4.6 Application of cellulose in an ocular drug delivery system.

S. no.	Type/form of cellulose	Applications	References
1.	Cellulose nanocrystals (CNC)	CNC nano-engineered smart gel-enabled medications provided the release of loaded drug over 25 h with excellent ophthalmic drug delivery system.	[140]
2.	Bacterial cellulose and bacterial cellulose/polycaprolactone composite (BC/PCL)	The results discovered a temperate inflammatory process that decreased from the 20th day in groups of study.	[141]

(*Continued*)

Table 4.6 Application of cellulose in an ocular drug delivery system. (*Continued*)

S. no.	Type/form of cellulose	Applications	References
3.	Microbial cellulose graft	Possessed corneal healing with potential as a useful material for injured corneal surfaces.	[142]
4.	Tosyl cellulose	1. Utilized for the development of novel water-soluble ampholytic cellulose derivatives and act as a potential carrier for controlled drug release.	[143]
5.	Bacterial cellulose scaffolds	These scaffolds with doxoribicin showed a decrease in cell viability of HT-29 when compared with free drug.	[144]
6.	Bacterial nanocellulose (BNC)	This possessed good properties as corneal bandage.	[145]
7.	Carboxyl methyl cellulose	Utilized with PHMB and reduced its chemical availability and bioactivity; had a potential cytoprotective effect on the ocular surface when used before lens insertion.	[146]
8.	Bacterial cellulose/ poly(vinyl alcohol)	Utilized in corneal transplantations as hydrogel composites for tissue-engineered corneal stroma; artificial cornea biomaterial.	[147]
9.	Bacterial cellulose (BC)	Utilized for manufacturing that can have a wide application for epithelial regeneration.	[148]
10.	Bacterial cellulose (BC)	Utilized for RPE that enhanced the adhesion and proliferation of RPE cells.	[149]

4.6 Applications in Skeletal Systems

4.6.1 Bone Regeneration

Bone acts as a vital backing structure, binding set for muscles, ligaments, tendons, and as mechanical supportive mediator for most important tissues [150]. The majority of bone fractures take place due to bone regeneration [151]. BC membranes have proved to be useful in bone regeneration, for bone defects. Bacterial cellulose (BC) due to its excellent biomaterial finds medical applications, for bone regeneration, as they endorse bone formation at site, with cost effective treatment.

Literature finds the study on how Fang *et al.* observed the HA/BC hydroxyapatite nanocomposites scaffolds, that helped cell propagation with differentiation for human bone marrow (hBMSC) [152]. Study on BC-HA composites showed high potential bone regeneration. Microporous bacterial cellulose scaffolds with size range between 300–500 μm pore sizes were successfully produced [153]. Researchers have developed and characterized BC-Hap composites, wherein the results depicted the crystals when partly substituted with carbonate, resembled normal bones. Thus, with this unique property, in tissue engineering they encompass place as scaffolds [154, 155].

4.6.2 Cartilage Invertebrate Disc

The major health issue in industrialized societies is the back pain. This leads to distress to patients as well as families, with its prevalence is incurring due to the change in the lifestyle [156]. This back pain is related to degeneration of the intervertebral disc. Thus, to mimic the structure of native tissues is really challenging, but that can be achieved by use of bacterial cellulose. The exceptional properties of BC with amalgamation of micro fluidic techniques can help to formulate an IVD at micro scale.

The research explained by Junchuan Yang *et al.* a reverse reconstruction structure of native IVD with bacterial cellulose (BC) nanofibers, was done and later the microchip was screened, where the results depicted that the BC layers possessed excellent structural and functional performance [157].

The study performed by Leonardo Millon *et al.* they prepared nanocomposites in combination of bacterial cellulose and poly(vinyl alcohol) nanofibers, they found that when BC was added to PVA, it ranged in good mechanical properties, elasticity, and proved to a replacement for articular cartilage replacement with intervertebral discs [158].

4.6.3 Meniscus Implants

With the excellent mechanical properties of BC, and in terms of biodegradability indicates a good scaffold for replacement of cartilage [159]. As per the study by Lopes *et al.*, they determined the wear and frictional properties of BC against bovine cartilage. BC showed low friction coefficient and preserved the mating surface, due to the property of plastic deformation, hence BC can act as a budding material for joints [160]. Martínez *et al.* devised BC with micro-channels, which resulted in the configuration of cells with collagen fibers with parallel orientation and with good strength, thus making it ideal for the replacement of knee meniscus [161].

As per the studies conducted by Ávila *et al.*, for substitute of cartilage, BC with non-resorbable 15% cellulose was devised, which had similar properties with that of host tissue in terms of mechanical strength [162].

Azuma *et al.* documented the properties of BC with poly(dimethyl acrylamide) and possessed similar properties with regard to mechanical properties for artificial cartilage [163].

4.6.4 Dental and Ligament Implants

Biomaterials have played a major role in the medical field. Though they are classified as natural and synthetic, the natural biomaterials possess the property of tissue biocompatibility. With this, these are now being explored in dentistry due to their unique properties like those of native tissues.

As per the scientific documentation reported by Taba Jr *et al.*, the cellulose membranes have been utilized as a barrier for periodontal GTR, with expanded polytetrafluoroethylene along with absorbable polymers [164].

The first ever use of BC in the dental field was when it was used for periodontal disease with guided tissue regeneration, and in 1992, Novaes reported the application of GTR in periodontal defect [165]. When an infection is settled in the dental pulp, due to pathogens, it results in the treatment of the dental root canal. The conventional methods include the use of chemical substances, electrochemically activated solutions, and ultrasonic energy, but they possess limitations and there are chances of recurrence. Hence, bacterial cellulose can be a good alternative for treatment, wherein paper points are made and used to dry the canal and help in the incorporation of medications. As per the study of Yoshino *et al.*, they explored that BC sheets had the possibility to be used for root canal treatment with their properties of biocompatibility, good tensile strength, etc. [166]. In another study, it has been reported that BC hydrogels are biocompatible with connective tissues, and as per the results reported by Chiaoprakobkij *et al.*,

Table 4.7 Application of cellulose for bone regeneration, cartilage in vertebrates, meniscus implants, and dental and ligament implants.

S. no.	Types/form of cellulose	Applications	References
1	Bacterial cellulose	Utilized for the regeneration of tissue, periodontal and mucosa	[168]
2	Cellulose-based hydroxy gel	Utilized for tissue growth and regeneration	[169]
3	Bacterial cellulose	Collagen implants used to regenerate meniscal tissue after partial meniscectomy	[170]
4	Poly(methyl methacrylate) (PMMA)	Polymer used for the fabrication of dental prostheses	[171]
5	Porous silk fibroin/ cellulose hydrogels	Used to mimic natural bone, with appropriate mechanical strength	[172]
6	Non-woven cellulose II fabrics	Helps in the development of cartilage and its repair	[173]
7	(CS/HA/Si-HPMC) hydrogel	Utilized for excellent ability for repair of joint cartilage	[174]
8	Methyl cellulose	Bone generation and periodontal and bone diseases	[175]
9	Nanocellulose	Utilized for soft tissue implants and cartilage replacement with wound healing	[176]
10	2. Amidatedcarboxymethyl cellulose hydrogel (CMCA)	3. Utilized as a potential compound for cartilage regeneration	[177]

they explained the nanofibrils in the form of a three-dimensional network as mucosal flaps for oral tissue regeneration [167]. Several skeletal-related applications of cellulose are depicted in Table 4.7.

4.7 Miscellaneous

BC is also reported to have potential for urinary cancer. The study done by Bodin *et al.* was on micro-porous BC scaffolds that were developed for use in urinary diversion [178, 179]. BC has also found its place for tympanic membrane (TM), which is related to hearing loss and chronic perforations. Kim *et al.* fabricated the nanofiber patch by the use of BC, basically a wound healing scaffold for TM perforations, and the patch was found to significantly increase the healing [179, 180].

Though BC finds wide applications in treatment, yet still more research can be initiated to identify its unidentified potentials. It will indeed have a wide application in the biomedical field. As BC is endowed with unique characteristics, it is one of the emerging biomaterials with diverse applications. Thus, the interest in BC can intensify good research and inspire in the development of new materials that will also have wide applications in the medical field.

4.8 Conclusion

Cellulose and its several forms gain tremendous applications in the biomedical field for possessing versatile characteristics. The abundant availability from various sources attracts the manufacturers to give highest priority to the cellulose derivatives in the biomedical field despite other materials. All the forms of cellulose such as CNC, BNC, and CNF have proven applications in the biomedical field. The major areas of application cover drug delivery, tissue engineering, wound dressing, and biological systems such as cardiovascular, ocular, skeletal, etc., whereas cellulose is widely utilized in support of other materials and finds application as antimicrobial agents, for health and hygiene, etc. Hence, cellulose is a versatile material for the biomedical field.

References

1. Khattak, S., Wahid, F., Liu, L.-P., Jia, S.-R., Chu, L.-Q., Xie, Y.-Y., Li, Z.-X., Zhong, C., Applications of cellulose and chitin/chitosan derivatives and composites as antibacterial materials: Current state and perspectives. *Appl. Microbiol. Biotechnol.*, 103, 5, 1989–2006, 2019.
2. Bilal, M. and Iqbal, H.M.N., Naturally-derived biopolymers: Potential platforms for enzyme immobilization. *Int. J. Biol. Macromol.*, 130, 462–482, 2019.
3. Hemmati, F., Jafari, S.M., Taheri, R.A., Optimization of homogenization-sonication technique for the production of cellulose nanocrystals from cotton linter. *Int. J. Biol. Macromol.*, 137, 374–381, 2019.
4. Nehra, P. and Chauhan, R.P., Eco-friendly nanocellulose and its biomedical applications: Current status and future prospect. *J. Biomater. Sci. Polym. Ed.*, 32, 1, 112–149, 2020, https://doi.org/10.1080/09205063.2020.1817706.
5. Joseph, B., Sagarika, V.K., Sabu, C., Kalarikkal, N., Thomas, S., Cellulose nanocomposites: Fabrication and biomedical applications. *J. Bioresour. Bioprod.*, 5, 4, 223–237, 2020.
6. Raghav, N., Sharma, M.R., Kennedy, J.F., Nanocellulose: A mini-review on types and use in drug delivery systems. *Carbohydr. Polym. Technol. Appl.*, 2, 100031, 2021.
7. Gumrah Dumanli, A., Nanocellulose and its composites for biomedical applications. *Curr. Med. Chem.*, 24, 5, 512–528, 2017, http://dx.doi.org/10.2174/0929867323666161014124008
8. Nicu, R., Ciolacu, F., Ciolacu, D.E., Advanced functional materials based on nanocellulose for pharmaceutical/medical applications. *Pharm.*, 13, 8, 1125, 2021.
9. Farooq, A., Patoary, M.K., Zhang, M., Mussana, H., Li, M., Naeem, M.A., Mushtaq, M., Farooq, A., Liu, L., Cellulose from sources to nanocellulose and an overview of synthesis and properties of nanocellulose/zinc oxide nanocomposite materials. *Int. J. Biol. Macromol.*, 154, 1050–1073, 2020.
10. Liu, L., Kerr, W.L., Kong, F., Characterization of lipid emulsions during *in vitro* digestion in the presence of three types of nanocellulose. *J. Colloid Interface Sci.*, 545, 317–329, 2019.
11. Pourmadadi, M., Rahmani, E., Shamsabadipour, A., Samadi, A., Esmaeili, J., Arshad, R., Rahdar, A., Tavangarian, F., Pandey, S., Novel carboxymethyl cellulose based nanocomposite: A promising biomaterial for biomedical applications. *Process Biochem.*, 130, 211–226, 2023.
12. Du, H., Liu, W., Zhang, M., Si, C., Zhang, X., Li, B., Cellulose nanocrystals and cellulose nanofibrils based hydrogels for biomedical applications. *Carbohydr. Polym.*, 209, 130–144, 2019.

13. Pötzinger, Y., Kralisch, D., Fischer, D., Bacterial nanocellulose: The future of controlled drug delivery? *Ther. Deliv.*, 8, 9, 753–761, 2017, https://doi.org/10.4155/tde-2017-0059.
14. Safari, J. and Zarnegar, Z., Advanced drug delivery systems: Nanotechnology of health design a review. *J. Saudi Chem. Soc.*, 18, 2, 85–99, 2014.
15. Trucillo, P., Drug carriers: Classification, administration, release profiles, and industrial approach. *Process*, 9, 3, 470, 2021.
16. Salama, A., Cellulose/calcium phosphate hybrids: New materials for biomedical and environmental applications. *Int. J. Biol. Macromol.*, 127, 606–617, 2019.
17. Ullah, H., Wahid, F., Santos, H.A., Khan, T., Advances in biomedical and pharmaceutical applications of functional bacterial cellulose-based nanocomposites. *Carbohydr. Polym.*, 150, 330–352, 2016.
18. Surendran, G. and Sherje, A.P., Cellulose nanofibers and composites: An insight into basics and biomedical applications. *J. Drug Deliv. Sci. Technol.*, 75, 103601, 2022.
19. Kabir, S.M.F., Sikdar, P.P., Haque, B., Bhuiyan, M.A.R., Ali, A., Islam, M.N., Cellulose-based hydrogel materials: Chemistry, properties and their prospective applications. *Prog. Biomater.*, 7, 3, 153–174, 2018.
20. Imlimthan, S., Correia, A., Figueiredo, P., Lintinen, K., Balasubramanian, V., Airaksinen, A.J., Kostiainen, M.A., Santos, H.A., Sarparanta, M., Systematic *in vitro* biocompatibility studies of multimodal cellulose nanocrystal and lignin nanoparticles. *J. Biomed. Mater. Res. Part A*, 108, 3, 770–783, 2020.
21. Löbmann, K. and Svagan, A.J., Cellulose nanofibers as excipient for the delivery of poorly soluble drugs. *Int. J. Pharm.*, 533, 1, 285–297, 2017.
22. Li, W., Lan, Y., Guo, R., Zhang, Y., Xue, W., Zhang, Y., *In vitro* and *in vivo* evaluation of a novel collagen/cellulose nanocrystals scaffold for achieving the sustained release of basic fibroblast growth factor. *J. Biomater. Appl.*, 29, 6, 882–893, 2014, http://dx.doi.org/10.1177/0885328214547091.
23. Mauricio, M.R., Da Costa, P.G., Haraguchi, S.K., Guilherme, M.R., Muniz, E.C., Rubira, A.F., Synthesis of a microhydrogel composite from cellulose nanowhiskers and starch for drug delivery. *Carbohydr. Polym.*, 115, 715–722, 2015.
24. Lin, N., Gèze, A., Wouessidjewe, D., Huang, J., Dufresne, A., Biocompatible double-membrane hydrogels from cationic cellulose nanocrystals and anionic alginate as complexing drugs codelivery. *ACS Appl. Mater. Interfaces*, 8, 11, 6880–6889, 2016.
25. Supramaniam, J., Adnan, R., Mohd Kaus, N.H., Bushra, R., Magnetic nanocellulose alginate hydrogel beads as potential drug delivery system. *Int. J. Biol. Macromol.*, 118, 640–648, 2018.
26. Åhlén, M., Tummala, G.K., Mihranyan, A., Nanoparticle-loaded hydrogels as a pathway for enzyme-triggered drug release in ophthalmic applications. *Int. J. Pharm.*, 536, 1, 73–81, 2018.

27. Laurén, P., Lou, Y.R., Raki, M., Urtti, A., Bergström, K., Yliperttula, M., Technetium-99m-labeled nanofibrillar cellulose hydrogel for *in vivo* drug release. *Eur. J. Pharm. Sci.*, 65, 79–88, 2014.
28. Yildir, E., Sjöholm, E., Preis, M., Trivedi, P., Trygg, J., Fardim, P., Sandler, N., Investigation of dissolved cellulose in development of buccal discs for oromucosal drug delivery. *Pharm Dev Technol.*, 23, 5, 520–529, 2017, https://doi.org/10.1080/10837450.2017.1397163.
29. Masruchin, N., Park, B.-D., Causin, V., Dual-responsive composite hydrogels based on TEMPO-oxidized cellulose nanofibril and poly(N-isopropylacrylamide) for model drug release. *Cellul.*, 25, 1, 485–502, 2017.
30. Pourmoazzen, Z., Sadeghifar, H., Chen, J., Yang, G., Zhang, K., Lucia, L., The morphology, self-assembly, and host-guest properties of cellulose nanocrystals surface grafted with cholesterol. *Carbohydr. Polym.*, 233, 115840, 2020.
31. Li, C., Evans, J., Wang, N., Guo, T., He, S., pH dependence of the chirality of nematic cellulose nanocrystals. *Sci. Rep.*, 9, 1, 1–7, 2019.
32. Yang, J., Dahlström, C., Edlund, H., Lindman, B., Norgren, M., pH-responsive cellulose–chitosan nanocomposite films with slow release of chitosan. *Cellul.*, 26, 6, 3763–3776, 2019.
33. Seabra, A.B., Bernardes, J.S., Fávaro, W.J., Paula, A.J., Durán, N., Cellulose nanocrystals as carriers in medicine and their toxicities: A review. *Carbohydr. Polym.*, 181, 514–527, 2018.
34. Fekete, T., Borsa, J., Takács, E., Wojnárovits, L., Synthesis of carboxymethylcellulose/starch superabsorbent hydrogels by gamma-irradiation. *Chem. Cent. J.*, 11, 1, 1–10, 2017.
35. Shao, W., Liu, H., Wang, S., Wu, J., Huang, M., Min, H., Liu, X., Controlled release and antibacterial activity of tetracycline hydrochloride-loaded bacterial cellulose composite membranes. *Carbohydr. Polym.*, 145, 114–120, 2016.
36. Karzar Jeddi, M. and Mahkam, M., Magnetic nano carboxymethyl cellulose-alginate/chitosan hydrogel beads as biodegradable devices for controlled drug delivery. *Int. J. Biol. Macromol.*, 135, 829–838, 2019.
37. Abo-Elseoud, W.S., Hassan, M.L., Sabaa, M.W., Basha, M., Hassan, E.A., Fadel, S.M., Chitosan nanoparticles/cellulose nanocrystals nanocomposites as a carrier system for the controlled release of repaglinide. *Int. J. Biol. Macromol.*, 111, 604–613, 2018.
38. Zainuddin, N., Ahmad, I., Zulfakar, M.H., Kargarzadeh, H., Ramli, S., Cetyltrimethylammonium bromide-nanocrystalline cellulose (CTAB-NCC) based microemulsions for enhancement of topical delivery of curcumin. *Carbohydr. Polym.*, 254, 117401, 2021.
39. Singhsa, P., Narain, R., Manuspiya, H., Bacterial cellulose nanocrystals (BCNC) preparation and characterization from three bacterial cellulose sources and development of functionalized BCNCs as nucleic acid delivery systems. *ACS Appl. Nano Mater.*, 1, 1, 209–221, 2017.
40. Gonzalez-Fernandez, T., Rathan, S., Hobbs, C., Pitacco, P., Freeman, F.E., Cunniffe, G.M., Dunne, N.J., McCarthy, H.O., Nicolosi, V., O'Brien, F.J.,

Kelly, D.J., Pore-forming bioinks to enable spatio-temporally defined gene delivery in bioprinted tissues. *J. Control. Release*, 301, 13–27, 2019.

41. Tang, K.Y., Heng, J.Z.X., Chai, C.H.T., Chan, C.Y., Low, B.Q.L., Chong, S.M.E., Loh, H.Y., Li, Z., Ye, E., Loh, X.J., Modified bacterial cellulose for biomedical applications. *Chem. Asian J.*, 17, 19, e202200598, 2022.
42. Karimian, A., Parsian, H., Majidinia, M., Rahimi, M., Mir, S.M., Samadi Kafil, H., Shafiei-Irannejad, V., Kheyrollah, M., Ostadi, H., Yousefi, B., Nanocrystalline cellulose: Preparation, physicochemical properties, and applications in drug delivery systems. *Int. J. Biol. Macromol.*, 133, 850–859, 2019.
43. Dong, S., Cho, H.J., Lee, Y.W., Roman, M., Synthesis and cellular uptake of folic acid-conjugated cellulose nanocrystals for cancer targeting. *Biomacromolecules*, 15, 5, 1560–1567, 2014.
44. Sheng, Y., Gao, J., Yin, Z.Z., Kang, J., Kong, Y., Dual-drug delivery system based on the hydrogels of alginate and sodium carboxymethyl cellulose for colorectal cancer treatment. *Carbohydr. Polym.*, 269, 118325, 2021.
45. Ehsanimehr, S., Najafi Moghadam, P., Dehaen, W., Shafiei-Irannejad, V., Synthesis of pH-sensitive nanocarriers based on polyacrylamide grafted nanocrystalline cellulose for targeted drug delivery to folate receptor in breast cancer cells. *Eur. Polym. J.*, 150, 110398, 2021.
46. Do, T.T.A., Grijalvo, S., Imae, T., Garcia-Celma, M.J., Rodríguez-Abreu, C., A nanocellulose-based platform towards targeted chemo-photodynamic/photothermal cancer therapy. *Carbohydr. Polym.*, 270, 118366, 2021.
47. Hussein, Y., Loutfy, S.A., Kamoun, E.A., El-Moslamy, S.H., Radwan, E.M., Elbehairi, S.E.I., Enhanced anti-cancer activity by localized delivery of curcumin form PVA/CNCs hydrogel membranes: Preparation and *in vitro* bioevaluation. *Int. J. Biol. Macromol.*, 170, 107–122, 2021.
48. Taokaew, S., Chiaoprakobkij, N., Siripong, P., Sanchavanakit, N., Pavasant, P., Phisalaphong, M., Multifunctional cellulosic nanofiber film with enhanced antimicrobial and anticancer properties by incorporation of ethanolic extract of Garcinia mangostana peel. *Mater. Sci. Eng. C*, 120, 111783, 2021.
49. Ning, L., You, C., Zhang, Y., Li, X., Wang, F., Synthesis and biological evaluation of surface-modified nanocellulose hydrogel loaded with paclitaxel. *Life Sci.*, 241, 117137, 2020.
50. Kim, Y.M., Lee, Y.S., Kim, T., Yang, K., Nam, K., Choe, D., Roh, Y.H., Cationic cellulose nanocrystals complexed with polymeric siRNA for efficient anticancer drug delivery. *Carbohydr. Polym.*, 247, 116684, 2020.
51. Alavi, M. and Rai, M., Recent progress in nanoformulations of silver nanoparticles with cellulose, chitosan, and alginic acid biopolymers for antibacterial applications. *Appl. Microbiol. Biotechnol.*, 103, 21, 8669–8676, 2019.
52. Alavi, M. and Rai, M., Recent advances in antibacterial applications of metal nanoparticles (MNPs) and metal nanocomposites (MNCs) against multi-drug-resistant (MDR) bacteria. *Expert Rev Anti Infect Ther*, 17, 6, 419–428, 2019, https://doi.org/10.1080/14787210.2019.1614914.

53. Kim, J.S., Kuk, E., Yu, K.N., Kim, J.-H., Park, S.J., Lee, H.J., Kim, S.H., Park, Y.K., Park, Y.H., Hwang, C.-Y., Kim, Y.-K., Lee, Y.-S., Jeong, D.H., Cho, M.-H., Antimicrobial effects of silver nanoparticles. *Nanomed. Nanotechnol. Biol. Med.*, 3, 1, 95–101, 2007.
54. Kalwar, K. and Shan, D., Antimicrobial effect of silver nanoparticles (AgNPs) and their mechanism – A mini review. *Micro Nano Lett.*, 13, 3, 277–280, 2018.
55. Liu, Y., Wang, S., Wang, Z., Yao, Q., Fang, S., Zhou, X., Yuan, X., Xie, J., The *in situ* synthesis of silver nanoclusters inside a bacterial cellulose hydrogel for antibacterial applications. *J. Mater. Chem. B*, 8, 22, 4846–4850, 2020.
56. Li, J., Cha, R., Mou, K., Zhao, X., Long, K., Luo, H., Zhou, F., Jiang, X., Nanocellulose-based antibacterial materials. *Adv. Healthc. Mater.*, 7, 20, 1800334, 2018.
57. Li, J., Kang, L., Wang, B., Chen, K., Tian, X., Ge, Z., Zeng, J., Xu, J., Gao, W., Controlled release and long-term antibacterial activity of dialdehyde nanofibrillated cellulose/silver nanoparticle composites. *ACS Sustain. Chem. Eng.*, 7, 1, 1146–1158, 2018.
58. Fei, P., Liao, L., Meng, J., Cheng, B., Hu, X., Song, J., Non-leaching antibacterial cellulose triacetate reverse osmosis membrane via covalent immobilization of quaternary ammonium cations. *Carbohydr. Polym.*, 181, 1102–1111, 2018.
59. Smiechowicz, E., Niekraszewicz, B., Kulpinski, P., Dzitko, K., Antibacterial composite cellulose fibers modified with silver nanoparticles and nanosilica. *Cellul.*, 25, 6, 3499–3517, 2018.
60. Araújo, I.M.S., Silva, R.R., Pacheco, G., Lustri, W.R., Tercjak, A., Gutierrez, J., Júnior, J.R.S., Azevedo, F.H.C., Figuêredo, G.S., Vega, M.L., Ribeiro, S.J.L., Barud, H.S., Hydrothermal synthesis of bacterial cellulose–copper oxide nanocomposites and evaluation of their antimicrobial activity. *Carbohydr. Polym.*, 179, 341–349, 2018.
61. Chen, J., Chen, C., Liang, G., Xu, X., Hao, Q., Sun, D., *In situ* preparation of bacterial cellulose with antimicrobial properties from bioconversion of mulberry leaves. *Carbohydr. Polym.*, 220, 170–175, 2019.
62. Abral, H., Ariksa, J., Mahardika, M., Handayani, D., Aminah, I., Sandrawati, N., Sugiarti, E., Muslimin, A.N., Rosanti, S.D., Effect of heat treatment on thermal resistance, transparency and antimicrobial activity of sonicated ginger cellulose film. *Carbohydr. Polym.*, 240, 116287, 2020.
63. Gao, G., Fan, H., Zhang, Y., Cao, Y., Li, T., Qiao, W., Wu, M., Ma, T., Li, G., Production of nisin-containing bacterial cellulose nanomaterials with antimicrobial properties through co-culturing Enterobacter sp. FY-07 and Lactococcus lactis N8. *Carbohydr. Polym.*, 251, 117131, 2021.
64. He, Y., Li, H., Fei, X., Peng, L., Carboxymethyl cellulose/cellulose nanocrystals immobilized silver nanoparticles as an effective coating to improve barrier and antibacterial properties of paper for food packaging applications. *Carbohydr. Polym.*, 252, 117156, 2021.

65. Califano, D., Patenall, B.L., Kadowaki, M.A.S., Mattia, D., Scott, J.L., Edler, K.J., Enzyme-functionalized cellulose beads as a promising antimicrobial material. *Biomacromolecules*, 22, 2, 754–762, 2021.
66. Demircan, D. and Zhang, B., Facile synthesis of novel soluble cellulose-grafted hyperbranched polymers as potential natural antimicrobial materials. *Carbohydr. Polym.*, 157, 1913–1921, 2017.
67. El-Naggar, M.E., Hasanin, M., Youssef, A.M., Aldalbahi, A., El-Newehy, M.H., Abdelhameed, R.M., Hydroxyethyl cellulose/bacterial cellulose cryogel dopped silver@titanium oxide nanoparticles: Antimicrobial activity and controlled release of Tebuconazole fungicide. *Int. J. Biol. Macromol.*, 165, 1010–1021, 2020.
68. Janpetch, N., Saito, N., Rujiravanit, R., Fabrication of bacterial cellulose-ZnO composite via solution plasma process for antibacterial applications. *Carbohydr. Polym.*, 148, 335–344, 2016.
69. Liakos, I.L., Iordache, F., Carzino, R., Scarpellini, A., Oneto, M., Bianchini, P., Grumezescu, A.M., Holban, A.M., Cellulose acetate - essential oil nanocapsules with antimicrobial activity for biomedical applications. *Colloids Surf. B Biointerfaces*, 172, 471–479, 2018.
70. Trache, D., Tarchoun, A.F., Derradji, M., Hamidon, T.S., Masruchin, N., Brosse, N., Hussin, M.H., Nanocellulose: From fundamentals to advanced applications. *Front. Chem.*, 8, 70, e392, 2020.
71. Shrivastav, P., Pramanik, S., Vaidya, G., Abdelgawad, M.A., Ghoneim, M.M., Singh, A., Abualsoud, B.M., Amaral, L.S., Abourehab, M.A.S., Bacterial cellulose as a potential biopolymer in biomedical applications: A state-of-the-art review. *J. Mater. Chem. B*, 10, 17, 3199–3241, 2022.
72. Curvello, R., Raghuwanshi, V.S., Garnier, G., Engineering nanocellulose hydrogels for biomedical applications. *Adv. Colloid Interface Sci.*, 267, 47–61, 2019.
73. Qian, H., Liu, J., Wang, X., Pei, W., Fu, C., Ma, M., Huang, C., The state-of-the-art application of functional bacterial cellulose-based materials in biomedical fields. *Carbohydr. Polym.*, 300, 120252, 2023.
74. Khan, S., Ul-Islam, M., Ullah, M.W., Zhu, Y., Narayanan, K.B., Han, S.S., Park, J.K., Fabrication strategies and biomedical applications of three-dimensional bacterial cellulose-based scaffolds: A review. *Int. J. Biol. Macromol.*, 209, 9–30, 2022.
75. Sharma, C. and Bhardwaj, N.K., Bacterial nanocellulose: Present status, biomedical applications and future perspectives. *Mater. Sci. Eng. C*, 104, 109963, 2019.
76. Bacakova, L., Pajorova, J., Bacakova, M., Skogberg, A., Kallio, P., Kolarova, K., Svorcik, V., Versatile application of nanocellulose: From industry to skin tissue engineering and wound healing. *Nanomater.*, 9, 2, 164, 2019.
77. Mahendiran, B., Muthusamy, S., Selvakumar, R., Rajeswaran, N., Sampath, S., Jaisankar, S.N., Krishnakumar, G.S., Decellularized natural 3D cellulose

scaffold derived from Borassus flabellifer (Linn.) as extracellular matrix for tissue engineering applications. *Carbohydr. Polym.*, 272, 118494, 2021.
78. Nouri-Felekori, M., Nezafati, N., Moraveji, M., Hesaraki, S., Ramezani, T., Bioorthogonal hydroxyethyl cellulose-based scaffold crosslinked via click chemistry for cartilage tissue engineering applications. *Int. J. Biol. Macromol.*, 183, 2030–2043, 2021.
79. Pour-Esmaeil, S., Sharifi-Sanjani, N., Khoee, S., Taheri-Qazvini, N., Biocompatible chemical network of α-cellulose-ESBO (epoxidized soybean oil) scaffold for tissue engineering application. *Carbohydr. Polym.*, 241, 116322, 2020.
80. Amaral, H.R., Wilson, J.A., do Amaral, R.J.F.C., Pasçu, I., de Oliveira, F.C.S., Kearney, C.J., Freitas, J.C.C., Heise, A., Synthesis of bilayer films from regenerated cellulose nanofibers and poly(globalide) for skin tissue engineering applications. *Carbohydr. Polym.*, 252, 117201, 2021.
81. Sofi, H.S., Akram, T., Shabir, N., Vasita, R., Jadhav, A.H., Sheikh, F.A., Regenerated cellulose nanofibers from cellulose acetate: Incorporating hydroxyapatite (HAp) and silver (Ag) nanoparticles (NPs), as a scaffold for tissue engineering applications. *Mater. Sci. Eng. C*, 118, 111547, 2021.
82. Janarthanan, G., Tran, H.N., Cha, E., Lee, C., Das, D., Noh, I., 3D printable and injectable lactoferrin-loaded carboxymethyl cellulose-glycol chitosan hydrogels for tissue engineering applications. *Mater. Sci. Eng. C*, 113, 111008, 2020.
83. Halib, N., Ahmad, I., Grassi, M., Grassi, G., The remarkable three-dimensional network structure of bacterial cellulose for tissue engineering applications. *Int. J. Pharm.*, 566, 631–640, 2019.
84. Huerta, R.R., Silva, E.K., El-Bialy, T., Saldaña, M.D.A., Clove essential oil emulsion-filled cellulose nanofiber hydrogel produced by high-intensity ultrasound technology for tissue engineering applications. *Ultrason. Sonochem.*, 64, 104845, 2020.
85. Hosseini, H., Zirakjou, A., Goodarzi, V., Mousavi, S.M., Khonakdar, H.A., Zamanlui, S., Lightweight aerogels based on bacterial cellulose/silver nanoparticles/polyaniline with tuning morphology of polyaniline and application in soft tissue engineering. *Int. J. Biol. Macromol.*, 152, 57–67, 2020.
86. Singh, B.N., Panda, N.N., Mund, R., Pramanik, K., Carboxymethyl cellulose enables silk fibroin nanofibrous scaffold with enhanced biomimetic potential for bone tissue engineering application. *Carbohydr. Polym.*, 151, 335–347, 2016.
87. Naseri-Nosar, M., Salehi, M., Hojjati-Emami, S., Cellulose acetate/poly lactic acid coaxial wet-electrospun scaffold containing citalopram-loaded gelatin nanocarriers for neural tissue engineering applications. *Int. J. Biol. Macromol.*, 103, 701–708, 2017.
88. Wei, Z., Pan, P., Hong, F.F., Cao, Z., Ji, Y., Chen, L., A novel approach for efficient fabrication of chitosan nanoparticles-embedded bacterial nanocellulose conduits. *Carbohydr. Polym.*, 264, 118002, 2021.

89. Aki, D., Ulag, S., Unal, S., Sengor, M., Ekren, N., Lin, C.C., Yılmazer, H., Ustundag, C.B., Kalaskar, D.M., Gunduz, O., 3D printing of PVA/hexagonal boron nitride/bacterial cellulose composite scaffolds for bone tissue engineering. *Mater. Des.*, 196, 109094, 2020.
90. Dutta, S.D., Hexiu, J., Patel, D.K., Ganguly, K., Lim, K.T., 3D-printed bioactive and biodegradable hydrogel scaffolds of alginate/gelatin/cellulose nanocrystals for tissue engineering. *Int. J. Biol. Macromol.*, 167, 644–658, 2021.
91. Tetik, H., Zhao, K., Shah, N., Lin, D., 3D freeze-printed cellulose-based aerogels: Obtaining truly 3D shapes, and functionalization with cross-linking and conductive additives. *J. Manuf. Process.*, 68, 445–453, 2021.
92. Wang, H.-K., Tsai, C.-H., Chen, K.-H., Tang, C.-T., Leou, J.-S., Li, P.-C., Tang, Y.-L., Hsieh, H.-J., Wu, H.-C., Cheng, C.-M., Cellulose-based diagnostic devices for diagnosing serotype-2 dengue fever in human serum. *Adv. Healthc. Mater.*, 3, 2, 187–196, 2014.
93. Schyrr, B., Pasche, S., Voirin, G., Weder, C., Simon, Y.C., Foster, E.J., Biosensors based on porous cellulose nanocrystal–poly(vinyl alcohol) scaffolds. *ACS Appl. Mater. Interfaces*, 6, 15, 12674–12683, 2014.
94. Yee, Y.C., Hashim, R., Yahya, A.R.M., Bustami, Y., Colorimetric analysis of glucose oxidase-magnetic cellulose nanocrystals (CNCs) for glucose detection. *Sensors*, 19, 11, 2511, 2019.
95. Esmaeili, C., Abdi, M.M., Mathew, A.P., Jonoobi, M., Oksman, K., Rezayi, M., Synergy effect of nanocrystalline cellulose for the biosensing detection of glucose. *Sensors*, 15, 10, 24681–24697, 2015.
96. Edwards, J.V., Prevost, N., Sethumadhavan, K., Ullah, A., Condon, B., Peptide conjugated cellulose nanocrystals with sensitive human neutrophil elastase sensor activity. *Cellul.*, 20, 3, 1223–1235, 2013.
97. Abbasi-Moayed, S., Golmohammadi, H., Hormozi-Nezhad, M.R., A nanopaper-based artificial tongue: A ratiometric fluorescent sensor array on bacterial nanocellulose for chemical discrimination applications. *Nanoscale*, 10, 5, 2492–2502, 2018.
98. Lv, P., Zhou, H., Mensah, A., Feng, Q., Wang, D., Hu, X., Cai, Y., Amerigo Lucia, L., Li, D., Wei, Q., A highly flexible self-powered biosensor for glucose detection by epitaxial deposition of gold nanoparticles on conductive bacterial cellulose. *Chem. Eng. J.*, 351, 177–188, 2018.
99. Baldikova, E., Pospiskova, K., Ladakis, D., Kookos, I.K., Koutinas, A.A., Safarikova, M., Safarik, I., Magnetically modified bacterial cellulose: A promising carrier for immobilization of affinity ligands, enzymes, and cells. *Mater. Sci. Eng. C*, 71, 214–221, 2017.
100. Ariaeenejad, S., Motamedi, E., Hosseini Salekdeh, G., Immobilization of enzyme cocktails on dopamine functionalized magnetic cellulose nanocrystals to enhance sugar bioconversion: A biomass reusing loop. *Carbohydr. Polym.*, 256, 117511, 2021.

101. Li, P., Ma, X.H., Jin, L., Chen, J., Dopamine-polyethyleneimine co-deposition cellulose filter paper for α-Glucosidase immobilization and enzyme inhibitor screening. *J. Chromatogr. B*, 1167, 122582, 2021.
102. Shin, S.K., Ko, Y.J., Jeong, D.W., Lee, M.E., You, S.K., Hyeon, J.E., Han, S.O., Enhanced production of polyhydroxybutyrate from syngas by using nanoscaled cellulose particles with a syngas-converting enzyme complex immobilized on Ralstonia eutropha. *J. Cleaner Prod.*, 285, 124903, 2021.
103. Almeida, A.P.C., Saraiva, J.N., Cavaco, G., Portela, R.P., Leal, C.R., Sobral, R.G., Almeida, P.L., Crosslinked bacterial cellulose hydrogels for biomedical applications. *Eur. Polym. J.*, 177, 111438, 2022.
104. Ul-Islam, M., Alhajaim, W., Fatima, A., Yasir, S., Kamal, T., Abbas, Y., Khan, S., Khan, A.H., Manan, S., Ullah, M.W., Yang, G., Development of low-cost bacterial cellulose-pomegranate peel extract-based antibacterial composite for potential biomedical applications. *Int. J. Biol. Macromol.*, 231, 123269, 2023.
105. Kujath, P. and Michelsen, A., Wounds - from physiology to wound dressing. *Dtsch. Arztebl.*, 105, 13, 239–248, 2008.
106. Robson, M.C., Steed, D.L., Franz, M.G., Wound healing: Biologic features and approaches to maximize healing trajectories. *Curr. Probl. Surg.*, 38, 2, 72–140, 2001.
107. Szycher, M. and Lee, S., Modern wound dressings: A systematic approach to wound healing. *Biomater. Appl.*, 7, 2, 142–213, 1992.
108. Dhivya, S., Padma, V., Santhini, E., Wound dressings - A review. *Biomedicine (Taipei)*, 5, 4, 22, 2015.
109. Tarnuzzer, R. and Schultz, G., Biochemical analysis of acute and chronic wound environments. *Wound Repair Regener.*, 4, 321–5, 1996.
110. Clark, R., Wound repair overview and general considerations. The molecular and cellular biology of wound repair. *Plenum*, 1, 3–5, 1996.
111. Dowsett, C. and Newton, H., Wound bed preparation: TIME in practice. *Wounds UK*, 1, 58–70, 2005.
112. Vanwijck, R., Surgical biology of wound healing. *Bull. Mem. Acad. R. Med. Belg.*, 156, 175–84, 2000.
113. Degreef, H., How to heal a wound fast. *Dermatol. Clin.*, 16, 365–75, 1998.
114. Hunt, T., Hopf, H., Hussain, Z., Physiology of wound healing. *Adv. Skin Wound Care*, 13, 6–11, 2000.
115. Fontana, J., Dsouza, A., Fontana, C., Acetobacter cellulose pellicle as a temporary skin substitute. *Appl. Biochem. Biotechnol.*, 24–25, 253–264, 1990.
116. Czaja, W., Young, J., Kawecki, M., The future prospects of microbial cellulose in biomedical applications. *Biomacromolecules*, 8, 1–12, 2002.
117. Wouk, F., Diniz, M., Círio, M., Santos, H., Baltazar, L., Membrana biológica Biofill-estudocomparativo com outros agentespromotores da cicatrizac, ão da peleemsuínos: Aspectosclínicos, histopatológicos e morfométricos. *Arch. Vet. Sci.*, 3, 31–37, 1998.

150. Glonek, T. and Levinson, E., A microbial cellulose graft for ocular surface defects. *ARVO*, 48, 787, 2007.
151. El-Sayed, N.S., El-Sakhawy, M., Hesemann, P., Brun, N., Kamel, S., Rational design of novel water-soluble ampholytic cellulose derivatives. *Int. J. Biol. Macromol.*, 114, 363–72, 2018 Jul 15.
152. Cacidedo, M.L., Leon, I.E., S., J., Modified bacterial cellulose scaffolds for localized doxorubicin release in human colorectal HT-29 cells. *Colloids Surf. B*, 140, 421–429, 2016.
153. Sales, I., Antin, J., Engorba, J., Bacterial nanocellulose as a corneal bandage material: A comparison with amniotic membrane. *Biomater. Sci.*, 8, 2921–2930, 2020.
154. Han, Y., Li, C., Cai, Q., Bao, X., Tang, L., Ao, H., Liu, J., Jin, M., Zhou, Y., Wan, Y., Liu, Z., Studies on bacterial cellulose/poly (vinyl alcohol) hydrogel composites as tissue-engineered corneal stroma. *Biomed. Mater.*, 15, 3, 035022, 2020.
155. Wang, J., Gao, C., Zhang, Y., Wan, Y., Preparation and *in vitro* characterization of BC/PVA hydrogel composite for its potential use as artificial cornea biomaterial. *Mater. Sci. Eng. C*, 30, 1, 214–8, 2010.
156. Anton-Sales, I., Beekmann, U., Laromaine, A., Roig, A., Kralisch, D., Opportunities of bacterial cellulose to treat epithelial tissues. *Curr. Drug Targets*, 20, 8, 808–822, 2019.
157. Goncalves, S., Padrao, J., Silva, J., Bacterial cellulose as a support for the growth of retinal pigment epithelium. *Biomacromolecules*, 16, 4, 1341–51, 2014.
158. Pajarinen, J., Lin, T., Gibon, E., Kohno, Y., Mesenchymal stem cell-macrophage crosstalk and bone healing. *Biomaterials*, 196, 80–89, 2019.
159. Naghib, M., Ansari, M., Pedram, A., Bioactivation of 304 stainless steel surface through 45S5 bioglass coating for biomedical applications. *Int. J. Electrochem. Sci.*, 7, 2890–2903, 2012.
160. Fang, B., Wan, Z., Tang, T., Proliferation and osteoblastic differentiation of human bone marrow stromal cells on hydroxyapatite/bacterial cellulose nanocomposite scaffolds. *Tissue Eng.—Part A*, 15, 5, 1091–1098, 2009.
161. Zaborowska, M., Bodin, A., Bäckdahl, H., Popp, J., Microporous bacterial cellulose as a potential scaffold for bone regeneration. *Acta Biomater.*, 6, 7, 2540–7, 2010.
162. Hong, L., Wang, L., Jia, R., Huang, Y., Gao, C., Wan, Z., Hydroxyapatite/bacterial cellulose composites synthesized via a biomimetic route. *Mater. Lett.*, 60, 1710–1713, 2006.
163. Wan, Z., Hong, L., Jia, R., Huang, Y., Zhu, Y., Wang, L., Synthesis and characterization of hydroxyapatite—Bacterial cellulose nanocomposites. *Compos. Sci. Technol.*, 66, 1825–1832, 2006.
164. Urban, P. and Roberts, S., Degeneration of the intervertebral disc (IVD). *Arthritis Res. Ther.*, 5, 3, 120–130, 2003.

165. Yang, J. and Wang, L., Reverse reconstruction and bioprinting of bacterial cellulose-based functional total intervertebral disc for therapeutic implantation. *Small*, 1702582, 1–9, 2017.
166. Millon, L., Oates, J., Wan, W., Compression properties of polyvinyl alcohol–bacterial cellulose nanocomposite. *J. Biomed. Mater. Res. B Appl. Biomater.*, 90, 2, 922–9, 2009.
167. Svensson, A., Nicklasson, E., Harrah, T., Panilaitis, B., Kaplan, D.L., Brittberg, M., Bacterial cellulose as a potential scaffold for tissue engineering of cartilage. *Biomaterials*, 26, 419–431, 2005.
168. Lopes, L., Machado, M., Castanheira, L., Granja, L., Gama, M., Dourado, F., Friction and wear behaviour of bacterial cellulose against articular cartilage. *Wear*, 271, 2328–2333, 2011.
169. Martínez, H., Brackmann, C., Enejder, A., Gatenholm, P., Mechanical stimulation of fibroblasts in micro-channeled bacterial cellulose scaffolds enhances production of oriented collagen fibers. *J. Biomed. Mater. Res. Part A*, 100, 948–957, 2012.
170. Ávila, H., Schwarz, S., Feldmann, M., Mantas, A., Von, A., Gatenholm, P., Biocompatibility evaluation of densified bacterial nanocellulose hydrogel as an implant material for auricular cartilage regeneration. *Appl. Microbiol. Biotechnol.*, 98, 7423–7435, 2014.
171. Azuma, C., Yasuda, K., Tanabe, Y., Taniguro, H., Kanaya, F., Nakayama, A., Biodegradation of high-toughness double network hydrogels as potential materials for artificial cartilage. *J. Biomed. Mater. Res. Part A*, 81A, 373–380, 2007.
172. Taba, Jr., Jin, Q., Sugai, V., Giannobile, V., Current concepts in periodontal bioengineering. *Orthod. Craniofac. Res.*, 8, 292–302, 2005.
173. Novaes, B. and Novaes, B., IMZ implants placed into extraction sockets in association with membrane therapy (Gengiflex) and porous hydroxyapatite: A case report. *Int. J. Oral. Maxillofac. Implants*, 7, 536–540, 1992.
174. Yoshino, A., Tabuchi, M., Uo, M., Applicability of bacterial cellulose as an alternative to paper points in endodontic treatment. *Acta Biomater.*, 9, 6116–6122, 2013.
175. Chiaoprakobkij, N., Sanchavanakit, N., Subbalekha, K., Pavasant, P., Phisalaphong, M., Characterization and biocompatibility of bacterial cellulose/alginatecomposite sponges with human keratinocytes and gingival fibroblasts. *Carbohydr. Polym.*, 853, 548–553, 2011.
176. Gutierrez, A., Osorio, M., Molina-Ramirez, C., Arboleda-Toro, D., Bacterial cellulose: A biomaterial with high potential in dental and oral applications. *Cellulose*, 19, 1–8, 2020.
177. Dutta, D., Patel, K., Lim, T., Functional cellulose-based hydrogels as extracellular matrices for tissue engineering. *J. Biol. Eng.*, 13, 1, 1–9, Dec 2019.
178. Bodin, A., Concaro, S., Brittberg, M., Gatenholm, P., Bacterial cellulose as a potential meniscus implant. *J. Tissue Eng. Regen. Med.*, 1, 5, 406–8, Sep 2007.

179. Juan, A., Rene, C., Germán, S., Antimicrobial poly (methyl methacrylate) with silver nanoparticles for dentistry: A systematic review. *Appl. Sci.*, 10, 11, 4007, 2020.
180. Burger, D., Beaumont, M., Rosenau, T., Porous silk fibroin/cellulose hydrogels for bone tissue engineering via a novel combined process based on sequential regeneration and porogen leaching. *Mol.*, 25, 21, 5097, 2020.

5

Biomedical Applications of Starch

Daud Hussain[1], Suhail Ayoub Khan[1], Annu[2] and Tabrez Alam Khan[1*]

[1]*Department of Chemistry, Jamia Millia Islamia, New Delhi, India*
[2]*Thin-Film and Materials Laboratory, School of Mechanical Engineering, Yeungnam University, Gyeongsan, Republic of Korea*

Abstract
Starch is a natural polymer which acquires several exceptional properties and some shortcomings simultaneously. There is a growing interest in the biomedical and pharmaceutical sectors for biocompatible and bioresorbable materials. Its functional and structural diversity make starch appropriate for numerous applications. Different approaches can be employed to modify and improve the properties of starch to promote its utilization for various biomedical and pharmaceutical purposes. These physical, chemical, or enzymatic changes regulate and adjust the properties of starches for various usages such as transdermal patches, drug carriers, injectables, tablet excipients, transient embolizants, scaffolds, wound dressing materials, and stents. Starches that are chemically modified have more reactive sites to carry bioactive compounds and are beneficial biocompatible carriers that are metabolized easily inside the human body. This chapter reviews the modifications of starch and its utilization in biomedical and pharmaceutical applications, either as excipients or as drug delivery systems. Furthermore, a few drug release mechanisms, which involve encapsulants, micro/nanoparticle hydrogels, and scaffolds, are discussed.

Keywords: Starch, biodegradable, biomedical, drug carriers, encapsulants, modified, scaffold, hydrogel

**Corresponding author*: takhan@jmi.ac.in

Annu (ed.) Biopolymers for Biomedical Applications, (129–150) © 2024 Scrivener Publishing LLC

5.2 Structure and Properties of Starch

The major reservoir of polysaccharides in plants is starch, which is present in a discrete granular form comprised of mainly two homopolymers of D-glucose amylose and amylopectin (Figure 5.1) Amylose is a linear α-D (1→4) polymer composed of glucan units, while amylopectin is a branched polymer that has a similar backbone structure as amylose but connected through α-1,6′-linked branch points, having the highest molecular weight 107–109 among all the known naturally occurring polymers [43, 44]. Starch chains accommodate a lot of hydroxyl groups. Evidently, the starch is hydrophilic in nature because the hydroxyl groups available on starch possibly display reactivity particularly for alcohols [45]. Starch has different magnitudes of amylose that range from 10% to 20% and amylopectin at 80–90% depending on the source [46]. Starch occurs generally in the form of discrete granules, and a subsequently strong intermolecular association is shown by hydrogen bonding of these granules formed through OH groups on the surface of granules, exhibiting hydrophilic properties due to hydrophilicity, and with the morphology and internal interaction of starch rapidly altered by water molecules. However, the hydrophilicity of starch could be utilized to intensify the rate of degradation for hydrophobic degradable polymers. The characterization of biomaterials based

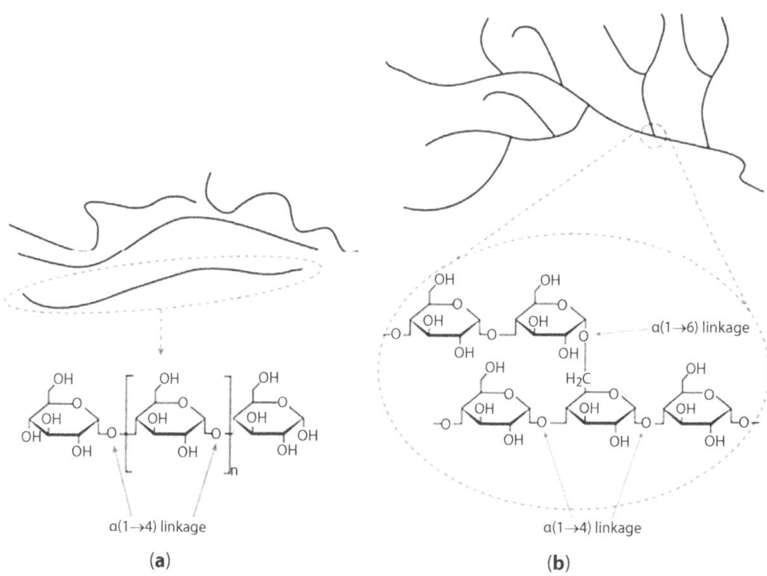

Figure 5.1 Molecular structure of starch: (a) amylose and (b) amylopectin [53].

on starch for usage in biomedical operations is significant because of their solubility differences, swelling, and surface features. The powdered starch is converted to gel for its utilization in pharmaceutical functions, and this conversion is accomplished by gelatinization. Subsequently, during the gelatinization process, amylopectin shows stabilizing properties, while amylose form gels and holds an efficient complex-forming tendency with lipids and other components [47]. Therefore, amylose and amylopectin are approved for several pharmaceutical and food applications. The properties of starch rely upon the chemical and physical features, such as granular size, amylose/amylopectin ration, distribution size, and mineral content. The amylose content of starch granules differs with their botanical source and is influenced by the type of soil and climatic conditions during growth.

Starch is completely biodegradable in a vast diversity of conditions. It can be hydrolyzed by enzymes or microorganism into glucose and then metabolized into CO_2 and H_2O [48]. It is significant to note that starch is again produced in plants by reutilizing carbon dioxide and sunshine. The processability of starch is very low, and it is also poor in its mechanical properties and dimensional stability for its products. Hence, the native form of starch is not utilized exactly. Therefore, starches can be categorized into (1) slowly digestible, (2) rapidly digestible, and (3) resistant starches (RS). This categorization relies upon the intrinsic features that influence the digestibility of starch. A few studies indicate the usage of RS as constituents with probable biomedical operations as drug delivery devices [49–52]. The potentiality indicated that chemical alterations of starches and their adaptability to dissolve them in aqueous solution provide the advancement of several forms of devices for drug delivery in the field of biomedical operations.

5.3 Modified Starches

Several chemical, physical, genetic, and enzymatic changes will alter and enhance the useful properties of starch to enable its usage for numerous intents, consenting its utilization as an excipient either alone or in concurrence with effective materials [54, 55]. Certain alterations can develop flow and compressibility performances for appropriate drug delivery devices.

Physical approaches of modification involve any treatment under the act of mechanics, pressure, or heat in wet or dry situation, involving fractionation. Variation by enzymes means the involvement of treatment of starch in existence of minute quantities of enzymes to attain water-soluble or moderately hydrolyzed goods. Chemical alteration includes the treatment of starch in occupancy of one or more chemical constituents.

5.3.1 Physical Modifications

The utmost common derivative of starch utilized in pharmaceutical production is pre-gelled starch. The granules of starch absorb a little amount of water after suspension in water or having been exposed to high humidity. Consequently, they are insoluble in cold water, which restricts its usage in a few products. Pre-gelled starch is acquired by gelatinizing and drying of the suspension. Gelatinization is a vital procedure in the modification of starch consisting of disorganized crystalline structures with the objective to ultimately propose a new form of arrangement in structure. It also leads to an alteration in the organization of granules. The order of crystallinity in starch granules is generally the basic feature that influences the functional properties [56, 57]. Within the starch granules, the collapse in the order of crystallinity exhibits inevitable variations in properties such as loss of birefringence, pasting, granule swelling, and starch solubility.

5.3.2 Enzyme Modification

Amylase and related enzymes are utilized to hydrolyze starch to maltose, glucose, and malto-oligosaccharides. Glucoamylase is an exo-acting hydrolase, with hydrolyzed (1→6)-branched points, and converts starch effectively to glucose [58]. The susceptibility of the hydrolyses of starches by enzymes has also been recommended to be influenced by the particular area's starch granules' surface, which declines the magnitude of enzyme binding, and eventually resulted in less hydrolyses in larger, in contrast to small, granules [59]. Usually, amylases degrade starch to a pertinent extent of polymerization (DP), and to influence the particular DP to process an exact kind of cyclodextrin, cyclodextrin glycosyl transferase is utilized [60]. The particular product attained depends upon the category and situation of the processing employed, and its features rely upon the resource of the starch utilized. The significance of cyclodextrins depends upon its capability to process compounds with hydrophobic materials, permitting the development of additional compounds with an extensive diversity of organic ingredients. This procedure signifies the utmost progressive technologies of molecular encapsulation, and it is treated as inexpensive and simple in relation with others [61].

5.3.3 Chemical Modification

The chemical changes in starch significantly influence their capability of arrangement under several conformations (i.e., random, coils helices) and suggest further methods to regulate their self-assemblage. Numerous materials for biomedical or pharmaceutical operations were acquired by complexation reactions by grafting of new functional groups or cross-linking. It is alluring to mention the reality that the efficacy of self-assemblage starch in helical fashion distinguishes entirely its behavior in comparison to synthetic polymers. Furthermore, the ratio between crystals and helices (ordered) and amorphous structures (i.e., disordered) of the scheme is a crucial component to acquire particular properties [62, 63]. The modified starches mostly display improved stability and paste clarity and increased resistance to retrogradation.

5.3.4 Cross-Linked Starch

Starch network structures are generally intensified by utilizing cross-linking. Lately, the idea of minor modifications may initiate major changes in self-assemblage excipient to access the means to develop, with a similar chemistry, absolutely various starch excipient substances [64, 65]—for example, contramid excipient for the continued release of drug was attained at low extended cross-linking degree (CLD), while a binder–disintegrant was acquired at a somewhat greater CLD [66–68].

Cross-linking creates intra- and inter-organic bonds which afford additional integrity of structure to reinforce and stabilize complexes, to promote resistance to low pH, greater temperature, and large shear, and to utilize as an agent for disintegrant and matrices in measured discharge formulations. Although they contain extra grimy gels and minimal water holding capability throughout aging [69], the only alteration was a minute difference in the level of cross-linking, though the physical and practical influence was main: less CLD permits sufficient changeability of chains to set up hydrogen linkages in addition to stabilizing the matrix, cross-linking bridges, restricting hydration, and improving the time of release. It was reported that the crystallinity of starch is associated to the CLD and influenced considerably the release time of drug [66].

5.3.5 Oxidized Starch

To develop new drug delivery systems, oxidized starches have been principally utilized. The comfort of chemical operation of the structure of starch,

encourage an accurate dose and has admirable disintegration properties. Furthermore, they afforded some encouraging outcomes as hydrophilic matrices for enhanced release.

5.5 Starch for Novel Drug Delivery

The application of starches has made considerable advancement in biomedical operations, involving innovative devices for drug delivery, due to its biodegradability, biocompatibility, inexpensiveness, and nontoxicity. Moreover, starch can safeguard bioactive compounds from oxidative or thermal degradation such as detrimental ecological circumstances frequently experienced throughout processing. In this system, drugs are carried to the targeted sites in the human body and then controls the drug release in the small intestine [83, 84]. The objective of these devices is uniform loading and enhancing the outcome of drugs. They intensify the bioavailability and hence control the continuous and/or constant concentrations of therapeutic staffs by safeguarding the short half-lives of bioactive compounds. Furthermore, they enhance patient relief since they might carry huge local concentrations, evading adverse impacts on untargeted sites and involving detrimental outcomes in the gastrointestinal tract (GIT), such as peptic ulcer, bleeding, and gastric erosion, along with increases in patient therapy by diminishing the drug taking repetition. These arrangements essentially have the succeeding features: (1) mechanical strength, because of the presence of hydrogen bonding in the starch chain preferring inter-chain interactions; (2) sorption of large quantities of drug during formation, because the distinctive and several interaction positions approve the interactions stated above including those supported between the active materials and starch chains; (3) no accidental drug release due to the firm feasibility of the absorption of materials from the sites of interaction, though a strong interaction in the matrix of the polysaccharide avoids or hampers any undesirable delivery; (4) smooth manufacturing as its flexibility permits starch to be applied in numerous kinds of procedures; and (5) the chances of sterilization [85, 86].

5.5.1 Drug Delivery Systems

The system of drug delivery comes from ancient Egypt, in the pattern of drug and pill formulations. Along the progress in the management of pharmaceuticals, practices have been upgraded with time—for example, solid devices such as powders developed into granules and its reciprocal systems, like tablets, dragees, coated tablets, and capsules. Presently, pharmaceutical advancement has capitalized in distinguished delivery devices, pointing at the operation of lesser dosages of drug with a consequential decrease in adverse impacts. Apparently, novel release schemes involve the production of micro/nanoparticles, encapsulants, and scaffolds that are formulated [87, 88].

5.6 Encapsulants

Encapsulation is employed for the stabilization, protection, and gradual discharge of materials. It is treated as a means to ensnare active materials inside a suitable carrier with the intent to safeguard and deliver to the targeted site at the right time. The wall or encapsulating materials mostly comprise of starch and its derivatives (like cyclodextrins), lipids, proteins, gums, or any arrangement of them. The external surface of these annular-designed particles is hydrophilic, and although they own a central hallow pit that is hydrophobic and capable to involve, additionally, a polar moiety for geometric affinity, the significance of molecular encapsulation is attained through incorporation of compound formation [60]. Several approaches were utilized for encapsulation like emulsification, nanoprecipitation, freeze-drying, extrusion, spray-drying, molecular addition, co-crystallization, fluidized bed coating, and coacervation [89, 90]. To substitute encapsulation or transporters, gum arabic powder and maltodextrin succinylated starches were employed [91]. The encapsulation procedure

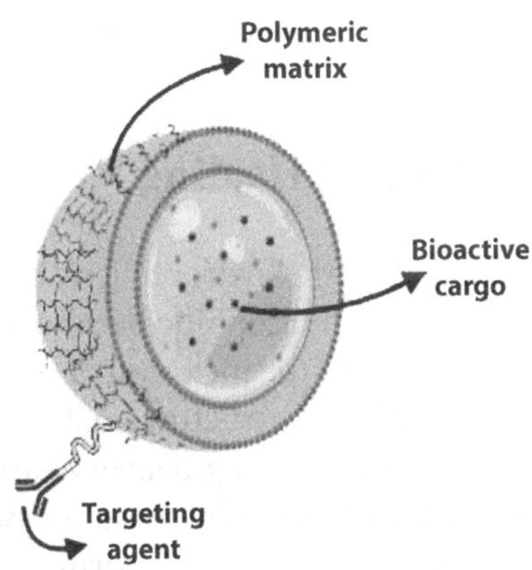

Figure 5.2 Schematic representation of a nanocapsule. A bioactive cargo is encapsulated into a core-shell polymeric matrix [92].

stabilizes the air and light-sensitive products. A bioactive cargo is encapsulated into a core-shell polymeric matrix as a nanocapsule as shown in Figure 5.2.

5.7 Micro/Nanoparticle Hydrogels

As a promising nanocarrier, amylose could be employed, as it can process bioactive molecular incorporated compounds that are susceptible to oxidation and heat by constituting a helix over ligands and therefore safeguarding and increasing the bioavailability to prepare them for appropriate delivery and constant discharge inside small intestine. The molecular incorporated complexes can considerably be extra effective in protecting responsive materials from detrimental processing situations than the prevailing conventional encapsulation methods [93]. Microspheres and/or nanospheres and/or microcapsules and polymeric nanocapsules are extensively utilized in controlling drug delivery systems. The approaches employed to load bioactive materials or drugs inside the matrices of polymers are as follows: (1) incorporation—in polymerization mechanism, the material is solubilized before the inclusion of a monomer that encourages the trapping of the encapsulated material into the matrix and (2) adsorption—addition of material into the medium while the matrix of polymer is previously established. Therefore, the material is sorbed mostly on the external surface of the substance [94–96]. Because of its storage stability, its great stability against swelling, large shear, greater temperature, and acidic situations, including its simple formation technique, the cross-linked starch microspheres were the most examined drug carriers, particularly in intranasal delivery devices [97–99].

5.8 Scaffolds for Wound Healing

An excellent substitute to resolve the hydrophilicity of starches is the grafting of a hydrophobic group on a hydrophilic backbone. The hydroxyl-rich backbone of polysaccharide grants the inclusion of several peculiar ligands to acquire polyfunctional colloidal structures. Chen *et al.* (2011) stated the usage of polymers as significant drug delivery carriers in colonic systems [100]. The digestion is prevented by these polysaccharides and can be poorly permeable in the upper GIT for the drug, although after arrival in the colon it can be assimilated by enzymes formed by colonic microbiota. They determined that the resistant starch acetate film-coated pellet is exceptional and a possible oral colon-directed delivery carrier. The protection of starch acetate digestion can be controlled by the substitution degree of starch acetate; consequently, starch acetate could be employed as a probable delivery system for colon-directed biomacromolecule drugs [101]. To assure its capabilities, a perfect kind of structure should essentially be biodegradable and biocompatible to offer a surface chemistry satisfactory for distinction, proliferation, and cellular adhesion to display mechanical endurance and improve the local biomechanics. Alongside this utilization, scaffolds can be categorized into the following groups:

- Osteoconductive scaffolds are formed from substances that support growth and cellular migration as well as certify the mechanical uphold required throughout bone regeneration [102].
- Carrier scaffolds are fabricated when autogenic osteogenic cells are introduced in the scaffold to encourage osteoinduction [103].

Similarly, this class also includes porous delivery devices. An appropriate recognition of the porosity necessities relies upon the final use of the substances. An additional condition for the porosity of scaffold is the interconnection of pores (the interconnection of pores is also termed as "mesoporosity") and for suitable nutrient transport and cellular migration to the newly produced tissues. For drug delivery systems, the interconnection of

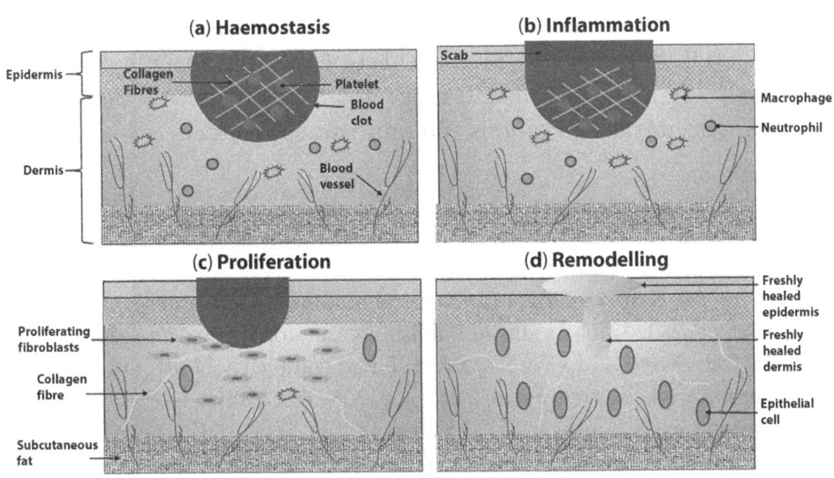

Figure 5.3 Schematic representation of the cutaneous wound healing [107].

pores provides a homogeneous distribution of drug inside the entire bulk of the scaffold. A recent study demonstrated that starch films could photo-inactivate 99.9% *Escherichia coli*, which is viable for wound healing [104]. The starch-mediated synthesis of chitosan/silver nanocomposite exhibited a wound gap closure by promoting cell migration in mouse fibroblast cells [105]. In a study, PVA/chitosan/starch produced nanofibrous mates which depicted the ability to accelerate wound healing [106]. The representation of wound healing is shown in Figure 5.3 [107].

5.9 Conclusions

Starch is one of the most appropriate polysaccharides recommending an extensive range of biomedical applications. Starch is commercially accessible in powder form, film, hydrogel, and sponges. In the form of fiber, starch attains greater attention from researchers working in this domain. It is cheap, biodegradable, and modified easily by physical or chemical approaches. Advancement in understanding the parameters influencing the functionality of starch and the outcomes of chemical alterations has authorized the starch industry to manufacture starches with appropriate and enhanced functional features. The application of starch in the pharmaceutical industry varies from ingredients to advanced drug delivery devices for the treatment of numerous diseases. Because of the biocompatibility features, versatility, and low immunogenicity of physical and chemical amendments, this polymer is broadly utilized in medicines and has encouraging usage for biomedical operations. Furthermore, manufacturers that alter its molecule to acquire appropriate nanocarriers have formed nanomaterials that have less adverse effects. Lastly, the flexibility of starch reflects its potential that grants its usage in the various areas of biomedical and pharmaceutical applications and futuristic substances.

References

1. Chen, Y.F., Kaur, L., Singh, J., Chemical modification of starch, in: *Starch in Food*, pp. 283–321, Woodhead Publishing, Sawston, United Kingdom, 2018.
2. Lu, D.R., Xiao, C.M., Xu, S.J., Starch-based completely biodegradable polymer materials. *Express Polym. Lett.*, *3*, 6, 366–375, 2009.
3. Masina, N., Choonara, Y.E., Kumar, P., du Toit, L.C., Govender, M., Indermun, S., Pillay, V., A review of the chemical modification techniques of starch. *Carbohydr. Polym.*, *157*, 1226–1236, 2017.

4. Puchongkavarin, H., Varavinit, S., Bergthaller, W., Comparative study of pilot scale rice starch production by an alkaline and an enzymatic process. *Starch-Stärke, 57*, 3-4, 134–144, 2005.
5. Mano, J.F. and Reis, R.L., Viscoelastic monitoring of starch-based biomaterials in simulated physiological conditions. *Mater. Sci. Eng.: A, 370*, 1-2, 321–325, 2004.
6. Garcia, M.A.V.T., Garcia, C.F., Faraco, A.A.G., Pharmaceutical and biomedical applications of native and modified starch: A review. *Starch-Stärke, 72*, 7-8, 1900270, 2020.
7. Pal, K., Banthia, A.K., Majumdar, D.K., Starch based hydrogel with potential biomedical application as artificial skin. *Afr. J. Biomed. Res., 9*, 1, 23–29. 2006.
8. Chakraborty, S., Sahoo, B., Teraoka, I., Miller, L.M., Gross, R.A., Enzyme-catalyzed regioselective modification of starch nanoparticles. *Macromolecules, 38*, 1, 61–68, 2005.
9. Pereira, C.S., Cunha, A.M., Reis, R.L., Vazquez, B., San Roman, J., New starch-based thermoplastic hydrogels for use as bone cements or drug-delivery carriers. *J. Mater. Sci.: Mater. Med., 9*, 12, 825–833, 1998.
10. Domb, A.J., Manor, N., Elmalak, O., Biodegradable bone cement compositions based on acrylate and epoxide terminated poly (propylene fumarate) oligomers and calcium salt compositions. *Biomaterials, 17*, 4, 411–417, 1996.
11. Kaur, L., Singh, N., Singh, J., Factors influencing the properties of hydroxypropylated potato starches. *Carbohydr. Polym., 55*, 2, 211–223, 2004.
12. Griffin, G.J.L., Starch polymer blends. *Polym. Degrad. Stab., 45*, 2, 241–247, 1994.
13. Pareta, R. and Edirisinghe, M.J., A novel method for the preparation of starch films and coatings. *Carbohydr. Polym., 63*, 3, 425–431, 2006.
14. Hemamalini, T. and Dev, V.R.G., Comprehensive review on electrospinning of starch polymer for biomedical applications. *Int. J. Biol. Macromol., 106*, 712–718, 2018.
15. Elvira, C., Mano, J.F., San Roman, J., Reis, R.L., Starch-based biodegradable hydrogels with potential biomedical applications as drug delivery systems. *Biomaterials, 23*, 9, 1955–1966, 2002.
16. Noreen, A., Akram, J., Rasul, I., Mansha, A., Yaqoob, N., Iqbal, R., Zia, K.M., Pectins functionalized biomaterials; a new viable approach for biomedical applications: A review. *Int. J. Biol. Macromol., 101*, 254–272, 2017.
17. Bikiaris, D., Prinos, J., Koutsopoulos, K., Vouroutzis, N., Pavlidou, E., Frangis, N., Panayiotou, C., LDPE/plasticized starch blends containing PE-g-MA copolymer as compatibilizer. *Polym. Degrad. Stab., 59*, 1-3, 287–291, 1998.
18. Mani, R. and Bhattacharya, M., Properties of injection moulded blends of starch and modified biodegradable polyesters. *Eur. Polym. J., 37*, 3, 515–526, 2001.
19. Liu, J., Willför, S., Xu, C., A review of bioactive plant polysaccharides: Biological activities, functionalization, and biomedical applications. *Bioact. Carbohydr. Diet. Fibre, 5*, 1, 31–61, 2015.

20. Muzzarelli, R.A.A. and Muzzarelli, C., Chitosan chemistry: Relevance to the biomedical sciences, in: *Polysaccharides I*, pp. 151–209, 2005.
21. Munarin, F. A. B. I. O. L. A., Tanzi, M.C., Petrini, P. A. O. L. A., Advances in biomedical applications of pectin gels. *Int. J. Biol. Macromol.*, *51*, 4, 681–689, 2012.
22. Liu, Z., Jiao, Y., Wang, Y., Zhou, C., Zhang, Z., Polysaccharides-based nanoparticles as drug delivery systems. *Adv. Drug Deliv. Rev.*, *60*, 15, 1650–1662, 2008.
23. Vasilyev, G., Vilensky, R., Zussman, E., The ternary system amylose-amylopectin-formic acid as precursor for electrospun fibers with tunable mechanical properties. *Carbohydr. Polym.*, *214*, 186–194, 2019.
24. Wadke, P., Chhabra, R., Jain, R., Dandekar, P., Silver-embedded starch-based nanofibrous mats for soft tissue engineering. *Surf. Interfaces*, *8*, 137–146, 2017.
25. Berens, A.M., Newman, S., Bhrany, A.D., Murakami, C., Sie, K.C., Zopf, D.A., Computer-aided design and 3D printing to produce a costal cartilage model for simulation of auricular reconstruction. *Otolaryngol.-Head Neck Surg.*, *155*, 2, 356–359, 2016.
26. Abd Karim, A., Norziah, M.H., Seow, C.C., Methods for the study of starch retrogradation. *Food Chem.*, *71*, 1, 9–36, 2000.
27. Kasemsuwan, T. and Jane, J., Location of amylose in normal starch granules. II. Locations of phosphodiester cross-linking revealed by phosphorus-31 nuclear magnetic resonance. *Cereal Chem.*, *71*, 3, 282–286, 1994.
28. Lenaerts, V., Moussa, I., Dumoulin, Y., Mebsout, F., Chouinard, F., Szabo, P., Marchessault, R., Cross-linked high amylose starch for controlled release of drugs: Recent advances. *J. Control. Release*, *53*, 1-3, 225–234, 1998.
29. García-González, C.A., López-Iglesias, C., Concheiro, A., Alvarez-Lorenzo, C., Biomedical applications of polysaccharide and protein based aerogels, in: *Biobased Aerogels*, pp. 295–323, 2018.
30. Silva, G.A., Coutinho, O.P., Ducheyne, P., Shapiro, I.M., Reis, R.L., The effect of starch and starch-bioactive glass composite microparticles on the adhesion and expression of the osteoblastic phenotype of a bone cell line. *Biomaterials*, *28*, 2, 326–334, 2007.
31. Wadke, P., Chhabra, R., Jain, R., Dandekar, P., Silver-embedded starch-based nanofibrous mats for soft tissue engineering. *Surf. Interfaces*, *8*, 137–146, 2017.
32. Silva, I., Gurruchaga, M., Goni, I., Fernández-Gutiérrez, M., Vázquez, B., Román, J.S., Scaffolds based on hydroxypropyl starch: Processing, morphology, characterization, and biological behavior. *J. Appl. Polym. Sci.*, *127*, 3, 1475–1484, 2013.
33. Craig, S.S., Maningat, C.C., Seib, P.A., Hoseney, R.C., Starch paste clarity. *Cereal Chem.*, *66*, 3, 173–182, 1989.

34. Marchessault, R.H., Ravenelle, F., Zhu, X.X. (Eds.), *Polysaccharides for Drug Delivery and Pharmaceutical Applications*, American Chemical Society, Washington, DC, 2006.
35. Yihui, L., Hang, H., Qing, Z., Jiangling, W., Ying, W., Zifu, L., Xiangliang, Y., Novel hydroxyethyl starch-paclitaxel conjugates based nanoparticles for enhanced antitumor effect. *Nanomed.: Nanotechnol. Biol. Med.*, *14*, 5, 1798, 2018.
36. Hadisi, Z., Nourmohammadi, J., Nassiri, S.M., The antibacterial and anti-inflammatory investigation of Lawsonia Inermis-gelatin-starch nano-fibrous dressing in burn wound. *Int. J. Biol. Macromol.*, *107*, 2008–2019, 2018.
37. Kim, H.R., Hermansson, A.M., Eriksson, C.E., Structural characteristics of hydroxypropyl potato starch granules depending on their molar substitution. *Starch-Stärke*, *44*, 3, 111–116, 1992.
38. López, O.V., Zaritzky, N.E., García, M.A., Physicochemical characterization of chemically modified corn starches related to rheological behavior, retrogradation and film forming capacity. *J. Food Eng.*, *100*, 1, 160–168, 2010.
39. Maningat, C.C. and Seib, P.A., RS 4-Type resistant starch: Chemistry, functionality and health benefits, in: *Resistant Starch: Sources, Applications and Health Benefits*, pp. 43–78, 2013.
40. Nugent, A.P., Health properties of resistant starch. *Nutr. Bull.*, *30*, 1, 27–54, 2005.
41. Zhu, J., Zhong, L., Chen, W., Song, Y., Qian, Z., Cao, X., Chen, W., Preparation and characterization of pectin/chitosan beads containing porous starch embedded with doxorubicin hydrochloride: A novel and simple colon targeted drug delivery system. *Food Hydrocolloids*, *95*, 562–570, 2019.
42. Olayinka, F.S., Olayinka, O.O., Olu-Owolabi, B.I., Adebowale, K.O., Effect of chemical modifications on thermal, rheological and morphological properties of yellow sorghum starch. *J. Food Sci. Technol.*, *52*, 12, 8364–8370, 2015.
43. Aberle, T., Burchard, W., Vorwerg, W., Radosta, S., Conformational contributions of amylose and amylopectin to the structural properties of starches from various sources. *Starch-Stärke*, *46*, 9, 329–335, 1994.
44. Hoover, R., Composition, molecular structure, and physicochemical properties of tuber and root starches: A review. *Carbohydr. Polym.*, *45*, 3, 253–267, 2001.
45. Hussain, D., Khan, S.A., Khan, T.A., Fabrication and characterization of mesoporous guar gum/$NiWO_4$ nanocomposite for efficient adsorption of phloxine B and crystal violet from aqueous solution and evaluation of its antioxidant activity. *Colloids Interface Sci. Commun.*, *44*, 100488, 2021.
46. Tomasik, P. and Schilling, C.H., Chemical modification of starch, in: *Advances in Carbohydrate Chemistry and Biochemistry*, vol. 59, pp. 175–403, 2004.
47. Singh, J. and Singh, N., Studies on the morphological and rheological properties of granular cold water soluble corn and potato starches. *Food Hydrocolloids*, *17*, 1, 63–72, 2003.

48. Primarini, D. and Ohta, Y., Some enzyme properties of raw starch digesting amylases from Streptomyces sp. No. 4. *Starch-Stärke, 52*, 1, 28–32, 2000.
49. Labelle, M.A., Ispas-Szabo, P., Mateescu, M.A., Structure-functions relationship of modified starches for pharmaceutical and biomedical applications. *Starch-Stärke, 72*, 7-8, 2000002, 2020.
50. Alves, N.M., Leonor, I.B., Azevedo, H.S., Reis, R.L., Mano, J.F., Designing biomaterials based on biomineralization of bone. *J. Mater. Chem., 20*, 15, 2911–2921, 2010.
51. Azevedo, H.S. and Reis, R.L., Encapsulation of α-amylase into starch-based biomaterials: An enzymatic approach to tailor their degradation rate. *Acta Biomater., 5*, 8, 3021–3030, 2009.
52. Santos, M.I., Fuchs, S., Gomes, M.E., Unger, R.E., Reis, R.L., Kirkpatrick, C.J., Response of micro-and macrovascular endothelial cells to starch-based fiber meshes for bone tissue engineering. *Biomaterials, 28*, 2, 240–248, 2007.
53. Ali, A., Xie, F., Yu, L., Liu, H., Meng, L., Khalid, S., Chen, L., Preparation and characterization of starch-based composite films reinfoced by polysaccharide-based crystals. *Compos. Part B: Eng., 133*, 122–128, 2018.
54. Bhuniya, S.P., Rahman, S., Satyanand, A.J., Gharia, M.M., Dave, A.M., Novel route to synthesis of allyl starch and biodegradable hydrogel by copolymerizing allyl-modified starch with methacrylic acid and acrylamide. *J. Polym. Sci. Part A: Polym. Chem., 41*, 11, 1650–1658, 2003.
55. Reis, A.V., Guilherme, M.R., Moia, T.A., Mattoso, L.H., Muniz, E.C., Tambourgi, E.B., Synthesis and characterization of a starch-modified hydrogel as potential carrier for drug delivery system. *J. Polym. Sci. Part A: Polym. Chem., 46*, 7, 2567–2574, 2008.
56. Lawal, M.V., Odeniyi, M.A., Itiola, O.A., The effect of thermal and chemical modifications of excipients on the compressional properties of paracetamol tablet formulations including maize, cassava and sweet potato starches as filler-binders. *J. Excip. Food Chem., 6*, 3, 927, 2016.
57. Rowe, R.C., Sheskey, P., Quinn, M., *Handbook of pharmaceutical excipients*, Libros Digitales-Pharmaceutical Press, London, UK, 2009.
58. Tester, R.F., Karkalas, J., Qi, X., Starch structure and digestibility enzyme-substrate relationship. *Worlds Poult. Sci. J., 60*, 2, 186–195, 2004.
59. Cottrell, J.E., Duffus, C.M., Paterson, L., Mackay, G.R., Properties of potato starch: Effects of genotype and growing conditions. *Phytochemistry, 40*, 4, 1057–1064, 1995.
60. Szejtli, J., Past, present and futute of cyclodextrin research. *Pure Appl. Chem., 76*, 10, 1825–1845, 2004.
61. Hedges, A.R., Shieh, W.J., Sikorski, C.T., Use of cyclodextrins for encapsulation in the use and treatment of food products, *Encapsulation and Controlled Release of Food Ingredients*, 6, 60–71, 1995.
62. Zhong, Y., Li, Z., Qu, J., Bertoft, E., Li, M., Zhu, F., ... & Liu, X. Relationship between molecular structure and lamellar and crystalline structure of rice starch. *Carbohydr. Polym.*, 258, 117616, 2021.

63. Fadzlina, Z.A.N., Karim, A.A., Teng, T.T., Physicochemical properties of carboxy-methylated sago (Metroxylon sagu) starch. *J. Food Sci.*, *70*, 9, C560–C567, 2005.
64. Labelle, M.A., Ispas-Szabo, P., Mateescu, M.A., Structure-functions relationship of modified starches for pharmaceutical and biomedical applications. *Starch-Stärke*, *72*, 7-8, 2000002, 2020.
65. Mateescu, M.A., Ispas-Szabo, P., Assaad, E., *Controlled drug delivery: The role of self-assembling multi-task excipients*, Elsevier, Sawston, United Kingdom, 2014.
66. Ispas-Szabo, P., Ravenelle, F., Hassan, I., Preda, M., Mateescu, M.A., Structure–properties relationship in cross-linked high-amylose starch for use in controlled drug release. *Carbohydr. Res.*, *323*, 1-4, 163–175, 1999.
67. Labelle, M.A., Ispas-Szabo, P., Mateescu, M.A., Structure-functions relationship of modified starches for pharmaceutical and biomedical applications. *Starch-Stärke*, *72*, 7-8, 2000002, 2020.
68. Dumoulin, Y., Clement, P., Mateescu, M.A., Cartilier, L., Cross-linked amylose as a binder/disintegrant in compressed tablets. *STP Pharma Sci.*, *4*, 5, 329–335, 1994.
69. Kim, M. and Lee, S.J., Characteristics of crosslinked potato starch and starch-filled linear low-density polyethylene films. *Carbohydr. Polym.*, *50*, 4, 331–337, 2002.
70. Kuakpetoon, D. and Wang, Y.J., Characterization of different starches oxidized by hypochlorite. *Starch-Stärke*, *53*, 5, 211–218, 2001.
71. Wen, N., Gao, C., Lü, S., Xu, X., Bai, X., Wu, C., Liu, M., Novel amphiphilic glucose-responsive modified starch micelles for insulin delivery. *RSC Adv.*, *7*, 73, 45978–45986, 2017.
72. Wang, P., Sheng, F., Tang, S.W., ud-Din, Z., Chen, L., Nawaz, A., Xiong, H., Synthesis and characterization of corn starch crosslinked with oxidized sucrose. *Starch-Stärke*, *71*, 5-6, 1800152, 2019.
73. Kulp, K. (Ed.), *Handbook of Cereal Science and Technology, Revised and Expanded*, CRC Press, Marcel Dekker, New York, 2000.
74. Okunlola, A. and Akingbala, O., Characterization and evaluation of acid-modified starch of Dioscorea oppositifolia (Chinese yam) as a binder in chloroquine phosphate tablets. *Braz. J. Pharm. Sci.*, *49*, 4, 699–708, 2013.
75. Khalid, G.M., Musa, H., Olowosulu, A.K., Evaluation of filler/binder properties of modified starches derived from plectranthusesculentus by direct compression in metronidazole tablet formulations. *Pharm. Anal. Acta*, *7*, 1, 74, 2016.
76. Wang, S. and Copeland, L., Effect of acid hydrolysis on starch structure and functionality: A review. *Crit. Rev. Food Sci. Nutr.*, *55*, 8, 1081–1097, 2015.
77. Odeku, O.A., Potentials of tropical starches as pharmaceutical excipients: A review. *Starch-Stärke*, *65*, 1-2, 89–106, 2013.
78. Builders, P.F., Nnurum, A., Mbah, C.C., Attama, A.A., Manek, R., The physicochemical and binder properties of starch from Persea americana Miller (Lauraceae). *Starch-Stärke*, *62*, 6, 309–320, 2010.

79. Zhang, Y.E. and Schwartz, J.B., Effect of diluents on tablet integrity and controlled drug release. *Drug Dev. Ind. Pharm.*, *26*, 7, 761–765, 2000.
80. Nachaegari, S.K. and Bansal, A.K., Coprocessed excipients for solid dosage forms. *Pharm. Technol. (2003)*, *28*, 1, 52–64, 2004.
81. Builders, P.F. and Arhewoh, M.I., Pharmaceutical applications of native starch in conventional drug delivery. *Starch-Stärke*, *68*, 9-10, 864–873, 2016.
82. Flores-Arriaga, J.C., de Jesús Pozos-Guillén, A., Escobar-García, D.M., Grandfils, C., Cerda-Cristerna, B.I., Cell viability and hemocompatibility evaluation of a starch-based hydrogel loaded with hydroxyapatite or calcium carbonate for maxillofacial bone regeneration. *Odontology*, *105*, 4, 398–407, 2017.
83. Kang, B., Okwieka, P., Schöttler, S., Seifert, O., Kontermann, R.E., Pfizenmaier, K., Landfester, K., Tailoring the stealth properties of biocompatible polysaccharide nanocontainers. *Biomaterials*, *49*, 125–134, 2015.
84. Noga, M., Edinger, D., Rödl, W., Wagner, E., Winter, G., Besheer, A., Controlled shielding and deshielding of gene delivery polyplexes using hydroxyethyl starch (HES) and alpha-amylase. *J. Control. Release*, *159*, 1, 92–103, 2012.
85. Breitenbach, J., Melt extrusion: From process to drug delivery technology. *Eur. J. Pharm. Biopharm.*, *54*, 2, 107–117, 2002.
86. Elvira, C., Mano, J.F., San Roman, J., Reis, R.L., Starch-based biodegradable hydrogels with potential biomedical applications as drug delivery systems. *Biomaterials*, *23*, 9, 1955–1966, 2002.
87. Konstantakos, S., Marinopoulou, A., Papaemmanouil, S., Emmanouilidou, M., Karamalaki, M., Kolothas, E., Karageorgiou, V., Preparation of model starch complex hydrogels. *Food Hydrocolloids*, *96*, 365–372, 2019.
88. Ahmad, M.Z., Akhter, S., Ahmad, I., Singh, A., Anwar, M., Shamim, M., Ahmad, F.J., *In vitro* and *in vivo* evaluation of Assam Bora rice starch-based bioadhesive microsphere as a drug carrier for colon targeting. *Expert Opin. Drug Delivery*, *9*, 2, 141–149, 2012.
89. Rostami, M., Yousefi, M., Khezerlou, A., Mohammadi, M.A., Jafari, S.M., Application of different biopolymers for nanoencapsulation of antioxidants via electrohydrodynamic processes. *Food Hydrocolloids*, *97*, 105170, 2019.
90. Tian, B., Hua, S., Liu, J., Cyclodextrin-based delivery systems for chemotherapeutic anticancer drugs: A review. *Carbohydr. Polym.*, *232*, 115805, 2020.
91. LeCorre, D., Vahanian, E., Dufresne, A., Bras, J., Enzymatic pretreatment for preparing starch nanocrystals. *Biomacromolecules*, *13*, 1, 132–137, 2012.
92. Vega-Vásquez, P., Mosier, N.S., Irudayaraj, J., Nanoscale drug delivery systems: From medicine to agriculture. *Front. Bioeng. Biotechnol.*, *8*, 79, 2020.
93. Marinopoulou, A., Papastergiadis, E., Raphaelides, S.N., Inclusion complexes of non-granular maize starch with fatty acids and ibuprofen. A comparative study of their morphology and structure. *Starch-Stärke*, *71*, 1-2, 1800100, 2019.

94. Fresta, M., Cavallaro, G., Giammona, G., Wehrli, E., Puglisi, G., Preparation and characterization of polyethyl-2-cyanoacrylate nanocapsules containing antiepileptic drugs. *Biomaterials, 17,* 8, 751–758, 1996.
95. Fang, Y.Y., Wang, L.J., Li, D., Li, B.Z., Bhandari, B., Chen, X.D., Mao, Z.H., Preparation of crosslinked starch microspheres and their drug loading and releasing properties. *Carbohydr. Polym., 74,* 3, 379–384, 2008.
96. Malafaya, P.B., Stappers, F., Reis, R.L., Starch-based microspheres produced by emulsion crosslinking with a potential media dependent responsive behavior to be used as drug delivery carriers. *J. Mater. Sci.: Mater. Med., 17,* 4, 371–377, 2006.
97. Mao, S., Chen, J., Wei, Z., Liu, H., Bi, D., Intranasal administration of melatonin starch microspheres. *Int. J. Pharm., 272,* 1-2, 37–43, 2004.
98. Pereswetoff-Morath, L., Microspheres as nasal drug delivery systems. *Adv. Drug Delivery Rev., 29,* 1-2, 185–194, 1998.
99. Li, B.Z., Wang, L.J., Li, D., Adhikari, B., Mao, Z.H., Preparation and characterization of crosslinked starch microspheres using a two-stage water-in-water emulsion method. *Carbohydr. Polym., 88,* 3, 912–916, 2012.
100. Chen, L., Pu, H., Li, X., Yu, L., A novel oral colon-targeting drug delivery system based on resistant starch acetate. *J. Controlled Release, 152,* e51–e52, 2011.
101. Singh, A.V. and Nath, L.K., Evaluation of acetylated moth bean starch as a carrier for controlled drug delivery. *Int. J. Biol. Macromol., 50,* 2, 362–368, 2012.
102. Luo, K., Gao, P., Yang, W., Lei, X., Wong, T.W., Ismail, A.F., Wang, L., Antibacterial polyurethane composite scaffolds for minimally invasive alveolar bone repair. *Appl. Mater. Today*, 31, 101752, 2023.
103. Wang, X., Sun, X., Ma, C., Zhang, Y., Kong, L., Huang, Z., Wang, P., Multifunctional AuNPs@ HRP@ FeMOF immune scaffold with a fully automated saliva analyzer for oral cancer screening. *Biosens. Bioelectron.*, 222, 114910, 2023.
104. Lopes, P., Joaquinito, A.S.M., Ribeiro, A., Moura, N.M., Gomes, A.T., Guerreiro, S.G., Gonçalves, I., Starch-based films doped with porphyrinoid photosensitizers for active skin wound healing. *Carbohydr. Polym.*, 313, 120894, 2023.
105. Muneeswaran, T., Maruthupandy, M., Mary, A.S., Vennila, T., Kaushik, R., Ramakritinan, C.M., Quero, F., Starch-mediated synthesis of chitosan/silver nanocomposites for antibacterial, antibiofilm and wound healing applications. *J. Drug Deliv. Sci. Technol.*, 84, 104424, 2023.
106. Adeli, H., Khorasani, M.T., Parvazinia, M., Wound dressing based on electrospun PVA/chitosan/starch nanofibrous mats: Fabrication, antibacterial and cytocompatibility evaluation and *in vitro* healing assay. *Int. J. Biol. Macromol.*, 122, 238–254, 2019.
107. Deng, X., Gould, M., Ali, M.A., A review of current advancements for wound healing: Biomaterial applications and medical devices. *J. Biomed. Mater. Res. Part B: Appl. Biomater.*, 110, 11, 2542–2573, 2022.

6

Biomedical Applications of Carrageenan

Hitesh Chopra[1] and Rupesh K. Gautam[2*]

[1]*Department of Biosciences, Saveetha School of Engineering, Saveetha Institute of Medical and Technical Sciences, Chennai, Tamil Nadu, India*
[2]*Department of Pharmacology, Indore Institute of Pharmacy, IIST Campus, Rau-Indore, M.P., India*

Abstract
Carrageenan is a sulfated linear polysaccharide comprising D-galactose and 3,6-anhydro D-galactose extracted from some red seaweeds of the Rhodophyceae family. Carrageenan is derived from these red seaweeds. This chapter's goal is to outline current carrageenan medication delivery system uses. In the pharmaceutical business, carrageenan has been studied as a polymer matrix in oral extended-release tablets, as a new extrusion assist for pellet manufacture, and as a carrier/stabilizer in micro/nanoparticle systems. Because of carrageenan's unique properties, such as its high negative charge and gelling, it has been employed as a gelling agent/viscosity enhancer for controlled drug release and long-term retention. It has also been employed to transport cells and regenerate tissue with medicinal macromolecules like carrageenan. Carrageenan's safety and other possible uses will be investigated in the near future.

Keywords: Carrageenan, drug delivery, pellet, marine, seaweeds

6.1 Introduction

Carrageen (CG) is a general word used to represent a broad collection of sulfated polysaccharide molecules found in the red algae cell wall matrix [1, 2]. These algae are all part of the Gigartinales order. From a commercial point of view, three major kinds of carrageen extracted are important: iota (β), kappa (μ), and lambda (μ). Carrageenan is widely utilized for textured functioning in a broad variety of culinary applications, mainly in gelling

**Corresponding author*: drrupeshgautam@gmail.com

Annu (ed.) Biopolymers for Biomedical Applications, (151–174) © 2024 Scrivener Publishing LLC

and viscosity. Processed meat, dairy, and desserts/jellies are the main food industries. It is also employed as a flocculating ingredient in beer and in certain applications, such as in air-cooling gels, toothpaste, and pet food [3–7]. All hydrocolloids have distinct characteristics and techniques and thus varied potential applications. Carrageenan possesses a certain number of characteristics that vary from other hydrocolloids and make them valuable, i.e., for the production of potassium and calcium ion gels, milk protein reactivity, thermally reversible gel formation, and synergistic behaviors [8–10].

The main seaweeds used to make CG are Chondrus, Eucheuma, Gigartina, and Hypnea [11–13]. Many CG manufacturers consider the extraction process' trade secrets, even though they often follow a similar pattern. The seaweed is dried quickly to minimize degradation before being processed. The algae are regularly cleaned to remove significant pollutants such as sand, salt, and marine life before a hot alkaline extraction process extracts the CG. When heated, CG crystallizes and turns into powder [14]. Extraction variables (temperature, pH, and time) and alkaline pretreatment variables (time) alter the chemical structure and gelling properties [15].

6.2 Structure of Carrageenan

Seaweed carrageenan is a linear hydrophilic polymer [16]. Carrageenan, a marine polysaccharide, has several beneficial properties [17]. Carrageenan is one of the most extensively used polysaccharides in food [18]. Based on structural variations between repeating disaccharide units (e.g., sulphate group number and position), carrageenan is divided into six types [17, 19]. Carrageenan is a sulfated galactan family (van); Greek letters prefix the main known backbone substitution patterns [20]. Enzymes and alkalis may produce the 3-anhydrobridge of galactose-6-sulfate [17]. Carrageenan can be refined or semi-refined during processing. Simple carrageenan extraction requires soaking in alcohol and gelling [21]. Carrageenan is classified as high molecular (or "undegraded") or low molecular (or "degraded").

Besides the carrageenans, there exist hybrid carrageenans. Marine algae create a variety of sulfated polysaccharides, including hybrid carrageenans. κ/ι hybrid carrageenans are natural polyelectrolyte copolymers comprised of both κ-carrageenan and ι-carrageenan repeating units, extensively used in the food business [22]. Aside from their great qualities, κ/ι hybrid carrageenans are gaining popularity due to their rising demand in culinary, biomedical, and pharmaceutical applications [23].

6.3 Biomedical Applications

6.3.1 In Oral Drug Delivery

Carrageenan, a novel pelletization aid, has greater formulation robustness and disintegration speed than non-disintegrating microcrystalline cellulose pellets. The pellets were enteric coated with Kollicoat(®) MAE 30 DP and Eudragit(®) NE 30 D and then crushed and embedded in silicified microcrystalline cellulose [21]. The effect of coating levels and pellet core types on drug release was studied. To make acid-resistant multiparticulate tablets, a proper coating thickness for carrageenan pellets was needed. However, unlike carrageenan pellets and pills, MCC pellets and tablets only released 20–24% of the medicine during the neutral period. The oblong or round punches used had no influence on the drug release from the tablets. Kollicoat(®) SR 30 D was used to coat carrageenan pellets with theophylline as a model medicine. Theophylline pellets produced with MCC needed less coating and has increased pore development [24].

Buchholcz *et al.* created modified-release monolithic matrix tablets containing API [25]. The pH 1.2 (gastric juice) dissolving ended at pH 4.5. The MCC/lactose ratio impacts the tablet's properties, especially its dissolution. To better understand how chitosan (CS) and three carrageenans affect medication release from matrix tablets (kappa, iota, and lambda-CG), the model drug used was trimetazidine hydrochloride. It was examined to better understand drug release mechanisms. They tested the drug release from CS–CG-based matrix at various pH levels and ionic strengths. The controlled-release drug carriers based on CS–CG show considerable promise [25].

Combining a marine polymer (chitosan or kappa-carrageenan) with a semisynthetic polymer (hydroxyl-propyl-methyl-cellulose) in solid compacted formulations resulted in full and sustained acyclovir release for 220–260 h [26]. These formulations may cling to the vaginal mucosa for 72 h. The type and number of polymers in the formulations affect the swelling results. Hydroxypropylmethyl cellulose swells more than chitosan and kappa-carrageenan, which have lower swelling values and are erodible in acid. Anhydrous calcium hydrogen phosphate, as a structural agent, modulates polymer swelling characteristics, decreases HPMC swelling, and inhibits the development of a mixed HPMC/chitosan gel. In addition to their favorable features of muco-adhesion and prolonged release of ACV, chitosan-containing formulations demonstrated a protective effect of chitosan against cell damage [26].

To provide safe and effective quantities of the antiviral medicines griffithsin (GRFT) and carrageenan, Lal *et al.* developed a self-administered vaginal rapid-dissolving insert manufactured by freeze-drying [27]. The lead formulation, FDI-024, comprises 4 mg GRFT, 15 mg CG, and excipients (the cryoprotectant sucrose and bulking agents dextran 40 and mannitol). The FDI is friable, firm, and stable for 6 months at 40°C/75% relative humidity. When mixed with simulated vaginal fluid (1 mL), it disintegrates in less than 60 s, producing a viscous semi-solid gel with mucoadhesive and spreading characteristics. In cell-based tests, the formulation maintains GRFT and CG antiviral efficacy against HIV-1 and HPV, respectively [27].

6.3.2 In Wound Healing

Boateng *et al.* developed polyethylene oxide and carrageenan-based films for wound healing using diclofenac as active ingredient. The films showed high absorbing ability with respect to wound fluid and showed significant mucoadhesion. The films also showed inhibition against *Staphylococcus aureus*, *Pseudomonas aeruginosa*, and *Escherichia coli* and was able to target inflammatory pain and swelling [28].

Pettinelli *et al.* prepared injectable hydrogel based on iota and kappa carrageenan locust bean gum and gelatin. The injectable hydrogel has shear thinning characteristics and can be injected using a minimally invasive technique. Hydrogel was able to grow fibroblasts and could support the tissue repair mechanism. Hydrogels released the vascular endothelial growth factors and released these in a sustained manner [29]. Another group of researchers also developed wound-healing hydrogel films encapsulating the chitosan-capped sulfur nanoparticles and grapeseed extract. The films thus formed have high UV barrier properties and swelling ratio. The films showed antibacterial action against *S. epidermis* and *E. coli* and high biocompatibility against the mouse fibroblasts cells [30]. Other researchers developed chitosan/carrageenan-based wound dressings using freeze-drying procedure for hemorrhage. The dressing showed blood coagulation implying the adhesion of red blood cells and platelets and causing the formation of thrombin, thus enhancing wound healing [31]. Nanoemulsions of kappa-carrageenan encapsulating astaxanthin and alpha-tocopherol were shown on streptozotocin-induced diabetic mice. The dressing showed enhanced anti-diabetic potential. The dressing showed enhanced restoration of body weight and reduced fasting blood glucose levels. The dressing superbly decreased the complications of diabetes mellitus [32]. Along with octenidine dihydrochloride and chitosan-treated serum, kappa-carrageenan hydrogel was developed. Octenidine acted as an antiseptic agent. The presence of chitosan-treated serum induced the migration of polymorphonuclear neutrophils and fibroblasts, easing the regeneration of tissue. An *in vivo* antibacterial study showed the inhibition of *S. aureus* growth [33]. Researchers developed electrospun nanofiber scaffolds composed of polyvinyl alcohol and iota-carrageenan doped with partially reduced grapheme oxide for wound healing. The histopathological evaluation showed re-epithelialization and the presence of yeast-induced hair follicles growth [34].

Tavakoli *et al.* developed sprayable bioadhesive wound healing dressing. A light-sensitive, cross-linkable nanocomposite hydrogel was developed (shown in Figure 6.1). Methacrylated kappa-carrageenan mimicking the natural glycosaminoglycan was used as hydrogel matrix with polydopamine-modified ZnO nanoparticles. The presence of L-glutamic acid accelerated wound healing. The hydrogels showed an effectual increase in blood clotting ability and biocompatibility [35]. Researchers developed carboxymethyl k-carrageenan using k-carrageenan and monochloroacetic acid. It was found that, as the degree of substitution increased from 0.42 to 1.09, the anticoagulant activity decreased. The polymer showed good antimicrobial properties against *E. coli* and *S. aureus* [36]. Similarly, cyclic β-(1→3) (1→6)glucan/carrageenan hydrogels encapsulating ciprofloxacin increased the porosity of hydrogels and increased cell adhesion and proliferation. The presence of carrageenan increased the drug loading

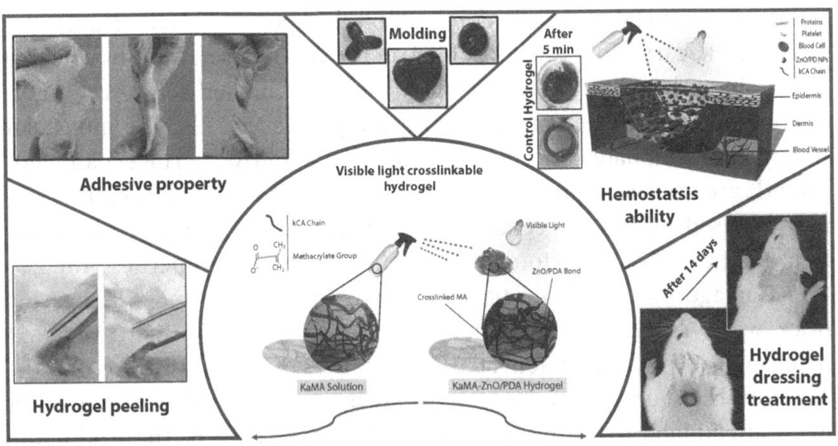

Figure 6.1 Multifunctional nanocomposite spray dressing of Kappa-carrageenan-polydopamine-modified ZnO/L-glutamic acid for diabetic wounds. Reproduced with permission from [32].

and aided in sustained release along with antibacterial action. An animal model showed accelerated wound healing action in a group treated with hydrogels [37].

Chandika *et al.* developed robust wound dressing of polyvinyl alcohol and methacrylate kappa-carrageenan hydrogels encapsulating chitooligosaccharide *via* freeze/thaw cycles and photo and chemical crosslinking. The hydrogel completely healed the wound by the 14th day as confirmed by full re-epithelialization and rapid skin modeling [38].

6.3.3 Anticancer Activity

Sulfated polysaccharide carrageenan, found in red seaweed, has been shown in studies to have anticancer effects. Cancer's uncontrolled growth is caused by a malfunctioning cell cycle regulator. In HeLa and HUVEC, carrageenans kappa (k-CO) and lambda (l-CO) were tested for cytotoxicity at dosages ranging from 250 to 2,500 g/mL. The cell cycle profile and progress of HeLa cells were examined using fluorescence ubiquitination-based cell cycle indicator probes. Using HUVECs as a test subject, no negative effects were seen from the presence of carrageenan (normal cells). If carrageenan interacts differently with each cell cycle, it may either halt or stop the HeLa cells from cycling [39].

Studies on seaweed extracts' carrageenans and disaccharide-alditols found promising anticancer properties (also known as carrabiitols). In the presence of carrageenan oligosaccharides and carrageenan itself, the LM2 tumor cells were completely eliminated. Carbabioses were studied using a carbabitol metabolite called carrabibitol. Apoptosis (cell death) was caused by κ-carrabiose, the most toxic of the carabinoses. It is possible that κ-carrabiose has anticancer properties since it inhibited cell-to-cell interaction and matrix interaction when given to cancer cells. The molecular weights and sulphate concentrations of most cytotoxic carrageenan compounds are modest. Antineoplasics and disaccharide units like carrabioses may be used with carrageenan as an adjuvant or a carrier in anticancer treatment to boost the cytotoxicity and antimetastatic properties [40].

λ-Carrageenan inhibited tumor growth in B16-F10 and 4T1-bearing mice while enhancing the tumor immune response by increasing M1 macrophages, DCs, and activated CD4+CD8+ T cells in the spleen. The TNF levels in tumors are also increased by -carrageenan, with the latter being attributable to the invading macrophages secreting it. The production of anti-OVA antibodies is aided using λ-carrageenan as an adjuvant in OVA-based cancer vaccines, according to studies. Studies on animals showed that carrageenan was not harmful; thus, it was approved for use in humans. Thus, carrageenan has the potential to be employed in cancer immunotherapy as an adjuvant as well as a potent anticancer medication [41].

The anticancer and immunotropic characteristics of the red sea algae *Chondrus armatus* and the breakdown products of LMW have been examined. Human monocytes were tested for their capacity to produce pro- and anti-inflammatory cytokines using the esophageal cancer cell lines KYSE30 and FLO1. High molecular weight carrageenans are more antimetabolic and less cytostatic than their low molecular weight breakdown products. Tumor necrosis factor and interleukin 1 were produced by carrageenans in human monocytes. The anti-inflammatory cytokine IL10 could only be obtained from LMW-carrageenan, which was also the most effective in inducing cytokine production in general [42].

These anticancer and antioxidant properties were studied using the enzyme κ-carrageenase, which was shown to be very effective in producing low molecular weight carrageenan oligosaccharides. Oligosaccharides are water soluble due to their low molecular weight and hydrophilic side chains (sulphate groups). Polysaccharides with low molecular weight and sulphate substitution may improve the solubility and antitumor effectiveness by increasing the molecular weight. Oligosaccharides had strong anticancer effects in Sarcoma 180; however, the results were dose dependent. In high dosages, the anticancer activity is less potent. For gelling and thickening carrageenan oligosaccharides, use this method since it produces the best dose-dependent curve. Tumor inhibition is enhanced by increasing the sulfonation degree. In terms of enhancing oligosaccharide tumor inhibition, the functional group sulphate is helpful. The anticancer action of oligosaccharides was also revealed to be substantially influenced by their molecular weight and enzyme breakdown [43]. For sulfated oligosaccharides, larger molecular weight fractions were more effective. It was fascinating to watch how oligosaccharide efficacy changed with sulfonation and molecular weight. The potency of oligosaccharides with added sulphate groups in their backbone was significantly improved when tested against tumor systems. It worked differently than naturally occurring anticancer oligosaccharides because of its mechanism of action. Changes in side chain polarity or kinetic properties affected the binding abilities of molecules on the cell surface. Sulfate modifications may assist oligosaccharides adhere to cellular surfaces by improving the polarity [44].

On MDA-MB-231 cells, ESC exhibits time- and concentration-dependent cytotoxic inhibitory effects [45]. Inhibiting ESC-treated cells induced apoptosis. Anti-tumor activity was shown with the use of the sulfated polysaccharide (PI-88), which prevented the death of cancer cells in the pancreatic islet of the mice model. ESC has been shown to have cell killing anti-proliferative effects. Many different types of cancer have been proven to benefit from apoptosis-inducing cytotoxic drugs [46]. A greater understanding of apoptosis and its signaling system may help in the development of novel cancer medicines [47]. ESC regulates caspase-8 in MDA-MB-231 cells, generating an extrinsic cascade of death, according to the results of this research. ESCs treated with caspase-3 and caspase-9 [48] show DNA breakage, nuclear condensation, and cell shrinkage [49–51]. The Bax and Bcl-2 proteins also impact apoptosis [52, 53]. By directly activating the transcription of the Bax gene, p53 promotes mitochondrial membrane permeabilization as shown in Figure 6.2 [54]. Increased p53 expression raises the Bax/Bcl-2 ratio, resulting in caspase activation and apoptosis [55, 56].

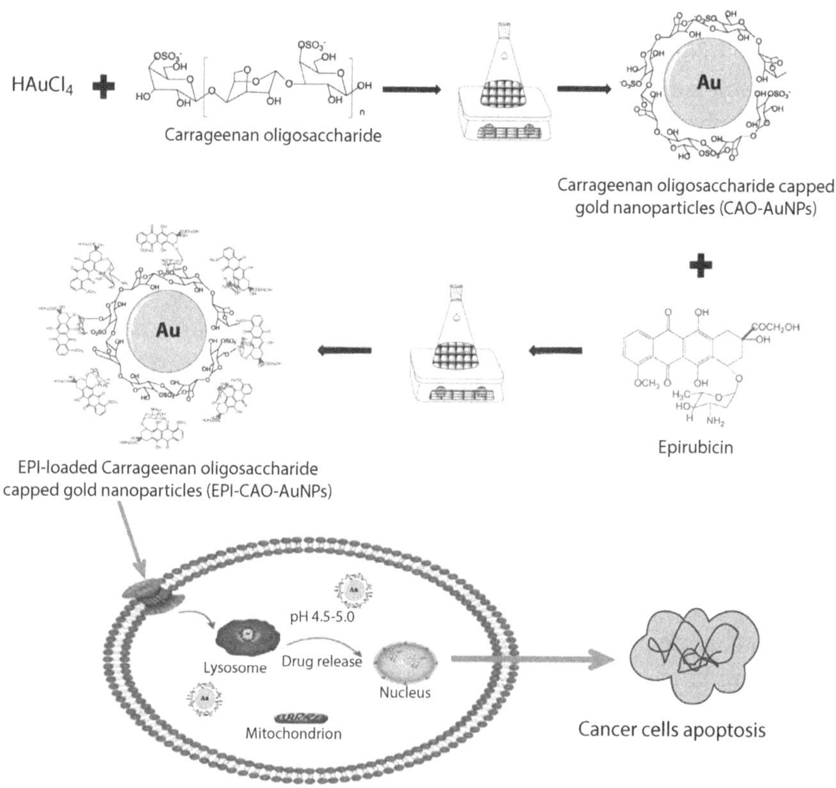

Figure 6.2 An intracellular endo/lysosomal EPI-CAO-AuNPs delivery system with pH-triggered drug release. Reproduced with permission from [54] under creative common license.

With the EPI-CAO-AuNPs system, the first pH-triggered anticancer drug release was made possible employing marine carrageenan CAO as a biocompatible nanomaterial [57]. EPI-CAO-AuNPs were found to be 12.5% and 94.3% spherical, homogenous, and drug-loading/entrapment efficient. If the pH of the surroundings was low (as opposed to high), the EPI-CAO-AuNPs released EPI more rapidly (as opposed to slower). During the G2/M phase of HCT-116 and HepG2 cells, we were able to trigger apoptosis utilizing EPI-CAO-AuNPs. More cancer cells died and necrosed when exposed to EPI-CAO-AuNPs. The authors propose a novel method of delivering personalized medicine by using marine polysaccharides as nanoparticles.

A water-soluble chemotherapeutic drug, mitoxantrone hydrochloride (MTO), may have increased the oral bioavailability and anticancer efficacy because of NLCCs' high loading ability and long release time. MTO was electrostatically encapsulated in NLCCs using carrageenan, a negative polymer. In *in vivo* pharmacokinetic investigations in rats, MTO-NLCCs had a 3.5 times greater apparent bioavailability than MTO-Sol. In MTT cytotoxicity tests, NLCCs greatly increased the anticancer efficacy against chemotherapy-resistant MCF-7/MX cells. MTO-NLCCs had a 9.2-fold greater relative cellular association than MTO-Sol because NLCCs reduced BCRP-mediated drug efflux in MCF-7/MX cells overexpressing BCRP. According to the endocytosis inhibition investigation, NLCCs entered MCF-7/MX cells by bypassing the BCRP-mediated MTO efflux. New NLCC compounds increased the encapsulation efficiency, oral bioavailability, and cytotoxicity against resistant breast cancer cells [58].

6.3.4 Anticoagulant Action

Researchers assessed the anticoagulant effect of sulfated polysaccharides (SPS) from Codium dwarkense Bergesen in K-carrageenan hypercoagulable albino Wistar rats. A single dose of K-carrageenan was able to induce hypercoagulability. SPS treatment and prevention decreased the microthrombi and structural damage in lung, liver, and mesentery histopathology [59].

Inhibiting cancer cell motility, invasion, and angiogenesis, HPSE degrades heparan sulphate (HS). Due to their structural similarity to HS, sulfated polysaccharides may be used to control these actions. Researchers used depolymerization and desulfation to create -carrageenan oligosaccharides (-CO). HPSE inhibition was primarily, but not entirely, linked to -CO substitution degree. Chain length and DS both inhibited factor Xa and thrombin IIa, two enzymes in the coagulation cascade. Because of this, it had the least anticoagulant effects of all -CO produced while inhibiting MDA-MB-231 breast cancer cell motility [60]. It was produced by alkali-treating native-carrageenan and then auto-hydrolyzing it for 0.5–72 h. Size-exclusion chromatography was used to compute the molecular weight. High molecular weight carrageenan anticoagulant action was equal to heparin, which was 2.5–4 times less active [61].

These composite hydrogels lower the endotoxin levels and bacterial load in septic blood. Less hemolysis, complement activation, platelet activation, and contact activation were seen in C-K hydrogels compared to raw chitosan hydrogels. The enhanced C2-K1 hydrogels removed 63.3% endotoxin from septic blood after 3 h of simulative hemoperfusion. The enhanced C2-K1 hydrogels decreased the *E. coli* and *S. aureus* burdens by 46% and 68.7% in bacterial cleaning tests [62].

Through the use of structural and functional group mimicry, researchers created a hemoperfusion adsorbent. A dual-network structure with improved mechanical properties and better control over swelling ratios was produced using post-crosslinking of AA and carrageenan phase inversion. The use of beads decreased protein adsorption, hemolysis, cytotoxicity, and complement activation as well as the complement system's overall activity. The experimental group's APTT, prothrombin time, and thrombin time were all 13 times greater than those of the control group. Self-anticoagulant and biocompatible beads were used to demonstrate the adsorption capacities for exogenous toxins (560.34 mg/g heavy metal ions) and endogenous toxins (18.15 mg/g LDL, 228.16 mg/g bilirubin, and 14.83 mM creatinine) [63].

Panlasigui and coauthors claim that the participants ate mixed carrageenans for 8 weeks [64]. Sokolova's study also used carrageenan as a dietary supplement for those with CVD [65]. For 20 days, the participants consumed 250 mL of a hybrid K-carrageenan (3:1) supplement [65]. After a daily intake of 100 mL of vegetable jelly for 60 days, we found a statistically significant reduction in TC and HDL-C concentrations. Female patients had lower TC and LDL-C values based on gender distribution and EG-60 days. Male patients, however, saw a rise in TG [66].

After 30 days, the TC levels fell by 7.6% and after 60 days by 5.3%. According to Sokolova *et al.* [65], supplementing with carrageenan reduced TC by 16.5% in cardiovascular patients, which may be attributable to a different research design. Panlasigui *et al.* [64] found that consuming carrageenan-rich foods at the three most essential meals (breakfast, lunch, and supper) reduced the TC levels by 33%. After 60 days, the LDL-C fraction concentration exhibited a tendency to decline, with female patients exhibiting a considerable drop. After 30 days, the TG concentration increased by 30.3%, and after 60 days it increased by 17.8%. It has not achieved the maximum reference levels [67]. While Panlasigui and coauthors reported a 32% drop in TG, Sokolova and coauthors reported an increase [64, 65]. However, in both sexes, poor cholesterol absorption is linked to increased TG and lower HDL [68]. Despite their many health advantages, carrageenans' mode of action is still unknown, and they are classed as long-chain carbohydrates [69, 70]. Carrageenans increase the volume and viscosity of the intestinal contents, increasing the digestion time and decreasing the efficiency [66].

That it can reduce cholesterol absorption and promote endogenous cholesterol production is what gives it its bioactive potential [68]. Some studies claim that when cholesterol absorption indicators decline, the serum triglyceride concentration rises and the HDL-C levels fall [71]. The blood LDL-C fraction values dropped when high endogenous cholesterol levels were observed, according to Matthan *et al.* [68]. The increased expression of the LDL-C receptor in the cell produces an increase in the cholesterol internalization rate, lowering 3-hydroxy-3-methylglutaryl-coenzyme A (HMG-CoA) reductase activity and so decreasing the endogenous cholesterol production [68]. Carrageenan consumption reduces cholesterol absorption, increasing TG levels and decreasing HDL-C. However, the demand for cholesterol for bile salt production caused by carrageenan excretion promotes endogenous cholesterol synthesis, lowering the serum LDL-C levels [72, 73].

6.3.5 Anti-Mutagenic Activity

The screening of carrageenan's antimutagenic characteristics has been a new study focus in recent years. Compounds having mutagenic potential may chemically interact with DNA and cause genetic damage, resulting in chronic and degenerative diseases. Cancer is caused by DNA damage that was not successfully repaired by cellular DNA reparative enzyme systems. Ineffective DNA repair causes mutations. Somatic cell mutations may cause cancer.

Thus, the development of preventive techniques employing chemoprotective chemicals on genetic material is required. Complex polysaccharides are regarded as one of the most promising sources of anticancer chemoprophylactic drugs. Researchers examined the ability of carrageenan type CSM-2 (Genuvisco®, CP Kelco) to protect cultivated meristematic cells of *Allium cepa* against methyl-sulfonylmethane [74]. The examined carrageenan has chemoprophylactic action at all the tested doses (5, 10, and

20 g/mL). The polysaccharide employed to prevent mutagenesis damage had a potency of 62.54% to 96.66% [75]. The authors of this study concluded that carrageenan is anti-mutagenic, that is, it can adsorb and inactivate DNA toxicants. They also believe carrageenan has bio-antimutagenic action, which means that it may modify DNA repair enzyme systems. It should be noted that carrageenan has some binding potential [76, 77]. Considering that carrageenan is a mixture of predominantly high molecular polysaccharides, we may deduce that carrageenan entering cells is improbable. A diet or functional food with antioxidant and antimicrobial properties may be created using its lack of direct mutagenesis effects [75].

6.3.6 For Vaccines

Immunotherapy for cancer uses monoclonal antibodies, adoptive cell transfer, and anticancer vaccines [78]. Its main goal is to stimulate an immune response that may kill cancer cells and create long-lasting immunity. To combat cancer, DC precursors are loaded with tumor antigens and matured into DCs. Maybe they are monocytes, HSPs, or DCs. The most frequent method is to extract DCs from PBMCs [79].

To provide or augment the immune response caused by vaccine antigens, adjuvants are chemicals that stimulate or alter the humoral or cellular immune response [80]. The CCR7 ligand promotes migration, and CD4+ T cells are stimulated because TLR ligands activate dendritic cells. To activate the immune system, researchers found the TLR ligand iquimod (TLR7 and TLR8) to be the most efficient [81, 82]. Scientists believe that CD4+ T cells with the CD40 receptor release more co-stimulating chemicals and cytokines, leading to the activation of CD8+ T cells, and IL-15, IL-7, and IL-12 increase T-cell survival and function [83]. Immunologic adjuvants include inorganic chemicals, bacterial metabolites, cytokines, and other substances that stimulate the immune system.

Polysaccharides generated from marine algae may be found in this group [84]. Li *et al.* studied the anticancer effects of a DC-based vaccine loaded with E6 and E7 proteins from HPV-16 [85]. Subcutaneously implanted TC-1 cells expressing HPV-16 E6 and E7 were tested in C57BL/6 mice. At 30 days later, the tumor nodes weighed three times less than the controls. In comparison to the control values, the adjuvant immunization with -carrageenan significantly reduced the tumor formation. Carrageenan had the same adjuvant effect as CpGODN 1826. The oncogenic HPV-specific CD8+ T-cell response increased, but regulatory T-cells (nTreg) and myeloid-derived suppressor cells were suppressed by both vaccinations. With an HPV-16 human papillomavirus E7 protein vaccine, carrageenan was employed as an adjuvant to generate antigen-specific immune responses and anticancer effects in mice. Carrageenan (Sigma) improved the TLR4 pathway activation and E7 peptide vaccination against E7 expressing tumors [86]. Given that carrageenan enhances DC maturation and cytokine generation, its immune-activating action is likely TLR4 signaling.

The efficacy of carrageenan adjuvant in preventive and therapeutic vaccines was studied in C57BL mice with tumor E.G7 line expressing ovalbumin (cell line E.G7-OVA). In the first study, the mice were immunized three times a week with ovalbumin (5 g) or carrageenan (100 g). The tumor volume was measured weekly after the last immunization. In the second experiment, the mice were subcutaneously injected with E.G7-OVA cells and then given the vaccine three times a week until the tumor reached 3 mm. The results showed that, compared to the control and OVA groups, the vaccine with carrageenan adjuvant significantly reduced the tumor growth. By the fourth week, the OVA group's average tumor volume was around seven times lower. The therapeutic vaccine having only ovalbumin had no effect on mice. In contrast, OVA/-carrageenan infusions slowed the tumor development by 30–40%. Six out of 10 mice inoculated with the preventive vaccine containing carrageenan had no tumor 40 days after the tumor cell injection, but all mice immunized with the control or ovalbumin had tumors. The total anti-OVA immunoglobulin G serum level was substantially greater in mice vaccinated with OVA/-carrageenan than in mice inoculated with OVA (2,000 and 8,000, respectively) a week after the final immunization [41].

Thus, pre-clinical and clinical investigations on -carrageenan may be proposed to assess its potential as an adjuvant in dendritic cell-based therapeutic vaccines.

6.4 Toxicity

Even though CG has been used in the food and pharmaceutical industries for many years, many people are still worried about its safety. Non-parenteral forms of CG are generally thought to be safe and non-irritating [14]. Despite this, it is well known that CG may cause inflammatory reactions. When used in pharmacodynamic studies of anti-inflammatory medicines, CG has recently been linked to damage to the intestine's mucous membrane and ulcerous colonitis in many animals when used for an extended period. These lesions include adenocarcinomas, anaplastic large cell lymphomas, and squamous cell carcinomas, all of which may be induced by CG. Colonic metaplasia occurred at 100% after 15 months of feeding on 5% degraded CGs [87]. As a result, CG safety assessment is critical.

6.5 Challenges, Conclusion, and Future Trends

Researchers are concentrating their efforts on developing a health-improving nutraceutical delivery system based on carrageenan because of its unique properties. However, there are a number of long-term issues relating to delivery system development and utilization that must be addressed. One possibility is a reduction in the bioavailability of the active ingredient(s).

Bioavailability determines the efficacy of bioactive compounds [88]. During system preparation and distribution, bioactive components might lose or change, reducing their bioavailability [89]. Increased availability of bioactive compounds requires practical solutions based on their properties, therefore this is crucial. Antioxidants, crosslinking agents, and the correct packaging materials may help protect bioactive substances in capsules [90].

Nutraceuticals are products derived from food sources that are purported to provide extra health benefits, in addition to the basic nutritional value found in foods [91–93]. They are defined as any substance that is a food or part of a food and provides medical or health benefits, including the prevention and treatment of chronic diseases [94–99].

Nutraceutical delivery systems based on carrageenan may potentially be impacted by interactions between many components. As delivery carriers, the performance of delivery trucks in actual food systems has to be investigated and proven [100]. Poor sensory perception (such as a foul taste) or interactions involving many components might reduce the nutritional value of bioactive substances. These errands must be finished right away.

Moreover, particle sizes may be enormous. Some established carrageenan-based bioactive component delivery methods are huge, limiting their implementation in many commercial goods in terms of look, taste, and nutrition [101]. Choosing adequate packing materials and crosslinking agents may help generate a more compact particle structure, which is beneficial for smaller sizes.

Lack of research on carrageenan metabolism and carrageenan-based nutraceutical delivery techniques that have been updated is also a concern. Colonic bacteria may desulfate carrageenan, resulting in a polymer with just a degradable galactan polymer backbone. Neither the intestinal fate nor the metabolic mechanism of carrageenan is well understood. Carrageenan's metabolic fate in the human body has to be extensively investigated. Due to modified carrageenan's increased properties, it will be interesting to observe whether it outperforms normal carrageenan-based delivery techniques. For this reason, carrageenan is still being studied to determine whether reducing its viscosity improves its suitability for delivery methods. A more effective extraction method is necessary to get higher-quality carrageenan.

References

1. Necas, J. and Bartosikova, L., Carrageenan: A review. *Vet. Med. (Praha)*, *58*, 187–205, 2013.
2. Pacheco-Quito, E.M., Ruiz-Caro, R., Veiga, M.D., Carrageenan: Drug delivery systems and other biomedical applications. *Mar. Drugs*, *18*, 1–39, 2020.
3. Hotchkiss, S., Brooks, M., Campbell, R., Philp, K., Trius, A., The use of carrageenan in food, in: *Carrageenans: Sources and Extraction Methods, Molecular Structure, Bioactive Properties and Health Effects*, 2016.
4. Falcó, I., Randazzo, W., Sánchez, G., López-Rubio, A., Fabra, M.J., On the use of carrageenan matrices for the development of antiviral edible coatings of interest in berries. *Food Hydrocolloids*, *92*, 74–85, 2019, doi: 10.1016/j.foodhyd.2019.01.039.
5. Tye, R.J., Industrial and non-food uses for carrageenan. *Carbohydr. Polym.*, *10*, 259–280, 1989.
6. Liao, Y.C., Chang, C.C., Nagarajan, D., Chen, C.Y., Chang, J.S., Algae-derived hydrocolloids in foods: Applications and health-related issues. *Bioengineered*, *12*, 3787–3801, 2021.
7. Dal Bosco Ducatti, R., da Anunciação, C.R., Sartori, V.C., Piva, M.C.B., Comunello, L., Tironi, S.P., Use of carrageenan for the reduction of deoxynivalenol contamination in wheat and barley kernels. *J. Biotechnol. Biodivers.*, *9*, 040–047, 2021, doi: 10.20873/jbb.uft.cemaf.v9n1.ducatti.
8. Chevenier, A., Jouanneau, D., Ficko-Blean, E., Carrageenan biosynthesis in red algae: A review. *Cell Surf.*, *9*, 100097, 2023, doi: 10.1016/J.TCSW.2023.100097.
9. Zampouni, K., Mouzakitis, C.K., Lazaridou, A., Moschakis, T., Katsanidis, E., Physicochemical properties and microstructure of bigels formed with gelatin and κ-carrageenan hydrogels and monoglycerides in olive oil oleogels. *Food Hydrocolloids*, *140*, 108636, 2023, doi: 10.1016/J.FOODHYD.2023.108636.
10. Shi, F., Chang, Y., Shen, J., Chen, G., Xue, C.A., Comparative investigation of anionic polysaccharides (sulfated fucan, ι-carrageenan, κ-carrageenan, and alginate) on the fabrication, stability, rheology, and digestion of multilayer emulsion. *Food Hydrocolloids*, *134*, 108081, 2023, doi: 10.1016/J.FOODHYD.2022.108081.
11. Rhein-Knudsen, N., Ale, M.T., Meyer, A.S., Seaweed hydrocolloid production: An update on enzyme assisted extraction and modification technologies. *Mar. Drugs*, *13*, 3340–3359, 2015.

12. Cunha, L. and Grenha, A., Sulfated seaweed polysaccharides as multifunctional materials in drug delivery applications. *Mar. Drugs*, *14*, 1–41, 2016.
13. Shukla, P.S., Borza, T., Critchley, A.T., Prithiviraj, B., Carrageenans from red seaweeds as promoters of growth and elicitors of defense response in plants. *Front. Mar. Sci.*, *3*, 1–9, 2016.
14. Rowe, R.C., Sheskey, P.J., dan, M.Q., *Handbook of pharmaceutical excipients*, Sixth Edition, Pharmaceutical Press, London, UK, vol. E.28, 2009.
15. Hilliou, L., Larotonda, F.D.S., Abreu, P., Ramos, A.M., Sereno, A.M., Gonçalves, M.P., Effect of extraction parameters on the chemical structure and gel properties of κ/ι-hybrid carrageenans obtained from mastocarpus stellatus. *Biomol. Eng.*, *23*, 201–208, 2006, doi: 10.1016/j.bioeng.2006.04.003.
16. Azevedo, G., Torres, M.D., Sousa-Pinto, I., Hilliou, L., Effect of pre-extraction alkali treatment on the chemical structure and gelling properties of extracted hybrid carrageenan from chondrus crispus and ahnfeltiopsis devoniensis. *Food Hydrocoll.*, *50*, 150–158, 2015, doi: 10.1016/j.foodhyd.2015.03.029.
17. Khotimchenko, M., Tiasto, V., Kalitnik, A., Begun, M., Khotimchenko, R., Leonteva, E., Bryukhovetskiy, I., Khotimchenko, Y., Antitumor potential of carrageenans from marine red algae. *Carbohydr. Polym.*, *246*, 1–17, 2020.
18. Bakry, A.M., Huang, J., Zhai, Y., Huang, Q., Myofibrillar protein with κ- or λ-carrageenans as novel shell materials for microencapsulation of tuna oil through complex coacervation. *Food Hydrocolloids*, *96*, 43–53, 2019, doi: 10.1016/j.foodhyd.2019.04.070.
19. Li, L., Ni, R., Shao, Y., Mao, S., Carrageenan and its applications in drug delivery. *Carbohydr. Polym.*, *103*, 1–11, 2014.
20. Knutsen, S.H., Myslabodski, D.E., Larsen, B., Usov, A.I.A., Modified system of nomenclature for red algal galactans. *Bot. Mar.*, *37*, 163–169, 1994, doi: 10.1515/botm.1994.37.2.163.
21. Alba, K. and Kontogiorgos, V., Seaweed polysaccharides (agar, alginate carrageenan), in: *Encyclopedia of Food Chemistry*, 2018.
22. Shen, Y.R. and Kuo, M.I., Effects of different carrageenan types on the rheological and water-holding properties of tofu. *LWT - Food Sci. Technol.*, *78*, 122–128, 2017, doi: 10.1016/j.lwt.2016.12.038.
23. Torres, M.D., Chenlo, F., Moreira, R., Structural features and water sorption isotherms of carrageenans: A prediction model for hybrid carrageenans. *Carbohydr. Polym.*, *180*, 72–80, 2018, doi: 10.1016/j.carbpol.2017.10.010.
24. Ghanam, D. and Kleinebudde, P., Suitability of κ-carrageenan pellets for the formulation of multiparticulate tablets with modified release. *Int. J. Pharm.*, *409*, 9–18, 2011, doi: 10.1016/j.ijpharm.2011.02.016.
25. Buchholcz, G., Kelemen, A., Sovány, T., Pintye-Hódi, K., Matrix tablets based on a carrageenan with the modified-release of sodium riboflavin 50-phosphate. *Pharm. Dev. Technol.*, *20*, 676–683, 2015, doi: 10.3109/10837450.2014.910810.
26. Pacheco-Quito, E.M., Ruiz-Caro, R., Rubio, J., Tamayo, A., Veiga, M.D., Carrageenan-based acyclovir mucoadhesive vaginal tablets for prevention of genital herpes. *Mar. Drugs*, *18*, 1–19, 2020, doi: 10.3390/md18050249.

27. Lal, M., Lai, M., Ugaonkar, S., Wesenberg, A., Kizima, L., Rodriguez, A., Levendosky, K., Mizenina, O., Fernández-Romero, J., Zydowsky, T., Development of a vaginal fast-dissolving insert combining griffithsin and carrageenan for potential use against sexually transmitted infections. *J. Pharm. Sci.*, *107*, 2601–2610, 2018, doi: 10.1016/j.xphs.2018.06.002.
28. Boateng, J.S., Pawar, H.V., Tetteh, J., Polyox and carrageenan based composite film dressing containing anti-microbial and anti-inflammatory drugs for effective wound healing. *Int. J. Pharm.*, *441*, 181–191, 2013, doi: 10.1016/j.ijpharm.2012.11.045.
29. Pettinelli, N., Rodríguez-Llamazares, S., Bouza, R., Barral, L., Feijoo-Bandín, S., Lago, F., Carrageenan-based physically crosslinked injectable hydrogel for wound healing and tissue repairing applications. *Int. J. Pharm.*, *589*, 1–9, 2020, doi: 10.1016/j.ijpharm.2020.119828.
30. Jaiswal, L., Shankar, S., Rhim, J.W., Carrageenan-based functional hydrogel film reinforced with sulfur nanoparticles and grapefruit seed extract for wound healing application. *Carbohydr. Polym.*, *224*, 1–10, 2019, doi: 10.1016/j.carbpol.2019.115191.
31. Biranje, S.S., Madiwale, P.V., Patankar, K.C., Chhabra, R., Bangde, P., Dandekar, P., Adivarekar, R.V., Cytotoxicity and hemostatic activity of chitosan/carrageenan composite wound healing dressing for traumatic hemorrhage. *Carbohydr. Polym.*, *239*, 1–31, 2020, doi: 10.1016/j.carbpol.2020.116106.
32. Shanmugapriya, K., Kim, H., Kang, H.W.A., New alternative insight of nanoemulsion conjugated with κ-carrageenan for wound healing study in diabetic mice: *In vitro* and *in vivo* evaluation. *Eur. J. Pharm. Sci.*, *133*, 236–250, 2019, doi: 10.1016/j.ejps.2019.04.006.
33. Muthuswamy, S., Viswanathan, A., Yegappan, R., Selvaprithiviraj, V., Vasudevan, A.K., Biswas, R., Jayakumar, R., Antistaphylococcal and neutrophil chemotactic injectable κ-carrageenan hydrogel for infectious wound healing. *ACS Appl. Bio Mater.*, *2*, 378–387, 2019, doi: 10.1021/acsabm.8b00625.
34. Gouda, M.H., Ali, S.M., Othman, S.S., A. Abd Al-Aziz, S., Abu-Serie, M.M., Elsokary, N.A., Elessawy, N.A., Novel scaffold based graphene oxide doped electrospun iota carrageenan/polyvinyl alcohol for wound healing and pathogen reduction: *In-vitro* and *in-vivo* study. *Sci. Rep.*, *11*, 1–11, 2021, doi: 10.1038/s41598-021-00069-0.
35. Tavakoli, S., Mokhtari, H., Kharaziha, M., Kermanpur, A., Talebi, A., Moshtaghian, J.A., Multifunctional nanocomposite spray dressing of kappa-carrageenan-polydopamine modified ZnO/L-glutamic acid for diabetic wounds. *Mater. Sci. Eng. C*, *111*, 1–17, 2020, doi: 10.1016/j.msec.2020.110837.
36. Fan, L., Wang, L., Gao, S., Wu, P., Li, M., Xie, W., Liu, S., Wang, W., Synthesis, characterization and properties of carboxymethyl kappa carrageenan. *Carbohydr. Polym.*, *86*, 1167–1174, 2011, doi: 10.1016/j.carbpol.2011.06.010.

37. Nair, A.V., Raman, M., Doble, M., Cyclic β-(1→3) (1→6) glucan/carrageenan hydrogels for wound healing applications. *RSC Adv.*, *6*, 98545–98553, 2016, doi: 10.1039/c6ra23386d.
38. Chandika, P., Kim, M.S., Khan, F., Kim, Y.M., Heo, S.Y., Oh, G.W., Kim, N.G., Jung, W.K., Wound healing properties of triple cross-linked poly (vinyl alcohol)/methacrylate kappa-carrageenan/chitooligosaccharide hydrogel. *Carbohydr. Polym.*, *269*, 1–11, 2021, doi: 10.1016/j.carbpol.2021.118272.
39. Prasedya, E.S., Miyake, M., Kobayashi, D., Hazama, A., Carrageenan delays cell cycle progression in human cancer cells *in vitro* demonstrated by FUCCI imaging. *BMC Complement. Altern. Med.*, *16*, 1–9, 2016, doi: 10.1186/s12906-016-1199-5.
40. Calvo, G.H., Cosenza, V.A., Sáenz, D.A., Navarro, D.A., Stortz, C.A., Céspedes, M.A., Mamone, L.A., Casas, A.G., Di Venosa, G.M., Disaccharides obtained from carrageenans as potential antitumor agents. *Sci. Rep.*, *9*, 1–13, 2019, doi: 10.1038/s41598-019-43238-y.
41. Luo, M., Shao, B., Nie, W., Wei, X.W., Li, Y.L., Wang, B.L., He, Z.Y., Liang, X., Ye, T.H., Wei, Y.Q., Antitumor and adjuvant activity of λ-carrageenan by stimulating immune response in cancer immunotherapy. *Sci. Rep.*, *5*, 1–12, 2015, doi: 10.1038/srep11062.
42. Cicinskas, E., Begun, M.A., Tiasto, V.A., Belousov, A.S., Vikhareva, V.V., Mikhailova, V.A., Kalitnik, A.A., *In vitro* antitumor and immunotropic activity of carrageenans from red algae chondrus armatus and their low-molecular weight degradation products. *J. Biomed. Mater. Res. - Part A*, *108*, 1–13, 2020, doi: 10.1002/jbm.a.36812.
43. Yuan, H. and Song, J., Preparation, structural characterization and *in vitro* antitumor activity of kappa-carrageenan oligosaccharide fraction from kappaphycus striatum. *J. Appl. Phycol.*, *17*, 7–13, 2005, doi: 10.1007/s10811-005-5513-8.
44. Hensel, A., γ-Propoxy-sulfo-lichenin, an antitumor polysaccharide derived from lichenin. *Pharm. Acta Helv.*, *70*, 25–31, 1995, doi: 10.1016/0031-6865(94)00048-Z.
45. Murad, H., Ghannam, A., Al-Ktaifani, M., Abbas, A., Hawat, M., Algal sulfated carrageenan inhibits proliferation of MDA-MB-231 cells via apoptosis regulatory genes. *Mol. Med. Rep.*, *11*, 2153–2158, 2015, doi: 10.3892/mmr.2014.2915.
46. Green, D.R. and Kroemer, G., The pathophysiology of mitochondrial cell death. *Science*. *305*, 5684, 626–629, 2004.
47. Hengartner, M.O., The biochemistry of apoptosis. *Nature*, *407*, 770–776, 2000.
48. Guessous, I., Cornuz, J., Paccaud, F., Lung cancer screening: Current situation and perspective. *Swiss Med. Wkly.*, *137*, 949–953, 2007.
49. Shah, S., Gapor, A., Sylvester, P.W., Role of caspase-8 activation in mediating vitamin E-induced apoptosis in murine mammary cancer cells. *Nutr. Cancer*, *45*, 236–246, 2003, doi: 10.1207/S15327914NC4502_14.

50. Riedl, S.J. and Shi, Y., Molecular mechanisms of caspase regulation during apoptosis. *Nat. Rev. Mol. Cell Biol.*, *5*, 897–907, 2004.
51. Yew, H.C., Nordin, F.J., Thiam, T.T., Azimahtol, H.L.P., Abdullah, N.R., Ismail, Z., Antiproliferative property and apoptotic effect of xanthorrhizol on MDA-MB-231 breast cancer cells. *Anticancer Res.*, *28*, 3677–3689, 2008.
52. Gao, Z., Shao, Y., Jiang, X., Essential roles of the Bcl-2 family of proteins in caspase-2-induced apoptosis. *J. Biol. Chem.*, *280*, 38271–38275, 2005, doi: 10.1074/jbc.M506488200.
53. Cheah, Y.H., Azimahtol, H.L.P., Abdullah, N.R., Xanthorrhizol exhibits antiproliferative activity on MCF-7 breast cancer cells via apoptosis induction. *Anticancer Res.*, *26*, 4527–4534, 2006.
54. Toshiyuki, M. and Reed, J.C., Tumor suppressor P53 is a direct transcriptional activator of the human bax gene. *Cell*, *80*, 293–299, 1995, doi: 10.1016/0092-8674(95)90412-3.
55. Rassouli, F.B., Matin, M.M., Iranshahi, M., Bahrami, A.R., Investigating the cytotoxic and apoptosis inducing effects of monoterpenoid stylosin *in vitro*. *Fitoterapia*, *82*, 742–749, 2011, doi: 10.1016/j.fitote.2011.03.005.
56. Luo, G., Guan, X., Zhou, L., Apoptotic effect of citrus fruit extract nobiletin on lung cancer cell line A549 *in vitro* and *in vivo*. *Cancer Biol. Ther.*, *7*, 966–973, 2008, doi: 10.4161/cbt.7.6.5967.
57. Chen, X., Han, W., Zhao, X., Tang, W., Wang, F., Epirubicin-loaded marine carrageenan oligosaccharide capped gold nanoparticle system for pH-triggered anticancer drug release. *Sci. Rep.*, *9*, 1–10, 2019, doi: 10.1038/s41598-019-43106-9.
58. Ling, G., Zhang, T., Zhang, P., Sun, J., He, Z., Nanostructured lipid-carrageenan hybrid carriers (NLCCs) for controlled delivery of mitoxantrone hydrochloride to enhance anticancer activity bypassing the BCRP-mediated efflux. *Drug Dev. Ind. Pharm.*, *42*, 1351–1359, 2016, doi: 10.3109/03639045.2015.1135937.
59. Golakiya, H.N., Naik, V.N., Hirapara, H.N., Mody, K.H., Goswami, A.P., Tripathi, C., Evaluation of anticoagulant effect of sulfated polysaccharide (SPS) from codium dwarkense børgesen in κ-carrageenan induced hypercoagulable state in wistar albino rats. *Acta Pol. Pharm. - Drug Res.*, *74*, 987–994, 2017.
60. Groult, H., Cousin, R., Chot-Plassot, C., Maura, M., Bridiau, N., Piot, J.M., Maugard, T., Fruitier-Arnaudin, I., Λ-Carrageenan oligosaccharides of distinct anti-heparanase and anticoagulant activities inhibit MDA-MB-231 breast cancer cell migration. *Mar. Drugs*, *17*, 1–15, 2019, doi: 10.3390/md17030140.
61. Saluri, K. and Tuvikene, R., Anticoagulant and antioxidant activity of lambda- and theta-carrageenans of different molecular weights. *Bioact. Carbohydr. Diet. Fibre*, *24*, 1–6, 2020, doi: 10.1016/j.bcdf.2020.100243.
62. Li, Y., Li, J., Shi, Z., Wang, Y., Song, X., Wang, L., Han, M., Du, H., He, C., Zhao, W. *et al.*, Anticoagulant chitosan-kappa-carrageenan composite hydrogel

sorbent for simultaneous endotoxin and bacteria cleansing in septic blood. *Carbohydr. Polym., 243*, 1–10, 2020, doi: 10.1016/j.carbpol.2020.116470.

63. Song, X., Wang, K., Tang, C.Q., Yang, W.W., Zhao, W.F., Zhao, C.S., Design of carrageenan-based heparin-mimetic gel beads as self-anticoagulant hemoperfusion adsorbents. *Biomacromolecules, 19*, 1966–1978, 2018, doi: 10.1021/acs.biomac.7b01724.
64. Panlasigui, L.N., Baello, O.Q., Dimatangal, J.M., Dumelod, B.D., Blood cholesterol and lipid-lowering effects of carrageenan on human volunteers. *Asia Pac. J. Clin. Nutr., 12*, 209–214, 2003.
65. Sokolova, E.V., Bogdanovich, L.N., Ivanova, T.B., Byankina, A.O., Kryzhanovskiy, S.P., Yermak, I.M., Effect of carrageenan food supplement on patients with cardiovascular disease results in normalization of lipid profile and moderate modulation of immunity system markers. *PharmaNutrition, 2*, 33–37, 2014, doi: 10.1016/j.phanu.2014.02.001.
66. Valado, A., Pereira, M., Caseiro, A., Figueiredo, J.P., Loureiro, H., Almeida, C., Cotas, J., Pereira, L., Effect of Carrageenans on Vegetable Jelly in Humans with Hypercholesterolemia. *Mar. Drugs, 18*, 1–11, 2020, doi: 10.3390/md18010019.
67. Graham, I., Atar, D., Borch-Johnsen, K., Boysen, G., Burell, G., Cifkova, R., Dallongeville, J., De Backer, G., Ebrahim, S., Gjelsvik, B. *et al.*, European guidelines on cardiovascular disease prevention in clinical practice Fourth Joint Task Force of the European Society of Cardiology and Other Societies on Cardiovascular Disease Prevention in Clinical Practice. *Ration. Pharmacother. Cardiol., 4*, 2375–2414, 2008, doi: 10.20996/1819-6446-2008-4-3-111-128.
68. Matthan, N.R., Zhu, L., Pencina, M., D'Agostino, R.B., Schaefer, E.J., Lichtenstein, A.H., Sex-specific differences in the predictive value of cholesterol homeostasis markers and 10-year cardiovascular disease event rate in framingham offspring study participants. *J. Am. Heart Assoc., 2*, 1–13, 2013, doi: 10.1161/JAHA.112.005066.
69. McKim, J.M., Food additive carrageenan: Part I: A critical review of carrageenan *in vitro* studies, potential pitfalls, and implications for human health and safety. *Crit. Rev. Toxicol., 44*, 211–243, 2014, doi: 10.3109/10408444.2013.861797.
70. Weiner, M.L., Nuber, D., Blakemore, W.R., Harriman, J.F., Cohen, S.M.A., 90-day dietary study on kappa carrageenan with emphasis on the gastrointestinal tract. *Food Chem. Toxicol., 45*, 98–106, 2007, doi: 10.1016/j.fct.2006.07.033.
71. Andrade, I., Santos, L., Ramos, F., Cholesterol absorption and synthesis markers in portuguese hypercholesterolemic adults: A cross-sectional study. *Eur. J. Intern. Med., 28*, 85–90, 2016, doi: 10.1016/j.ejim.2015.10.022.
72. Wang, W., Onnagawa, M., Yoshie, Y., Suzuki, T., Binding of bile salts to soluble and insoluble dietary fibers of seaweeds. *Fish. Sci., 67*, 1169–1173, 2001, doi: 10.1046/j.1444-2906.2001.00376.x.

73. Sokolova, E.V., Kravchenko, A.O., Sergeeva, N.V., Davydova, V.N., Bogdanovich, L.N., Yermak, I.M., Effect of carrageenans on some lipid metabolism components *in vitro*. *Carbohydr. Polym.*, *230*, 1–15, 2020, doi: 10.1016/j.carbpol.2019.115629.
74. Fedel-Miyasato, L.E.S., Formagio, A.S.N., Auharek, S.A., Kassuya, C.A.L., Navarro, S.D., Cunha-Laura, A.L., Monreal, A.C.D., Vieira, M.C., Oliveira, R.J., Antigenotoxic and antimutagenic effects of schinus terebinthifolius raddi in allium cepa and swiss mice: A comparative study. *Genet. Mol. Res.*, *13*, 3411–3425, 2014, doi: 10.4238/2014.April.30.2.
75. Nantes, C.I., Pesarini, J.R., Mauro, M.O., Monreal, A.C.D., Ramires, A.D., Oliveira, R.J., Evaluation of the antimutagenic activity and mode of action of carrageenan fiber in cultured meristematic cells of Allium cepa. *Genet. Mol. Res.*, *13*, 9523–9532, 2014, doi: 10.4238/2014.November.12.1.
76. Khotimchenko, M.Y., Kolenchenko, E.A., Khotimchenko, Y.S., Khozhaenko, E.V., Kovalev, V.V., Cerium binding activity of different pectin compounds in aqueous solutions. *Colloids Surf. B Biointerfaces*, *77*, 104–110, 2010, doi: 10.1016/j.colsurfb.2010.01.015.
77. Khotimchenko, Y.S., Khozhaenko, E.V., Khotimchenko, M.Y., Kolenchenko, E.A., Kovalev, V.V., Carrageenans as a new source of drugs with metal binding properties. *Mar. Drugs*, *8*, 1106–1121, 2010, doi: 10.3390/md8041106.
78. Topalian, S.L., Weiner, G.J., Pardoll, D.M., Cancer immunotherapy comes of age. *J. Clin. Oncol.*, *29*, 480–489, 2011.
79. Constantino, J., Gomes, C., Falcão, A., Neves, B.M., Cruz, M.T., Dendritic cell-based immunotherapy: A basic review and recent advances. *Immunol. Res.*, *65*, 798–810, 2017.
80. Yi, C., He, Y., Xia, H., Zhang, H., Zhang, P., Review and perspective on adjuvant and neoadjuvant immunotherapies in NSCLC. *OncoTargets Ther.*, *12*, 7329–7336, 2019, doi: 10.2147/OTT.S218321.
81. Adams, S., O'Neill, D.W., Nonaka, D., Hardin, E., Chiriboga, L., Siu, K., Cruz, C.M., Angiulli, A., Angiulli, F., Ritter, E. *et al.*, Immunization of malignant melanoma patients with full-length NY-ESO-1 protein using TLR7 agonist imiquimod as vaccine adjuvant. *J. Immunol.*, *181*, 776–784. 2008, doi: 10.4049/jimmunol.181.1.776.
82. Di Sciullo, P., Menay, F., Cocozza, F., Gravisaco, M.J., Waldner, C.I., Mongini, C., Systemic administration of imiquimod as an adjuvant improves immunogenicity of a tumor-lysate vaccine inducing the rejection of a highly aggressive T-cell lymphoma. *Clin. Immunol.*, *203*, 154–161, 2019, doi: 10.1016/j.clim.2019.04.013.
83. RL, S., S, B., N, B., Dendritic cell-based immunotherapy. *Cell Res.*, *27*, 74–95, 2017, doi: 10.1038/CR.2016.157.
84. Del Giudice, G., Rappuoli, R., Didierlaurent, A.M., Correlates of adjuvanticity: A review on adjuvants in licensed vaccines. *Semin. Immunol.*, *39*, 14–21, 2018.

85. Li, J., Aipire, A., Li, J., Zhu, H., Wang, Y., Guo, W., Li, X., Yang, J., Liu, C., λ-carrageenan improves the antitumor effect of dendritic cellbased vaccine. *Oncotarget*, *8*, 29996–30007, 2017, doi: 10.18632/oncotarget.15610.
86. Zhang, Y.Q., Tsai, Y.C., Monie, A., Hung, C.F., Wu, T.C., Carrageenan as an adjuvant to enhance peptide-based vaccine potency. *Vaccine*, *28*, 5212–5219, 2010, doi: 10.1016/j.vaccine.2010.05.068.
87. Tobacman, J.K., Review of harmful gastrointestinal effects of carrageenan in animal experiments. *Environ. Health Perspect.*, *109*, 983–994, 2001.
88. Gonçalves, R.F.S., Martins, J.T., Duarte, C.M.M., Vicente, A.A., Pinheiro, A.C., Advances in nutraceutical delivery systems: From formulation design for bioavailability enhancement to efficacy and safety evaluation. *Trends Food Sci. Technol.*, *78*, 270–291, 2018, doi: 10.1016/J.TIFS.2018.06.011.
89. Wang, W., Liu, F., Gao, Y., Quercetagetin loaded in soy protein isolate–κ-carrageenan complex: Fabrication mechanism and protective effect. *Food Res. Int.*, *C*, 31–40, 2016, doi: 10.1016/J.FOODRES.2016.02.012.
90. Wei, Z. and Huang, Q., Assembly of protein-polysaccharide complexes for delivery of bioactive ingredients: A perspective paper. *J. Agric. Food Chem.*, *67*, 1344–1352, 2019, doi: 10.1021/acs.jafc.8b06063.
91. Chopra, H., Sangeeta, Garg, G., Mineral nanoencapsulation, in: *Handbook Nanoencapsulation*, pp. 189–209, 2023.
92. Frenkel, Y., Cherno, V., Kostenko, H., Chopra, H., Gautam, R.K., Kostenko, V., Dietary supplementation with resveratrol attenuates serum melatonin level, pro-inflammatory response and metabolic disorder in rats fed high-fructose high-lipid diet under round-the-clock lighting. *Pathophysiol.*, *30*, 37–47, 2023, doi: 10.3390/PATHOPHYSIOLOGY30010005.
93. Chopra, H., Sangeeta, Elshikh, M.S., Introduction to the role of fruits in nutraceutical and functional foods, in: *Fruits Their Roles Nutraceuticals Funct. Foods*, pp. 1–28, 2023.
94. Chopra, H., Bibi, S., Goyal, R., Gautam, R.K., Trivedi, R., Upadhyay, T.K., Mujahid, M.H., Shah, M.A., Haris, M., Khot, K.B. *et al.*, Chemopreventive potential of dietary nanonutraceuticals for prostate cancer: An extensive review. *Front. Oncol.*, *12*, 1–25, 2022, doi: 10.3389/fonc.2022.925379.
95. Mustafa, F., Chopra, H., Baig, A.A., Avula, S.K., Kumari, S., Mohanta, T.K., Saravanan, M., Mishra, A.K., Sharma, N., Mohanta, Y.K., Edible mushrooms as novel myco-therapeutics: Effects on lipid level, obesity, and BMI. *J. Fungi*, *8*, 1–21, 2022.
96. Bhattacharya, T., Soares, G.A.B.E., Chopra, H., Rahman, M.M., Hasan, Z., Swain, S.S., Cavalu, S., Applications of phyto-nanotechnology for the treatment of neurodegenerative disorders. *Mater.*, *15*, 804, 2022, doi: 10.3390/MA15030804.

97. Chopra, H., Mishra, A.K., Baig, A.A., Mohanta, T.K., Mohanta, Y.K., Baek, K.H., Narrative review: Bioactive potential of various mushrooms as the treasure of versatile therapeutic natural product. *J. Fungi*, *7*, 1–30, 2021, doi: 10.3390/jof7090728.
98. Walia, V., Kaushik, D., Mittal, V., Kumar, K., Verma, R., Parashar, J., Akter, R., Rahman, M.H., Bhatia, S., Al-Harrasi, A. *et al.*, Delineation of neuroprotective effects and possible benefits of antioxidantstherapy for the treatment of Alzheimer's diseases by targeting mitochondrial-derived reactive oxygen species: Bench to bedside. *Mol. Neurobiol.*, *59*, 657–680, 2022, doi: 10.1007/s12035-021-02617-1.
99. Singla, R.K., Behzad, S., Khan, J., Tsagkaris, C., Gautam, R.K., Goyal, R., Chopra, H., Shen, B., Natural kinase inhibitors for the treatment and management of endometrial/uterine cancer: Preclinical to clinical studies. *Front. Pharmacol.*, *13*, 1–15, 2022, doi: 10.3389/fphar.2022.801733.
100. Gómez-Mascaraque, L.G., Llavata-Cabrero, B., Martínez-Sanz, M., Fabra, M.J., López-Rubio, A., Self-assembled gelatin-ι-carrageenan encapsulation structures for intestinal-targeted release applications. *J. Colloid Interface Sci.*, *517*, 113–123, 2018, doi: 10.1016/j.jcis.2018.01.101.
101. Ma, D., Tu, Z.C., Wang, H., Zhang, Z., McClements, D.J., Microgel-in-microgel biopolymer delivery systems: Controlled digestion of encapsulated lipid droplets under simulated gastrointestinal conditions. *J. Agric. Food Chem.*, *66*, 3930–3938, 2018, doi: 10.1021/acs.jafc.8b00132.

7

Biomedical Applications of Gums

Suhail Ahmed[1]*, Nagender Singh[2] and Annu[3†]

[1]*Department of Textile and Fibre Engineering, Indian Institute of Technology Delhi, New Delhi, India*

[2]*The Technological Institute of Textile and Sciences (TIT&S), Bhiwani, Haryana, India*

[3]*Thin-Film and Materials Laboratory, School of Mechanical Engineering, Yeungnam University, Gyeongsan, Republic of Korea*

Abstract

Today life expectancy and healthy life expectancy rates have increased tremendously because of advances and the latest development in the medical sector. However, the medical sector still uses sustainable materials to develop medical products. Therefore, gum hydrocolloids are obtained from sustainable origins. Gums are promptly accessible, eco-friendly, inexpensive, and non-hazardous. Based on origin, gums might be sorted as plant exudate gums, microbial gums, and seed gums. Gum hydrocolloids usually have numerous uses in the biomedical field, in contrast to synthetic materials, on account of their novel edifices and functionalities. Their matrices are favored to support medication conveyance since they are edible, harmless, and biodegradable. This chapter provides a contemporary overview of the biomedical use of gum to provide context for further research.

Keywords: Biomedical, bionanocomposites, biopolymers, gum, drug delivery

7.1 Introduction

Gums are hydrocolloids found in almost all biospheres on the planet, in plants, animals, and various microscopic organisms [1, 2], and are hydrophilic. They have a large number of hydroxyl groups that are normally

**Corresponding author*: alsuhail007@gmail.com

†*Corresponding author*: annuchem92@gmail.com

Annu (ed.) Biopolymers for Biomedical Applications, (175–198) © 2024 Scrivener Publishing LLC

arranged lateral to the molecule's backbone, taking into account the chelation of mono- and di-valent ions, fusing the polymer webs simultaneously and forming compound macro-structures. Gums are most likely distinct polysaccharides with glycosidic linkage and complex structures. Gums often consist of a variety of polysaccharides with intricate structures and significant molecular mass [1]. Gums are successful water adsorbents that might be solubilized by water to a more substantial or smaller degree. The water is held within the microstructure by hydrogen and the cavities created by the intricate molecular arrangement, as is evident from the numerous -OH groups [2]. The significant gums accessible are gum arabic (GA), xanthan gum (XG), tragacanth gum (TG), and guar gum (GG). A broad examination of different parts of these gum polysaccharides has been done, incorporating their accessibility, chemical structures, molecular weight appropriation, and food and non-food applications [3]. These gums have a complex chemical composition that varies based on age and source. Therefore, expecting to provide specified main structural formulations for such biodegradable polymers is unrealistic. Exudate gums have various applications, especially in the food industry. However, there are numerous non-food uses as well [4].

In recent years, many articles have been available on gum in biomedical applications [5, 6]. Be that as it may, no particular survey has investigated gum-based materials for biomedical applications. In this article, the new advancements in gums and their subordinates as medication conveyance are examined exhaustively. The application of these resources as transdermal, protein, anti-hypertensive, and colon explicit drug delivery systems has received particular emphasis. In addition, various techniques for the utilization and characteristics of the chemically altered forms intended for the associations and conveyance of drug delivery were additionally talked about.

7.2 Physicochemical Properties of Gums

The broad utilization of gums in various fields has prompted the improvement of products with new advancements in assembling different material frameworks through alterations in the mechanical and biochemical properties or structure of these biomaterials. The typical polymers' extraordinary synthetic and organic qualities have been exceptionally viable in many fields like medication, nutrition, pharmacology, agribusiness, and natural and compound designing [6]. A few gums are biocompatible, and numerous researchers have persistently attempted to utilize them in planning clinical frameworks or clinical gadgets [7]. Such natural polymers have been thought of because of their incredible biocompatibility, biodegradability, nontoxicity, reasonableness, and simplicity of openness when contrasted with engineered polymers. Various kinds of normal polysaccharides, like gum tragacanth (GT), thickener, dextran, chitosan, GG, and so forth, have been used for drug uses as nanofibers and nanoparticles [8, 100, 101]. The high molar mass of the gum exudates can be framed by substances like arabinose, xylose, galactose, galacturonic corrosive, and rhamnose. They are capable of emulsification, thickening, maintaining moisture, and gelling. Due to their high degree of flexibility, polysaccharides are frequently used in food preparation [9].

The *Cyamopsis tetragonoloba* endosperm seeds are used to make guar gum. Galactomannan, a collection of polysaccharides, makes up this substance [10]. Figure 7.1 depicts the structure of the latter, which is composed of (14)-D-mannose (b-D-mannopyranose), with linkages built of D-galactose (a-D-galactopyranose) units connected by 1–6 links. The composition of GG is 80% galactomannan, 12% water, 5% protein, 2% acidic insoluble ash, 0.7% ash, and 0.7% fat [11]. Guar gum is utilized as an additive in food, drug, beauty care products, and material ventures, just as grease in oil well drilling. There are lots of hydroxyl groups in this polymer. Guar gum forms hydrogen bonds with water when introduced, giving solutions with a significant thickness and thickening. Due to its excellent water swellability, safety, and affordability, it is also used to assemble hydrophilic grid tablets [12–14].

Xanthan gum is the utmost well-known bacteriological gum and orders an overwhelming majority of the business gum market on account of its accessibility and excellent rheologic properties. The XG is a branched

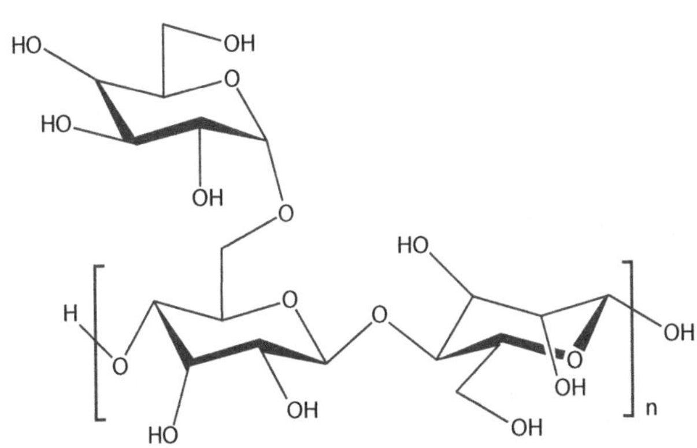

Figure 7.1 Structure of guar gum [11].

polysaccharide made by the fermentation process of bacteria. Its side-chains are composed of a trisaccharide made of d-glucuronic acid (-1,2), d-mannose (-1,4), and d-mannose, which are connected to the backbone's different glucose remains by -1,3 linkages, two of which are schematically shown in Figure 7.2 [15, 16]. It possesses unique qualities such as extreme pseudo-pliancy and high viscosity even at low concentrations. It has outstanding strength and solvency in harsh and corrosive circumstances and maintains thermal stability at high temperatures. The fermentation of the Gram-negative bacteria forms this microbial gum *Xanthomonas campestris*, precisely like the many types of *Xanthomonas* that contaminate particular plants [17, 18].

Gum arabic is a complex polysaccharide that is attained as a combination of potassium, magnesium, and calcium salts. It has branches and is neutrally acidic [9]. Figure 7.3 depicts the side chains composed of 2 to 5 b-D-galactopyranosyl units connected with 1–3 links and attached to the prime chain *via* 1–6 links [19]. The high-molecular-weight component of the gum is hypothesized to consist of massive starch blocks with a subatomic mass of around 2.5 * 105 Da connected only to a polypeptide amino acid chain. The GA is a helical polysaccharide derived from Acacia Seyal

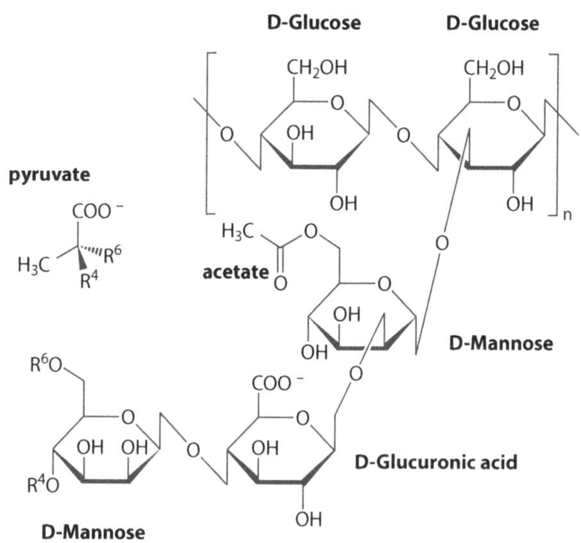

Figure 7.2 Representation of the xanthan repeating unit's chemical composition. The three counter-ions that result during fermentation most frequently are Na+, K+, and Ca^{2+} [16].

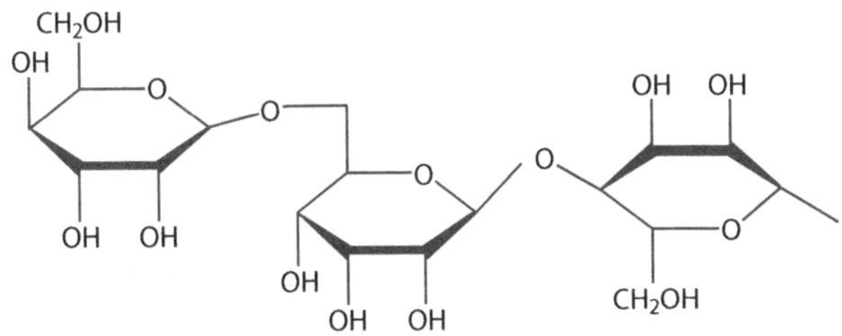

Figure 7.3 Structure of gum arabic [21].

and Acacia Senegal stems and has extraordinary underlying physical and material qualities. As a result, it is frequently used in food and drug applications [20–22].

Gum tragacanth is a heterogeneous, complex, densely branched polymer that often exhibits a mildly acidic appearance in calcium, potassium, and magnesium salt. Around 8.4 * 105 Da would be its molecular mass. The composition of the gum derived from several *Astragalus* species varies noticeably. Since *Astragalus* is the most prevalent variety within the Leguminosae family, this variation is not typical. It occurs worldwide in tropical regions and has more than 100 subdivisions made up of about 2,000 species [23–25]. The majority of GT is composed of a complicated mixture of branching acidic hetero-polysaccharides comprising d-galacturonic acid. Some additional sugars created during hydrolysis are d-xylose, d-galactose, l-arabinose, and l-fucose (6-deoxy-l-galactose). As seen in Figure 7.4, bassorin (arabinogalactan) appears almost neutral since it mainly contains d-galacturonic acid methyl ester rather than the acid groups [24]. The two components of GT are tragacanthin, which is water-solvent, and tragacanthin is corrosive, which is insoluble in water but can develop and form a gel. Methoxyl groups and proteinaceous material are present in only minor amounts in the two parts, with the water-solvent division having a more significant concentration of the latter. The component of the tragacanth corrosive that is water-swellable has a molecular structure like a bar and a high subatomic weight. The central chain comprises 1,4-connected D-galactose deposits, whereas the side chains are composed of 1,3-linked d-xylose units. Tragacanthin, an arabinogalactan that dissolves in water, has a circular atomic structure and is impartial and thoroughly extended. The appended chains of 1,2-, 1,3-, and 1,5-connected l-arabinose are most likely attached to a main of 1,3- and 1,6-connected D-galactose. GT is primarily utilized in the pharmaceutical and food industries [26–33].

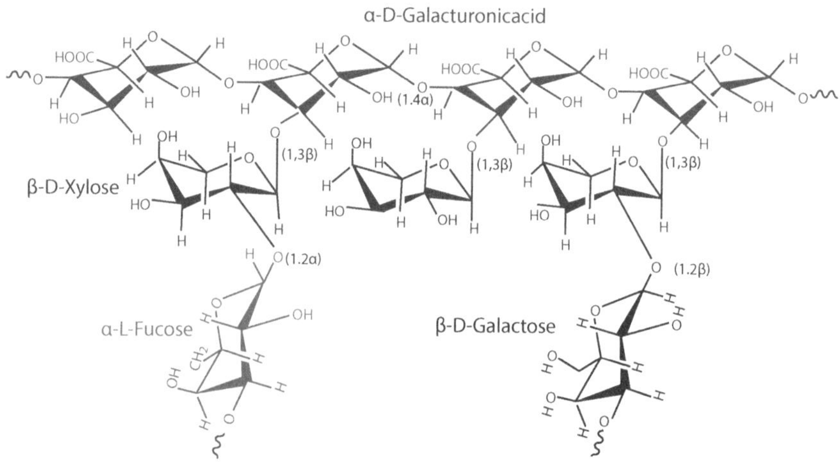

Figure 7.4 Structure of tragacanth gum polysaccharide [24].

7.3 Biomedical Applications of Guar Gum

The utilization of regular polysaccharides and their subsidiaries in biomedical and controlled release drug medication has been prospering. In tablet formulations, guar gum serves as a binder and a degrading agent. It is a thickener, stabilizer, suspender, and emulsifying agent in fluid formulations. The properties of guar gum can also be changed by getting fluctuating solvency, swelling, and film framing capacity. In this way, since the most recent couple of years, a parcel of examination studies has been done for the improvement of GG and its subordinates as anti-hypertensive, protein, colon-specific, and transdermal medication conveyance frameworks [34].

Guar gum polysaccharide, due to its adaptability to acquire an advantageous medication discharge profile, cost-viability, and wide administrative acknowledgment, has been broadly utilized for designated drug conveyance in different structures, for example, oral-controlled medication conveyance frameworks, coatings, nanoparticles, hydrogels, and tablet grids. Due to its inherent benefits of protection from separation in the stomach's acidic pH, gelling to lessen the release of medication, and vulnerability to enzymatic and microorganisms' degradation in the gastrointestinal tract, guar gum has been developed into a valuable water-soluble grid for the slow release of oral medications.

Dexamethasone and other reducing medicines, including 5-amino salicylic acid, indomethacin, metronidazole, albendazole, and mebendazole, have been successfully released into the colon in a precise, controlled manner using guar gum as a lattice. In addition, the colon-designated arrival of GG-bound tablets was established in healthy volunteers using gamma scintigraphy with a focus on tracer like technetium 99m-DTPA, which demonstrated the most excellent drug release in the colon because of enzymatic debasement. In pressure-covered tablets, guar gum has been demonstrated as a compelling medication discharge retardant to release metronidazole and 5-fluorouracil at a lower pH. The last measure of medication was delivered in reenacted gastro-digestive liquids, though the most effective medication was delivered in the presence of rodent cecal substance medium [35–41].

In order to effectively induce cross-linking and the growth of silver nuclei by reduction reaction through a quick and straightforward process for anti-cancer drug release and antibacterial applications, Ramasubba RP *et al.* created a dual-sensitive composite hydrogel network from GG-grafted poly-acrylamidoglycolic acid (graft polymer) utilizing sodium borohydride [42]. Rishikesh Deka and colleagues synthesized an intelligent hydrogel and hydrogel composite made of polyvinyl alcohol (PVA) and GG. The catalytic and antibacterial abilities of the intelligent hydrogel composite made them acceptable for biological applications [43]. The biomaterial was created by Piyali *et al.* by synthesizing carboxyl-methyl-guar-gum (CMGG) from indigenous GG, grafting it by gelatin to create CMGG-g-gelatin, and combining it together with curcumin. The prepared CMGG-g-gelatin inhibits both Gram-negative and Gram-positive bacteria [44]. Various published work has shown the use of GG and its derivatives for biomedical purposes [45–47]. Bio-polymeric hydrogels based on GG were created by Hina Daud *et al.* for the slow release of anti-hypertensive medications [48]. In order to create a polyelectrolyte complex for drug administration, Zunaira and colleagues used the solution casting approach to cross-link chitosan, guar gum, and polyvinyl pyrrolidone [49].

7.4 Biomedical Applications of Xanthan Gum

Xanthan gum is extensively used in water-based paints, food, oil recovery, personal care products, agribusiness, and other industries. Because of its common characteristics like non-Newtonian behavior, high consistency at low concentrations, low affectability of thickness to saltiness modifications, safety from mechanical corrosion, and better durability and eco-friendly properties, XG is used in food and other industries. The electro-spinning technique was used to organize XG nanofibers in a formic acid-corrosive environment without the use of copolymers. The polymer concentrates as the sub-atomic and rheological compliances of XG are studied. An electro-spinning thickener (1.5 to 2.5 w/v% in formic corrosive) was employed to produce nanofibers of xanthan. Since sub-atomic compliance and low shear pressure are caused by hydrogen retention, XG has a better consistency due to its diminishing shear conduct. The fundamental problems with XG are chain entanglement and poor gelling qualities in a wet atmosphere, prohibiting effortless electro-spinning. However, it has been discovered that XG has reversible shear diminishing behavior in formic corrosive, which may successfully circumvent the precautions mentioned above and generate the required rheological properties.

Additionally, the power stream record worth of XG in formic corrosive is 0.761 to 0.236 for convergences of 0.5 to 2.5 (w/v%), suggesting that the declining shear conduct needed to produce an electric field for the continuous electro-spinning of XG is necessary. The conductivity of XG configurations in formic corrosive was further investigated, and it was shown that increasing the XG centralization led to a more excellent arrangement conductivity due to more ionic charges, which improved XG's usefulness. A higher electrical conductivity increases the electrostatic repulsion and surface potential, prolonging the electro-spun-persistent fiber arrangement. Additionally, formic corrosive, which has a high dielectric consistency (57.9), increases the framework's charges in a way that is beneficial for the efficient electro-spinning of XG [16, 50–57].

Shekarforoush *et al.* organized XG–chitosan's nanofibers in formic corrosive with a high level of sound in liquid media and employed them for the embodiment and advancement with the pH-active curcumin (CU). The findings suggested that these XG–chitosan electro-spun nanofibers might be used as a transporter for hydrophobic bioactive formulations. Electro-spinning has been used to create crossover nanofibers based on XG and chitosan that carry CU. Due to oppositely charged polyelectrolytes, the nanosized strands organize ionically connected electro-spun nanofibers in a salt configuration (contrarily charge of XG and emphatically charge of chitosan). The digestibility of the CU in the hydride fibers was higher than that of free CU administration. A nanodevice has been created for the oral distribution of hydrophobic treatment specialists to the digestive tract. The electro-spinning method was used to create xanthan–chitosan nanofibers as CU's model and delivery system [58, 59]. Various works have been published on the biomedical application of xanthan gum [60–64].

In order to promote wound healing, Andreia Alves *et al.* created and examined a thermo-reversible hydrogel made of konjac glucomannan and xanthan gum in various doses (1% and 2% w/v) and proportions (50/50 and 60/400) [65]. Luciana and colleagues created an inventive aerogel using a xanthan gum polyelectrolyte/chitosan complex to explore its possible use in the medical field [66]. Beatrice Piola *et al.* created a cross-linked 3D printable hydrogel with gelatin, xanthan gum, and biocompatible natural polymers to be used as a wound dressing and a scaffold for cell growth.

The growth of human keratinocytes and fibroblasts *in vitro* over 14 days on gelatin/xanthan gum-printed hydrogels demonstrated the materials' biocompatibility [67]. Hydrogels have been created by using the freeze/thaw cycles technique with xanthan gum and poly(vinyl alcohol) in the existence of oxalic acid as a cross-linker. Gram-positive and Gram-negative bacteria were both resistant to the produced hydrogel's antibacterial properties [68]. Gelatin methacrylate and xanthan gum were combined with N-halamine/TiO_2 to create a 3D antibacterial wound dressing. The finding showed that the prepared dressings might significantly speed up the wound healing process, as shown in a mouse model, which indicates that this will be an excellent candidate for wound therapy [69].

Song *et al.* fabricated a multifunctional hydrogel using xanthan gum and polyacrylamide. According to the *in vivo* results, the produced hydrogel dressings considerably speed up the healing of wounds in a mouse model [70]. Winters *et al.* successfully synthesize the polystyrene sulfonate nanoparticles using hyaluronic acid as a stabilizer *via* an oxidative mini-emulsion polymerization technique for biomedical applications [71]. A porous tri-component framework made of xanthan, graphene oxide, and hydroxyapatite has been created for bone tissue creation. The *in vitro* results demonstrated the potential usefulness of the manufactured scaffold materials as a biomaterial for bone tissue engineering [72].

7.5 Biomedical Applications of Gum Arabic

A mixture of GA and sodium alginate created a unique hydrogel that relied on sundew adhesive and used Ca salt as a cross-connecting specialist. This hydrogel was then examined for use in the transportation of drugs and the healing of wounds [73]. Excellent biocompatibility has been achieved in the development of poly(dialdehyde)-corrected gum arabic with a melatonin- and collagen-based combination framework. By using sodium iodate to oxidize GA, poly(dialdehyde) was created, and then a function of the hybrid complex's amine was added. Applications for this substance include the treatment of wounds and *in vitro* and *in vivo* regeneration of tissue. The poly(dialdehyde)/gum arabic high content has shown outstanding qualities (elevated warm security, porosity, expanding limit, and biocompatible and biodegradable attributes) appropriate for biomedical fields [74].

After being formed using primaquine, an oxidized variety of GA was readied for regulated delivery in leishmaniasis or jungle fever. Schiff's initial work confirmed the complex interaction between the amine and GA aldehyde groups. When the productivity of stacking medications, form security, enlargement performance, and other factors were evaluated, the pre-assembled complex gel outperformed the dextran-based pharmaceutical delivery materials. The construction of unique hydrogel twisted swaths for the sluggish arrival of gentamicin (an anti-infection treatment) and to nurture the beneficial injury abilities has been attributed to the amalgamation of hydrogels utilizing gum arabic and cross-connected poly-acrylic corrosive carbopol. A number of variables have been considered, including cell reinforcement movement, blood similarity, muco-adhesion, oxygen/water fume porousness, antimicrobial activity, bacteriological infiltration, and mechanical properties (burst and elasticity, versatility, unwinding, and collapsing perseverance) of the hydrogel. Tests on wound healing in Swiss Balb C strain light-skinned rats revealed that, in contrast to painful injuries, wounds treated with hydrogel dressings resulted in quick wound twisting along with the arrangement of all-around produced fibroblasts and blood vessels. The biological properties showed that hydrogel films are mucoadhesive and anti-cancer while being non-thrombogenic and non-hemolytic. These wound-healing hydrogel layers are resistant to microbes even though permeable to oxygen and moisture. The assimilation of large volumes of liquid by hydrogel dressings was proved by the replicated liquid take-up in injuries and the release of anti-toxins in the injury fluids. The fluids from a reproduced wound they can consume are limited to 8.772–0.184 g/g of film. A Fickian dispersion component, which included a recreated wound fluid, in the investigations, led to the release of a gentamicin medicine from the dressings. As a result, given their ability to deliver the anti-toxins in a controlled manner, they are appropriate for mild to severe radiating injuries. For medicinal applications for the delivery of anti-cancer drugs, Ghazal *et al.* created a hybrid gum arabic aldehyde–bovine serum albumin nanogels by utilizing the inverse mini-emulsion method and the Schiff base reaction [75].

In order to supply nanocomposite for desired biomedical applications, Francesca *et al.* created biocompatible blended gum arabic/polyvinyl-alcohol matrix-like nanofibers for *in vitro* cell growth [76]. Gum arabic, a natural polysaccharide, was used to create polyaniline (PANI) nanoparticles, which were then tested for biocompatibility. When exposed to a protein-rich media, the GA successfully created the colloidal PANI–GA dispersions, demonstrating compatibility with the biological environment [77].

Silver nanoparticles were created synthetically by Mysoon *et al.* According to the research, AgNPs effectively combatted the oral infection by *S. mutans* [78]. Mohamed Hasanin and colleagues created a biocompatible polymer (gum arabic/gelatin) functionalized on graphene-oxide nanosheets *via* green biosynthesis. The created system showed a fresh amalgamation of gum arabic/gelatin and graphene-oxide for efficient anti-cancer therapy and tailored medication delivery [79]. Shahin and colleagues have utilized gum arabic aldehyde and PVA to create pH-sensitive, self-healing hydrogels. The outcomes showed that the hydrogel platformed a self-healing nanocomposite for drug administration at pH 7.4 [83].

Selenium nanorods (Se-NRs) were created by Mahin Velayati *et al.* using a natural, straightforward biological process that included gum arabic as a stabilizer, ascorbic acid as a reducing agent, and sodium selenite salt as the source of selenium [80]. The first unique Mn-doped CuS (Mn-CuS) nanoprisms with a high surface area and various Mn concentrations (0%, 1%, 3%, and 5%) were prepared by Mustafa M. *et al.* using the two-phase colloidal approach. According to the findings, GA@Mn-CuS 1% nano-prisms successfully destroyed cancer cells [81]. To increase medication release and alizarin penetration inside biofilm cells, chitosan/gum arabic-coated liposomes alizarin nanocarriers were created. As a result of the prolonged release of alizarin from NCs, CGL-Alz NCs had an antibiofilm impact against biofilms produced by *C. Albicans* or *S. aureus* and increased the alizarin diffusion inside biofilms [82].

7.6 Biomedical Applications of Gum Tragacanth

The accumulation of gum tragacanth to calcium alginate led to the formation of globules with an enormous potential for encasing bone cells. High cell multiplication, manageability, and qualification were studied in bone cells embodied by gum tragacanth–calcium alginate dots, suggesting that these cells may have uses in tissue engineering in addition to more modern medication preparations [83, 84]. A biocompatible GT-PVP-based hydrogel was developed with ammonium persulfate as an initiator and N,N′-methylenebisacrylamide as a cross-linker. The antifungal drug fluconazole, which has been shown to have excellent blood similarity and mucoadhesive qualities and is non-thrombogenic, was fused with the GT-PVP hydrogel to test its ability to release medication [85].

Electro-spinning has been used to produce poly(L-lactic corrosive)/GT (PLLA/GT) filaments that are 3% (w/v) and in a variety of ratios (100:0, 75:25, and 50:50). Nanofibers with PLLA/GT adjustments and a 75:25 ratio demonstrated balanced characteristics. They were used to test the feasibility of using these frameworks as substrates for nerve recovery using an *in vitro* culture of nerve cells. Electro-spun nanofibers were produced using PVA and GT in deionized water arrangements at numerous weight ratios [86]. FTIR analysis confirmed the hydrogen holding between PVA and GT for the optimal PVA/GT proportion (40/60). It was discovered that human fibroblast cells adhered to the GT/PVA nanofiber frameworks well and proliferated there. The GT/PVA NF demonstrated cell practicality and natural resemblance according to MTT tests. These NFs' biocompatibility and antibacterial qualities showed that they are suitable as an injury dressing since they can shield the affected area from the elements and prevent disease and dehydration. It has been reported that ciprofloxacin-stacked GT-based hydrogels, where GT was combined with engineered polymers containing acrylic corrosive, PVP, ammonium persulfate, and a cross-connecting specialist (N,N'- methylenebisacrylamide), have improved the delivery of diverticulitis medication. The pH influenced the hydrogel's mechanical stability and organizational structure, resulting in the surprise pH-responsive and mucoadhesive resources that may be used for spot-explicit medicine conveyance applications. Singh *et al.* accounted for GT and poly(acrylic corrosive)-based polymer hydrogels for amoxicillin drug delivery frameworks [85].

Gum tragacanth (6%, w/v), acrylic corrosive (0.874 mol/L), ammonium persulfate (0.0263 mol/L), and N,N'- methylenebisacrylamide (0.0389 mol/L) were used to create the hydrogel. There is some flexibility in the relationships between the artificial response bounds and the underlying qualities. The distribution profile of pharmaceuticals *via* polymeric networks was examined using a variety of models that focused on discharge energy. Carriers made of tragacanth gum have become popular drug delivery systems with particular uses. A functionalized TG was combined with a synthetic copolymer called poly(methyl methacrylate-alt-maleic anhydride)-g-polycaprolactone to create an amphiphilic co-network hydrogel that responds to pH. The hydrogel's potential for *in vitro* drug release was examined [87].

To create a hydrogel for ciprofloxacin distribution, TG was cross-linked with poly(vinylpyrrolidone)/acrylic acid [88]. In an experiment, tragacanth gum/collagen composites were employed to release the anti-inflammatory and vasodilator pentoxifylline. The findings show that adding gelatin to the TG matrix resulted in a porous structure that promotes more effective drug release [89].

Various studies have suggested the application of TG in medical applications [90–93]. The first effort to create random and aligned composite scaffolds from gum tragacanth/poly(l-lactic acid) was made by Marziyeh *et al.* The cell culture experiments with aligned (PLLA/GT 75:25) nanofibers revealed improved neurite outgrowth, increased proliferation, and better cellular phenotype [94]. In a different work, the author created curcumin (Cur)-loaded PCL/GT scaffold layers, which allowed for the regulated release of curcumin for more than 20 days [99]. Through cytotropic gelation, Didem *et al.* created a scaffold for tissue engineering applications using chitosan and gum tragacanth [95].

Marjan and colleagues have created a honey-infused nanofiber utilizing ethyl cellulose and gum tragacanth as a powerful antibacterial wound treatment. The outcomes showed appropriate antibacterial capabilities, better antioxidant action, enhanced degrading proficiency, improved mechanical characteristics, and proper cell growth, proliferation, and attachment compared to NIH-3 T3 fibro-blast cells [96]. Polyvinyl alcohol and gum tragacanth, two artificial and natural polymers, combined with graphene oxide that contained tetracycline were used by Mohadese *et al.* to create electro-spun nanofiber scaffolds for a transdermal drug delivery method [97]. For the administration of tetracycline hydrochloride (TCH), Salar *et al.* created a polymeric nanocomposite made of GT, PVA, and molybdenum disulfide (MoS_2). The antibacterial test showed that the nanofiber structure's ability to suppress bacterial growth is enhanced by the synergistic effect of MoS_2 and THC nanosheets [98]. Zein, GT, PLA, and TCH were mixed in various zein/GT percentages to create a unique drug delivery mat. The outcome demonstrates that the scaffold had sufficient degradability and antibacterial characteristics for skin tissue engineering [99].

7.7 Conclusion and Future Perspective

Drug nanocarriers must be durable over time and in a variety of electrolytic and pH settings, in addition to having specific biological properties. An additional advantage of utilizing gums for synthesis, depending on the type of metal particles used, is that the resulting nanostructures are more resistant to pH fluctuations and electrolyte expansion than typical nano-compounds, which are inhibited by citrate and borohydride—for instance, after being exposed to strong or noteworthy light for a short period at room temperature, Au NPs mixed with gum do not show any apparent molecular size and shape change.

Since gum polysaccharides are not polymers that conduct electricity, they cannot be used as biosensors. They can, however, be utilized to biosynthesize metal and metal oxide nanoparticles for use in detection applications, such as the biosynthesizing of silver nanoparticles for use in gas sensors for the detection of smelling salts. Biosensors, which must be protected and biocompatible, are one type of sensor used in biomedicine that can be helped by green science. The majority of pharmaceuticals and biomolecules face the challenges of low water solvency, accessibility, bioreactivity in the body, and soundness during capacity. Gum exemplification has two advantages over conventional conveyance that are valuable for medicines and biomolecules. It is probably the greatest technique to deal with making sure that drugs and biomolecules are unstable and dormant in the physiological environment before they reach their respective receptors. The strength and bioactivity of the substances mentioned above can also be increased by using proper security in conjunction with a successful vehicle.

Additionally, these mixes may be delivered gradually and under regulated conditions. In addition to the safety and controlled delivery for biomolecules, the biodegradability of gum-based composites in comparison to inorganic-based nanocarriers like carbon/metal-based transporters or synthetic polymer-based nanodevices is an advantage. There are not many publications on the use of gum nanocomposites as biosensors, even though they are expected to be contenders for chemical immobilization and biocompatibility.

This hole might be partially filled by changing the gums' synthetic design and using the right conductive nanoparticles, such as metal NPs and carbon-based combinations. In various contemporary applications, including the food and therapeutic industries, gum-biobased materials have been employed as thickeners and emulsifiers to manage surface and microstructure and improve the consistency and strength of the outcomes. The hydrogels based on gum nanocomposites have also been used to remove metal particles and colors from water. They have been combined with inorganic/natural nanomaterials in biomedical applications such as bone tissue recovery, medication delivery, tissue engineering, wound dressings, and wound healing.

References

1. Ahmad, S., Ahmad, M., Manzoor, K., Purwar, R., Ikram, S., A review on latest innovations in natural gums based hydrogels: Preparations & applications. *Int. J. Biol. Macromol.*, 136, 870-890, 2019, https://doi.org/10.1016/j.ijbiomac.2019.06.113.
2. Nussinovitch, A., *Plant gum exudates of the world: Sources, distribution, properties, and applications*, Routledge Taylor and Francis, New York, United States, 2009.
3. Karmakar, K., Application of natural gum as a binder in modern drug delivery. *J. Anal. Pharm. Res.*, 3, 00061, 2016, https://doi.org/10.15406/japlr.2016.03.00061.
4. Jani, G.K., Shah, D.P., Prajapatia, V.D., Jain, V.C., Gums and mucilages: Versatile excipients for pharmaceutical formulations. *Asian J. Pharm. Sci.*, Article ID 138951001, 2009.
5. Kulal, P. and Badalamoole, V., Efficient removal of dyes and heavy metal ions from waste water using Gum ghatti-graft-poly(4-acryloylmorpholine) hydrogel incorporated with magnetite nanoparticles. *J. Environ. Chem. Eng.*, 8, 104207, 2020, https://doi.org/10.1016/j.jcce.2020.104207.
6. Padil, V.V.T., Wacławek, S., Černík, M., Varma, R.S., Tree gum-based renewable materials: Sustainable applications in nanotechnology, biomedical and environmental fields. *Biotechnol. Adv.*, 36, 1984–2016, 2018, https://doi.org/10.1016/j.biotechadv.2018.08.008.
7. Nazarzadeh Zare, E., Makvandi, P., Tay, F.R., Recent progress in the industrial and biomedical applications of tragacanth gum: A review. *Carbohydr. Polym.*, 212, 450–467, 2019, https://doi.org/10.1016/j.carbpol.2019.02.076.
8. Korrapati, P.S., Karthikeyan, K., Satish, A., Krishnaswamy, V.R., Venugopal, J.R., Ramakrishna, S., Recent advancements in nanotechnological strategies in selection, design and delivery of biomolecules for skin regeneration. *Mater. Sci. Eng. C*, 67, 747–765, 2016, https://doi.org/10.1016/j.msec.2016.05.074.
9. Barak, S., Mudgil, D., Taneja, S., Exudate gums: Chemistry, properties and food applications - A review. *J. Sci. Food Agric.*, 100, 2828–2835, 2020, https://doi.org/10.1002/JSFA.10302.
10. Cheetham, N.W.H. and Mashimba, E.N.M., Conformational aspects of xanthan—Galactomannan gelation. Further evidence from optical-rotation studies. *Carbohydr. Polym.*, 14, 17–27, 1990, https://doi.org/10.1016/0144-8617(90)90004-C.
11. Prabaharan, M., Prospective of guar gum and its derivatives as controlled drug delivery systems. *Int. J. Biol. Macromol.*, 49, 117–124, 2011, https://doi.org/10.1016/J.IJBIOMAC.2011.04.022.
12. Nayak, A.K., Hasnain, M.S., Pal, K., Banerjee, I., Pal, D., Gum-based hydrogels in drug delivery, in: *Biopolymer-Based Formulations: Biomedical and Food Applications*, pp. 605–645, 2020, https://doi.org/10.1016/B978-0-12-816897-4.00025-4.

13. Thombare, N., Jha, U., Mishra, S., Siddiqui, M.Z., Guar gum as a promising starting material for diverse applications: A review. *Int. J. Biol. Macromol.*, 88, 361–372, 2016, https://doi.org/10.1016/J.IJBIOMAC.2016.04.001.
14. Liu, Z., Jiao, Y., Wang, Y., Zhou, C., Zhang, Z., Polysaccharides-based nanoparticles as drug delivery systems. *Adv. Drug Deliv. Rev.*, 60, 1650–1662, 2008, https://doi.org/10.1016/J.ADDR.2008.09.001.
15. Kumar, A., Rao, K.M., Han, S.S., Application of xanthan gum as polysaccharide in tissue engineering: A review. *Carbohydr. Polym.*, 180, 128–144, 2018, https://doi.org/10.1016/J.CARBPOL.2017.10.009.
16. Petri, D.F.S., Xanthan gum: A versatile biopolymer for biomedical and technological applications. *J. Appl. Polym. Sci.*, 132, 42035, 2015, https://doi.org/10.1002/APP.42035.
17. Dave, P.N. and Gor, A., Natural polysaccharide-based hydrogels and nanomaterials: Recent trends and their applications, in: *Handbook of Nanomaterials for Industrial Applications*, pp. 36–66, 2018, https://doi.org/10.1016/B978-0-12-813351-4.00003-1.
18. Ahmad, S., Ahmad, M., Manzoor, K., Purwar, R., Ikram, S., A review on latest innovations in natural gums based hydrogels: Preparations & applications. *Int. J. Biol. Macromol.*, 136, 870–890, 2019, https://doi.org/10.1016/J.IJBIOMAC.2019.06.113.
19. Dror, Y., Cohen, Y., Yerushalmi-Rozen, R., Structure of gum arabic in aqueous solution. *J. Polym. Sci. B Polym. Phys.*, 44, 3265–3271, 2006, https://doi.org/10.1002/POLB.20970.
20. Hamouda, A., Factors affecting the quality of Acacia senegal gums, 2017, http://hdl.handle.net/10034/620895 (accessed August 26, 2022).
21. Annu, and Ahmed, S., Advanced green materials: An overview, in: *Advanced Green Materials: Fabrication, Characterization and Applications of Biopolymers and Biocomposites*, pp. 1–13, 2021, https://doi.org/10.1016/B978-0-12-819988-6.00001-X.
22. Dave, P.N. and Gor, A., Natural polysaccharide-based hydrogels and nanomaterials: Recent trends and their applications, in: *Handbook of Nanomaterials for Industrial Applications*, pp. 36–66, 2018, https://doi.org/10.1016/B978-0-12-813351-4.00003-1.
23. Ehsan, M., Yazdi, T., Nazarnezhad, S., Mousavi, S.H., Amiri, M.S., Darroudi, M., Baino, F., Kargozar, S., Gum tragacanth (GT): A versatile biocompatible material beyond borders. *Molecules*, 26, 1510, 2021, https://doi.org/10.3390/MOLECULES26061510.
24. Nazarzadeh Zare, E., Makvandi, P., Tay, F.R., Recent progress in the industrial and biomedical applications of tragacanth gum: A review. *Carbohydr. Polym.*, 212, 450–467, 2019, https://doi.org/10.1016/J.CARBPOL.2019.02.076.
25. Nejatian, M., Abbasi, S., Azarikia, F., Gum tragacanth: Structure, characteristics and applications in foods. *Int. J. Biol. Macromol.*, 160, 846–860, 2020, https://doi.org/10.1016/J.IJBIOMAC.2020.05.214.

26. Mohammadifar, M.A., Musavi, S.M., Kiumarsi, A., Williams, P.A., Solution properties of targacanthin (water-soluble part of gum tragacanth exudate from Astragalus gossypinus). *Int. J. Biol. Macromol.*, 38, 31–39, 2006, https://doi.org/10.1016/J.IJBIOMAC.2005.12.015.
27. Mohammed, A.S.A., Naveed, M., Jost, N., Polysaccharides; classification, chemical properties, and future perspective applications in fields of pharmacology and biological medicine (a review of current applications and upcoming potentialities). *J. Polym. Environ.*, 29, 8, 2359–2371, 2021, https://doi.org/10.1007/S10924-021-02052-2.
28. Koyyada, A. and Orsu, P., Natural gum polysaccharides as efficient tissue engineering and drug delivery biopolymers. *J. Drug Deliv. Sci. Technol.*, 63, 102431, 2021, https://doi.org/10.1016/J.JDDST.2021.102431.
29. Ahmad, S., Ahmad, M., Manzoor, K., Purwar, R., Ikram, S., A review on latest innovations in natural gums based hydrogels: Preparations & applications. *Int. J. Biol. Macromol.*, 136, 870–890, 2019, https://doi.org/10.1016/J.IJBIOMAC.2019.06.113.
30. Verbeken, D., Dierckx, S., Dewettinck, K., Exudate gums: Occurrence, production, and applications. *Appl. Microbiol. Biotechnol.*, 63, 1, 10–21, 2003, https://doi.org/10.1007/S00253-003-1354-Z.
31. Padil, V.V.T., Wacławek, S., Černík, M., Varma, R.S., Tree gum-based renewable materials: Sustainable applications in nanotechnology, biomedical and environmental fields. *Biotechnol. Adv.*, 36, 1984–2016, 2018, https://doi.org/10.1016/J.BIOTECHADV.2018.08.008.
32. Rahimi, M., Noruzi, E.B., Sheykhsaran, E., Ebadi, B., Kariminezhad, Z., Molaparast, M., Mehrabani, M.G., Mehramouz, B., Yousefi, M., Ahmadi, R., Yousefi, B., Ganbarov, K., Kamounah, F.S., Shafiei-Irannejad, V., Kafil, H.S., Carbohydrate polymer-based silver nanocomposites: Recent progress in the antimicrobial wound dressings. *Carbohydr. Polym.*, 231, 115696, 2020, https://doi.org/10.1016/j.carbpol.2019.115696.
33. Mohammadinejad, R., Kumar, A., Ranjbar-Mohammadi, M., Ashrafizadeh, M., Han, S.S., Khang, G., Roveimiab, Z., Recent advances in natural gum-based biomaterials for tissue engineering and regenerative medicine: A review. *Polymers*, 12, 176, 2020, https://doi.org/10.3390/POLYM12010176.
34. Prabaharan, M., Prospective of guar gum and its derivatives as controlled drug delivery systems. *Int. J. Biol. Macromol.*, 49, 117–124, 2011, https://doi.org/10.1016/J.IJBIOMAC.2011.04.022.
35. Krishnaiah, Y.S.R., Karthikeyan, R.S., Satyanarayana, V., A three-layer guar gum matrix tablet for oral controlled delivery of highly soluble metoprolol tartrate. *Int. J. Pharm.*, 241, 353–366, 2002, https://doi.org/10.1016/S0378-5173(02)00273-9.
36. Prabaharan, M., Review paper: Chitosan derivatives as promising materials for controlled drug delivery. *J. Biomater. Appl.*, 23, 5–36, 2008, https://doi.org/10.1177/0885328208091562.

37. Prabaharan, M. and Jayakumar, R., Chitosan-graft-β-cyclodextrin scaffolds with controlled drug release capability for tissue engineering applications. *Int. J. Biol. Macromol.*, 44, 320–325, 2009, https://doi.org/10.1016/J.IJBIOMAC.2009.01.005.
38. Liu, Z., Jiao, Y., Wang, Y., Zhou, C., Zhang, Z., Polysaccharides-based nanoparticles as drug delivery systems. *Adv. Drug Deliv. Rev.*, 60, 1650–1662, 2008, https://doi.org/10.1016/j.addr.2008.09.001.
39. Sinha, V.R. and Kumria, R., Polysaccharides in colon-specific drug delivery. *Int. J. Pharm.*, 224, 19–38, 2001, https://doi.org/10.1016/S0378-5173(01)00720-7.
40. Cheetham, N.W.H. and Mashimba, E.N.M., Conformational aspects of xanthan—Galactomannan gelation. Further evidence from optical-rotation studies. *Carbohydr. Polym.*, 14, 17–27, 1990, https://doi.org/10.1016/0144-8617(90)90004-C.
41. Lee, J.W., Park, J.H., Robinson, J.R., Bioadhesive-based dosage forms: The next generation. *J. Pharm. Sci.*, 89, 850–866, 2000, https://doi.org/10.1002/1520-6017(200007)89:7<850::aid-jps2>3.0.co;2-g.
42. Palem, R.R., Shimoga, G., Rao, K.S.V.K., Lee, S.H., Kang, T.J., Guar gum graft polymer-based silver nanocomposite hydrogels: Synthesis, characterization and its biomedical applications. *J. Polym. Res.*, 27, 1–20, 2020, https://doi.org/10.1007/S10965-020-2026-8/FIGURES/9.
43. Deka, R., Sarma, S., Patar, P., Gogoi, P., Sarmah, J.K., Highly stable silver nanoparticles containing guar gum modified dual network hydrogel for catalytic and biomedical applications. *Carbohydr. Polym.*, 248, 116786, 2020, https://doi.org/10.1016/J.CARBPOL.2020.116786.
44. Manna, P.J., Mitra, T., Pramanik, N., Kavitha, V., Gnanamani, A., Kundu, P.P., Potential use of curcumin loaded carboxymethylated guar gum grafted gelatin film for biomedical applications. *Int. J. Biol. Macromol.*, 75, 437–446, 2015, https://doi.org/10.1016/J.IJBIOMAC.2015.01.047.
45. Sarmah, J.K., Dutta, A., Sarmah, S., Ankaleswar, B., Guar gum nanoparticles: A new paradigm in biomedical applications, in: *Polysaccharide Nanoparticles: Preparation and Biomedical Applications*, pp. 119–144, 2022, https://doi.org/10.1016/B978-0-12-822351-2.00003-6.
46. Nazarzadeh Zare, E., Makvandi, P., Borzacchiello, A., Tay, F.R., Ashtari, B., Padil, V.T.V., Antimicrobial gum bio-based nanocomposites and their industrial and biomedical applications. *Chem. Commun.*, 55, 14871–14885, 2019, https://doi.org/10.1039/C9CC08207G.
47. Koyyada, A. and Orsu, P., Natural gum polysaccharides as efficient tissue engineering and drug delivery biopolymers. *J. Drug Deliv. Sci. Technol.*, 63, 102431, 2021, https://doi.org/10.1016/J.JDDST.2021.102431.
48. Daud, H., Ghani, A., Iqbal, D.N., Ahmad, N., Nazir, S., Muhammad, M.J., Hussain, E.A., Nazir, A., Iqbal, M., Preparation and characterization of guar gum based biopolymeric hydrogels for controlled release of anti-hypertensive

drug. *Arab. J. Chem.*, 14, 103111, 2021, https://doi.org/10.1016/J.ARABJC.2021.103111.

49. Ghauri, Z.H., Islam, A., Qadir, M.A., Gull, N., Haider, B., Khan, R.U., Riaz, T., Development and evaluation of pH-sensitive biodegradable ternary blended hydrogel films (chitosan/guar gum/PVP) for drug delivery application. *Sci. Rep.*, 11, 1, 1–10, 2021, https://doi.org/10.1038/s41598-021-00452-x.
50. Sworn, G., Xanthan gum, in: *Handbook of Hydrocolloids*, pp. 833–853, Elsevier, Amsterdam, Netherlands, 2021, https://doi.org/10.1016/B978-0-12-820104-6.00004-8.
51. Patel, J., Maji, B., Narayana Moorthy, N.S.H., Maiti, S., Xanthan gum derivatives: Review of synthesis, properties and diverse applications. *RSC Adv.*, 10, 27103–27136, 2020, https://doi.org/10.1039/D0RA04366D.
52. Hamcerencu, M., Desbrieres, J., Popa, M., Khoukh, A., Riess, G., New unsaturated derivatives of Xanthan gum: Synthesis and characterization. *Polym. (Guildf)*, 48, 1921–1929, 2007, https://doi.org/10.1016/J.POLYMER.2007.01.048.
53. Kumar, A., Rao, K.M., Han, S.S., Application of xanthan gum as polysaccharide in tissue engineering: A review. *Carbohydr. Polym.*, 180, 128–144, 2018, https://doi.org/10.1016/J.CARBPOL.2017.10.009.
54. Wei, B., Flow characteristics of three enhanced oil recovery polymers in porous media. *J. Appl. Polym. Sci.*, 132, 41598, 2015, https://doi.org/10.1002/APP.41598.
55. Petri, D.F.S. and de Queiroz Neto, J.C., Identification of lift-off mechanism failure for salt drill-in drilling fluid containing polymer filter cake through adsorption/desorption studies. *J. Pet. Sci. Eng.*, 70, 89–98, 2010, https://doi.org/10.1016/J.PETROL.2009.10.002.
56. Hui, Y., Corke, H., de Leyn, I., Nip, W., Cross, N., *Bakery products: Science and technology*, John Wiley and Sons, New Jersey, USA, 2008, https://doi.org/10.1002/9781118792001.ch28.
57. Rinaudo, M., Polyelectrolytes derived from natural polysaccharides, monomers, in: *Polymers and Composites from Renewable Resources*, pp. 495–516, 2008, https://doi.org/10.1016/B978-0-08-045316-3.00024-7.
58. Shekarforoush, E., Ajalloueian, F., Zeng, G., Mendes, A.C., Chronakis, I.S., Electrospun xanthan gum-chitosan nanofibers as delivery carrier of hydrophobic bioactives. *Mater. Lett.*, 228, 322–326, 2018, https://doi.org/10.1016/J.MATLET.2018.06.033.
59. Shekarforoush, E., Ajalloueian, F., Zeng, G., Mendes, A.C., Chronakis, I.S., Electrospun xanthan gum-chitosan nanofibers as delivery carrier of hydrophobic bioactives. *Mater. Lett.*, 228, 322–326, 2018, https://doi.org/10.1016/J.MATLET.2018.06.033.
60. Shende, P. and Pathan, N., Potential of carbohydrate-conjugated graphene assemblies in biomedical applications. *Carbohydr. Polym.*, 255, 117385, 2021, https://doi.org/10.1016/J.CARBPOL.2020.117385.

61. Patel, J., Moorthy, N.S.H.N., Maiti, S., Xanthan-based nanomaterials for drug delivery applications, in: *Biopolymer-Based Nanomaterials in Drug Delivery and Biomedical Applications*, pp. 275–292, 2021, https://doi.org/10.1016/B978-0-12-820874-8.00014-2.
62. Petri, D.F.S., Xanthan gum: A versatile biopolymer for biomedical and technological applications. *J. Appl. Polym. Sci.*, 132, 42035, 2015, https://doi.org/10.1002/APP.42035.
63. Abu Elella, M.H., Goda, E.S., Gab-Allah, M.A., Hong, S.E., Pandit, B., Lee, S., Gamal, H., Rehman, A.U., Yoon, K.R., Xanthan gum-derived materials for applications in environment and eco-friendly materials: A review. *J. Environ. Chem. Eng.*, 9, 104702, 2021, https://doi.org/10.1016/J.JECE.2020.104702.
64. Priya, S., Batra, U., S.R.N., Sharma, S., Chaurasiya, A., Singhvi, G., Polysaccharide-based nanofibers for pharmaceutical and biomedical applications: A review. *Int. J. Biol. Macromol.*, 218, 209–224, 2022, https://doi.org/10.1016/J.IJBIOMAC.2022.07.118.
65. Alves, A., Miguel, S.P., Araujo, A.R.T.S., de J. Valle, M.J., Navarro, A.S., Correia, I.J., Ribeiro, M.P., Coutinho, P., Xanthan gum–konjac glucomannan blend hydrogel for wound healing. *Polymers*, 12, 99, 2020, https://doi.org/10.3390/POLYM12010099.
66. Tomé, L.I.N., Reis, M.S., de Sousa, H.C., Braga, M.E.M., Chitosan-xanthan gum PEC-based aerogels: A chemically stable PEC in scCO2. *Mater. Chem. Phys.*, 287, 126294, 2022, https://doi.org/10.1016/J.MATCHEMPHYS.2022.126294.
67. Piola, B., Sabbatini, M., Gino, S., Invernizzi, M., Renò, F., 3D bioprinting of gelatin–xanthan gum composite hydrogels for growth of human skin cells. *Int. J. Mol. Sci.*, 23, 539, 2022, https://doi.org/10.3390/IJMS23010539.
68. Enache, A.A., Serbezeanu, D., Vlad-Bubulac, T., Ipate, A.M., Suflet, D.M., Drobotă, M., Barbălată-Mândru, M., Udrea, R.M., Rîmbu, C.M., Tunable properties via composition modulations of poly(vinyl alcohol)/xanthan gum/oxalic acid hydrogels. *Materials*, 15, 2657, 2022, https://doi.org/10.3390/MA15072657.
69. Yang, Z., Ren, X., Liu, Y., Multifunctional 3D printed porous GelMA/xanthan gum based dressing with biofilm control and wound healing activity. *Mater. Sci. Eng.: C*, 131, 112493, 2021, https://doi.org/10.1016/J.MSEC.2021.112493.
70. Tang, S., Gong, Z., Wang, Z., Gao, X., Zhang, X., Multifunctional hydrogels for wound dressings using xanthan gum and polyacrylamide. *Int. J. Biol. Macromol.*, 217, 944–955, 2022, https://doi.org/10.1016/J.IJBIOMAC.2022.07.181.
71. Winters, C., Zamboni, F., Beaucamp, A., Culebras, M., Collins, M.N., Synthesis of conductive polymeric nanoparticles with hyaluronic acid based bioactive stabilizers for biomedical applications. *Mater. Today Chem.*, 25, 100969, 2022, https://doi.org/10.1016/J.MTCHEM.2022.100969.
72. Vanpeene, M., Rajesh, R., Ravichandran, Y.D., Kuo, Y.C., Gure, G., Biomimetic graphene oxide-xanthan gum-hydroxyapatite composite scaffold for bone

tissue engineering. *Chemistry Africa*, 6, 1–8, 2022, https://doi.org/10.1007/S42250-022-00368-7.

73. Sun, L., Huang, Y., Bian, Z., Petrosino, J., Fan, Z., Wang, Y., Park, K.H., Yue, T., Schmidt, M., Galster, S., Ma, J., Zhu, H., Zhang, M., Sundew-inspired adhesive hydrogels combined with adipose-derived stem cells for wound healing, *ACS Appl. Mater. Interfaces*, 8, 2423–2434, 2016, https://doi.org/10.1021/ACSAMI.5B11811.
74. Murali, R., Thanikaivelan, P., Cheirmadurai, K., Melatonin in functionalized biomimetic constructs promotes rapid tissue regeneration in Wistar albino rats. *J. Mater. Chem. B*, 4, 5850–5862, 2016, https://doi.org/10.1039/C6TB01221C.
75. Bashiri, G., Shojaosadati, S.A., Abdollahi, M., Synthesis and characterization of Schiff base containing bovine serum albumin-gum arabic aldehyde hybrid nanogels via inverse miniemulsion for delivery of anti-cancer drug. *Int. J. Biol. Macromol.*, 170, 222–231, 2021, https://doi.org/10.1016/J.IJBIOMAC.2020.12.150.
76. Serio, F., da Cruz, A.F., Chandra, A., Nobile, C., Rossi, G.R., D'Amone, E., Gigli, G., del Mercato, L.L., de Oliveira, C.C., Electrospun polyvinyl-alcohol/gum arabic nanofibers: Biomimetic platform for *in vitro* cell growth and cancer nanomedicine delivery. *Int. J. Biol. Macromol.*, 188, 764–773, 2021, https://doi.org/10.1016/j.ijbiomac.2021.08.069.
77. Gonçalves, J.P., de Oliveira, C.C., da Silva Trindade, E., Riegel-Vidotti, I.C., Vidotti, M., Simas, F.F., *In vitro* biocompatibility screening of a colloidal gum Arabic-polyaniline conducting nanocomposite. *Int. J. Biol. Macromol.*, 173, 109–117, 2021, https://doi.org/10.1016/J.IJBIOMAC.2021.01.101.
78. Al-Ansari, M.M., Al-Dahmash, N.D., Ranjitsingh, A.J.A., Synthesis of silver nanoparticles using gum Arabic: Evaluation of its inhibitory action on Streptococcus mutans causing dental caries and endocarditis. *J. Infect. Public Health*, 14, 324–330, 2021, https://doi.org/10.1016/J.JIPH.2020.12.016.
79. Hasanin, M., Taha, N.F., Abdou, A.R., Emara, L.H., Green decoration of graphene oxide Nano sheets with gelatin and gum Arabic for targeted delivery of doxorubicin. *Biotechnol. Rep.*, 34, e00722, 2022, https://doi.org/10.1016/J.BTRE.2022.E00722.
80. Velayati, M., Hassani, H., Sabouri, Z., Mostafapour, A., Darroudi, M., Biosynthesis of Se-Nanorods using Gum Arabic (GA) and investigation of their photocatalytic and cytotoxicity effects. *Inorg. Chem. Commun.*, 128, 108589, 2021, https://doi.org/10.1016/J.INOCHE.2021.108589.
81. Muhsen, M.M., Al-Jawad, S.M.H., Taha, A.A., Gum Arabic-modified Mn-doped CuS nanoprisms for cancer photothermal treatment. *Chem. Pap.*, 1, 1–18, 2022, https://doi.org/10.1007/S11696-022-02364-0.
82. Raj, V., Kim, Y., Kim, Y.G., Lee, J.H., Lee, J., Chitosan-gum arabic embedded alizarin nanocarriers inhibit biofilm formation of multispecies microorganisms. *Carbohydr. Polym.*, 284, 118959, 2022, https://doi.org/10.1016/J.CARBPOL.2021.118959.

83. Kulanthaivel, S., Agarwal, T., Sharan Rathnam, V.S., Pal, K., Banerjee, I., Cobalt doped nano-hydroxyapatite incorporated gum tragacanth-alginate beads as angiogenic-osteogenic cell encapsulation system for mesenchymal stem cell based bone tissue engineering. *Int. J. Biol. Macromol.*, 179, 101–115, 2021, https://doi.org/10.1016/J.IJBIOMAC.2021.02.136.
84. Kulanthaivel, S., Agarwal, T., Rathnam, V.S.S., Pal, K. and Banerjee, I., Cobalt doped nano-hydroxyapatite incorporated gum tragacanth-alginate beads as angiogenic-osteogenic cell encapsulation system for mesenchymal stem cell based bone tissue engineering, *Int. J. Biol. Macromol.*, 179, 101–115, 2021.
85. Singh, B. and Sharma, V., Influence of polymer network parameters of tragacanth gum-based pH responsive hydrogels on drug delivery. *Carbohydr. Polym.*, 101, 928–940, 2014, https://doi.org/10.1016/J.CARBPOL.2013.10.022.
86. Ranjbar-Mohammadi, M., Bahrami, S.H., Joghataei, M.T., Fabrication of novel nanofiber scaffolds from gum tragacanth/poly(vinyl alcohol) for wound dressing application: *In vitro* evaluation and antibacterial properties. *Mater. Sci. Eng.: C*, 33, 4935–4943, 2013, https://doi.org/10.1016/J.MSEC.2013.08.016.
87. Hemmati, K., Masoumi, A., Ghaemy, M., pH responsive tragacanth gum and poly(methyl methacrylate-co-maleic anhydride)-g-poly(caprolactone) conetwork microgel for *in vitro* quercetin release. *Polym. (Guildf)*, 59, 49–56, 2015, https://doi.org/10.1016/J.POLYMER.2014.12.050.
88. Singh, B. and Sharma, V., Crosslinking of poly(vinylpyrrolidone)/acrylic acid with tragacanth gum for hydrogels formation for use in drug delivery applications. *Carbohydr. Polym.*, 157, 185–195, 2017, https://doi.org/10.1016/J.CARBPOL.2016.09.086.
89. Gupta, M. and Chatterjee, S., Quantified differentiation of surface topography for nano-materials as-obtained from atomic force microscopy images. *J. Mater. Eng. Perform.*, 27, 2734–2740, 2018, https://doi.org/10.1007/S11665-018-3360-4/FIGURES/7.
90. Koyyada, A. and Orsu, P., Natural gum polysaccharides as efficient tissue engineering and drug delivery biopolymers. *J. Drug Deliv. Sci. Technol.*, 63, 102431, 2021, https://doi.org/10.1016/J.JDDST.2021.102431.
91. Putro, J.N., Lunardi, V.B., Soetaredjo, F.E., Yuliana, M., Santoso, S.P., Wenten, I.G., Ismadji, S., A review of gum hydrocolloid polyelectrolyte complexes (PEC) for biomedical applications: Their properties and drug delivery studies. *Processes*, 9, 1796, 2021, https://doi.org/10.3390/PR9101796.
92. Mahendhran, K. and Ramanathan, M., Biopolymer-based nanomaterials for biomedical applications: Biomedical applications of electrospun nanofibers, *Handbook of Research on Nano-Strategies for Combatting Antimicrobial Resistance and Cancer*, IGI Global Publishing, Pennsylvania, USA, pp. 29-55, 2021. pp. 29–55, https://doi.org/10.4018/978-1-7998-5049-6.CH002.

93. Nazarzadeh Zare, E., Makvandi, P., Tay, F.R., Recent progress in the industrial and biomedical applications of tragacanth gum: A review. *Carbohydr. Polym.*, 212, 450–467, 2019, https://doi.org/10.1016/J.CARBPOL.2019.02.076.
94. Ranjbar-Mohammadi, M., Prabhakaran, M.P., Bahrami, S.H., Ramakrishna, S., Gum tragacanth/poly(l-lactic acid) nanofibrous scaffolds for application in regeneration of peripheral nerve damage. *Carbohydr. Polym.*, 140, 104–112, 2016, https://doi.org/10.1016/J.CARBPOL.2015.12.012.
95. Demir, D., Uğurlu, M.A., Ceylan, S., Sakım, B., Genç, R., Bölgen, N., Assessment of chitosan: Gum tragacanth cryogels for tissue engineering applications. *Polym. Int.*, 71, 1109–1118, 2022, https://doi.org/10.1002/PI.6372.
96. Ghorbani, M., Ramezani, S., Rashidi, M.R., Fabrication of honey-loaded ethylcellulose/gum tragacanth nanofibers as an effective antibacterial wound dressing. *Colloids Surf. A Physicochem. Eng. Asp.*, 621, 126615, 2021, https://doi.org/10.1016/J.COLSURFA.2021.126615.
97. Abdoli, M., Sadrjavadi, K., Arkan, E., Zangeneh, M.M., Moradi, S., Zangeneh, A., Shahlaei, M., Khaledian, S., Polyvinyl alcohol/Gum tragacanth/graphene oxide composite nanofiber for antibiotic delivery. *J. Drug Deliv. Sci. Technol.*, 60, 102044, 2020, https://doi.org/10.1016/J.JDDST.2020.102044.
98. Electro-spun nanofiber patch based on gum tragacanth/polyvinyl alcohol/molybdenum disulfide composite for tetracycline delivery and their inhibitory effect on Gram+ and Gram– bacteria. *J. Mol. Liq.*, 334, 115989, 2021, https://doi.org/10.1016/J.MOLLIQ.2021.115989.
99. Ghorbani, M., Mahmoodzadeh, F., Yavari Maroufi, L., Nezhad-Mokhtari, P., Electrospun tetracycline hydrochloride loaded zein/gum tragacanth/poly lactic acid nanofibers for biomedical application. *Int. J. Biol. Macromol.*, 165, 1312–1322, 2020, https://doi.org/10.1016/J.IJBIOMAC.2020.09.225.
100. Annu, Ahmed, S., Nirala, R.K., Kumar, R., Ikram, S., Green synthesis of chitosan/nanosilver hybrid bionanocomposites with promising antimicrobial, antioxidant and anticervical cancer activity. *Polym. Polym. Compos.*, 29, 9_suppl, S199–S210, 2021.
101. Annu, Ali, A., Ahmed, S., Eco-friendly natural extract loaded antioxidative chitosan/polyvinyl alcohol based active films for food packaging. *Heliyon*, 7, 3, e06550, 2021.

8

Biomedical Applications of Cyclodextrin

Iftkhar Ahmad[1], Abu Taha[2], Annu[3*], Athar Adil Hashmi[2†] and Saiqa Ikram[1‡]

[1]*Bio/Polymer Research Laboratory, Department of Chemistry, Jamia Millia Islamia, New Delhi, India*

[2]*Bioinorganic Research Laboratory, Department of Chemistry, Jamia Millia Islamia, New Delhi, India*

[3]*Thin-Film and Materials Laboratory, School of Mechanical Engineering, Yeungnam University, Gyeongsan, Republic of Korea*

Abstract

Biopolymers are always preferred for making tools used in biomedical applications due to their biocompatibility, biodegradability, and ability to mimic the extracellular matrix of tissues, making them ideal for biomedical applications. Cyclodextrins, natural cyclic oligosaccharides with hydrophobic cavities and hydrophilic outer surfaces, are preferred for making biomedical tools due to their high solubility, biocompatibility, and non-toxicity. Their unique structure makes them suitable for dissolving hydrophobic molecules in water and biological systems, making them widely employed in biomedical applications such as drug formulation, gene delivery, tissue engineering, cancer treatment, medical device fabrication, biosensor, tissue engineering, and bio-imaging. Cyclodextrin has a well-established toxicological profile and is safe for human consumption in food and medicine. This chapter explores the various biomedical applications of cyclodextrins.

Keywords: Cross-linked, tissue engineering, drug delivery, cancer, polypyrrole, biomarker

Abbreviations

CD	Cyclodextrin
PPy	polypyrrole

**Corresponding author*: annuchem92@gmail.com

†*Corresponding author*: ahashmi@jmi.ac.in

‡*Corresponding author*: sikram@jmi.ac.in

Annu (ed.) Biopolymers for Biomedical Applications, (199–218) © 2024 Scrivener Publishing LLC

PEI polyethylenimine
RNA ribonucleic acid
siRNA short interference RNA
RNAi interference RNA
DA dopamine
DNA deoxyribonucleic acid
DMSO dimethyl sulphaoxide
AIBN azobisisobutyronitrile
AEG allyl glycidyl ether
FR folate receptor
PS photosensitizer

8.1 Introduction

Cyclodextrins (CDs) are naturally occurring cyclic oligosaccharides formed of glucopyranose units linked together by α-1,4-glycosidic bond.

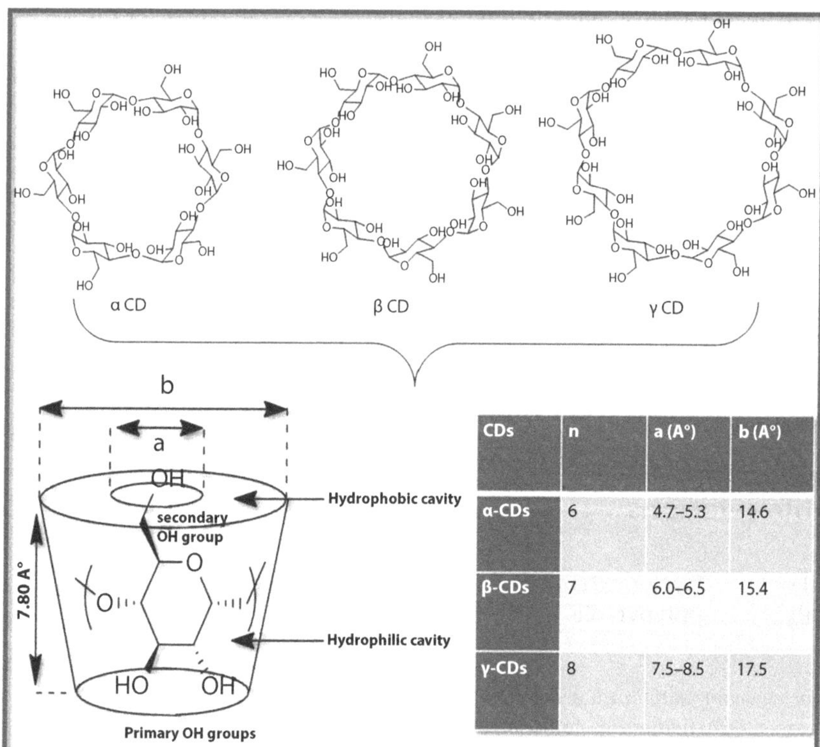

CDs	n	a (A°)	b (A°)
α-CDs	6	4.7–5.3	14.6
β-CDs	7	6.0–6.5	15.4
γ-CDs	8	7.5–8.5	17.5

Figure 8.1 Structure of α, β, and ϒ cyclodextrin.

The most commonly occurring CDs are alpha (α), beta (β), and gamma (γ). α, β, and γ-CD have six, seven, and eight glucopyranose units, respectively [1]. The outer periphery of CDs has a hydroxyl (OH) group, which makes it hydrophilic, while the inner cavity is lipophilic. This property allows CDs to make complexes in the water of hydrophobic molecules and moieties [2]. CDs are biocompatible, biodegradable, and natural oligosaccharides used in various biomedical technologies [3, 4]. CDs have been utilized as ligands in the creation of metal–organic frameworks, which have found numerous applications in areas such as drug delivery, catalysis, sensors, and adsorption [5, 6]. CDs have been widely used in industry and academia as functional compounds with molecular recognition because different CDs have different and specific cavity sizes [1]. In recent times, derivatives of CDs have been synthesized to achieve desired results such as improved interaction of guest with host, good solubility, and enhanced stability of CD conjugate [7]. Figure 8.1 shows the structure of various CDs.

8.2 Biomedical Applications of Cyclodextrin

CDs were used as an excipient, solubilizing, and stabilizing agent for many years. CDs are natural oligomers, so they are biocompatible and non-toxic. Due to these properties, CDs were explored as a scaffold material in drug delivery, gene delivery, biosensor, and wound healing. At the end of the 20th century, scientists began to explore the medicinal application of CDs in drug delivery, gene delivery, and wound healing [8]. CDs with other natural or synthetic polymers are used to make highly cross-linked polymers called hydrogel, which can absorb large amounts of water. CD hydrogels have proven useful in drug delivery and wound healing [9, 10]. CDs having a hydrophobic cavity is utilized to make an inclusion compound with hydrophobic molecules insoluble in water. Various hydrophobic drugs are delivered to a specific site using this property of CDs [11]. Other molecules can act as host molecules for drugs such as pillar arenes, crown ether, calixarenes, and cucurbiturils. CDs are widely used due to their non-toxicity, high-water solubility, facile modification, and abundant availability.

Moreover, the side effect of the drug is reduced due to the shielding effect of cyclodextrin on the drug [12]. Not only in drug delivery but folate-appended CDs are also used in delivery systems for DNA and short-interference RNA (siRNA) (gene delivery) [13]. Some researchers have also used CD-based biosensors, biomarkers, and antibacterial composite materials [14–16]. Research is still going on with CDs in biomedical applications,

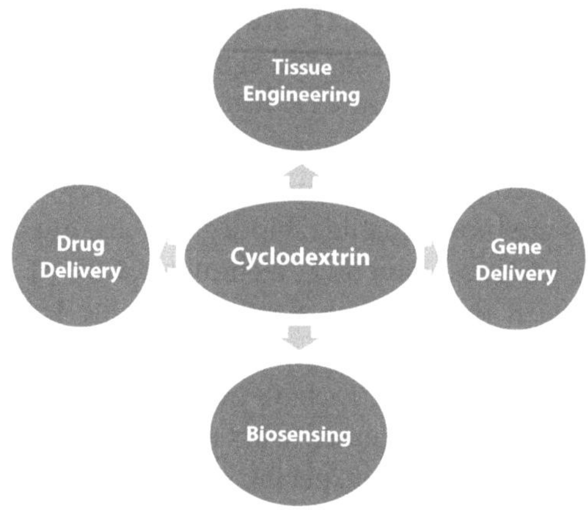

Figure 8.2 Biomedical applications of cyclodextrins.

mainly in drug delivery, gene delivery tissue engineering, and biosensors [11]. The biomedical applications of cyclodextrins is shown in Figure 8.2.

8.2.1 CDs as a Scaffold in Tissue Engineering

Tissue engineering is a very complex and important essential discipline that aims to develop a material that can act as a scaffold for tissue renewal and mimic natural support. The material which is extensively used to fabricate such scaffolds is polypyrrole (PPy). Pyrrole is a naturally occurring five-membered ring compound used to make PPy. PPy is extensively used as a scaffold material in tissue engineering due to its high stability, biocompatibility, easy preparation, structure tunability, and ability to adhere and proliferate different types of cells [17, 18]. It was also found that PPy shows a negative test for cell toxicity and hemolysis and no adverse effect on cell viability [19]. CDs are combined with other polymers to form a high cross-link network-like structure. These cross-link network structures with CDs are hydrophilic and are called hydrogels [20]. Hydrogels are semi-solid three-dimension-like structures dispersed in a large volume of liquid medium. Hydrogels with tuned pore structures are used in tissue engineering and drug delivery for the controlled release of different drug sizes at a particular site [21]. Some other natural polymers are also used with CDs to make cross-linked polymer chains for scaffolds in tissue engineering, like chitosan, polycaprolactone, and polyhydroxy butyrate [22–24]. In order to make a polymeric chain of CDs, different monomeric derivatives are used. Some examples of its monomeric structure are depicted in Figure 8.3.

Linear polymer is converted into cross-linked polymer hydrogel by chemical modification using a cross-linking agent. These cross-linked

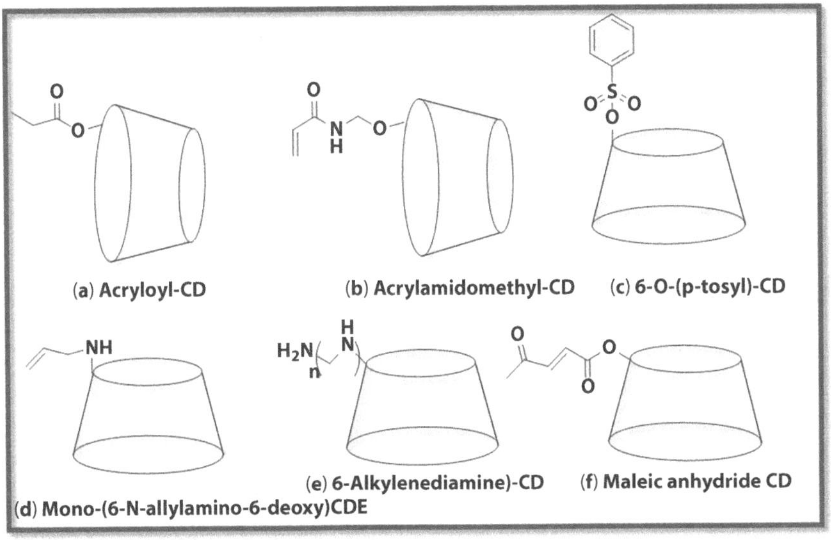

Figure 8.3 Structure of the monomeric derivative of cyclodextrins used for making a polymer chain.

polymeric materials are used extensively in tissue engineering as a template. Biodegradable nanogels have been synthesized by cross-linking between polylactic acid and vinyl cyclodextrin monomer with 1-allyloxy-2,3-epoxy propane, AGE, as shown in Figure 8.4 [25]. These are used as active filling materials for soft and hard tissue damage. The hydrogel may be loaded with other compounds, such as drugs, and growth hormones, which are released to their surroundings and help regenerate new cells and fill damaged tissue. The pH of the environment and degree of cross-linking also affect the scaffold material for drug release and their use in tissue engineering [22].

8.2.2 CDs in Drug Delivery

According to the US National Institutes of Health, drug delivery is defined as the "formulation of a device that enables the introduction of therapeutic substances in the body and improves efficiency and safety by controlling the rate, time, and place of drug release in the body" [26]. As mentioned, CDs have a hydrophilic outer surface and a hydrophobic inner core that enable them to form a stable inclusion compound of hydrophobic drug molecules in the water system. Natural CDs are more resistant to non-enzymatic hydrolysis than linear oligosaccharides. CDs are also

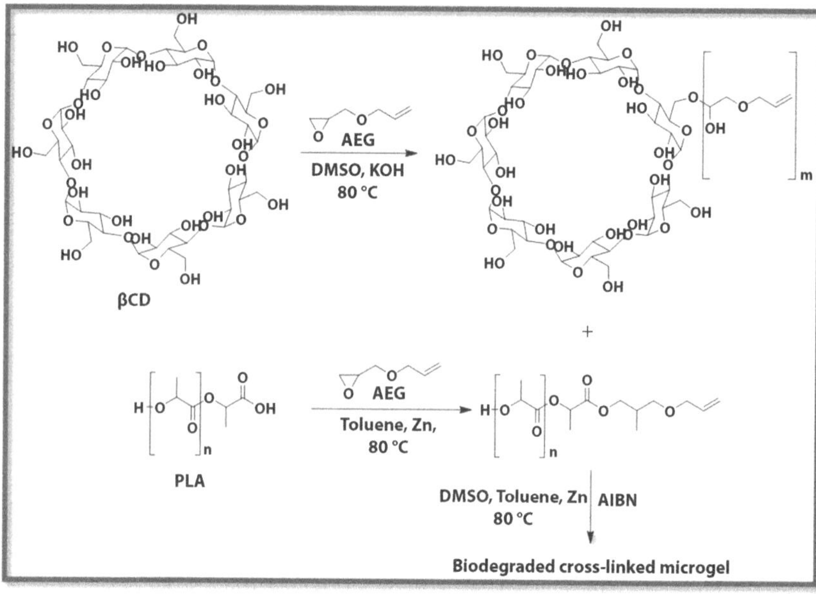

Figure 8.4 Synthesis of cross-linked CD polymer nanogel.

not digested by salivary and pancreatic amylase. Due to this property, the CD–drug conjugate remains stable in the upper gastrointestinal tract until it reaches the colon region. Many bacteria that are present in the colon region breaks down cyclodextrin and causes a rapid drug release. These properties of CDs make them a useful carrier for the colon region. Most anticancer drugs are poorly soluble in water, so CDs carry these drug molecules to the colon region to treat colonic cancer [27, 28].

Folate receptor (FR) is a type of transmembrane glycoprotein present in the cell. Generally, FR is overexpressed in tumor cells and not detected in normal cells. Folate can be used as a targeting ligand due to the high affinity between folate and its receptor. Thus, β-CD was modified as FR and made a drug conjugate to the delivery drug at the desired site. Like FR, biotin receptors are overexpressed on tumor cells compared to normal cells. The biotin–CD system was used as a carrier for targeted paclitaxel drug delivery. Some polypeptides are also found to be targeting ligands. Cyclic polypeptide containing arginine–glycine–aspartate works as a targeting ligand. They have a strong affinity with receptors present in cancer cells. Carboxymethyl-β-CD, polyamidoamine as dendrimer, and polyethyleneglycol as building blocks were used for the delivery of doxorubicin as shown in Figure 8.5 [12].

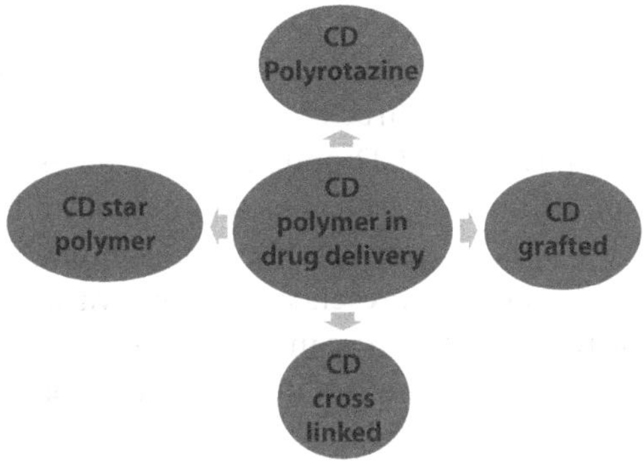

Figure 8.5 A different form of cyclodextrin polymer in drug delivery.

Currently, there are four classes of CD polymers for drug delivery:

(1) In this type of combination, CD-based polymers are combined with a polymer such that it does not escape from the polymer chain as both ends are locked with a large group.
(2) Grafted CD polymers: here CDs are patched on polymer or some other substance that works as a host for many substances.
(3) Cross-linked CD polymers area type of copolymer and CDs work as one of the monomers with some other monomer. The cross-linked polymer work as a hydrogel and is used in the healing of wounds and drug delivery.
(4) CD-based star polymers: polymer chains are attached to CDs and form multi-armed star-like structures. These CD polymers have diverse structures, qualities, applications, and unique superiority. Thus, they all have the potential to work as drug delivery agents [29].

The drug concentration peaks soon after it is taken and then reduces sharply, which is not suitable for the bio-accessibility of the drug. Thus, different drug delivery agents are used to carry the drug molecule and release it to a specific site in a controlled manner to check the bio-accessibility and maintain a balanced drug concentration. The advantages of CD polymers as drug carriers are outstanding. The pores of CDs in CD polymers are occupied with guest polymers and are not accessible for the drug to be included. However, the drugs can be directly attached to the CD polymers to form the polymer prodrug or physically enveloped by CD polymers.

The threaded CDs on CD polymers have unique effects on drug loading and release properties [26, 30]. Thus, CDs can be used for drug delivery in two formats: (1) the drugs are directly linked to CDs or (2) the drugs are embedded in pores formed by CD polymers. The pores in CD polymers and the fabrication method are critical factors for controlled drug release. The entangled CDs on CD polymer have outstanding effects on drug release to the specific site. The presence of many hydroxyl groups on the CD ring causes the release of the drug in a controlled manner. The mobility of the entangled CDs (sliding and rotation) is also another factor for the bioconjugation of CD polymers with the biological receptors. CDs, being a natural product, can degrade biologically. Some other physical factors, such as light, pH, and temperature, are also responsible for the controlled release of the drug to its specific site [31].

8.2.3 CDs in Cancer Treatment

Cancer is the second leading cause of death after cardiovascular disease worldwide [32]. About 277 types of cancer have been discovered; some of them are very common, like breast, lung, prostate, colon, or rectal cancer. Various kinds of cancer treatment depend on many parameters such as the stage of cancer, type of cancer, and age and health of the patient. Some common treatments for cancer are chemotherapy, surgery, and radiation therapy. The treatment of cancer may involve a single therapy or a combination of therapies depending upon the type and stage of cancer. Surgery is used to treat and diagnose many types of cancer in which the whole or some parts of the tumor are removed when cancer is localized. In radiation therapy, high-energy beams kill cancer cells or stop their growth, cell division, and proliferation. Chemotherapy uses an anticancer drug to kill or stop the growth of cancer cells. All three treatments have some adverse side effects and are not always practical for cancer treatment. Scientists have developed other cancer treatment therapies that have more effectiveness and a reduced side effect, such as hormone therapy, immunotherapy, gene therapy, cryotherapy, stem cell transplant, and thermal and photodynamic therapy [33].

Photodynamic therapy (PDT) uses three component combinations for cancer treatment: (1) a photoactive molecule or photosensitizer (PS), (2) a light source generally in the visible range, and (3) naturally occurring oxygen molecule in the triplet state ($^3O_2$). PDT cancer treatment involves the delivery of PS to the cancer patient, followed by the excitation of PS using visible light. The excited PS reacts with triplet molecular oxygen, producing reactive oxygen species, including singlet oxygen molecule ($^1O_2$), responsible for cancer cell damage [33]. The PSs are divided into four categories (first-, second-, third-, and fourth-generation PSs) based on their chronological development. There are various PSs, but most of them are from porphyrinoid groups.

Among the problems faced by many PSs are their low solubility in water and tissue fluid, causing low bioavailability and reduced delivery efficiency of PSs. CDs are used to enhance the solubility and delivery of PS in cancer tissue [34].

Kolarova *et al.* explored the phototoxicity of two PSs: meso-tetrakis(4-sulphonatophenyl)porphyrine (TPPS4) and its zinc complex (ZnTPPS4) on cancer cells in the presence and absence of CD derivatives. Their experiment found that both PSs, when conjugated with CD, showed lower cytotoxicity and higher phototoxicity than PSs alone [35]. Paul and his co-worker studied the conjugation of chlorin e6 (Ce6) into the 2-hydroxypropyl-β-cyclodextrin (HP-β-CD) cavity. The researchers showed that the conjugation of Ce6 with HP-β-CD increased the Ce6 solubility, decreased its aggregation in water, and enhanced its singlet oxygen molecule ($^1O_2$) yield at the pH tumor environment compared to the free Ce6 [36]. Yankovsky *et al.* also studied two modified forms of β-CD, i.e., methyl-β-cyclodextrin (Me-β-CD) and HP-β-CD on the distribution of meta-tetra(hydroxyphenyl)chlorin (mTHPC). The author found that β-CD prevents the aggregation of mTHPC after ingestion into the blood and enhances diffusion through the biological system. It was also found that the addition of β-CD affects the distribution of mTHPC and enhances the photocytotoxicity of effect toward cancer cells [37].

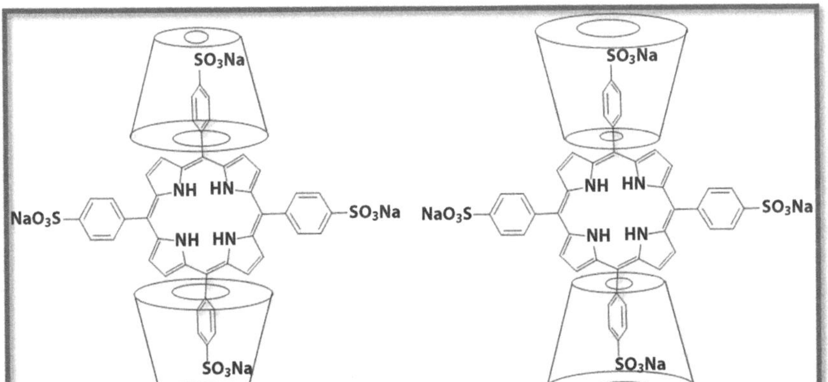

Figure 8.6 Cyclodextrin (CD)–PS inclusion complexes between β-CD and γ-CD with a porphyrinoid.

The CD inclusion complex between CD and porphyrin is synthesized by encapsulation of the aryl substituent of PS into a large cavity of CD. The cavity of α-CD is tiny; hence inclusion compounds of α-CD are not found, whereas β- and γ-CD generate inclusion complexes [38]. Figure 8.6 shows the typical structure of the inclusion complex between β-CD and γ-CD with porphyrin [39].

Králová *et al.* synthesized two inclusion complexes of perfluorinated porphyrin (P) with one ($P(\beta\text{-CD})_1$) and two ($P(\beta\text{-CD})_2$) β-CD units, as shown in Figure 8.7. The *in vitro* PDT effect of these two complexes was performed on different cell lines, and the inclusion complex $P(\beta\text{-CD})_1$ was more efficient than $P(\beta\text{-CD})_2$. The *in vivo* studies showed that $P(\beta\text{-CD})_2$ was most efficient and totally inhibited tumor growth [40].

Ikeda *et al.* described the effect of inclusion complexes 1–6 of trimethyl-β-CD (TMe-β-CD), as shown in Figure 8.8. It is formed in a 2:1 ratio, and their PDT effect was studied. Complexes 2–4 were evaluated for their PDT effect due to their photostability and ability to generate $^1O_2$. The

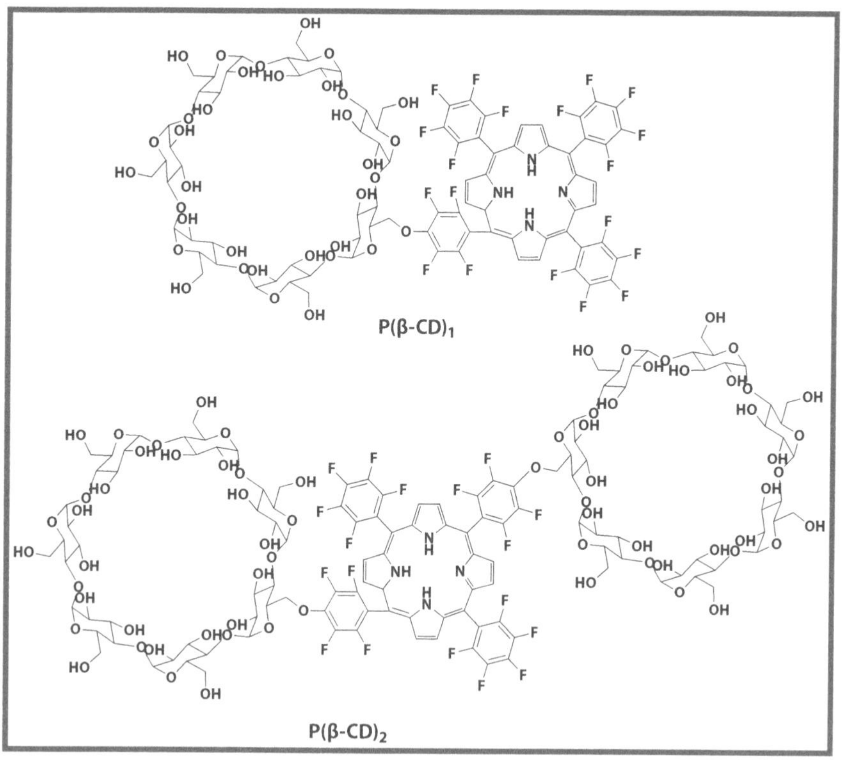

Figure 8.7 Structure of P(β-CD)1 and P(β-CD)2.

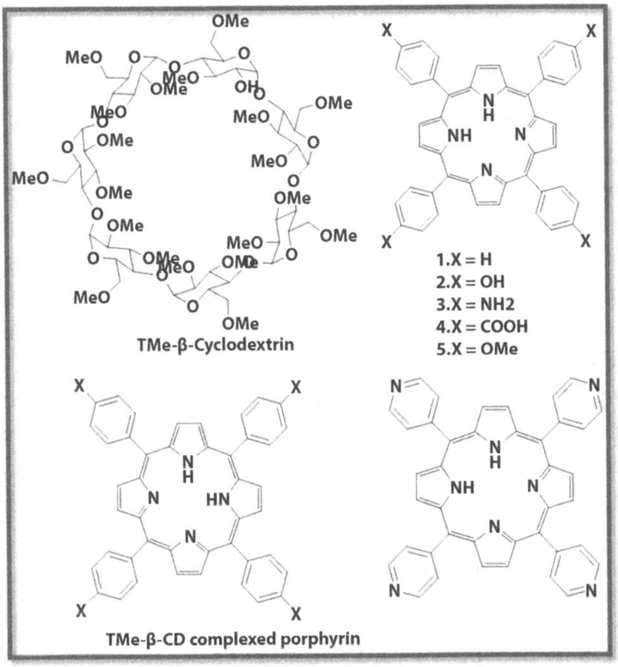

Figure 8.8 Chemical structure and schematic design of the trimethyl-β-CD (TMe-β-CD) complexed with porphyrin derivatives.

inclusion complexes generated $^1O_2$ in the order 3 > 2 ≈ 4. PDT activity was found in the order 3 > 2 ≫ 4, which suggests that PDT activity was due to the higher uptake of inclusion complexes 2 and 3 than $^1O_2$ formation [41].

8.2.4 CDs in Gene Delivery

Traditionally, CD polymer properties were utilized to deliver low molecular weight proteins, peptides, and drug molecules. Later on, CD was explored to deliver oligonucleotide, and it was found that CD enhanced the absorption and resistance to nuclease due to binding and interaction [42]. Initially, cationic polymer polyethyleneimine (PEI) was used for gene delivery. The use of CD for gene delivery was started in 1999. The linear cationic polymer containing CDs has been used in gene delivery, and Davis and co-workers reported the first cationic polymer containing β-CD for gene delivery. They developed such polymer with modified CD and studied their effect on gene delivery [43].

The cationic polymer containing CDs shows lower cytotoxicity. The most crucial feature of a cationic polymer containing CD is that the polyplex formed between the polymer and DNA can be further modified as the polymer contains many CD moieties. A cationic polymer such as polyethyleneimine and polyamidoamine was modified by grafting with CD on polymer and explored for gene delivery. All CD-containing polymers showed reduced cytotoxicity [44]. Gene therapy is used to transfer genetic material into the cell. Recently, interference RNA has been explored in cancer treatment by silencing genes causing uncontrolled cell growth. Various polymers containing cationic polymers, i.e., PEI, and polysaccharides, i.e., CDs and chitosan, have been explored for siRNA distribution. CDs were used as DNA plasmid carrier in 1999, and after decades, it is still considered an appropriate and potential gene carrier for cancer treatment [45]. CDs are considered a favorable carrier among various carriers of genetic material due to their biocompatible nature, hydrophilic surface, hydrophobic cavity, and a hydroxyl functional group on the CD ring that provides a modification of the functional group [39].

8.2.5 CDs in Biosensor and Biomarker

Biosensors are a tool used in the health sector, designed for fast detection, analysis, and diagnosis. Biosensors contain biomolecules or synthetic materials with cross unique sensing capacity for the analyte. The first biosensor used in the health sector for glucose was made up of polyacrylamide gel. Since then, the polymer has been explored to design biosensors as a polymer that can be easily modified and tuned for different types of receptors [46]. CDs, natural oligomers, got lots of attention in designing biosensors due to their unique ability to form inclusion compounds with guest molecules. Additionally, CDs are non-toxic, cost-effective, biodegradable, and functionalized easily. The amphipathic nature of CDs with a hydrophobic cavity can easily host many hydrophobic molecules of appropriate sizes and suitable functional groups [47, 48].

Dopamine (DA) is a kind of neurotransmitter whose concentration in the blood plays an essential role in the proper functioning of the central nervous, cardiovascular, and endocrine systems. Low levels of DA cause disorders such as dementia and schizophrenia. Hence, it became necessary to detect DA for diagnostic applications. DA belongs to o-diphenol, and its oxidation is carried out by an enzyme called laccase. In order to determine DA, an enzyme-based biosensor is used to detect DA. Due to robust host–guest interactions, β-cyclodextrin has been used to manufacture various

DA sensors. Harley *et al.* proposed a PPy film doped with sulfonated β-CD for the selective determination of DA. Tan *et al.* have prepared a nanocomposite of β-CD and graphene sheet for the nanomolar detection of DA in the presence of ascorbic acid [49].

Sulfur-containing agrochemicals and amino acids are also detected using functionalized CD polymer-derived gold nanoprobe. β-Cyclodextrin, along with sodium dodecyl sulphate, is used to detect cholesterol. β-Cyclodextrin capped with the gold particle is used as a catechol biosensor and also to inhibit ibuprofen. β-Cyclodextrin is used as a biosensor for various biomolecules such as DA, serotonin, ascorbic acid, and uric acid [50].

8.2.6 Cyclodextrin-Based Stimulus-Responsive Biomaterial Systems

Traditional methods of administering drugs result in unpredictable plasma concentrations at the intended location, causing fluctuations that can lead to toxicity and related adverse reactions following the drug administration. A superlative medication transport mechanism preserves the requisite therapeutic level in the aimed tissue, with negligible or zero exposure to the systemic or non-targeted tissues, thus circumventing any undesirable secondary outcomes. Novel smart polymers featuring sophisticated responsive drug delivery systems have been engineered to facilitate controlled drug release *via* the activation of environmental cues, encompassing physical stimuli such as temperature, light, electrical impulses, magnetic fields, and ultrasonic waves as well as chemical stimuli including pH, ionic strength, and redox potential in addition to biological triggers such as enzymes, inflammatory signals, and biomolecular markers [51]. The inclusion chemistry of cyclodextrins and the resulting supramolecular assemblies serve the fundamental purpose of augmenting the solubility and bioavailability of drugs. By leveraging their hydrophobic cavity and ability to form inclusion complexes with guest molecules, these assemblies function as carriers that enhance the therapeutic profile of drugs [52]. Furthermore, the utilization of supramolecular assemblies comprising cyclodextrins and polymers that are responsive to external stimuli yields intelligent and innovative systems, as illustrated in Figure 8.9. In this context, the association between host and guest molecules for drug delivery can be activated *via* externally regulated stimuli, affording the benefits of drug entrapment, precise and localized release, and improved physicochemical and biopharmaceutical properties [47].

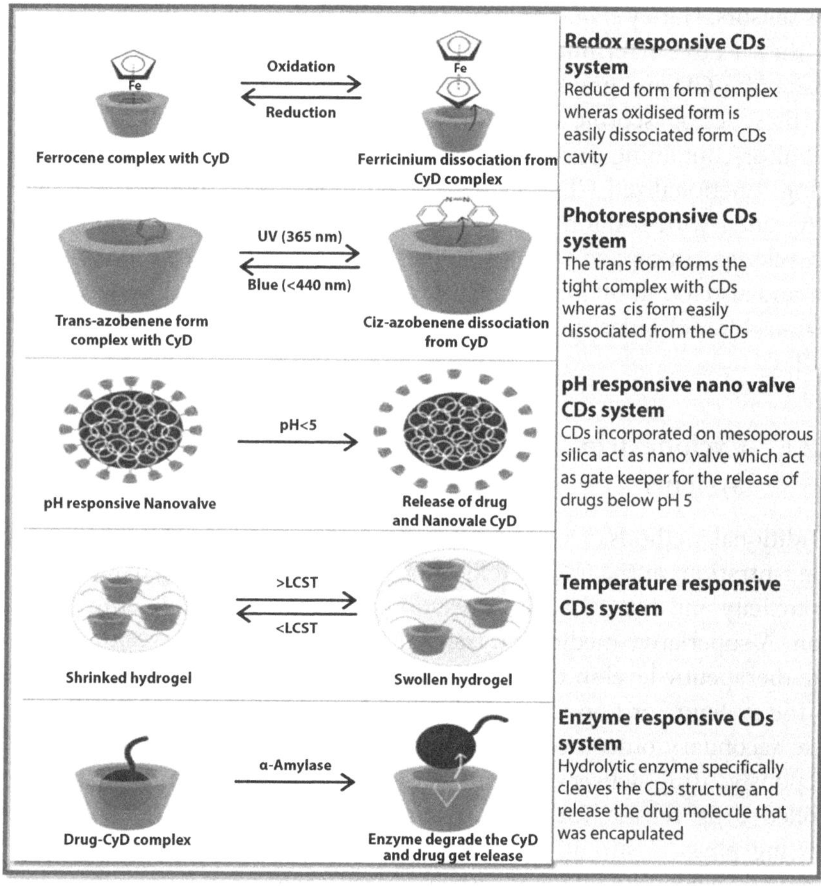

Figure 8.9 Cyclodextrin-based supramolecular inclusion complexes are induced by various stimuli and exhibit guest molecule discharging. Reproduced with permission from Wiley [47].

8.3 Future Prospects

Various chitosan aspects of CDs, such as tissue engineering, drug delivery, gene delivery, and biosensor, are discussed in this chapter. CDs show a biocompatible nature and are used as a green polymer to prepare different tools for biomedical applications. CDs were mostly used with some other polymer or nanomaterials for making these tools. In the coming times, we may see a more precise tool for drug delivery, more bio-mimicking scaffold material, and sensing capacity of the biomolecule in very trace amounts. β-CD is most explored among CDs; we hope to see α-CD and γ-CD for various biomedical applications in future. CDs are not used alone for biomedical application but as a combination of a CD oligomer with other polymers and nanomaterials by doing some chemical modifications. In the coming time, we will see a simple preparation method for these biomaterials of CDs. There are immense possibilities for the chemical modification of the hydroxyl group on CD rings, matching our need for coming applications and challenges. There is still much work to do to make valuable tools for biomedical applications with more precision and a higher degree of effectiveness. As the demand for biocompatible and biodegradable materials is increasing daily, we will see extensive the use of CDs in various biomedical applications.

8.4 Conclusion

Cyclodextrins have proven to be a highly versatile and valuable material in biomedical applications due to their biocompatibility, biodegradability, and unique structural properties. They can be chemically modified to tune their properties according to specific biomedical applications. The use of cyclodextrins with other natural or synthetic polymers and nanomaterials has further expanded their applications in tissue engineering, drug delivery, and gene therapy. Although β-CD is the most widely studied among the various cyclodextrins, research on α-CD and γ-CD is also gaining momentum. However, further research is required to explore the full potential of cyclodextrins in various biomedical applications, including their toxicity profile, pharmacokinetics, and long-term biodegradability. With the increasing demand for biocompatible and biodegradable materials in biomedical applications, cyclodextrins are likely to play a significant role in the future of regenerative medicine, drug delivery, and biomedical device development.

References

1. Shahiwala, A., Cyclodextrin conjugates for colon drug delivery. *J. Drug Deliv. Sci. Technol.*, 55, 101448, Feb. 2020, doi: 10.1016/j.jddst.2019.101448.
2. Choi, J.H. *et al.*, Preparation and characterization of an injectable dexamethasone-cyclodextrin complexes-loaded gellan gum hydrogel for cartilage tissue engineering. *J. Control. Release*, 327, 747–765, Nov. 2020, doi: 10.1016/j.jconrel.2020.08.049.

3. Xu, W. *et al.*, Design of cyclodextrin-based functional systems for biomedical applications. *Front. Chem.*, 9, 2021, [Online]. Available: https://www.frontiersin.org/articles/10.3389/fchem.2021.635507.
4. Chen, Y., Mensah, A., Wang, Q., Li, D., Qiu, Y., Wei, Q., Hierarchical porous nanofibers containing thymol/beta-cyclodextrin: Physico-chemical characterization and potential biomedical applications. *Mater. Sci. Eng. C*, 115, 111155, Oct. 2020, doi: 10.1016/j.msec.2020.111155.
5. Shahzaib, A. *et al.*, Ultrarapid and highly efficient reduction of nitroaromatic compounds using cyclodextrin MOF. *Catal. Commun.*, 174, 106569, Nov. 2022, doi: 10.1016/j.catcom.2022.106569.
6. Shahzaib, A. *et al.*, One pot synthesis of cyclodextrin MOF as a promising heterogeneous catalyst for the reduction of nitroaromatic compounds and azo dyes. *Res. Chem. Intermed.*, 49, 2507–2528, 2023, doi: 10.1007/s11164-023-04986-9.
7. Li, T., Guo, R., Zong, Q., Ling, G., Application of molecular docking in elaborating molecular mechanisms and interactions of supramolecular cyclodextrin. *Carbohydr. Polym.*, 276, 118644, Jan. 2022, doi: 10.1016/j.carbpol.2021.118644.
8. Moriya, T. *et al.*, Potent inhibitory effect of a series of modified cyclodextrin sulfates on the replication of HIV-1 *in vitro*. *J. Med. Chem.*, 34, 7, 2301–2304, Jul. 1991, doi: 10.1021/jm00111a055.
9. Escobar, K., Garrido-Miranda, K.A., Pulido, R., Naveas, N., Manso-Silván, M., Hernandez-Montelongo, J., Coatings of cyclodextrin/citric-acid biopolymer as drug delivery systems: A review. *Pharmaceutics*, 15, 1, 2023, doi: 10.3390/pharmaceutics15010296.
10. He, F. *et al.*, Highly stretchable and tough alginate-based cyclodextrin/Azo-polyacrylamide interpenetrating network hydrogel with self-healing properties. *Carbohydr. Polym.*, 296, 256, 117595, Mar. 2021, doi: 10.1016/j.carbpol.2020.117595.
11. Mohandoss, S. *et al.*, Ultrasonic-assisted efficient synthesis of inclusion complexes of salsalate drug and β-cyclodextrin derivatives for potent biomedical applications. *J. Mol. Liq.*, 319, 114358, Dec. 2020, doi: 10.1016/j.molliq.2020.114358.
12. Zhang, D., Lv, P., Zhou, C., Zhao, Y., Liao, X., Yang, B., Cyclodextrin-based delivery systems for cancer treatment. *Mater. Sci. Eng. C*, 96, 872–886, Mar. 2019, doi: 10.1016/j.msec.2018.11.031.
13. Ceborska, M., Folate appended cyclodextrins for drug, DNA, and siRNA delivery. *Eur. J. Pharm. Biopharm.*, 120, 133–145, Nov. 2017, doi: 10.1016/j.ejpb.2017.09.005.
14. Kapan, B., Kurbanoglu, S., Esenturk, E.N., Soylemez, S., Toppare, L., Electrochemical catechol biosensor based on β-cyclodextrin capped gold nanoparticles and inhibition effect of ibuprofen. *Process Biochem.*, 108, 80–89, Sep. 2021, doi: 10.1016/j.procbio.2021.06.004.

15. Shrivastava, S., Bairagi, P.K., Verma, N., Spermine biomarker of cancerous cells voltammetrically detected on a poly(β-cyclodextrin) - Electropolymerized carbon film dispersed with Cu-CNFs. *Sens. Actuators B Chem.*, 313, 128055, Jun. 2020, doi: 10.1016/j.snb.2020.128055.
16. Sharif, N., Golmakani, M.T., Hajjari, M.M., Aghaee, E., Ghasemi, J.B., Antibacterial cuminaldehyde/hydroxypropyl-β-cyclodextrin inclusion complex electrospun fibers mat: Fabrication and characterization. *Food Packag. Shelf Life*, 29, 100738, 2021, doi: https://doi.org/10.1016/j.fpsl.2021.100738.
17. Huang, Z.-B., Yin, G.-F., Liao, X.-M., Gu, J.-W., Conducting polypyrrole in tissue engineering applications. *Front. Mater. Sci.*, 8, 1, 39–45, Mar. 2014, doi: 10.1007/s11706-014-0238-8.
18. Guo, B. and Ma, P.X., Conducting polymers for tissue engineering. *Biomacromolecules*, 19, 6, 1764–1782, Jun. 2018, doi: 10.1021/acs.biomac.8b00276.
19. Wang, X. *et al.*, Evaluation of biocompatibility of polypyrrole *in vitro* and *in vivo*. *J. Biomed. Mater. Res.*, 68A, 3, 411–422, Mar. 2004, doi: 10.1002/jbm.a.20065.
20. Lukášek, J. *et al.*, Cyclodextrin-polypyrrole coatings of scaffolds for tissue engineering. *Polymers*, 11, 3, 459, Mar. 2019, doi: 10.3390/polym11030459.
21. Moya-Ortega, M.D., Alvarez-Lorenzo, C., Concheiro, A., Loftsson, T., Cyclodextrin-based nanogels for pharmaceutical and biomedical applications. *Int. J. Pharm.*, 428, 1–2, 152–163, May 2012, doi: 10.1016/j.ijpharm.2012.02.038.
22. Prabaharan, M. and Jayakumar, R., Chitosan-graft-β-cyclodextrin scaffolds with controlled drug release capability for tissue engineering applications. *Int. J. Biol. Macromol.*, 44, 4, 320–325, May 2009, doi: 10.1016/j.ijbiomac.2009.01.005.
23. Lee, J. *et al.*, Development of poly(ε-caprolactone) scaffold loaded with simvastatin and beta-cyclodextrin modified hydroxyapatite inclusion complex for bone tissue engineering. *Polymers*, 8, 2, 49, Feb. 2016, doi: 10.3390/polym8020049.
24. Vogel, R., Tändler, B., Häussler, L., Jehnichen, D., Brünig, H., Melt spinning of poly(3-hydroxybutyrate) fibers for tissue engineering using α-cyclodextrin/polymer inclusion complexes as the nucleation agent. *Macromol. Biosci.*, 6, 9, 730–736, Sep. 2006, doi: 10.1002/mabi.200600116.
25. Lu, D., Yang, L., Zhou, T., Lei, Z., Synthesis, characterization and properties of biodegradable polylactic acid-β-cyclodextrin cross-linked copolymer microgels. *Eur. Polym. J.*, 7, 44, 2140–2145, 2008, doi: 10.1016/j.eurpolymj.2008.04.014.
26. Chavda, V.P., Nanobased nano drug delivery, in: *Applications of Targeted Nano Drugs and Delivery Systems*, pp. 69–92, Elsevier, Amsterdam Netherlands, 2019, doi: 10.1016/B978-0-12-814029-1.00004-1.

27. Uekama, K., Design and evaluation of cyclodextrin-based drug formulation. *Chem. Pharm. Bull. (Tokyo)*, 52, 8, 900–915, 2004, doi: 10.1248/cpb.52.900.
28. Uekama, K., Hirayama, F., Arima, H., Recent aspect of cyclodextrin-based drug delivery system. *J. Incl. Phenom. Macrocycl. Chem.*, 56, 1, 3–8, Oct. 2006, doi: 10.1007/s10847-006-9052-y.
29. Liu, Z., Ye, L., Xi, J., Wang, J., Feng, Z., Cyclodextrin polymers: Structure, synthesis, and use as drug carriers. *Prog. Polym. Sci.*, 118, 101408, Jul. 2021, doi: 10.1016/j.progpolymsci.2021.101408.
30. Donoso-González, O. *et al.*, Functionalization of gold nanostars with cationic β-cyclodextrin-based polymer for drug co-loading and SERS monitoring. *Pharmaceutics*, 13, 2, 2021, doi: 10.3390/pharmaceutics13020261.
31. Duan, Q., Shi, J., Zhou, L., Zhang, B., Wang, X., Sang, S., pH-responsive and sustained release drug delivery system of BSA coated CDs-DOX. *J. Mol. Struct.*, 1248, 131358, Jan. 2022, doi: 10.1016/j.molstruc.2021.131358.
32. Naghavi, M. *et al.*, Global, regional, and national age-sex specific mortality for 264 causes of death, 1980–2016: A systematic analysis for the Global Burden of Disease Study 2016. *Lancet*, 390, 10100, 1151–1210, Sep. 2017, doi: 10.1016/S0140-6736(17)32152-9.
33. Macdonald, I.J. and Dougherty, T.J., Basic principles of photodynamic therapy. *J. Porphyr. Phthalocyanines*, 05, 02, 105–129, Feb. 2001, doi: 10.1002/jpp.328.
34. Tiwari, G., Tiwari, R., Rai, A.K., Cyclodextrins in delivery systems: Applications. *J. Pharm. Bioallied Sci.*, 2, 2, 72, Apr. 2010, doi: 10.4103/0975-7406.67003.
35. Kolárová, H., Mosinger, J., Lenobel, R., Kejlová, K., Jírová, D., Strnad, M., *In vitro* toxicity testing of supramolecular sensitizers for photodynamic therapy. *Toxicol. In Vitro*, 17, 5–6, 775–778, Oct. 2003, doi: 10.1016/S0887-2333(03)00094-8.
36. Paul, S., Heng, P.W.S., Chan, L.W., pH-dependent complexation of hydroxypropyl-beta-cyclodextrin with chlorin e6: Effect on solubility and aggregation in relation to photodynamic efficacy. *J. Pharm. Pharmacol.*, 68, 4, 439–449, Apr. 2016, doi: 10.1111/jphp.12535.
37. Yankovsky, I. *et al.*, Inclusion complexation with β-cyclodextrin derivatives alters photodynamic activity and biodistribution of meta-tetra(hydroxyphenyl)chlorin. *Eur. J. Pharm. Sci.*, 91, 172–182, Aug. 2016, doi: 10.1016/j.ejps.2016.06.012.
38. Moradi, S., Barati, A., Salehi, E., Tonelli, A.E., Hamedi, H., Preparation and characterization of chitosan based hydrogels containing cyclodextrin inclusion compounds or nanoemulsions of thyme oil. *Polym. Int.*, 68, 11, 1891–1902, Nov. 2019, doi: 10.1002/pi.5899.
39. Singh, R.P. *et al.*, Self-assembled cationic β-cyclodextrin nanostructures for siRNA delivery. *Mol. Pharm.*, 16, 3, 1358–1366, Mar. 2019, doi: 10.1021/acs.molpharmaceut.8b01307.

40. Kralova, J., Synytsya, A., Pouckova, P., Koc, M., Dvorak, M., Kral, V., Novel porphyrin conjugates with a potent photodynamic antitumor effect: Differential efficacy of mono- and bis-beta-cyclodextrin derivatives *in vitro* and *in vivo*. *Photochem. Photobiol.*, 82, 2, 432–438, Apr. 2006, doi: 10.1562/2005-05-06-RA-516.
41. Ikeda, A., Satake, S., Mae, T., Ueda, M., Sugikawa, K., Shigeto, H., Funabashi, H. and Akio Kuroda, A., Photodynamic activities of porphyrin derivative–cyclodextrin complexes by photoirradiation. *ACS Med. Chem. Lett.*, 8, 555–559, 2017.
42. Redenti, E., Pietra, C., Gerloczy, A., Szente, L., Cyclodextrins in oligonucleotide delivery. *Adv. Drug Deliv. Rev.*, 53, 2, 235–244, Dec. 2001, doi: 10.1016/S0169-409X(01)00230-7.
43. Gonzalez, H., Hwang, S.J., Davis, M.E., New class of polymers for the delivery of macromolecular therapeutics. *Bioconjug. Chem.*, 10, 6, 1068–1074, Nov. 1999, doi: 10.1021/bc990072j.
44. Arima, H., Kihara, F., Hirayama, F., Uekama, K., Enhancement of gene expression by polyamidoamine dendrimer conjugates with α-, β-, and γ-cyclodextrins. *Bioconjug. Chem.*, 12, 4, 476–484, Jul. 2001, doi: 10.1021/bc000111n.
45. Mousazadeh, H., Pilehvar-Soltanahmadi, Y., Dadashpour, M., Zarghami, N., Cyclodextrin based natural nanostructured carbohydrate polymers as effective non-viral siRNA delivery systems for cancer gene therapy. *J. Control. Release*, 330, 1046–1070, Feb. 2021, doi: 10.1016/j.jconrel.2020.11.011.
46. Holzinger, M., Buzzetti, P.H.M., Cosnier, S., Polymers and nano-objects, a rational combination for developing health monitoring biosensors. *Sens. Actuators B Chem.*, 348, 130700, Dec. 2021, doi: 10.1016/j.snb.2021.130700.
47. Wankar, J., Kotla, N.G., Gera, S., Rasala, S., Pandit, A., Rochev, Y.A., Recent advances in host–guest self-assembled cyclodextrin carriers: Implications for responsive drug delivery and biomedical engineering. *Adv. Funct. Mater.*, 30, 44, 1909049, Oct. 2020, doi: 10.1002/adfm.201909049.
48. Monti, P. *et al.*, Use of β-cyclodextrin as enhancer of ascorbic acid rejection in permselective films for amperometric biosensor applications. *Talanta*, 186, 53–59, Aug. 2018, doi: 10.1016/j.talanta.2018.04.034.
49. Hua, Z., Qin, Q., Bai, X., Wang, C., Huang, X., β-Cyclodextrin inclusion complex as the immobilization matrix for laccase in the fabrication of a biosensor for dopamine determination. *Sens. Actuators B Chem.*, 220, 1169–1177, Dec. 2015, doi: 10.1016/j.snb.2015.06.108.
50. Bae, J. *et al.*, A succinct review of refined chemical sensor systems based on conducting polymer–cyclodextrin hybrids. *J. Ind. Eng. Chem.*, 79, 19–28, Nov. 2019, doi: 10.1016/j.jiec.2019.06.051.
51. Yuan, Y., Nie, T., Fang, Y., You, X., Huang, H., Wu, J., Stimuli-responsive cyclodextrin-based supramolecular assemblies as drug carriers. *J. Mater. Chem. B*, 10, 13, 2077–2096, 2022, doi: 10.1039/D1TB02683F.

52. Agnes, M., Pancani, E., Malanga, M., Fenyvesi, E., Manet, I., Implementation of water-soluble cyclodextrin-based polymers in biomedical applications: How far are we? *Macromol. Biosci.*, 22, 8, 2200090, Aug. 2022, doi: 10.1002/mabi.202200090.

9

Biomedical Applications of Dextran

Sonia Sharma and Jagadeesh Chandra Bose K*

Chandigarh University, Gharuan, Mohali, Punjab, India

Abstract

Dextran is a complex polysaccharide, generally the extract of bacteria, having a broad range of biomedical applications due to its the chemical structure which can be easily modified and excellent stable nature. It can enhance the property of any biomaterial with which it is made complex. Various fabricated Dex-QDs are studied for the application of bioanalysis and imaging. Dex and its derivatives in the form of nanoparticles can do successful insulin delivery in the glucose-rich medium. Antimicrobial and anticancer properties are shown by Dex-hesperetin/hesperidin conjugates. It is proven to be effective against stomach and breast cancer cell lines. Due to its high stability, doctors are using it as an anesthetic adjuvant that can be retained in the injected body part for more than 24 h to give long-term relief to the patient during and after surgery with less medication. Dextran, in the form of dextran/chitosan hydrogels and Au@DEX/PVA nanoparticles, shows its effectiveness in healing wounds. Its anti-adhesion ability has also been proven by analyzing sheep after spinal surgery. It does the detection of Hg^{2+} ions with high specificity and stability. Damage to the patient after myocardial infarction can be recovered by the cell delivery of stem cells in the patient's heart with the help of CS/Dex/β-GP hydrogels. All these biomedical applications will be discussed in detail in this chapter.

***Keywords*:** Hydrogels, dextran, nanocomposite, wound, toxicity

9.1 Introduction

Synthesis of dextran can be done using sucrose (substrate) by the lactic acid bacteria of different genera such as *Pediococcus*, *Weissella*, *Leuconostoc*, *Lactobacillus*, and *Streptococcus* [1, 2]. The number of branches and chain

**Corresponding author*: jcboseuibtlab@gmail.com

Annu (ed.) Biopolymers for Biomedical Applications, (219–252) © 2024 Scrivener Publishing LLC

styles completely depends upon the nutrition provided, fermentation technique, and the strains being used to synthesize [3]. Its polymerization, monosaccharide composition, and glycosidic linkage will determine its biological activities [4]. There are many benefits of dietary fibers, such as promotion of the growth of beneficial gut microbiota, reduction in the risk of colorectal cancer, prevention of constipation, lowering of blood cholesterol, reduction in bowel transit time, and production of short-chain fatty acids. Behind every benefit, probiotics are responsible, which provide the best intestinal environment to the host because probiotics promote health benefits by stimulating probiotic microorganisms. The animal body is not capable of digesting dietary fibers such as polysaccharides, oligosaccharides, and their derivatives, whereas the digestive enzymes of the animal body do not attack dextran which has the backbone α-(1→6) glycosidic. Dextran with prebiotic properties is produced by *Lactobacillus plantarum*, *Weissella confuses*, *Weissella cibaria*, *Pediococcus pentosaceus*, and many other marine lactic-producing bacteria [5–8]. Dextran does not get hydrolyzed by pancreatic amylase, a stomach acidic medium. Furthermore, dextran is an important source of carbon for *Bifidobacterium bifidum* [9]. Production of sourdough is done by the fermentation of yeast and LAB. Dextran, which is extracted from different microbial sources, is being used in the baking of pre-dough, i.e., sourdough [10]. If we talk about food items, dextrans show a good impact on the quality of bread due to its high molecular weight and less branched structure, which give moisture as well as softness to the bread and also increase the volume of the bread by 12% [11]. Different techniques such as NMR and FTIR are performed for the characterization of dextran. Many types of research have been done to find out more and more prebiotic applications. It has been found that it shows versatility in different techniques such as imaging and bioanalysis. The development of dextran ligands has been successfully done, which are used in the proper functioning of colloidal semiconductor quantum dots. Dextran increases the aggregation and adhesive properties of bacterial strains. It has a wide range of biomedical significance which are evident through different research such as drug delivery with dextran to treat many diseases and enhancement of the antibacterial and anticancer properties of the compound with which it makes complex. Specific detection of mercury ions can be done with the help of highly stable dextran, increasing the efficiency of anesthesia by improving its retention period in body, wound healing with hydrogels, efficient recovery of heart health after a myocardial infarction, and many more. Such a wide range of applications are due to the property of dextran which can be easily modified and its extraordinary stable chemical structure as shown in Figure 9.1 [12–21].

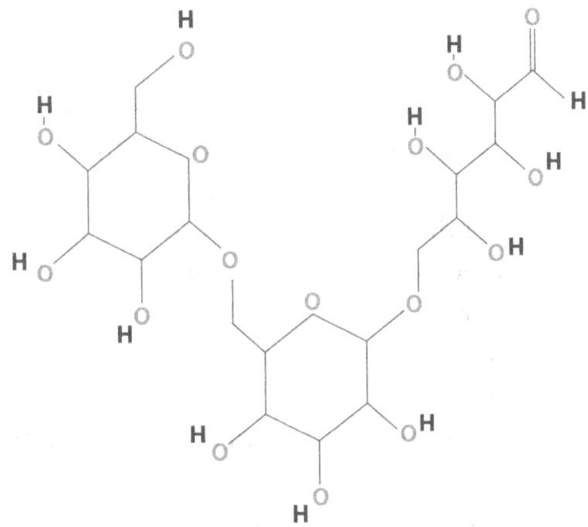

Figure 9.1 Chemical structure of dextran [22].

9.2 Biomedical Applications of Dextran

9.2.1 Bioanalysis and Imaging With Dextran

Combining nanoparticles and dextran for different applications is very successful and popular. The latest research has been discussed here, which concludes that Dex-QDs are highly efficient in imaging and bioanalysis. Various aspects have been studied to evaluate the efficiency of Dex-QDs with different modifications. Different ligands have been synthesized by the alteration of dextrans such as D-t-DHLA (altering with terminal dithiol group), D-p-API (with multiple pendant imidazole groups), and D-p-DHLA and D-p-DHLAm (with multiple pendant dithiol groups). Imidazole and dithiol are the inorganic parts of the quantum dots that are bound with the various anchoring groups. The main requirement for the modified dextrans are 1-(3-aminopropyl) imidazole (API) and dihydrolipoic acid (DHLA), both are derivatives of lipoic acid. $-NH_2$ (amine-functionalized dextrans) and -CHO (half oxidized) are the two intermediates of dextran ligand. It is reported that denser and thicker dextran coating is produced by the terminal anchoring group as compared to pendant anchoring groups. The attachment of quantum dots and multiple pendant anchoring groups shows good stability, whereas the imidazole groups are less tightly bound with the QDs as compared to the dithiol groups. They reduce the yield of quantum due to reoxidation. Due to its large size, good colloidal stability will be provided with dextran (high molecular weight). To predict the limitation and benefits of dextran, different types of ligands are studied. Due to less modification, it is very challenging to characterize dextran. The specific properties of dextran are preserved by pendant or terminal. Furthermore, expected observations have been recorded after 2,4,6-trinitrobenzene-sulfonic acid and different biological tests like the ninhydrin test. After performing the mentioned test, the positive result of dextran or its derivatives has been recorded after the modifications have been done with hexamethylenediamine (HMDA). Treatment of lipoic acid with HMDA will give a negative result for dextran. DHLA-modified dextran shows positive results for the thiol groups, which are concluded after Ellman's test.

Dextran shows its solubility only in aqueous buffers so that hydrophobic quantum dots get functionalized with dextran in the non-polar solvent. It gets functionalized in two different steps. Firstly, there is the transfer of the aqueous phase and exchange of ligands with histidine (His) for the formation of His-QDs, and then in the second step, His ligands are aqueously exchanged with the dextran ligands. Dextran's reducing end should be functionalized with HMDA for the terminal modifications. For the production of dithiolane-functionalized dextran, aminated dextran undergoes ethanol precipitation, which is further dried and treated with lipoic acid–succinimidyl. Oxidization (partial) of dextran with sodium (meta) periodate has been done for the pendant modifications. Thus, the ligands of dextran can be characterized by many characterization techniques such as FTIR, colorometric assays, and NMR. There are two steps involved in Dex-QDs synthesis. In the first step, *via* the exchange of ligands with histidine, there is the transfer of QDs (organic phase) into the aqueous phase. The second step is incubating (3 h at 60°C) His-QDs with dextran ligands. After that, spin filtration technique is being used for the purification of Dex-QDs.

Therefore, for the functionalization of colloidal semiconductor QDs, dextran ligands have been synthesized in the labs. In each ligand, there is a different use of pendant dithiol anchoring groups and terminal dithiol anchoring group, different bindings of dextran and QDs, and also a different molecular weight from each other. It is thus concluded that every ligand have their own benefits as well as limitations. Furthermore, after analyzing Dex-QDs, it has been revealed that it has a negligible impact on cell viability and specificity in binding with cells and protein is very low, which is suitable for multiple conjugation strategies and high colloidal stability. It is also useful in the assembly of polyhistidine-tagged peptides, covalent coupling, and in preparing immunoconjugates. Applications of Dex-QDs have been proved by an *in vitro* fluorescence immunoassay for EPO, selective immunolabeling of breast cancer cells, and FRET-based assays of proteolytic activity. Therefore, it is concluded that imaging and bioanalysis can be efficiently done by Dex-QDs [12, 23–28].

9.2.2 Drug Delivery Systems

For drug delivery, a nanoscale drug delivery system has been synthesized by natural as well as synthetic material to resolve medical and health issues. Synthetic nanomaterial with dextran is introduced to enhance its therapeutic effects as well as bioavailability and to reduce its limitations. In the encapsulation and delivery of pharmaceutical ingredients, natural polymers (proteins, nucleic acids, and polysaccharides) are very efficient, and they are also non-toxic, non-immunogenicity and biodegradable. Various delivery systems are synthesized, such as self-assembled micelles, nano-emulsions, nanoparticles, nanocrystals, and nano-complexes based on dextran, chitosan, heparin, alginate, and starch. However, researchers find delivery systems based on dextran's attractiveness because dextran is applicable in many fields such as biomedicine, nutraceutical, pharmaceutical, and food science. The fluid present in the intestine is pulled out and sent to the plasma with the help of colloid osmotic pressure. This pressure is produced when dextran replaces albumin. Dextran 40 is applicable in reducing the aggregation of erythrocytes and blood viscosity to improve blood flow. Parenteral nutrition and expansion in blood volume can be commonly done with the help of dextran 60 and intravenous solution. Dextran derivatives, micelles, dextran conjugates, and nano-emulsions worked as nanocarriers for the application of nanomedicine. Different types of bonding and interactions are there for self-assembly such as hydrophobic, van der Waals interactions, and noncovalent bonds. The self-assembled micelle has a hydrophobic tail and a hydrophilic head. When in liquid colloid, micelle gets the aggregate dispersed. The micelle's stability will get more stable in aqueous medium due to the hydrophilic shells. Furthermore, the room has been prepared for loading drugs or any nutrients by the hydrophobic core of the micelle. In recent researches, polymeric micelle synthesized by amphiphilic copolymers due to chemical modification *via* esterification and etherification is gathering special attention. On the backbone of dextran, it has a huge number of hydroxyl groups, due to which chemical modification has to be done for the synthesis of dextran derivatives. Mini-emulsions, hydrogels, spray-dried particles, and micelles are all efficient delivery systems based on dextran. It is evident through many studies that they are applicable in treating cancer due to their release mechanism, therapeutic nature, and physicochemical properties. Eco-friendly synthesis methods have been accepted by the researcher because, after clinical trials, it has been concluded that dextran made up of chemical treating methods causes serious issues such as hepatotoxicity, thrombocytopenia, etc. For consuming active pharmaceutical products through dextran-based nanoparticles orally, toxicity tests should be done through intravenous injections. Until then, more studies need to be done to be 100% sure about the good mucus permeability and transepithelial and GI stability because taking the drug orally is the most common and acceptable method [29–46]. Dextran-based nanoparticles for delivering insulin are depicted in Figure 9.2.

Insulin therapy nowadays is becoming very demanding due to the common disease diabetes mellitus. For oral insulin delivery, biopolymer-based nanocarriers are synthesized [48, 49]. Research have been done to synthesize the PLGA-based nanocarrier (dextran-poly (lactide-co-glycolic acid)) so that they are more biocompatible and biodegradable. As shown in Figure 9.2, *in vitro* studies have been done, in which nanocarriers reach and release insulin in the GI tract. The hypoglycemic effect is being analyzed in diabetic rats, after giving a high dose of insulin-loaded nanoparticles according to the body weight approximate (100 IU/kg). Furthermore, dextran-based delivery systems (pH/glucose-responsive) are synthesized by the immobilization of glucose on acryloyl cross-linked dialdehyde dextran nanoparticles and analyzed by *in vitro* experiments. It is observed that, at pH 7.4, insulin is released more efficiently when coming under a high glucose concentration medium. After that, nanocarriers degraded automatically when the pH get low due to the breakdown of glucose into gluconic acid [50, 51].

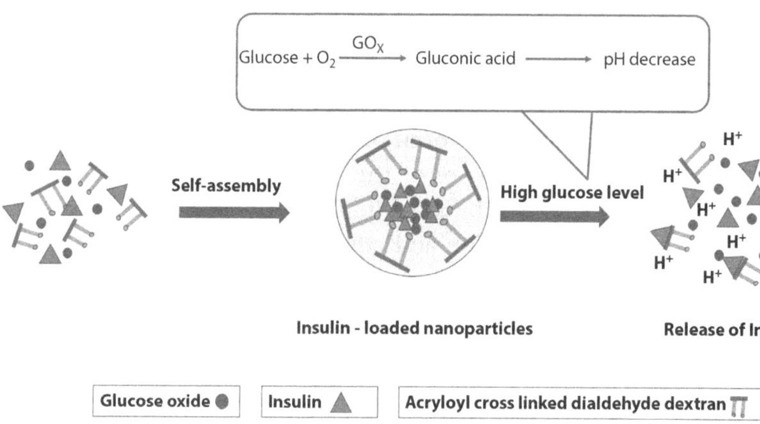

Figure 9.2 Dextran-based nanoparticles for delivering insulin [47].

9.2.3 Anticancer and Antimicrobial Conjugates

Due to the fact that microorganism contamination in food items is the most common problem that occurs during food distribution, packaging, production and storage [52], foodborne diseases spread and approximately 76 million of the population get ill, and more than 5,000 people died as estimated by the Centre for Disease Control and Prevention [53]. For the preservation of the food items, additives (antimicrobial and antioxidant) are to be used [54]. People are attracted to natural additives because they work as therapeutic agents for preventing many diseases, whereas synthetic additives have many side effects but are still used in massive amounts due to their cheap prices [55]. Cancer is also the deadliest disease all over the world, and in the treatment of cancer, plant-based materials are highly efficient [56]. Special attention is directed toward citrus fruits that contain flavonoids as the main content, due to which it has various applications such as anti-inflammatory, prevention from bone loss, antioxidant, antimicrobial, anticancer, and antiallergic [55–58]. Hesperidin and neo-hesperidin are the natural sources of flavonoids that can be found in the organic waste of citrus fruits [59]. Hesperidin is a powerful antioxidant found in orange peel in a large amount [60]. It has been evident from many studies that hesperetin is antioxidant, anticancer, antimicrobial, antidiabetic, and cardioprotective, but the clinical application is limited because of its low water solubility. For biomedical applications, flavonoid–polymer conjugates are introduced by scientists to increase flavonoid life and increase water solubility.

For the formation of dextran–hesperetin conjugates, dextran is used as the main component because of its high stability, ease of modification, cheap rate, and commercial availability. This conjugate is highly used in the pharmaceutical and food industries. The method for synthesis of dextran–hesperetin has been discussed; peels of oranges are dried (shady place) and then converted into powdered form by grinding. It can be stored for a long time at 4°C. After isolating hesperidin, it is purified [60]. Refluxing with petroleum ether has to be done for at least 4 h to remove fat from the powder. Furthermore, it is mixed with 150 ml methanol (approximately 12 h). Then, the rotary evaporator is used to dry out the methanolic extract (after filtering the methanol + hesperidin powder). Then, it is kept in the refrigerator overnight after adding acetic acid (50 ml) to increase the pH. On the next day, recrystallization is to be done with the help of DMSO/H_2O for purifying hesperidin (melting point, 260°C). Then, 5% H_2SO_4 is added with 10 g of hesperidin overnight. Furthermore, cooling down the resulting solution to room temperature will let the precipitation out of hesperetin. After washing the hesperetin precipitates with water, these were washed with acetone and water for additional purification. Finally, hesperetin (yellow) was obtained. Upon mixing dextran with hesperetin, the synthesis of dextran–hesperetin conjugate is done. Characterization of chemical structure is confirmed by the technique ^{1}H-NMR. This conjugate is supposed to be effective against cancer cell lines and various pathogens. The reason behind introducing dextran with hesperetin is to multiply the effectiveness of the antibacterial and anticancerous properties of hesperetin. The structure of hesperidin and hesperetin, respectively, is depicted in Figure 9.3.

The dextran–hesperetin conjugate is studied against *S. typhimurium* and *L. monocytogenes* (pathogenic bacteria). It has been observed that if Dex-HT is used in high concentrations, it will kill the bacteria, especially *L. monocytogenes*. Furthermore, Dex-HT is also capable of suppressing down the cell proliferation of bacteria observed after treatment for 72 h. Hence, it is evident from the various studies that Dex-HT can stop the proliferation and kill the bacteria. Gram-negative bacteria are less sensitive to Dex-HT and HD, whereas gram-positive bacteria are highly sensitive. It is reported that Dex-HD is effective in killing *Helicobacter pylori* as well as *Aeromonas hydrophila* [61, 62].

In one study, Dex-HT with breast cancer cell line, stomach cancer cell line, and normal fibroblast cells (for reference) were examined separately. It has been observed that it decreases the division ability of cancer cells (by initiating G1-phase cell cycle arrest and apoptosis process), whereas different concentrations give different results. That means that cell proliferation stops when treated with a higher amount of Dex-HT and *vice versa*. It has been concluded that Dex-HT is highly active against AGS and MCF-7 cells (even in low concentration), whereas with fibroblast cells it does not show

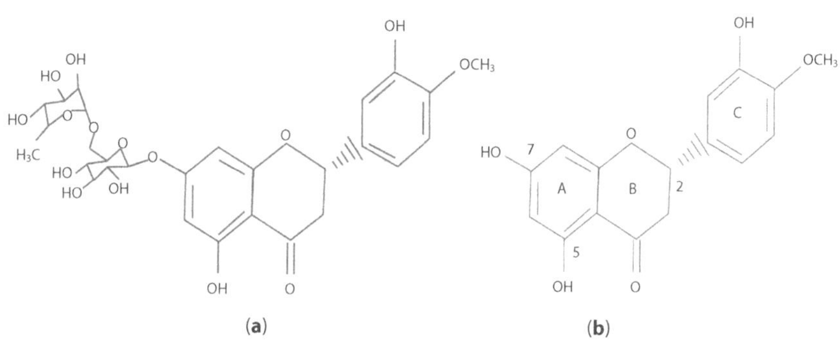

Figure 9.3 (a) Hesperidin. (b) Hesperetin [21].

any noticeable changes. Therefore, we can say that Dex-HT has high specificity toward cancer cells. Food and pharmaceutical companies should also accept this for its various biomedical applications as it is very effective, cheap, and highly stable [21, 61, 63–73].

9.2.4 Dextran as a Local Anesthetic Adjuvant

Why a Local Anesthetic Is Dangerous?
For many years until now, local anesthetics are given with the adjuvant (adrenaline applied with lidocaine), which causes no pain effect to the patient (analgesia), but it has been observed that lidocaine with adrenaline does not give many effective results. This is because blood vessels expand when they come in contact with lidocaine (i.e., called vasodilation) due to which the absorption level of adrenaline in the bloodstream increases and due to which the risk of high heart rate as well as blood pressure increases, which ultimately leads to lethal cardiac dysrhythmia. On the other hand, anesthesia with halothane is also proven dangerous, as it makes patients sensitive to adrenaline. The toxicity of lidocaine leads to severe conditions, i.e., convulsions, and a high concentration of lidocaine is also accumulated in the brain, which directly damages the central nervous system [74–76]. Therefore, after knowing all the major side effects of lidocaine–adrenaline, it is evident that it was harmful to us. The reason behind adding dextran (low molecular weight) into the lidocaine–adrenaline mixture is to decrease the absorption level of adrenaline in the bloodstream. The lidocaine–adrenaline–dextran mixture is a landmark finding, as it is proven to be safe in comparison with local anesthesia [18, 74]. Analgesia caused after the use of dextran (with a local anesthetic) has been studied, and their result has been discussed [77, 78]. When dextran is used in combination with lidocaine–adrenaline, it forms a complex with it which is water-soluble. It has viscous fluidity, so it is absorbed slowly in the bloodstream [79, 80]. The analgesia effect remains for a longer time as dextran maintains the pH [81, 82].

9.2.5 Increased Retention Time at the Targeted Area

The work efficiency of local anesthesia with dextran depends upon which type of procedure is opted and which nerve block has to be done—for example, if we talk about interfacial compartment nerve block, it becomes difficult with the conventional landmark method when an abdominal part has to be treated with anesthesia because it needs anesthesia in a high amount on an accurate point. It is clinically acceptable due to the advancement of ultrasound techniques. In the case of compartment block, to generate and maintain the analgesia effect, there is a need to give anesthesia through injection for a longer time. In this kind of case, dextran plays a very significant role by increasing its retention power. Therefore, we can say that, for interfacial compartment nerve block, dextran (low mol. wt) as anesthesia is most effective. Different cases of quadratus lumborum (QL) block have been studied to justify that dextran and local anesthesia can increase the retention period [18]. During surgery, dextran + ropivacaine is injected in the QL muscle of many patients of age approximately 68 years for approximately 8 h. Both sides of the interfacial space are injected with a saline solution of 8% dextran (low mol wt.) and 0.1% of ropivacaine and morphine (2 mg). Remifentanil and desflurane are used to maintain the anesthesia effect. For pain relief, after an interval of every 8 h, 50 mg intravenous flurbiprofen with acetaminophen is given to the patient. On the first night of surgery, no pain killer is given to the patient. The recovery rate of the patients has been observed on the next day of the surgery, such that they have less amount of pain and are capable of walking for approximately 20 m. No side effects are observed from injecting (low mol. wt) dextran. In fact, after examining the injected area through ultrasound, it is revealed that anesthesia is present over there in some amount until such time. It is evident from this study that low molecular dextran can hold anesthesia for a long time at the injected site.

For caudal block, this same mixture is also effective. Many anesthesiologists think that the caudal block in adults is very challenging and have many aftereffects, but this procedure is easily done by an ultrasound-guided technique. This is also accepted in other procedures such as lower abdominal and urological as well as gynecological [18, 83]. For brachial plexus nerve block, some other forms of anesthesia are also being studied, such as liposomal bupivacaine. It is highly effective but shows aftereffects of its use, and it is expensive at the same time. Thus, it is evident that low molecular dextran remains the best adjuvant in comparison with any other because it is easily modified, stable, and cheap as well as clinically and therapeutically safe to be used [18, 84, 85].

9.2.6 Treatment of Myocardial Infarction

Myocardial infarction (myocardial necrosis) is a very common heart disease that is caused when there is persistent ischemia and acute hypoxia in the coronary arteries [86–88]. When a patient has a heart attack, treatment like coronary intervention drugs and thrombolytic therapy may save the patient and gradually decrease the size of infract to recover heart health [89–91]. During myocardial infarction, a massive amount of non-regenerative cardiomyocytes are lost, so a patient cannot get fully recovered from it. One of the treatments given to a myocardial infarction patient is a heart transplant in serious cases, but transplantation of the heart has various disadvantages such as a higher risk of body rejection, costly procedures, and very few donor sources. If the surgery went successfully, post-operative complications will take place. In this new era, stem cells have shown great results for the regeneration of cardiac cells [92, 93]. Many successful types of research have been done, and many are going on stem cells. Implantation of stem cells will be used to replace cardiomyocyte cells that get affected during MI [94]. Some of the major examples of stem cells are human embryonic cells [95], induced pluripotent cells [96], and umbilical cord mesenchymal stem cells [97]. These cells have an immense ability to make cardiomyocytic cells under certain conditions. As there are so many advantages of stem cell therapy, there are also some disadvantages to this therapy, such as less efficient cell differentiation, low survival, and retention rate [98–100]. It is reported that, after 1 day of injecting, approximately 10% of cells are retained in the MI area. Biomaterials were used to fabricate hydrogels so that the desired cells that have to be implanted will be encapsulated with it *via* tissue engineering to overcome all the limitations by making cells more efficient. Hydrogels can be synthetic or biological which can regenerate cardiac cells during the transplantation of cells. By this method, cell escape can be prevented, which is majorly caused by venous reflux and heartbeats so that their retention time can be improved. When folic acid-derived hydrogels are provided in a 3D workplace, the survival rate of pluripotent cells can be increased. These cells can give involuntary support to the infected area by making the ventricular wall thick. As carriers of chemokines, genes, protein, and live cells are being used as bioactive agents for the treatment of heart cell regrowth, cartilage repair, bone regeneration, and repair of the spinal cord. During severe surgeries, hydrogels give relief to the patient by healing the wound quickly, and this method is very cost-effective. Chitosan and dextran are said to be the most suitable, being biodegradable and biocompatible and having very little toxicity; therefore, they can be used as a good source for tissue engineering. They have various other applications like delivery of drug, use in food and cosmetics, immune adjuvant, etc. Hydrophobic interactions get to increase with CS due to the thermosensitive hydrogels made up of β-GP as it is less cytotoxic and can be applied for tissue engineering of cardiac tissues, adipose, bone, and cartilages. For fabricating chitosan/dextran/β-glycerophosphate hydrogel, dextran (2%) at 37°C is introduced with CS/β-GP.

After that, efficient cell growth and more gelation time have been observed. Due to these excellent properties, these gels (CS/DEX/β-GP) are highly used for cell delivery to graft desired cells for treating myocardial infarctions [101].

9.2.7 Cell Delivery Efficiency

As shown in Figures 9.4 and 9.5, stem cells (MSCs) and fabricated hydrogels (CS/Dex/β-GP) are injected in the myocardial patient's heart. With the help of the potency of MSCs, the formation and proliferation of new cells began to make new heart cell layers in the infract area, whereas

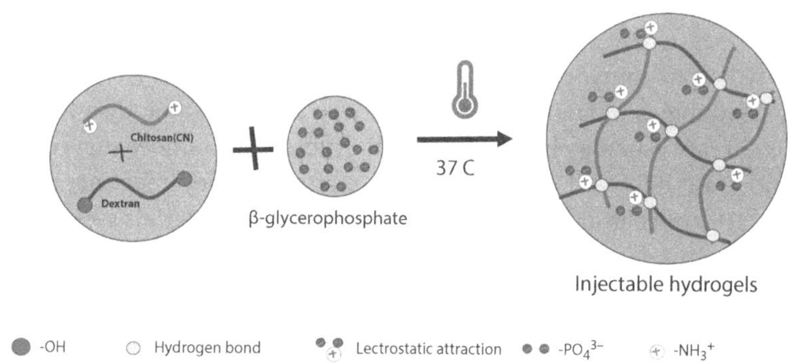

Figure 9.4 Fabrication of CS/Dex/β-GP hydrogel [19].

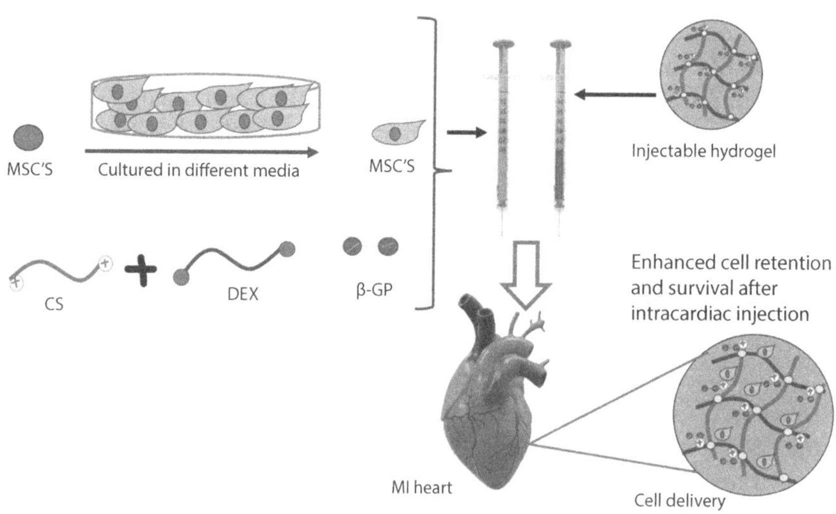

Figure 9.5 Encapsulation of UCMSCs in CS/Dex/β-GP hydrogel [19].

hydrogel (CS/Dex/β-GP) works as a cell delivery agent. Moreover, due to the high stability and antimicrobial and wound healing properties of dextran, it becomes the best choice for treating myocardial infarction. This whole study proves that hydrogels CS/DEX/β-GP are used as vehicle for cell delivery. Thus, in this way, it is capable to treat a damaged heart after myocardial infarction. Hydrophobic interactions and electrostatic attractions are used for the fabrication of CS/0.5-2.0DEX/β-GP as well as CS/β-GP. These gels have biodegradable properties, are injectable in surgical parts, and have good viability, and their gelatine period is controllable. The viability of CS/1.0DEX/β-GP as well as CS/β-GP is checked by *in vitro* experiments, after which the active proliferation of HUVECs and UCMSCs has been recorded. It is also concluded from past research that hydrogel (CS/1.0DEX/β-GP) also induce cell signaling pathways by activating p-ERK1/2 and p-Akt. Hence, CS/1.0DEX/β-GP is efficient in cell delivery for treating various diseases [19, 102–121].

9.2.8 Post-Laminectomy Spinal Sheep Model

Complications (scar formation and adhesion) occurring post-surgery is common. The pain level is around 5%–30% after a spinal cord surgery [122]. Adhesions between the nerve root and peridural fibrosis occur, which cause weakness in the lower limbs. Radicular pains are some post-surgery consequences called failed back surgery syndrome [123, 124]. Crustaceans are the major source for the collection of homeostatic polysaccharide dextran gel [125]. It has been reported by various clinical trials done on humans as well as animals that the anti-adhesion property of the dextran–chitosan gel enhances the anti-adhesion efficacy during endoscopic sinus surgery [126–128].

Before general anesthesia has been given to sheep, they were put on fasting for 12 h. In the jugular vein of the sheep, diazepam and ketamine were injected and then anesthesia has been given to the sheep before laminectomy was performed. On the laminectomy site, chitosan/dextran gel (CD) has been applied. After rinsing with saline solution, all exposed layers were bound and closed, and the sheep was left for approximately 3 months and then the recovery rate was checked [129, 130]. At first, the sheep was monitored on the basis of change in food intake habit, temperature, bleeding at the operation sites, grinding of teeth, ground pawing, etc. The strength of the hind limbs of sheep was checked. To minimize post-operational pain, the sheep is injected with adrenaline and bupivacaine on the site of the surgery. Intramuscular clavulox was injected in the sheep at 1 day before and after the surgery. It has been researched that CD gels show more accurate results than saline solution and gel forms in reducing adhesions. This similar type of experiment was previously performed on rats. As they are of a small size, it is thus very challenging to do the whole procedure and a post-surgery analysis on them. As in the sheep, the area is increased, and there is more space to work on; hence, the chance of post-operation complications and damage to the spinal cord is very less. For the thoracic and lumbar spine experiment, it has been proved that sheep would be the best test model [131–134]. When bonding of fibrosis and dura takes place, adhesions happen during laminectomy. On the surgery site, post-operative local hematoma is replaced by fibrous tissues. Due to this collaboration of dura and fibroblast, post-laminectomy adhesions happen [135]. In this way, it has been observed in all the sheep that new fiber tissues and new bone are developing with the help of dextran/chitosan gel. The gel also inhibits the growth of fibroblastic proliferation without hampering the epithelial cells. These gels, when applied on the outermost layers, act as mechanical barriers until they break down or get dissolved within the tissue [127].

Thrombin can also be used for adhesions, but as of now, every anti-adhesion method transmits a disease or causes allergic reaction and anti-body formation in the patient, whereas dextran/chitosan is used as an anti-adhesive agent as it shows no toxicity. The efficiency of anti-adhesive agents (CD gels) is also tested in brain surgeries. After which, great results have been analyzed [128, 136]. No successful results of clinical trials were observed with the use of chitosan gel without dextran. On the other hand, chitosan/dextran gels start to be used in human surgeries due to its stability and anti-adhesion and homeostasis properties [16, 122, 126, 137].

9.2.9 Wound Healing

The dynamic interactions between various compounds of the extracellular matrix have various cell types, resulting in the healing of a wound. In most cases, wounds are stitched up to bind the ruptured walls of the wound to prevent it from (microorganisms) environmental contact so that the growth factors get increased to make the wound heal fast [138, 139]. Modern techniques for wounds to heal are more significant and quicker than traditional ones. In modern healing techniques, the wound is covered in such materials that reduce the healing time and are easy to use. Wound healing is a challenge because it regulates the body temperature [140–142]. Wound dressings are of different types: polymeric, cryogels, fabrics, etc. However, every material used in wound dressing has its own limitations. Some of them are not biocompatible, produce toxins, cost-effective, has less airflow, and make wounds dry by decreasing the moisture intake. If we use bio-friendly wound dressings, they are less costly and do not cause any type of toxicity. To fabricate bio-friendly dressings, heparin, collagen, keratin, chitin, hyaluronic acid, polyethene, alginic acid, and, most importantly, dextran should be used because dextran enhances the property of any biomaterial with which it is made complex [143–145]. In various studies, it has been proven to be efficient for burn victims and patients having fractures [20].

9.2.10 Dextran With Nanoparticles

Here for wound healing dextran is encapsulated with gold nanoparticles. In *in vivo* studies of wound healing, there are so many challenges that occur, i.e., difficulties in normal tissue repair and cellular motion for bacterial tenderness and immunological trails. The outermost layer of the skin is the largest and fastest healing part of the human body. The interface between the outside environment and the skin is connected with peel. The peel gets renewed from time to time and also has self-repairing properties. When wound dressings are not properly sterilized, it increases the chances of infection and deposition of exudate on the wound site [146–148]. In this new era, nanoparticles are widely used in almost all fields, and it has shown tremendous results in biomedical fields of study. Nanoparticles can be synthesized and easy to modify with any biomaterial. They are biocompatible, biodegradable, and have little or no toxicity. When nanoparticles and dextran are combined together, they show enhanced antimicrobial activity and proliferation of new cells on the wound site [15, 149]. By observing limitations from other wound healing methods, researchers pay attention towards the DEX/PVA nanoparticles of montmorillonite clay. Studies are also going on with chitin–hyaluronic acid-coated gold nanoparticles to resist the drug used in wound healing. In addition to that, chitin shows promising results in the treatment of injury. A study has proved excellent results of wound healing with alginate-coating DEX/PVA with Au nanoparticles. It has antibacterial property and shows positive fluid adsorption.

DEX–polyvinyl alcohol nanoclusters are fabricated with the encapsulation of gold nanoparticles for the formulation of wound dressing. DEX/PVA nanoparticles have gone through various experimental levels to test its limitations. To check the size, shape, and structure of nanoparticles, scanning electron microscopy has been employed. Nanoparticles made with gold encapsulation and without gold encapsulation are analyzed to identify the better one. DIP coating system was used to make nanoparticles for wound dressing to produce bandages. A small piece of cotton cloth was dehydrated and dissolved in DEX/PVA and Au@DEX/PVA solution for 1 day at a specific temperature of 60°C. After 24 h, their moisture content was tested. They were carefully weighed and then dehydrated in hot-air oven for approximately 5 h. The piece of cotton fabric was weighed again and was ready to use.

Further experimentation was done on rats of age 4. They were categorized into three groups (six rats in each group). Group 1 is taken as the control. In group 2, the rats were treated with DEX/PVA. In group 3, the rats were treated with Au@DEX/PVA. The skin of rats has been removed very precisely without causing any pain to the animals. By using tongs and scissors, a cut was made on the inner layer of skin of about 1.5 cm. The gold nanoparticles used for this study are 10–50 nm in size. The wound was then covered with freshly prepared cotton fabric as bandage, and the rats were observed for 16 days. After 16 days, the bandages that are coated with gold nanoparticles show much more anti-bacterial properties.

When gold is used at a lower concentration, it is non-toxic and efficient enough to be used on humans. On the 16th day, 100% recovery rate has been observed in group 3 rats that were treated with Au@DEX/PVA. Gold nanoparticles support DEX/PVA for wound healing. No harmful effect was observed on the exposed tissues of the wound site. This study and the survey conducted on rats suggest that using gold nanoparticles with DEX/PVA can reduce damage and can be used as wound dressing for the easy, early, and quick healing of wounds [15, 150–153].

9.2.11 Dextran Hydrogels

Physically and chemically cross-linked 3D polymers retain water to form hydrogels. The various properties of this polymers, such as physical, biological, and chemical, are easily modified. Both synthetic as well as natural polymers can be used to fabricate hydrogels. Some of the examples of natural polymers are dextran, gelatine, collagen, chitosan, chitin, and gelatine. These polymers have great renewable properties, higher biocompatibility, and biodegradability. Cell growth and its function are influenced by the polymer due to its resemblance to the extracellular matrix. The most important factor in the healing of a wound is pH control. To subdue fluctuations that occur in the pH of the cells, dextran has been widely used. For the synthesis of hydrogel, derivatives of dextran, i.e., dextran sodium salt (DSS) is used as crosslinkers for the physical as well as chemical synthesis of hydrogels. DSS also helps in increasing the number of healthy cells at the wounded site; therefore, it is a good source of material for wound dressing [154–156]. Some therapeutic active components or agents, such as vitamins, flavonoids, polysaccharides, and peptides, can be utilized to speed up wound healing [157]. In medical and pharmacological investigations over 30 years, quercetin (Que), which belongs to the flavonoid family, has been one of the most important dietary antioxidants. Ascorbic acid (vitamin C) has a significant effect on the resurfacing of wounds with new cells and the production of collagen. Ascorbic acid (AA) boosts up the immune system and increases the antibody response and density of new cell growth, resulting in the wound being able to heal fast. When wounds are exposed to the air, AA protects the wound from various outside infections and radical damage because of its anti-bacterial, anti-viral, and anti-oxidant properties [158, 159]. An experiment has been performed in which a hydrogel system was established by a physical interaction between DSS polymer and chitosan (CH). Afterwards, Que and AA are loaded into the hydrogel system. The hydrogels were set aside for some time, and results were further examined. On the cell culture plates, the effects of CH-DSS hydrogel, Que, and AA were studied. In total, 1 mg CH was dissolved in 100 ml of acetic acid (1%) and 2 mg DSS in 100 ml of purified water. The three-dimensional hydrogel systems were formed by combining 1% (w/v) CH and 2% (w/v) DSS solutions in a 2:3 ratio. Hydrogels were produced, frozen at -80°C overnight, and then dried and lyophilized for around 2 days.

The water absorption capacity of the hydrogel is significantly more than its own weight, and it can vary depending on the circumstances. Temperature and pH influence the swelling, and the characterization of hydrogels was investigated according to it. By virtue of their moisture content and oxygen circulation potential, the hydrogels' water-holding capability promotes wound healing. The adsorption levels of the wound dressings get increased when the dressing is wet; therefore, hydrogels should always remain moist. If the dressing material is not properly moisturized, it gets difficult to be removed from over the wound, which causes pain or may worsen the condition of the wound, such that the new cells that are developed on the surface of the wound will be lost and thus the healing process will be delayed. pH plays an integral part for the swelling or adsorption of water content inside hydrogel. The maximum level of swelling is seen when pH is at 7.4 [160, 161]. The behavior of hydrogels' swelling and de-swelling properties depends on the level of the pH. This is because adsorption on the surface of hydrogels depends upon its hydrophobic and hydrophilic chains, and pH changes the interaction between the molecules and the hydrogen bonds. Physical interactions are used to fabricate the CH and DSS hydrogel system which is made up of a 3D polymeric network with high absorption capacity.

It is natural, is easy to fabricate, is cost-effective, and could be used as wound dressing material. AA and Que were added into 3D hydrogels to boost up the recovery rate of wounds. AA-loaded DSS-CH hydrogel shows more promising results for the production of a new cell layer on the surface of wounds. There are no toxic effects reported yet [20, 162].

9.2.12 Dextran Nanocomposite

Utilization of mercury has been increasing due to industries, in medicine, and in domestic areas, due to which the environment is being polluted [163]. Mercury spread in the environment is rising daily due to anthropogenic activities and becomes a threat to humans as well as the environment as it is entering our food chain directly or indirectly. Degrading it completely is nearly impossible as it is a heavy metal [164–167]. Exposure to mercury can be experienced in different industries, such as in pharmaceuticals and gold mining, through oral and dermal routes as well as by direct inhalation. Drastic symptoms have been noted, such as nephrotoxicity, affected neurological activity, and immune toxicity [168–174]. It is important to check the quality of drinking water as mercury content in water should be less than 10 nM as reported by U.S.E.P.A. [165]. It is compulsory to have a more specific and sensitive detector for detecting Hg^{2+}. There are different detection methods and techniques such as atomic fluorescence spectrometry, HPLC, ion-selective electrode, flame photometry, and atomic absorption/emission spectrometry, but these methods come with some limitations, i.e., very time-consuming and costly as well and need advanced instrumentation [17, 175–182]. Therefore, it is necessary to introduce methods to detect Hg^{2+}, i.e., specific and rapid as well as cheap. A method, i.e., Au nanoparticle-based colorimetric sensors, for detecting mercury is specific, have surface plasmon resonance property, and stable, but at the same time this method is highly expensive due to gold's prices. To overcome this limitation, metal nanoparticles have been introduced as they are easy to modify chemically as well as physically. Bimetallic nanoparticles are considered to be more beneficial due to their wide range of application, i.e., unique optical, electrical, antibacterial, and catalytic properties, full cell fabrication, and colorimetric sensors [183–189]. Dextran is used in the fabrication of nanocomposite for the purpose of *in situ* stabilization, effective reducer, biocompatible, and non-toxic to make a nanocomposite pocket as well as eco-friendly. Pollutants from azo dye are very challenging to remove physically as well as chemically [190–192]. For the photo-degradation of organic pollutants, some researchers give preferences to photocatalytic oxidation [193], whereas silver nanoparticles are effective against many bacteria, fungi, and some other microbes as silver ions are able to inhibit protein synthesis by denaturing ribosomes to kill microbes [194–196]. Combining Cu and Ag metals together give many benefits such as having antimicrobial and photocatalytic activity, suppression of the activity of cancer cells, and cost-effective [197]. In one study, Cu-sensitized Ag-dextran NCs are synthesized for the specific and sensitive detection of Hg^{2+} in liquid medium. Following are the steps for the synthesis of Cu-sensitized Ag-dextran nanocomposite: (1) 10 ml (1 mM of copper sulphate) and 10 ml (1 mM of silver nitrate) has been mixed, (2) sodium borohydride is mixed as a reducing agent, (3) furthermore, in 30 ml of chilled (2 mM) sodium borohydride mixed with dextran (2 ml), (4) in the dark room, a solution of copper sulphate and silver nitrate added in the above-mentioned prepared mixture, and (5) the appearance of a brownish-yellow color means that the fabrication of a Cu-sensitized Ag-dextran nanocomposite is completed [17]. After synthesis, the nanocomposite has been tested on many different samples to check the specificity and sensitivity on sewage water and purposely spiked water with mercury ions. A paper probe was used to detect the presence of mercury ions in the sample. In the aqueous solution of synthesized nanocomposite, a paper (Whatmann no. 1 filter paper) was used to dip in it and left for drying out. Different metal ions as well as mercury ions from dilution of 1 nm to 1 mM were poured on the paper strips. A specific color has been observed with the spiked water. Different metal ions give different color variations to the strip but give the same color repeatedly in the presence of mercury, which is evident of its specificity and sensitivity. A real water sample study was also done, which also gave accurate results. These strips were used again after almost 30 days to detect Hg^{2+}. It was observed

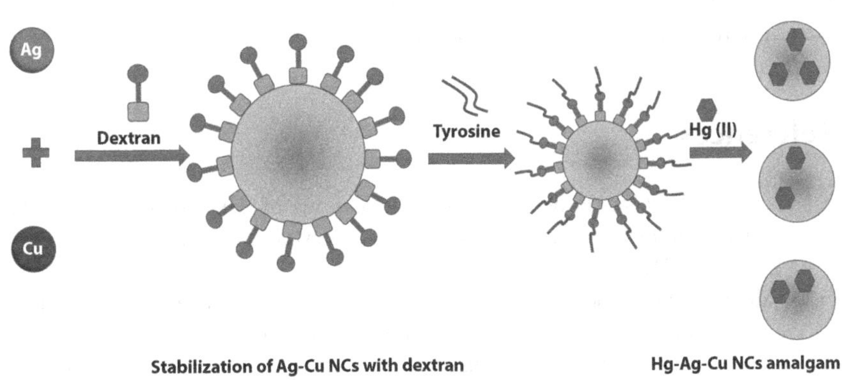

Figure 9.6 Detecting mercury with Cu-sensitized Ag–dextran nanocomposite [17].

that there was no change in the result accuracy, which means that it was highly stable as well. In Figure 9.6, detecting mercury with Cu-sensitized Ag-dextran nanocomposite is shown [17, 198].

9.3 Conclusion

Different biomedical applications have been concluded by different studies such as the application of Dex-QDs that has been proven by an *in vitro* fluorescent immunoassay for EPO, selective immunolabeling of breast cancer cells, and FRET-based assays of proteolytic activity. Dextran shows its significance in bioanalysis and imaging technique. Moreover, insulin has been released in the GI tract with the help of acryloyl cross-linked dialdehyde dextran nanoparticles more efficiently when coming under a high glucose concentration medium after drug delivery nanocarriers are degraded automatically when the pH gets low due to the breakdown of glucose into gluconic acid. Furthermore, the fabrication of Dex-HT is proven to be highly active against AGS and MCF-7 cells (even in low concentration), whereas with fibroblast cells it does not show any noticeable changes. Low molecular weight dextran is the best adjuvant in comparison with any other because it is easily modified, stable, and cheap as well as increases the retention time of anesthesia in the injected area, whereas in the case of myocardial infarction, CS/DEX/β-GP hydrogel and MSCs (stem cell) were injected in the heart, and it has been observed that recovery of the damage caused to the heart was very efficient. Here two applications of dextran together work for recovering heart damage, i.e., cell delivery and wound healing. Chitosan/dextran gels are being started to be used in human surgeries due to their anti-adhesion and homeostasis properties. The detection of mercury ions can be successfully and specifically done by the Cu-sensitized Ag-dextran nanocomposite.

References

1. Baruah, R. *et al.*, Functional food applications of dextran from Weissella cibaria RBA12 from pummelo (Citrus maxima). *Int. J. Food Microbiol.*, 242, 124–131, 2017.
2. Falconer, D.J., Mukerjea, R., Robyt, J.F., Biosynthesis of dextrans with different molecular weights by selecting the concentration of Leuconostoc mesenteroides B-512FMC dextransucrase, the sucrose concentration, and the temperature. *Carbohydr. Res.*, 346, 2, 280–284, 2011.

3. Heinze, T. *et al.*, Functional polymers based on dextran, in: *Polysaccharides II*, pp. 199–291, 2006.
4. Delattre, C. and Vijayalakshmi, M., Monolith enzymatic microreactor at the frontier of glycomic toward a new route for the production of bioactive oligosaccharides. *J. Mol. Catal. B: Enzym.*, 60, 3-4, 97–105, 2009.
5. Kothari, D., Tingirikari, J.M.R., Goyal, A., *In vitro* analysis of dextran from Leuconostoc mesenteroides NRRL B-1426 for functional food application. *Bioact. Carbohydr. Diet. Fibre*, 6, 2, 55–61, 2015.
6. Thebaudin, J. *et al.*, Dietary fibres: Nutritional and technological interest. *Trends Food Sci. Technol.*, 8, 2, 41–48, 1997.
7. Brennan, C.S. and Cleary, L.J., The potential use of cereal (1→ 3, 1→ 4)-β-D-glucans as functional food ingredients. *J. Cereal Sci.*, 42, 1, 1–13, 2005.
8. Das, D. and Goyal, A., Potential probiotic attributes and antagonistic activity of an indigenous isolate Lactobacillus plantarum DM5 from an ethnic fermented beverage "Marcha" of North Eastern Himalayas. *Int. J. Food Sci. Nutr.*, 65, 3, 335–344, 2014.
9. Hongpattarakere, T. *et al.*, *In vitro* prebiotic evaluation of exopolysaccharides produced by marine isolated lactic acid bacteria. *Carbohydr. Polym.*, 87, 1, 846–852, 2012.
10. Kajala, I. *et al.*, Rye bran as fermentation matrix boosts *in situ* dextran production by Weissella confusa compared to wheat bran. *Appl. Microbiol. Biotechnol.*, 100, 8, 3499–3510, 2016.
11. Lacaze, G., Wick, M., Cappelle, S., Emerging fermentation technologies: Development of novel sourdoughs. *Food Microbiol.*, 24, 2, 155–160, 2007.
12. Rees, K. *et al.*, Dextran-functionalized semiconductor quantum dot bioconjugates for bioanalysis and imaging. *Bioconjugate Chem.*, 31, 3, 861–874, 2020.
13. Besrour-Aouam, N. *et al.*, The role of dextran production in the metabolic context of Leuconostoc and Weissella Tunisian strains. *Carbohydr. Polym.*, 253, 117254, 2021.
14. Pacelli, S. *et al.*, Dextran-polyethylene glycol cryogels as spongy scaffolds for drug delivery. *Int. J. Biol. Macromol.*, 166, 1292–1300, 2021.
15. Xia, C. *et al.*, A feasible strategy of fabricating of gold-encapsulated dextran/polyvinyl alcohol nanoparticles for the treatment and care of wound healing. *J. Cluster Sci.*, 33, 2179–2187, 2021.
16. Rajiv, S. *et al.*, Chitosan Dextran gel as an anti adhesion agent in a postlaminectomy spinal sheep model. *J. Clin. Neurosci.*, 40, 153–156, 2017.
17. Kokilavani, S. *et al.*, Development of multifunctional Cu sensitized Ag-dextran nanocomposite for selective and sensitive detection of mercury from environmental sample and evaluation of its photocatalytic and antimicrobial applications. *J. Mol. Liq.*, 321, 114742, 2021.
18. Tsuchiya, M., New application of low-molecular weight dextran as local anesthetic adjuvant for ultrasound-guided nerve block, in: *Topics in Regional Anesthesia*, IntechOpen, 2021.

19. Ke, X. *et al.*, An injectable chitosan/dextran/β-glycerophosphate hydrogel as cell delivery carrier for therapy of myocardial infarction. *Carbohydr. Polym.*, 229, 115516, 2020.
20. Mansuroğlu, B. *et al.*, Chitosan/dextran sulphate sodium hydrogels for wound healing material: Preparation, characterisation and *in vitro* evaluation. *Mater. Technol.*, p, 1–8, 2020.
21. Zare, M. *et al.*, Dextran–hesperetin conjugate as a novel biocompatible medicine for antimicrobial and anticancer applications. *J. Polym. Environ.*, 29, 3, 811–820, 2021.
22. https://pubchem.ncbi.nlm.nih.gov/compound/dextran.
23. Hildebrandt, N. *et al.*, Energy transfer with semiconductor quantum dot bioconjugates: A versatile platform for biosensing, energy harvesting, and other developing applications. *Chem. Rev.*, 117, 2, 536–711, 2017.
24. Wang, W. *et al.*, Photoligation of an amphiphilic polymer with mixed coordination provides compact and reactive quantum dots. *J. Am. Chem. Soc.*, 137, 16, 5438–5451, 2015.
25. Petryayeva, E., Jeen, T., Algar, W.R., Optimization and changes in the mode of proteolytic turnover of quantum dot–peptide substrate conjugates through moderation of interfacial adsorption. *ACS Appl. Mater. Interfaces*, 9, 36, 30359–30372, 2017.
26. Jeen, T. and Algar, W.R., Mimicking cell surface enhancement of protease activity on the surface of a quantum dot nanoparticle. *Bioconjugate Chem.*, 29, 11, 3783–3792, 2018.
27. Liu, Y., Zhao, N., Xu, F.-J., pH-responsive degradable dextran-quantum dot nanohybrids for enhanced gene delivery. *ACS Appl. Mater. Interfaces*, 11, 38, 34707–34716, 2019.
28. Rahman, M.M. *et al.*, Recent advancements of nanoparticles application in cancer and neurodegenerative disorders: At a glance. *Biomed. Pharmacother.*, 153, 113305, 2022.
29. Alibakhshi, M.A. *et al.*, Picomolar fingerprinting of nucleic acid nanoparticles using solid-state nanopores. *ACS Nano*, 11, 10, 9701–9710, 2017.
30. Deng, L. *et al.*, Double-layered protein nanoparticles induce broad protection against divergent influenza A viruses. *Nat. Commun.*, 9, 1, 1–12, 2018.
31. Hu, Q. and Luo, Y., Chitosan-based nanocarriers for encapsulation and delivery of curcumin: A review. *Int. J. Biol. Macromol.*, 179, 125–135, 15 May 2021.
32. Qin, Y. *et al.*, Preparation of bioactive polysaccharide nanoparticles with enhanced radical scavenging activity and antimicrobial activity. *J. Agric. Food Chem.*, 66, 17, 4373–4383, 2018.
33. Luo, Y., Food colloids binary and ternary nanocomplexes: Innovations and discoveries. *Colloids Surf. B: Biointerfaces*, 196, 111309, 2020.
34. Luo, Y. and Wang, Q., Recent development of chitosan-based polyelectrolyte complexes with natural polysaccharides for drug delivery. *Int. J. Biol. Macromol.*, 64, 353–367, 2014.

35. Di Martino, A. *et al.*, Folic acid-chitosan-alginate nanocomplexes for multiple delivery of chemotherapeutic agents. *J. Drug Delivery Sci. Technol.*, 47, 67–76, 2018.
36. Fan, Y. *et al.*, Physicochemical stability and *in vitro* bioaccessibility of β-carotene nanoemulsions stabilized with whey protein-dextran conjugates. *Food Hydrocolloids*, 63, 256–264, 2017.
37. Hao, Y. *et al.*, Preparation of starch nanocrystals through enzymatic pretreatment from waxy potato starch. *Carbohydr. Polym.*, 184, 171–177, 2018.
38. Zhang, H. *et al.*, Stromal cell-derived factor-1α-encapsulated albumin/heparin nanoparticles for induced stem cell migration and intervertebral disc regeneration *in vivo*. *Acta Biomater.*, 72, 217–227, 2018.
39. Shen, Y. *et al.*, Synthesis of dextran/Se nanocomposites for nanomedicine application. *Mater. Chem. Phys.*, 109, 2-3, 534–540, 2008.
40. Semyonov, D. *et al.*, Enzymatically synthesized dextran nanoparticles and their use as carriers for nutraceuticals. *Food Funct.*, 5, 10, 2463–2474, 2014.
41. Duan, H.-L. *et al.*, Preparation of immunomagnetic iron-dextran nanoparticles and application in rapid isolation of E. coli O157: H7 from foods. *World J. Gastroenterol.: WJG*, 11, 24, 3660, 2005.
42. Deng, W. *et al.*, Green preparation process, characterization and antitumor effects of doxorubicin–BSA–dextran nanoparticles. *Macromol. Biosci.*, 10, 10, 1224–1234, 2010.
43. Gruber, U.F., Dextran and the prevention of postoperative thromboembolic complications. *Surg. Clin. North Am.*, 55, 3, 679–696, 1975.
44. Klotz, U. and Kroemer, H., Clinical pharmacokinetic considerations in the use of plasma expanders. *Clin. Pharmacokinet.*, 12, 2, 123–135, 1987.
45. Zinderman, C.E., Landow, L., Wise, R.P., Anaphylactoid reactions to Dextran 40 and 70: Reports to the United States Food and Drug Administration, 1969 to 2004. *J. Vasc. Surg.*, 43, 5, 1004–1009, 2006.
46. Danhauser-Riedl, S. *et al.*, Phase I clinical and pharmacokinetic trial of dextran conjugated doxorubicin (AD-70, DOX-OXD). *Invest. New Drugs*, 11, 2, 187–195, 1993.
47. Hu, Q., Lu, Y., Luo, Y., Recent advances in dextran-based drug delivery systems: From fabrication strategies to applications. *Carbohydr. Polym.*, 264, 117999, 15 July 2021.
48. He, Z. *et al.*, Scalable fabrication of size-controlled chitosan nanoparticles for oral delivery of insulin. *Biomaterials*, 130, 28–41, 2017.
49. Hu, Q. and Luo, Y., Recent advances of polysaccharide-based nanoparticles for oral insulin delivery. *Int. J. Biol. Macromol.*, 120, 775–782, 2018.
50. Alibolandi, M. *et al.*, Dextran-b-poly (lactide-co-glycolide) polymersome for oral delivery of insulin: *In vitro* and *in vivo* evaluation. *J. Controlled Release*, 227, 58–70, 2016.
51. Jamwal, S. *et al.*, New glucose oxidase-immobilized stimuli-responsive dextran nanoparticles for insulin delivery. *Int. J. Biol. Macromol.*, 123, 968–978, 2019.

52. Silveira, S.M.d. *et al.*, Chemical composition and antimicrobial activity of essential oils from selected herbs cultivated in the South of Brazil against food spoilage and foodborne pathogens. *Ciênc. Rural*, 42, 7, 1300–1306, 2012.
53. Diagnosis and management of foodborne illnesses, a primer for physicians and other health care professionals.recommendations and reports 53 no. RR-4, Centers for Disease Control and Prevention (CDC) Morbidity and mortality weekly report, 2004.
54. Garcia-Lomillo J, G.-S.M., Pino-Garcia, R.D., Rivero-Perez, M.D., Muniz-Rodriguez, P., Antioxidant and anti microbial properties of wine by-products and their potential uses in the food industry. *J. Agric. Food Chem.*, 62, 52, 12595–12602, 2014.
55. Celikel, N. and Kavas, G., Antimicrobial properties of some essential oils against some pathogenic microorganisms. *Czech J. Food Sci.*, 26, 3, 174, 2008.
56. Bishayee, A. and Sethi, G., Bioactive natural products in cancer prevention and therapy: Progress and promise, in: *Seminars in Cancer Biology*, Elsevier, 2016.
57. Middleton, E., Kandaswami, C., Theoharides, T.C., The effects of plant flavonoids on mammalian cells: Implications for inflammation, heart disease, and cancer. *Pharmacol. Rev.*, 52, 4, 673–751, 2000.
58. Marzouk, B., Characterization of bioactive compounds in Tunisian bitter orange (Citrus aurantium L.) peel and juice and determination of their antioxidant activities. *BioMed. Res. Int.*, 2013, 2013.
59. Lahmer, N. *et al.*, Hesperidin and hesperitin preparation and purification from Citrus sinensis peels. *Der Pharma Chem.*, 7, 2, 1–4, 2015.
60. Chaudhri, V.K. *et al.*, Isolation and characterization of hesperidin from dried orange peel. *Asian J. Ethnopharmacol. Med. Foods*, 22, 2016.
61. Lee Y, H.J., Nam, S., Moon, S., Lee, S., Enzymatic bio conversion of citrus hesperidin by aspergillus sojaenaringinase: Enhanced solubility of hesperetin-7-O-glucoside with *in vitro* inhibition of human intestinal maltase, HMG-CoA reductase, and growth of Helicobacter pylori. *Food Chem., 135*, 2253–2259, 2012.
62. Abuelsaad, A.S. *et al.*, Antimicrobial and immunomodulating activities of hesperidin and ellagic acid against diarrheic Aeromonas hydrophila in a murine model. *Life Sci.*, 93, 20, 714–722, 2013.
63. Trivedi, P.P., K.S., Tripathi, D.N., Jena, G.B., Cardiopro tective efects of hesperetin against doxorubicin-induced oxidative stress and DNA damage in rat. *Cardiovasc. Toxicol.*, 11, 3, 215–225, 2011.
64. Selvraj, A., J.P., Namasivayam, N., Efect of hesperetin, a citrus favonoid, on bacterial enzymes and carcinogen induced aberrant crypt foci in colon cancer rats: A dose-dependent study. *J. Pharm. Pharmacol.*, 60, 10, 1385–1392, 2008.
65. C.J., Antioxidant and neuroprotective efects of hesperi din and its aglycone hesperetin. *Arch. Pharm. Res.*, 29, 699–706, 2006.

66. Galati EM, T.A., Kirjavainen, S., Forestieri, A.M., Rossitto, M.M.A., Biological efects of hesperidin, a citrus favonoid. (Note III): Antihypertensive and diuretic activity in rat. *Farmaco*, 51, 219–221, 1996.
67. Mandalari G, B.R., Bisignano, G., Trombetta, D., Saija, A., Faulds CB, G.M., Narbad, A., Antimicrobial activ ity of favonoids extracted from bergamot (Citrus bergamiaRisso) peel, a byproduct of the essential oil industry. *J. Appl. Microbiol.*, 103, 2056–2064, 2007.
68. Moon SH, L.J., Kim, K.T., Park, Y.S., Nah, S.Y., Ahn, D.U.K., Paik, H.D., Antimicrobial efect of 7-O-Butylnaringenin, a novel favonoid, and various natural favonoids against helicobacterpy lori strains. *Int. J. Environ. Res. Public Health*, 10, 11, 5459–5469, 2013.
69. Qing-yi, M.A., C.-t.C., Xiao-yan, J., Wei-xia, Y., Studies on extraction of hesperidin and other active components from citrus peels and their antimicrobial efects. *J. Food Sci.*, 12, 345, 2004.
70. Giuseppe C, O.V., Hampel, S., Umile, G.S., Nevio, P., Francesca, I., Incorporation of carbon nanotubes into a gelatin–catechin conjugate: Innovative approach for the preparation of anticancer materials. *Int. J. Pharm.*, 446, 176–182, 2013.
71. Modak B, T.R., Wilkens, M., Urzua, A., Antibacterial activity of compounds isolated of the resinous exudate from Heli otropiumsinuatum on phytopathogenic bacteria. *J. Chil. Chem. Soc.*, 49, 1, 1–3, 2004.
72. Vittorio, O., C.G., Iemma, F.E., Turi, G.N.D., Acchetti, E., Curcio, B.S.M., Funel, N., Parisi, O.I., Puoci, F., Picci, N., Dextran catechin conjugate: A potential treatment against the pancreatic ductal adenocarcinoma. *Pharm. Res.*, 29, 2601–2614, 2012.
73. Shamala TR, P.M., Preliminary studies on the production of high and low viscosity dextran by leuconostoc spp. *Process Biochem.*, 30, 237–241, 1995.
74. Ueda, W., Hirakawa, M., Mori, K., Acceleration of epinephrine absorption by lidocaine. *Anesthesiology (Philadelphia)*, 63, 6, 717–720, 1985.
75. Yokoyama, M. *et al.*, Modification of intravenous lidocaine-induced convulsions by epinephrine in rats. *Can. J. Anaesth.*, 40, 3, 251–256, 1993.
76. Takahashi, R. *et al.*, Epinephrine increases the extracellular lidocaine concentration in the brain: A possible mechanism for increased central nervous system toxicity. *J. Am. Soc. Anesthesiologists*, 105, 5, 984–989, 2006.
77. Ito, E. *et al.*, Anesthetic duration of lidocaine with 10% dextran is comparable to lidocaine with 1: 160 000 epinephrine after intraosseous injection in the rabbit. *Oral. Surg. Oral. Med. Oral. Pathol. Oral. Radiol. Endod.*, 104, 3, e26–e31, 2007.
78. Simpson, P., Hughes, D., Long, D., Prolonged local analgesia for inguinal herniorrhaphy with bupivacaine and dextran. *Ann. R. Coll. Surg. Engl.*, 64, 4, 243, 1982.
79. Åberg, G., Friberger, P., Sydnes, G., Studies on the duration of local anaesthesia: A possible mechanism for the prolonging effect of dextran on the

duration of infiltration anaesthesia. *Acta Pharmacol. Toxicol.*, 42, 2, 88–92, 1978.
80. Hassan, H. *et al.*, Effects of adjuvants to local anaesthetics on their duration: I. Studies of dextrans of widely varying molecular weight and adrenaline in rat infraorbital nerve block. *Acta Anaesthesiol. Scand.*, 29, 4, 375–379, 1985.
81. Covino, B.G., The pharmacology of local anesthetic agents. *Anesthesiology*, 6–10, 1986.
82. Tsuchiya, M., Use of dextran in regional anesthesia, in: *Treatments, Mechanisms, and Adverse Reactions of Anesthetics and Analgesics*, pp. 15–26, Elsevier, 2022.
83. Tsuchiya, M. *et al.*, Ultrasound-guided single shot caudal block anesthesia reduces postoperative urinary catheter-induced discomfort. *Minerva Anestesiol.*, 79, 12, 1381–8, 2013.
84. Hejazi, M.S. *et al.*, Evaluation of volatile and intravenous anesthetics, effects on the threshold of neuroresponse telemetry and the threshold of acoustically evoked stapedial reflex in children undergoing cochlear implant surgery. *J. Anaesthesiol. Clin. Pharmacol.*, 34, 2, 177, 2018.
85. Skolnik, A. and Gan, T.J., New formulations of bupivacaine for the treatment of postoperative pain: Liposomal bupivacaine and SABER-Bupivacaine. *Expert Opin. Pharmacother.*, 15, 11, 1535–1542, 2014.
86. Arevalo, H.J. *et al.*, Arrhythmia risk stratification of patients after myocardial infarction using personalized heart models. *Nat. Commun.*, 7, 1, 1–8, 2016.
87. Lee, D.J. *et al.*, An injectable sulfonated reversible thermal gel for therapeutic angiogenesis to protect cardiac function after a myocardial infarction. *J. Biol. Eng.*, 13, 1, 1–13, 2019.
88. Majka, M. *et al.*, Concise review: Mesenchymal stem cells in cardiovascular regeneration: Emerging research directions and clinical applications. *Stem Cells Transl. Med.*, 6, 10, 1859–1867, 2017.
89. García-Prieto, J. *et al.*, Neutrophil stunning by metoprolol reduces infarct size. *Nat. Commun.*, 8, 1, 1–15, 2017.
90. Singhanat, K. *et al.*, Roles of melatonin and its receptors in cardiac ischemia–reperfusion injury. *Cell. Mol. Life Sci.*, 75, 22, 4125–4149, 2018.
91. Yang, H.Y. *et al.*, Predictors of in-hospital mortality in Korean patients with acute myocardial infarction. *Chonnam Med. J.*, 55, 1, 40–46, 2019.
92. Cahill, T.J., Choudhury, R.P., Riley, P.R., Heart regeneration and repair after myocardial infarction: Translational opportunities for novel therapeutics. *Nat. Rev. Drug Discovery*, 16, 10, 699–717, 2017.
93. Ma, J. *et al.*, Exosomes derived from AKt-modified human umbilical cord mesenchymal stem cells improve cardiac regeneration and promote angiogenesis via activating platelet-derived growth factor D. *Stem Cells Transl. Med.*, 6, 1, 51–59, 2017.
94. Ottersbach, A. *et al.*, Improved heart repair upon myocardial infarction: Combination of magnetic nanoparticles and tailored magnets strongly increases engraftment of myocytes. *Biomaterials*, 155, 176–190, 2018.

95. Kraehenbuehl, T.P. *et al.*, Human embryonic stem cell-derived microvascular grafts for cardiac tissue preservation after myocardial infarction. *Biomaterials*, 32, 4, 1102–1109, 2011.
96. Chow, A. *et al.*, Human induced pluripotent stem cell-derived cardiomyocyte encapsulating bioactive hydrogels improve rat heart function post myocardial infarction. *Stem Cell Rep.*, 9, 5, 1415–1422, 2017.
97. Nascimento, D.S. *et al.*, Human umbilical cord tissue-derived mesenchymal stromal cells attenuate remodeling after myocardial infarction by proangiogenic, antiapoptotic, and endogenous cell-activation mechanisms. *Stem Cell Res. Ther.*, 5, 1, 1–14, 2014.
98. Fan, Z. *et al.*, An injectable oxygen release system to augment cell survival and promote cardiac repair following myocardial infarction. *Sci. Rep.*, 8, 1, 1–22, 2018.
99. Follin, B. *et al.*, Retention and functional effect of adipose-derived stromal cells administered in alginate hydrogel in a rat model of acute myocardial infarction. *Stem Cells Int.*, 2018, 2018.
100. Vu, T.D. *et al.*, An autologous platelet-rich plasma hydrogel compound restores left ventricular structure, function and ameliorates adverse remodeling in a minimally invasive large animal myocardial restoration model: A translational approach: Vu and Pal "Myocardial Repair: PRP, Hydrogel and Supplements. *Biomaterials*, 45, 27–35, 2015.
101. Sun, Q. *et al.*, A self-sustaining antioxidant strategy for effective treatment of myocardial infarction. *Adv. Sci.*, 10, 5, 2204999, 2023.
102. Hou, D. *et al.*, Radiolabeled cell distribution after intramyocardial, intracoronary, and interstitial retrograde coronary venous delivery: Implications for current clinical trials. *Circulation*, 112, 9_supplement, I–150-I-156, 2005.
103. Ciuffreda, M.C. *et al.*, Synthetic extracellular matrix mimic hydrogel improves efficacy of mesenchymal stromal cell therapy for ischemic cardiomyopathy. *Acta Biomater.*, 70, 71–83, 2018.
104. Dong, R. *et al.*, Self-healing conductive injectable hydrogels with antibacterial activity as cell delivery carrier for cardiac cell therapy. *ACS Appl. Mater. Interfaces*, 8, 27, 17138–17150, 2016.
105. Hao, T. *et al.*, Injectable fullerenol/alginate hydrogel for suppression of oxidative stress damage in brown adipose-derived stem cells and cardiac repair. *ACS Nano*, 11, 6, 5474–5488, 2017.
106. Dorsey, S.M. *et al.*, MRI evaluation of injectable hyaluronic acid-based hydrogel therapy to limit ventricular remodeling after myocardial infarction. *Biomaterials*, 69, 65–75, 2015.
107. Chan, A.T. *et al.*, Hyaluronic acid-serum hydrogels rapidly restore metabolism of encapsulated stem cells and promote engraftment. *Biomaterials*, 73, 1–11, 2015.
108. Cai, L., Dewi, R.E., Heilshorn, S.C., Injectable hydrogels with *in situ* double network formation enhance retention of transplanted stem cells. *Adv. Funct. Mater.*, 25, 9, 1344–1351, 2015.

109. Parisi-Amon, A. *et al.*, Protein-engineered injectable hydrogel to improve retention of transplanted adipose-derived stem cells. *Adv. Healthc. Mater.*, 2, 3, 428–432, 2013.
110. Li, H. *et al.*, Folic acid derived hydrogel enhances the survival and promotes therapeutic efficacy of iPS cells for acute myocardial infarction. *ACS Appl. Mater. Interfaces*, 10, 29, 24459–24468, 2018.
111. Gwon, K., Kim, E., Tae, G., Heparin-hyaluronic acid hydrogel in support of cellular activities of 3D encapsulated adipose derived stem cells. *Acta Biomater.*, 49, 284–295, 2017.
112. Xu, G. *et al.*, Injectable biodegradable hybrid hydrogels based on thiolated collagen and oligo (acryloyl carbonate)–poly (ethylene glycol)–oligo (acryloyl carbonate) copolymer for functional cardiac regeneration. *Acta Biomater.*, 15, 55–64, 2015.
113. Ciocci, M. *et al.*, Injectable silk fibroin hydrogels functionalized with microspheres as adult stem cells-carrier systems. *Int. J. Biol. Macromol.*, 108, 960–971, 2018.
114. Hasan, A. *et al.*, Injectable hydrogels for cardiac tissue repair after myocardial infarction. *Adv. Sci.*, 2, 11, 1500122, 2015.
115. Cai, X. *et al.*, Transglutaminase-catalyzed preparation of crosslinked carboxymethyl chitosan/carboxymethyl cellulose/collagen composite membrane for postsurgical peritoneal adhesion prevention. *Carbohydr. Polym.*, 201, 201–210, 2018.
116. Sharma, S. *et al.*, Enhanced immune response against pertussis toxoid by IgA-loaded chitosan–dextran sulfate nanoparticles. *J. Pharm. Sci.*, 101, 1, 233–244, 2012.
117. Vecchione, R. *et al.*, Tunable stability of monodisperse secondary O/W nano-emulsions. *Nanoscale*, 6, 15, 9300–9307, 2014.
118. Yang, Y. *et al.*, Isolation and characterization of dextran produced by Leuconostoc citreum NM105 from manchurian sauerkraut. *Carbohydr. Polym.*, 133, 365–372, 2015.
119. Ahmad, U. *et al.*, Chitosan based thermosensitive injectable hydrogels for controlled delivery of loxoprofen: Development, characterization and *in-vivo* evaluation. *Int. J. Biol. Macromol.*, 129, 233–245, 2019.
120. Huang, C.-L. *et al.*, Development of chitosan/β-glycerophosphate/glycerol hydrogel as a thermosensitive coupling agent. *Carbohydr. Polym.*, 147, 409–414, 2016.
121. Yang, Y. *et al.*, Degradation behaviors of nerve guidance conduits made up of silk fibroin *in vitro* and *in vivo*. *Polym. Degrad. Stab.*, 94, 12, 2213–2220, 2009.
122. Diamond, M.P. *et al.*, Reduction of postoperative adhesions by N, O-carboxymethylchitosan: A pilot study. *Fertil. Steril.*, 80, 3, 631–636, 2003.
123. Kayaoglu, C., Calikoğlu, C., Binler, S., Re-operation after lumbar disc surgery: Results in 85 cases. *J. Int. Med. Res.*, 31, 4, 318–323, 2003.

124. Pheasant, H.C., Sources of failure in laminectomies. *Orthop. Clin. North Am.*, 6, 1, 319–329, 1975.
125. Liu, G. *et al.*, Synthesis and characterization of chitosan/dextran-based hydrogels for surgical use, in: *Macromolecular Symposia*, Wiley Online Library, 2009.
126. Valentine, R. *et al.*, The efficacy of a novel chitosan gel on hemostasis and wound healing after endoscopic sinus surgery. *Am. J. Rhinol. Allergy*, 24, 1, 70–75, 2010.
127. Athanasiadis, T. *et al.*, Effects of a novel chitosan gel on mucosal wound healing following endoscopic sinus surgery in a sheep model of chronic rhinosinusitis. *Laryngoscope*, 118, 6, 1088–1094, 2008.
128. Rajiv, S. *et al.*, The efficacy and safety of chitosan dextran gel in a burr hole neurosurgical sheep model. *Acta Neurochir.*, 155, 7, 1361–1366, 2013.
129. Klopp, L.S. *et al.*, Use of polylactide resorbable film as a barrier to postoperative peridural adhesion in an ovine dorsal laminectomy model. *Neurosurg. Focus*, 16, 3, 1–9, 2004.
130. Richards, P.J. *et al.*, Reduction in postlaminectomy epidural adhesions in sheep using a fibrin sealant-based medicated adhesion barrier. *J. Biomed. Mater. Res. Part B: Appl. Biomater.: An Official Journal of the Society for Biomaterials, The Japanese Society for Biomaterials, and The Australian Society for Biomaterials and the Korean Society for Biomaterials*, 92, 2, 439–446, 2010.
131. Drespe, I.H. *et al.*, Animal models for spinal fusion. *Spine J.*, 5, 6, S209–S216, 2005.
132. Farrokhi, M.R. *et al.*, The effect of methylene blue on peridural fibrosis formation after laminectomy in rats: An experimental novel study. *Spine J.*, 11, 2, 147–152, 2011.
133. Martini, L. *et al.*, Sheep model in orthopedic research: A literature review. *Comp. Med.*, 51, 4, 292–299, 2001.
134. Zhou, H. *et al.*, A new *in vivo* animal model to create intervertebral disc degeneration characterized by MRI, radiography, CT/discogram, biochemistry, and histology. *Spine*, 32, 8, 864–872, 2007.
135. LaRocca, H. and Macnab, I., The laminectomy membrane: Studies in its evolution, characteristics, effects and prophylaxis in dogs. *J. Bone Joint Surg. Br.*, 56, 3, 545–550, 1974.
136. Moore, M.L. *et al.*, The effect of amniotic tissue on spinal interventions: A systematic review. *Int. J. Spine Surg.*, 17, 1, 32–42, 2023.
137. Xu, R. *et al.*, Clinical study on chitosan in prevention of knee adhesion after patellar operation. *Zhongguo Xiu Fu Chong Jian Wai Ke Za Zhi= Zhongguo Xiufu Chongjian Waike Zazhi= Chin. J. Reparative Reconstr. Surg.*, 16, 4, 240–241, 2002.
138. Yalman, V. and Laçin, N.T., Development of humic acid and alginate-based wound dressing and evaluation on inflammation. *Mater. Technol.*, 34, 12, 705–717, 2019.

139. Sarabahi, S., Recent advances in topical wound care. *Indian J. Plast. Surg.*, 45, 02, 379–387, 2012.
140. Zhao, J. *et al.*, Polyester-based nanoparticles for nucleic acid delivery. *Mater. Sci. Eng.: C*, 92, 983–994, 2018.
141. Westmeier, D. *et al.*, Boosting nanotoxicity to combat multidrug-resistant bacteria in pathophysiological environments. *Nanoscale Adv.*, 2, 11, 5428–5440, 2020.
142. Alt, V. *et al.*, An *in vitro* assessment of the antibacterial properties and cytotoxicity of nanoparticulate silver bone cement. *Biomaterials*, 25, 18, 4383–4391, 2004.
143. Lambrechts, S.A. *et al.*, Photodynamic therapy for Staphylococcus aureus infected burn wounds in mice. *Photochem. Photobiol. Sci.*, 4, 7, 503–509, 2005.
144. Liu, H. *et al.*, A functional chitosan-based hydrogel as a wound dressing and drug delivery system in the treatment of wound healing. *RSC Adv.*, 8, 14, 7533–7549, 2018.
145. Sonamuthu, J. *et al.*, MMP-9 responsive dipeptide-tempted natural protein hydrogel-based wound dressings for accelerated healing action of infected diabetic wound. *Int. J. Biol. Macromol.*, 153, 1058–1069, 2020.
146. Shivaram, A., Bose, S., Bandyopadhyay, A., Understanding long-term silver release from surface modified porous titanium implants. *Acta Biomater.*, 58, 550–560, 2017.
147. Molnar, T. *et al.*, Associations between serum L-arginine and ficolins in the early phase of acute ischemic stroke–a pilot study. *J. Stroke Cerebrovasc. Dis.*, 29, 8, 104951, 2020.
148. Sener, G. *et al.*, Injectable, self-healable zwitterionic cryogels with sustained microRNA-cerium oxide nanoparticle release promote accelerated wound healing. *Acta Biomater.*, 101, 262–272, 2020.
149. Chen, P. *et al.*, Comparative assessment of early versus delayed surgery to treat proximal femoral fractures in elderly patients: A systematic review and meta-analysis. *Int. J. Surg.*, 68, 63–71, 2019.
150. Jeelani, H., N.T., Afroze, D., Rashid, F., Association of Paraoxonase1 enzyme and its genetic single nucleotide polymorphisms with cardio-metabolic and neurodegenerative diseases. *Gene Rep.*, 20, 100775, 2020.
151. Aerden, L.K., S.W., Houthoofd, S., Matricali, G.A., Reviving the debate: Articular cartilage preservation during disarticulation at the lower limb? A systematic review. *Foot Ankle Surg.*, 27, 246, 2020.
152. Rosselle, L. *et al.*, An 'on-demand'photothermal antibiotic release cryogel patch: Evaluation of efficacy on an *ex vivo* model for skin wound infection. *Biomater. Sci.*, 8, 21, 5911–5919, 2020.
153. Behnke, M. *et al.*, Ethoxy acetalated dextran nanoparticles for drug delivery: A comparative study of formulation methods. *Int. J. Pharm.: X*, 5, 100173, 2023.

154. Solomon, L. *et al.*, The dextran sulphate sodium (DSS) model of colitis: An overview. *Comp. Clin. Path.*, 19, 3, 235–239, 2010.
155. Crescenzi, V. *et al.*, Novel hydrogels via click chemistry: Synthesis and potential biomedical applications. *Biomacromolecules*, 8, 6, 1844–1850, 2007.
156. Biological evaluation of medical devices. Tests for *in vitro* cytotoxicity, 1999, SO B. 10993-5.
157. Ibrahim, N.I. *et al.*, Wound healing properties of selected natural products. *Int. J. Environ. Res. Public Health*, 15, 11, 2360, 2018.
158. De Souza, K., Bassani, V., Schapoval, E., Influence of excipients and technological process on anti-inflammatory activity of quercetin and Achyrocline satureioides (Lam.) DC extracts by oral route. *Phytomedicine*, 14, 2-3, 102–108, 2007.
159. Harris, C.L. and Fraser, C., Institutionalized elderly: The effects on wound healing. *Ostomy/Wound Manage.*, 50, 10, 54–63, 2004.
160. Li, Y. *et al.*, *In situ* crosslinkable hydrogels formed from modified starch and O-carboxymethyl chitosan. *RSC Adv.*, 5, 38, 30303–30309, 2015.
161. Kumar, A. *et al.*, Biodegradable hydrogel-based biomaterials with high absorbent properties for non-adherent wound dressing. *Int. Wound J.*, 14, 6, 1076–1087, 2017.
162. Le Guilcher, C. *et al.*, Engineered human liver based on pullulan-dextran hydrogel promotes mice survival after liver failure. *Mater. Today Bio*, 19, 100554, April 2023.
163. Jan, A.T. *et al.*, Mercury pollution: An emerging problem and potential bacterial remediation strategies. *World J. Microbiol. Biotechnol.*, 25, 9, 1529–1537, 2009.
164. Eagles-Smith, C.A. *et al.*, Modulators of mercury risk to wildlife and humans in the context of rapid global change. *Ambio*, 47, 2, 170–197, 2018.
165. Fisher, J.F. and Organization, W.H., *Elemental mercury and inorganic mercury compounds: Human health aspects*, World Health Organization, 2003.
166. Zheng, P. *et al.*, A gold nanohole array based surface-enhanced Raman scattering biosensor for detection of silver (I) and mercury (II) in human saliva. *Nanoscale*, 7, 25, 11005–11012, 2015.
167. Aragay, G., Pons, J., Merkoçi, A., Recent trends in macro-, micro-, and nanomaterial-based tools and strategies for heavy-metal detection. *Chem. Rev.*, 111, 5, 3433–3458, 2011.
168. Clarkson, T.W. and Magos, L., The toxicology of mercury and its chemical compounds. *Crit. Rev. Toxicol.*, 36, 8, 609–662, 2006.
169. Steckling, N. *et al.*, Mercury exposure in female artisanal small-scale gold miners (ASGM) in Mongolia: An analysis of human biomonitoring (HBM) data from 2008. *Sci. Total Environ.*, 409, 5, 994–1000, 2011.
170. Tchounwou, P.B. *et al.*, Environmental exposure to mercury and its toxicopathologic implications for public health. *Environ. Toxicol.: An Int. J.*, 18, 3, 149–175, 2003.

171. Hawley, D.M., Hallinger, K.K., Cristol, D.A., Compromised immune competence in free-living tree swallows exposed to mercury. *Ecotoxicology*, 18, 5, 499–503, 2009.
172. Crowe, W. *et al.*, Mercury as an environmental stimulus in the development of autoimmunity–a systematic review. *Autoimmun. Rev.*, 16, 1, 72–80, 2017.
173. Steuerwald, U. *et al.*, Maternal seafood diet, methylmercury exposure, and neonatal neurologic function. *J. Pediatr.*, 136, 5, 599–605, 2000.
174. Basu, N. *et al.*, An interspecies comparison of mercury inhibition on muscarinic acetylcholine receptor binding in the cerebral cortex and cerebellum. *Toxicol. Appl. Pharmacol.*, 205, 1, 71–76, 2005.
175. Karunasagar, D., Arunachalam, J., Gangadharan, S., Development of a 'collect and punch'cold vapour inductively coupled plasma mass spectrometric method for the direct determination of mercury at nanograms per litre levels. *J. Anal. At. Spectrom.*, 13, 7, 679–682, 1998.
176. Fong, B.M.W. *et al.*, Determination of mercury in whole blood and urine by inductively coupled plasma mass spectrometry. *J. Anal. Toxicol.*, 31, 5, 281–287, 2007.
177. Kuswandi, B. *et al.*, Development of a disposable mercury ion-selective optode based on trityl-picolinamide as ionophore. *Anal. Chim. Acta*, 591, 2, 208–213, 2007.
178. Yang, Q. *et al.*, Direct detection of mercury in vapor and aerosol from chemical atomization and nebulization at ambient temperature: Exploiting the flame atomic absorption spectrometer. *J. Anal. At. Spectrom.*, 20, 8, 760–762, 2005.
179. Suddendorf, R.F., Watts, J.O., Boyer, K., Simplified apparatus for determination of mercury by atomic absorption and inductively coupled plasma emission spectroscopy. *J. Assoc. Off. Anal. Chem.*, 64, 5, 1105–1110, 1981.
180. Ichinoki, S., Kitahata, N., Fujii, Y., Selective determination of mercury (II) ion in water by solvent extraction followed by reversed-phase HPLC. *J. Liq. Chromatogr. Relat. Technol.*, 27, 11, 1785–1798, 2004.
181. Nevado, J.J.B. *et al.*, Determination of mercury species in fish reference materials by gas chromatography-atomic fluorescence detection after closed-vessel microwave-assisted extraction. *J. Chromatogr. A*, 1093, 1-2, 21–28, 2005.
182. Fan, A. *et al.*, Direct colorimetric visualization of mercury (Hg2+) based on the formation of gold nanoparticles. *Talanta*, 82, 2, 687–692, 2010.
183. Wang, J. and Tian, B., Screen-printed stripping voltammetric/potentiometric electrodes for decentralized testing of trace lead. *Anal. Chem.*, 64, 15, 1706–1709, 1992.
184. Li, C. *et al.*, One-pot controllable synthesis of Au@Ag heterogeneous nanorods with highly tunable plasmonic absorption. *Chem. Mater.*, 25, 13, 2580–2590, 2013.

185. Yang, Y. *et al.*, Transformation of Ag nanocubes into Ag–Au hollow nanostructures with enriched Ag contents to improve SERS activity and chemical stability. *ACS Appl. Mater. Interfaces*, 6, 5, 3750–3757, 2014.
186. Zhang, J. *et al.*, Colorimetric iodide recognition and sensing by citrate-stabilized core/shell Cu@Au nanoparticles. *Anal. Chem.*, 83, 10, 3911–3917, 2011.
187. Ganguly, M. *et al.*, Green synthesis and reversible dispersion of a giant fluorescent cluster in solid and liquid phase. *Langmuir*, 29, 34, 10945–10958, 2013.
188. Dutta, S. *et al.*, Facile synthesis of bimetallic Au-Pt, Pd-Pt, and Au-Pd nanostructures: Enhanced catalytic performance of Pd-Pt analogue towards fuel cell application and electrochemical sensing. *Electrochim. Acta*, 180, 1075–1084, 2015.
189. Sankar, M. *et al.*, Designing bimetallic catalysts for a green and sustainable future. *Chem. Soc. Rev.*, 41, 24, 8099–8139, 2012.
190. Zangeneh, H. *et al.*, Photocatalytic oxidation of organic dyes and pollutants in wastewater using different modified titanium dioxides: A comparative review. *J. Ind. Eng. Chem.*, 26, 1–36, 2015.
191. Xu, J., Chen, M., Fu, D., Study on highly visible light active Bi-doped TiO2 composite hollow sphere. *Appl. Surf. Sci.*, 257, 17, 7381–7386, 2011.
192. Paul, T., Dodd, M.C., Strathmann, T.J., Photolytic and photocatalytic decomposition of aqueous ciprofloxacin: Transformation products and residual antibacterial activity. *Water Res.*, 44, 10, 3121–3132, 2010.
193. Kokilavani, S. *et al.*, Citrate functionalized Ag NPs-polyethylene glycol nanocomposite for the sensitive and selective detection of mercury (II) ion, photocatalytic and antimicrobial applications. *Physica E: Low-Dimens. Syst. Nanostruct.*, 124, 114335, 2020.
194. Deng, Q. *et al.*, Ag nanoparticle decorated nanoporous ZnO microrods and their enhanced photocatalytic activities. *ACS Appl. Mater. Interfaces*, 4, 11, 6030–6037, 2012.
195. Koduru, J.R. *et al.*, Phytochemical-assisted synthetic approaches for silver nanoparticles antimicrobial applications: A review. *Adv. Colloid Interface Sci.*, 256, 326–339, 2018.
196. Xiang, Q. *et al.*, Microwave-hydrothermal preparation and visible-light photoactivity of plasmonic photocatalyst Ag-TiO2 nanocomposite hollow spheres. *Chem.–Asian J.*, 5, 6, 1466–1474, 2010.
197. Bernard, V. *et al.*, AgCu bimetallic nanoparticles under effect of low intensity ultrasound: The cell viability study *in vitro*. *J. Cancer Res.*, 6, 2014, 2014.
198. Patra, S., Purohit, S.S., Swain, S.K., *In vivo* fluorescence non-enzymatic glucose sensing technique for diabetes management by CQDs incorporated dextran nanocomposites in human blood serums. *Microchem. J.*, 190, 108646, 2023.

10

Biomedical Applications of Pullulan

Sudipta Roy[1]*, Swaraj Biswas[2], Soumalya Chakraborty[3], Sanjay Kumar Das[4] and Richita Mondal[5]

[1]Department of Pharmacology, Bengal School of Technology (A College of Pharmacy), Sugandha, Hooghly, West Bengal, India
[2]Department of Veterinary Biochemistry, West Bengal University of Animal and Fishery Sciences, Kolkata, West Bengal, India
[3]National Institute of Pharmaceuticals Education and Research (NIPER), Mohali, Punjab, India
[4]Institute of Pharmacy, Jalpaiguri, West Bengal, India
[5]Department of Pharmaceutics, Jadavpur University, West Bengal, India

Abstract
Pullulan is an anaerobically produced water-soluble glucan gum derived from carbohydrates and is obtained from a fungus named *Aureobasidium pullulans*. It was first discovered by Bauer in 1938 and was named "pullulans" by Bender in the year 1959. Pullulan can alternatively be called 1,6-linked maltotriose and is a linearly structured homopolysaccharide of glucopyranose. Pullulan possesses a GRAS status with sufficient safety profile. It has several biological advantages since it is non-toxic, non-mutagenic, biodegradable, non-carcinogenic, and hemocompatible. Due to high structural flexibility and possibility of modulating the properties of pullulan by easy chemical derivitization reactions, pullulan has attracted significant interest in biomedical science as a useful biopolymer for diverse applications including targeted drug delivery, gene delivery, tissue engineering, vaccination, medical imaging, film-forming agents, molecular chaperons, capsule coating, etc. Pullulan has been used widely in the field of pharmaceuticals due to its high solubility and inevitable biological characteristics. This chapter aims to provide information regarding the biomedical applications of pullulan as well as its sources and its physicochemical and biological properties.

***Keywords*:** Pullulan, biomedical applications, drug delivery, tissue engineering, imaging, molecular chaperones

**Corresponding author*: sudiptaroy1229@gmail.com

Annu (ed.) Biopolymers for Biomedical Applications, (253–286) © 2024 Scrivener Publishing LLC

10.1 Introduction

Recently, biomasses, a renewable class of material, have drawn a significant research interest among materials scientists and mainly consist of different classes of biopolymers. Biopolymers are obtained from diverse sources, including, plant, animals, and microbial sources and possess distinct physicochemical and functional properties. Chemically, they may belong to the class of proteins or polysaccharides or sometimes lipids or composite material [1]. The biopolymers obtained from microbial sources may be intracellular, extracellular, or structural, depending on their morphological localization. Bacterial cellulose, levan, pullulan, kefiran, xanthan, gellan, haloferax exopolysaccharides, polyhydroxyalkanoates, etc., are some examples of microbial biopolymers [2]. Pullulan is a linear homopolysaccharide of glucopyranose subunits, more specifically a maltotriose, with α-(1→6) linkage and produced by the fungus *Aureobasidium pullulans* (often called black yeast due to its property of forming a black pigment—melanin) [3]. Other microorganisms that produce pullulan includes *Tremella mesenterica, Cytaria harioti, Cytaria darwinii, Teloschistes flavicans, Rhodotorula bacarum*, and *Cryphonectria parasitica* [4]. Bernier, in the year 1958, first mentioned about pullulan. The chemical structure of pullulan was first revealed by Bender *et al.* in 1959. The commercial manufacturing of pullulan was initiated in 1976 by Hayashibara Company, located in Okayama, Japan. The commercial production of pullulan film was started by the same company in 1982 [5]. The biosynthesis of pullulan is dependent on the activity of mainly three enzymes, i.e., α-phosphoglucose mutase, uridine 5′-diphosphate glucose (UDPG) pyrophosphorylase and glucosyltransferase [5, 6] and can be done in two steps: (1) after initial acetone drying of the cells of microbes, pullulan chain formation reaction is initiated by UDPG in the presence of adenosine 5'-triphosphate. UDPG attaches a D-glucose residue to the lipid through a phosphoester bridge, which ultimately forms a lipid-linked structure, and (2) these lipid-linked glucose molecules then participate in a reaction with isomaltosyl to yield pyranosyl or isopanosyl residue, which undergo polymerization reaction to yield a pullulan chain. [4]. Pullulan is a white- to off-white-colored powder with a smooth surface [7]. It is an odorless, tasteless, and edible polymer. No evidence of carcinogenicity of pullulan has been found. Depending on the source, its molecular weight varies from 4.5×10^4 to 6×10^5 Da, and its molecular formula is $(C_6H_{10}O_5)_n$ [8]. Pullulan does not possess a sharp melting temperature, mainly because of its amorphous nature. Melting of pullulan occurs in a range of temperature from 250°C to 300°C [9, 10]. At a higher temperature, pullulan undergoes degradation, without release of any toxic gaseous substance [11]. A high concentration of hydroxyl groups in the main chain imparts the distinct inherent physiological attributes of pullulan [4]. Pullulan possesses a GRAS status with a sufficient safety profile. In Japan, pullulan has been safely used since the last 20 years in food processing and as a pharmaceutical diluent [12]. Due to the high structural flexibility and the possibility of modulating the properties of pullulan by easy chemical derivitization reactions, pullulan has attracted significant interest in biomedical science as a useful biopolymer for diverse applications, including targeted drug delivery, gene delivery, tissue engineering, vaccination, capsule coating, etc. [4]. The high aqueous solubility of native pullulan and its easily available hydrophobic derivatives led to its popular use as a drug delivery carrier for both the immediate and controlled release of bioactives. Pullulan derivatized with hydrophobic functional groups is mainly used as a controlled-release drug delivery carrier [13].

The current book chapter aims to provide comprehensive information regarding the biomedical applications of pullulan as well as captures different relevant aspects including sources, physicochemical, and biological properties of pullulan.

10.2 Sources of Pullulan

The yeast-like fungus *Aureobasidium pullulans* is the primary source of pullulan, which is previously referred to as *Pulluria pullulans* [14, 15]. *A. pullulans* is commonly found in ecological niches like forest soil and fresh and sea water and as a saprophyte on wood, litter, decaying leaf, animal tissue, and other plants [16]. As a polymorphic fungus, at the beginning, it grows like yeast, and after that, it divides in its filamentous form—thus producing pullulan [17–19]. *A. pullulans* is also called "black yeast", and it can also produce blackish to dark-green-pigment melanin. Thus, due care should be taken during the production and extraction of pullulan in order to avoid the interference of melanin [20, 21]. Two varieties of *A. pullulans* are present, i.e., var. pullulans and var. aubasidani. *A. pullulans* var. pullulans strain of the fungi can only produce pullulan using ammonium sulphate as the nitrogen source, but *A. pullulans* var. aubasidani strain used sodium nitrate as a nitrogen source for the production of aubasidan exopolysaccharide [22, 23]. Moreover, *A. pullulans* can also produce β glucans and different types of enzymes [24, 25]. Apart from *A. pullulans*, pullulan has also been produced aerobically from multiple strains of *Aureobasidium* spp., such as *A. melanogenum* and *A. mousonni* [26]. In addition to this,

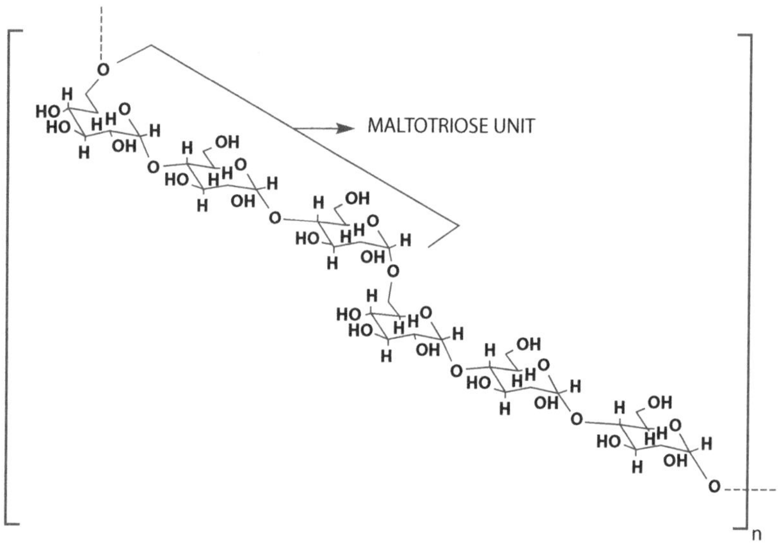

Figure 10.1 Structure of pullulan [17].

pullulan can also be isolated from *Cytaria darwinii* [27], *Cytaria harioti* [28], *Cryphonectria parasitica* [29], and a saprophytic fungus *Termella mesenterica* [30].

Besides these, there are different fungal sources of pullulan like *Teloschistes falvicans* [31], *Cryphonectria parasitica* (Murrill) [32], and *Eurotium cheyalieri* [33]. Pullulan can also be extracted from *Rhodotorula bacarum* found in Chinese plant leaves [34]. Pullulan is also produced from *Aspergillus japonicas* isolated from a culinary leaf sample and *Rhodosporidium paludigenum* isolated from fresh leaves as well as decaying leaves [35, 36]. Pullulan produced from these microbial cultures could use different carbon sources such as jaggery, corn steep liquor, de-oiled jatropha seed cake [37], cassava baggase [38], Asian palm kernel, jack fruit seed [39, 40], glucose, sucrose, maltose, lactose [17, 41], peat hydrolysate [42], and rice hull [43]. The structure of pullulan is depicted in Figure 10.1.

10.3 Properties of Pullulan

Pullulan is a highly water-soluble polysaccharide which forms a stable, viscous, adhesive, transparent, and colorless solution. Pullulan is insoluble in organic solvent like methanol and acetone, but it is soluble in dimethyl sulfoxide and dimethyl formamide [16, 44, 45]. It is highly hygroscopic, flexible, bearing elastic properties, and also stable in the presence of metal ions. Pullulan is structurally highly flexible and thermally stable. It can withstand a temperature of 250–280°C. A pullulan solution has low viscosity and is soluble in the pH range of 5–7 [46, 47]. It has good mechanical strength, foam retention, and adhesive properties. Besides these, it also has a film-forming ability and a fine adhesive property [48].

Pullulan is a biopolymer having biodegradability and biocompatibility. It is edible, non-mutagenic, non-carcinogenic, non-immunogenic, and non-toxic. It is devoid of odor and taste [46, 49–51]. All of these properties of pullulan have been utilized in the fields of food, pharmaceutical, cosmetics, and biomedicine.

10.4 Biomedical Applications

Polymers of natural origin have been used for a long time in biomedical sciences as they are biocompatible, non-toxic, and cost-effective. Pullulan has been extensively used in the field of pharmaceuticals and food industry. Apart from these, it has been applied in drug delivery, gene delivery, tissue engineering, wound healing, molecular imaging, vaccination, etc.

10.4.1 Drug Delivery

Pullulan has been used as a medium of drug delivery for several medical conditions. Various derivatives of pullulan, for example, carboxymethyl pullulan, cholesterol-bearing pullulan, pullulan succinylate, pullulan acetate, and pullulan amine, are widely used for drug release. Due to the presence of the excellent hemocompatibility properties of pullulan, it has shown promising results in the treatment of tumors. Because of the presence of discontinuous epithelial cells in the blood vessel linings of tumor tissues, the passages of nanoparticles into the tissues are enabled [51–53]. The derivatives of pullulan were studied for conjugating with protein for the drug delivery systems. The cholesterol-modified pullulan (CHP) has been conjugated with human recombinant erythropoietin and glucagon-like peptide-1-fluorescein-labeled insulin for sustained protein drug delivery into the blood [54] as well as CHP-doxorubicin to reduce its adverse effects in the liver [55]. The study on doxorubicin (DOX) was performed by Li *et al.*, and they reported that pullulan can be added with DOX of 30 wt% and nanoparticle of 50 to 110 nm. The amount of DOX conjugated with pullulan nanoparticle was found to be used as a better drug delivery system [55]. CHP, when complexed with docetaxel, markedly lowers the pleural tumor cells' growth in the lungs of a murine model and also lowers the inhibitory concentration of the prepared drug pullulan complex [56]. Mitoxantrone, another anti-cancer drug, when added to cholesterol-modified pullulan, has shown more than 90% drug release in tumor cells and has thus proven to be effective in the treatment of cancer [57]. The epirubicin-conjugated CHP has been reported to improve the half-life of the modified conjugate in female Wistar albino rats [58].

Pullulan acetate (PA) has been conjugated with drugs like clonazepam, adriamycin, naproxen, and doxorubicin. These conjugates have shown improved properties like sustained release of clonazepam [59] and strong adsorption between the drug adriamycin and hepatocellular carcinoma (HepG2) cells [60]. PA-conjugated nanoparticles also showed enhanced pH sensitivity and thus influences the release of both naproxen and doxorubicin [61, 62]. Similarly, microsphere formulations were prepared by the conjugation of pullulan succinate (PS) with drugs like diclofenac and antimicrobial enzyme lysozyme. At pH 7.4, particularly in the intestinal fluid, the PS microspheres swelled 20 times, thus triggering the busting of microspheres and the release of NSAID diclofenac in the intestine [63] and, at 28 times swelling, causing an effective release of the lysozyme [64]. Another modification of pullulan, carboxymethyl pullulan (CMP), after conjugation with doxorubicin, forms a doxorubicin–peptide–CMP conjugate. Nogusa *et al.*, in the year 2000, reported enhanced antitumor properties in mouse reticulosarcoma model with reduced toxicity [65]. Pluronic-modified pullulan nanogels with doxorubicin have shown antitumor effects and have potential to inhibit tumor growth [66]. All trans retinoic acid–pullulan have been conjugated again with doxorubicin, and nanogels were prepared, which exerted a controlled drug release pattern [67]. Additionally, doxorubicin conjugated with pullulan-g-poly(L-lactide) showed sustained drug release on tumur cells [68]. Folate-decorated maleilated pullulan nanoparticles formed with the drug doxorubicin have been reported to enhance the cellular uptake in human ovarian cancer cell line [69].

Oxidized pullulan (OP) has been reported to specifically target the carcinoma cells of the liver as reported by Balasso *et al.* [70]. Pullulan has been also widely utilized in the delivery of lipophilic drugs. Modified pullulan can be utilized as a polymeric stabilizer as well as emulsifier for lipophilic drugs by conjugating it with poly(ε-caprolactone) and beclomethasone dipropionate. The conjugate also exhibited delayed photodegradation of beclomethasone dipropionate by UVC [71]. In addition to these biomedical applications, pullulan has also been experimented to be used as dry powder inhalers for lung diseases [72]. Chitosan-grafted dihydrocaffeic acid and oxidized pullulan-conjugated hydrogels exhibited efficient drug release in the mucosal region and may be an ideal candidate for colon cancer treatment [73]. A chitosan–pullulan–silver nanocomposite was developed as a carrier system for the efficient release of the drug moxifloxacin. Moxifloxacin was highly effective against methicillin-resistant *Staphylococcus aureus* and *Pseudomonas aeruginosa* due to its prolonged release from the nanocomposites. This prolonged release occurred due to the presence of pullulan in the nanocomposites [74]. Huang *et al.* developed paclitaxel drug-loaded cross-linked pullulan nanoparticles [75]. These nanocomposites were efficiently distributed inside the liver cells and displayed prolonged retention time. This can act as an efficient delivery system for the treatment of liver cancer.

10.4.2 Gene Delivery System

Pullulan is regarded safe by the Food and Drug Administration in the United States and has GRAS status [76]. Genetic disorders, which are a result of defective genes, can be cured by gene therapy [77]. Pullulan is advantageous over other polysaccharides as it is biocompatible and heamocompatible and can travel across the blood–brain barrier and reach the brain. Being inert and non-toxic, it does not show any adverse effects on the host or does not undergo gene mutation and is thus used as a promising carrier in gene delivery systems for gene therapies [78]. The entrapment of genes inside a pullulan can protect the modified pullulan formulation from DNase degradation and will show improved effects on the targeted organs. Lee *et al.*, in the year 2004, explained how cholesterol-bearing pullulan is used for the targeted delivery of genes having a hydrophobic nature [79]. Pullulans, when modified by the addition of one cholesterol group, results into an inside hydrophobic core, leading to the improved self-aggregation properties of the modified pullulan [4, 80]. Diethylamino-ethylamine pullulan (DEAEP) is another modified pullulan which has the capability to protect genes from DNase enzyme degradation and thus can be formulated in either 3D-matrices or tabular forms [81]. Folate modification of pullulan has proven to increase the efficacy of gene transfection and gene slicing [82]. Pullulan spermine has been reported to support the release of dopamine from the brain and is effective in the treatment of Parkinson's disease [83]. Further addition of lipopolysaccharides to the pullulan spermine delivery system leads to an increase in the efficiency of pullulan as a gene delivery tool [84, 85]. Pullulan 3D matrices with DEAE and dendronized pullulan with guanidine are also promising gene delivery systems [86, 87]. Pullulan hydrogels can be used in the treatment of cancer because of its high loading efficiency with plasmid DNA, which results into sustained release of DNA to the cancer cells [88]. Another modification of pullulan is polyethylenimine pullulan (PEIP) and is advantageous in tumor cell targeting due to low side effects [89]. Liver targeting with the desired genes can also be carried out with the help of cationic polyethylenimine pullulan formulation [88]. Modified pullulan is also reported for protein delivery systems (e.g., truncated HER2-147) for therapy of the immune system [90]. Pullulan derivative formulations, for example, succinylated pullulan [91], pullulan-g-poly(L-lysine) [92], and pullulan-protamine [93], are reported to be useful carriers for gene delivery at targeted sites. Table 10.1 describes

Table 10.1 Modified pullulan formulations and their applications in gene delivery [11].

Modified pullulan	**Gene(s)**	**Target organs**	**Applications**
DEAEP (diethylamino-ethylamine pullulan)	Plasmid DNA	Vascular smooth muscle cells	Demonstrates high affinity for plasmid and enhances DNA transfection to the vascular smooth muscle cells
	Embryonic alkaline phosphatase plasmid	Vascular smooth muscle cells	Exhibits significant gene transfer with 150-fold increase in activity and shows no toxicity to normal cells
	Small interfering RNA	Vascular cells	Successful implantation of drug-eluting pullulan-coated stents loaded with siRNA in rabbit balloon-injured carotid arteries and induces an uptake of siRNA into the arterial wall with a lower activity of pro-matrix

(*Continued*)

Table 10.1 Modified pullulan formulations and their applications in gene delivery [11]. (*Continued*)

Modified pullulan	Gene(s)	Target organs	Applications
PEIP (polyethylenimine pullulan)	Wild-type human p53	Tumor cells	Betaine-modified PEIP nanoplexes with higher molecular weight PEI (10 kDa) possess excellent transfection efficacy for p53 gene and 90% cell death C6 rat glioma cells
	Wild-type human p53	Tumor cells	Ascorbic acid-modified PEIP nanoplexes show high transfection efficacy for p53 gene in C6 rat glioma cells and 96% cell death
	Small interfering RNA	Liver	Increases fluorescence level in liver; reduces gene toxicity and lowers the mortality rate in Balb/c mice
	Small interfering RNA; plasmid DNA	Tumor cells	DNA entrapped in folate-modified PEIP complex shows higher gene transfection in the presence of serum in HeLa cells with folate receptor in normal (FR+); siRNA complexed with folate-modified PEIP displays highly efficient gene silencing in HeLa cells with folate receptor in normal (FR+)

(*Continued*)

Table 10.1 Modified pullulan formulations and their applications in gene delivery [11]. (*Continued*)

Modified pullulan	Gene(s)	Target organs	Applications
	Tumor-suppressor gene p53	Tumor cells	Deoxycholic acid-grafted PEIP nanomicelles have good DNA-binding ability with efficient transport of genes p53 and higher cytotoxicity against human breast cancer (MCF-7) cells
PS (pullulan–spermine)	Plasmid DNA expressing green fluorescent protein wild-type human p53	Tumor cells	Iron oleate-modified PS shows lower magnetic saturation due to presence of organic compounds and 92% survival rate during cytotoxicity tests in Uppsala malignant glioma (U87) cells as glioblastoma cells
	Cytomegalovirus plasmid wild-type human p53; cytomegalovirus plasmid-β galactosidase	T24 cells of human bladder cancer	Successful transfection of plasmid DNA and 10-fold increase in gene expression level; Significantly reduces cytotoxicity and proliferation of T24 cells
	Mammalian expression vector expressing notch intracellular domain	Bone marrow stromal cells	Successfully transcribes exogenous plasmid genes and expresses as proteins in BMSCs cytoplasm with least cell death

(*Continued*)

Table 10.1 Modified pullulan formulations and their applications in gene delivery [11]. (*Continued*)

Modified pullulan	Gene(s)	Target organs	Applications
PPA (pullulan–protamine)	Methotrexate plasmid; DNA expressing green fluorescent protein	Tumor cells	Complete distribution of gene occurs in hepatocellular carcinoma (HepG2) bearing nude mice
	Renilla luciferase control vector plasmid, cytomegalovirus plasmid	Liver	Increases the gene transfection as well as the cellular uptake rate in hepatocellular carcinoma (HepG2) cells
	Wild-type human p53	Tumor cells	High efficiency to deliver plasmid inside cell nucleus, cellular uptake capability in C6 cells and facilitates high gene expression
CHP (cholesterol-modified pullulan)	Carbonic anhydrase B	Not Specified	Traps heat-denatured carbonic anhydrase B and prevents their aggregations by host-guest interaction of cholesteryl group and cyclodextrin
	Carbonic anhydrase B	Not Specified	Prevents thermal aggregation of heat-denatured protein by selective trapping and thermoresponsive controlled association in proteins

(*Continued*)

Table 10.1 Modified pullulan formulations and their applications in gene delivery [11]. (*Continued*)

Modified pullulan	Gene(s)	Target organs	Applications
	Insulin	Not Specified	Nanogels have ability to trap and release protein by host-guest interaction and shows high chaperone-like activity for refolding of chemically denatured protein
PH (pullulan hydrogel)	pBUDLacZ plasmid	Tumor cells	Gene transfection in common mammalian production cells
SP (succinylated pullulan)	Lysozyme	Not Specified	Improves the enzyme stability; Shows long term gene delivery at slow rate
PGPLL [pullulan-g-poly(L-lysine)]	Enhanced green fluorescent protein plasmid with N terminal fusion	Tumor cells	Successful gene transfection in human embryonic kidney (HEK293), hepatocellular carcinoma (HepG2) and keratin-forming tumor (KB) cell lines

the various modified pullulan formulations that have proven to be effective in targeted gene deliveries [11].

10.4.3 Tissue Engineering and Wound Healing

Wound injury is a common day-to-day problem leading to the breakdown of cellular and anatomic functions of a living tissue or organ. Among the various types of wounds (abrasion, laceration, burns, cutaneous, chemical, and punctured), burn injury is the most common type of traumatic injury leading to long hospitalizations and costly wound and scar

Table 10.2 Modified pullulan formulations and their applications in tissue engineering [98].

Application	Cell type treated	System used	Growth factors
Wound dressing	NIH 3T3 mouse embryonic fibroblast cells	Pullulan-chitosan composite nanofibers	Tannic acid
Wound healing	Rat skin cells	Pullulan film	-
Wound healing	Full thickness defect on rat dermal cells	Cholesterol-bearing pullulan nanogels	Prostaglandin E1
Wound dressing and antibacterial effect	*In vitro* drug release using phosphate buffer saline	Carboxymethyl pullulan hydrogels	Gentamicin
Early cutaneous wound healing	Full thickness transverse incisions on skin cells	Pullulan–collagen composite hydrogels	-
High-oxidative-stress wound healing	Anterior and posterior full-thickness transverse incisions on skin cells	Pullulan hydrogels	Mesenchymal stem cells
Non-load-bearing bone formation	Pre-osteoblast cell line	Pullulan-based composite scaffold	nHA and poly(3-hydroxybutyrate)
Bone formation	Skull cells	Cholesterol and acrylol-bearing pullulan nanogels	Bone morphogenetic protein and recombinant human fibroblast growth factor
Vascular endothelial cell proliferation	Human umbilical vein endothelial cells	Carboxymethyl pullulan membranes	Heparin
Vascular cell engineering	Rabbit vascular smooth muscle cells	Pullulan hydrogels	-

treatments [94]. Wound healing progresses in an orderly fashion through overlapping phases of processes which include induction of inflammation by the initial injury followed by granulation tissue formation and re-epithelization and finally ECM deposition and remodeling with wound contraction [95]. Such a process of healing the damaged tissues by creating a niche environment that is suitable for their differentiation, survival as well as functional achievement is called tissue engineering. The continuous supply of nutrients and growth factors enhances the differentiation of cells and thus the functional aspects of the cell [96, 97]. An artificial structure of a three-dimensional scaffold provides the nutrients required for the process of tissue engineering [48]. Pullulan has good swelling properties because of its hydrophilic nature and thus possesses the ability to form hydrogels with water rapidly [98, 99]. The prepared hydrogels have better ability to adjust with various physiological fluctuations and environmental factors like temperature, pH, ionic strength, polarity, mechanical properties, etc. For these contributing factors, pullulan has been proven to be a better candidate for its use in tissue engineering [100, 161]. Capsule preparation using pullulan, sodium alginate, and glass ceramics exhibited improved osteoblast proliferation as well as fibroblasts in host tissue, thus enabling the regeneration of bone tissue [101]. Pullulan films have shown good wound healing properties when applied on rat skin cells, whereas pullulan–collagen composite hydrogels may be used for early cutaneous wound healing [102, 103]. Three-dimensional scaffolds prepared from pullulan gels caused wound contraction and re-epithelialization of wound cells effectively in mice [104]. Table 10.2 describes the pullulans along with their applications that are being used in tissue engineering [98].

10.4.4 Medical Imaging

Medical imaging technique may be defined as the visual representation of a body's interior by labeling the body parts and/or inner cells with fluorescence dyes. It is a highly powerful tool to reveal the internal function of organs or tissues as well as to diagnose and treat diseases of medical importance. Quantum dots (QDs) are the nanometer-sized semiconductor nanocrystals that can be used as fluorescent probes to visualize the cellular environment. These QDs have properties like high luminescence, broad excitation spectra, narrow emission spectra, bright fluorescence, high photostability, low toxicity, and biocompatibility [105–109]. Quantum dots can enter inside the cell through endocytosis. The endocytosis of QDs is very low, and for bioimaging, there should be enough amount of detectable QDs inside the cells. Thus, poor tissue penetration depth and aqueous solubility and stability are the disadvantages of QDs [110]. Previously, cationic liposome has been used as a carrier molecule for QDs to overcome the poor tissue penetration properties, but liposomes can form large-sized aggregates inside the cells, which limit their use in bioimaging [111, 112]. In order to make QDs more biocompatible, different polysaccharide molecules have been linked to it. Pullulan is the one of them that can be conjugated with QDs for efficient fine imaging. Hasegawa *et al.* have developed cholesterol-conjugated pullulan and amine-modified cholesterol-bearing pullulan nanogel [113]. This nanogel can act as a carrier to deliver the QDs into the cells. The size of the nanoparticles was 38 nm. The study showed that the fluorescence intensity of the cholesterol-modified amino pullulan–QD nanoparticle was 3.4 times higher than the liposome–QD complex. The cells can uptake cholesterol pullulan-conjugated QD nanoparticles. Therefore, introduction of a cationic amino group on the backbone of cholesterol–pullulan enhances the cellular uptake, and the QDs were a better fluorescent probe for medical imaging than the conventional liposomes. Contrast-enhanced ultrasonography is an emerging tool in ultrasound imaging technique where gas-filled echogenic microbubbles are used as a contrast agent. In a study, a gas-generating calcium carbonate/pul-PCB (CPP) nanoparticle was developed by Chen *et al.* [114]. This CPP hybrid was used to detect the hepatoma through contrast-enhanced ultrasound imaging. CPP nanoparticles concentrated in tumor cells and under an acidic tumor environment of echogenic CO_2 bubbles produce enhanced ultrasound signals. CPP hybrid nanoparticles can generate six times of contrast enhancement than the normal imaging technique in hepatoma-bearing mice.

Pullulan could also be used as a novel contrast agent for magnetic resonance imaging (MRI). By using TEMPO reaction, a negatively charged carboxyl group was introduced into the backbone of pullulan. Then, carboxylated pullulan was further used for the coating of magnetic Fe_3O_4 nanoparticles. Therefore, carboxylated pullulan-embedded iron oxide nanoparticles of 3 nm in size can be used as a magnetic contrast agent for MRI. Upon reaching the liver, this contrast agent had a lower degree of diffusion compared to the conventional one. It could be the agent of choice where low diffusive property is suitable like a damaged blood vessel barrier of the affected organ or region [115]. Pullulan conjugated with iron oxide nanoparticles is a promising contrast agent for MRI labeling of mesenchymal stem cells [116].

Pullulan could be easily used as a liver targeting agent due to the presence of asialoglycoprotein receptor. Carbon dots were helpful in renal clearance. Using the property of pullulan and carbon dots, zerovalent iron-based nanoparticles stabilized by pullulan and coated with carbon dots were developed and analyzed as a possible contrast agent for MRI. On injection in Wister rats, it can produce a significant increase of signal in the liver where it resides for more than 30 min before getting eliminated *via* the kidney. This contrast agent also has the property of fluorescence which can be utilized in live animal imaging [117]. Yim *et al.* developed a liver-specific contrast agent, pullulan-conjugated gadolinium diethylene triamine pentaacetate (Gd-DTPA-pullulan), for delayed MRI [118]. It was preferentially used in hepatocarcinoma. It can last up to 24 h within the body and would not produce any toxicity. Pullulan-based cationic polymeric nanoparticles such as ethanolamine-functionalized poly(glycidyl methacrylate) and gadolinium were developed by Huang *et al.* as a good MRI agent [119]. Technetium-99m-radiolabeled pullulan and dextran and conjugated with fucoidan microparticles were developed for abdominal aortic aneurysm. Using SPECT radiotracing method, microparticles were injected in rats in order to estimate the human residence time [120]. Sentinel node navigation surgery can be used in case of gastric cancer. Infrared dye 900 conjugated with cholesterol-bearing pullulan, i.e., near-infrared polymer nanogels (NIR-PNG), was used as a tracer in sentinel nodes, and the dispersion pattern was studied using NIR imaging. This NIR-PNG nanogel

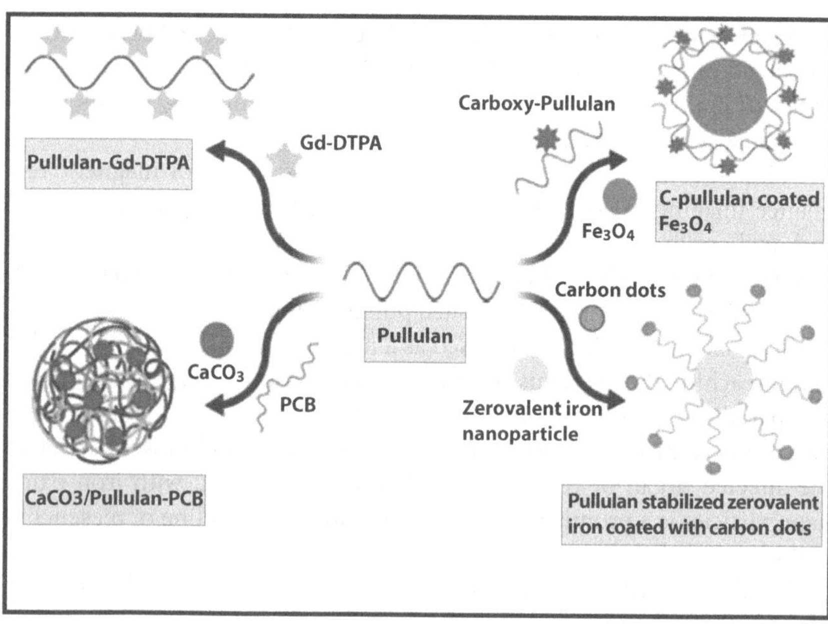

Figure 10.2 Various pullulan conjugates for imaging applications [122].

shows least dispersion and longer duration in sentinel nodes and could be used as a good tracer for sentinel node navigation surgery in gastric cancer [121]. Figure 10.2 illustrates various types of pullulan conjugates for biomedical application [122].

10.4.5 Vaccination

In order to prevent infectious diseases, it is essential to provoke immune responses by the process of vaccination. Most of the vaccines available in the market worldwide are injectables, with a very few that can be administered by oral or nasal route. Pullulan, when conjugated with chitosan, the cationic pullulan nanogels, has exhibited a promising nasal vaccine candidate for the prevention of various respiratory ailments [123]. A modified version of pullulan, cholesteryl group-bearing pullulan where a cholesterol group is attached to the backbone of pullulan, is widely used for vaccine delivery. A nanogel was synthesized bearing a non-toxic recombinant portion of *Clostridium botulinum* type A neurotoxin with cholesteryl group-bearing pullulan. The resultant cationic nanogel improved its adherence to the mucosal surface of the nasal epithelium which is negatively charged. It was observed that botulinum type A neurotoxin was retained on the mucosal surface for a long time, and a rapid release of the vaccine started within an hour and continued for 12 h. This nasal vaccine delivery system exhibited active immune responses, and high titers of serum and secretory antibodies were obtained [124, 125]. CHP nanogels were also used to deliver ghrelin conjugated with pneumococcal surface protein A (PspA) through the intranasal route using cyclic-di-GMP as an adjuvant. An increase in serum IgG antibodies was observed along with reduction of obesity in mice. Thus, the deposition of fat got attenuated, and an increase in energy expenditure was reported [126]. Kong *et al.* tested the efficacy of CHP-nanogel-bearing PspA to combat infections and tested them on murine models. Xen10 strain *Streptococcus pneumonia* infection, which can be lethal, was challenged with a nanogel vaccine, and it was observed that there was a marked decrease in the colonization of bacteria as well as upper and lower respiratory tract invasion. A specific rise in the levels of nasal and bronchial IgA, serum IgG, and Th17 cytokine production in the nasal and systemic–mucosal surfaces was reported, but the vaccine was not able to cross the blood–brain barrier, and this ensured the safety of the vaccine [127]. Similarly, Fukuyama *et al.* used CHP-nanogels for nasal administration in non-human primates. They reported long-term retention of the nanogels in the nasal cavity without its deposition in either CNS or olfactory bulb. A high amount of serum antibodies and PspA-specific IgG antibodies was also found. Along with these, increased levels of miR-181a and miR-326 in the respiratory tract tissues and plasma of the experimental non-human primates were reported [128]. Pullulan composites with TNF-α also have exhibited superior humoral and cell-mediated responses in case of influenza virus B [129]. Pullulan also takes part in the storage of vaccines, especially live vaccines.

10.4.6 Film-Forming Agent

The dosage form of a drug molecule could be tablet, capsule, syrup, elixir, paste, ointment, powder, emulsion, suspension, *etc.* Different dosage forms are introduced for the rapid disintegration and distribution of drug molecules at the site of action within the body. Film coating improves the ingestion ease of capsules and the tablet's appearance and covers the unpleasant taste of drug [130]. It helps in targeted drug delivery and could reduce the tablet processing time. Various film coating polymeric agents like gelatine, agar, amylase, modified starch pullulan, polyvinyl alcohol, and hydroxypropyl methylcellulose has been widely used. Pullulan ester and/or ether could be used as a film coating agent to increase the strength and shelf-life of capsules. Pullulan could be easily dissolved and forms a gel with water. Water-soluble neutral pullulan was also used in capsule coating, which forms a low-oxygen, permeable, solid film [131, 132]. Pullulan-based strip films may be effective for the rapid release of poorly water-soluble class II drugs of the Biopharmaceutical Classification System [133].

Pullulan and its derivatives can be used in oral consumable films. Various antimicrobial agents like essential oils thymol, methanol, eucalyptol, methyl salicylate, saliva-stimulating agents, deodorizing agents, and taste-masking agents (amberlite), i.e., sweeteners, flavors, and citric acid, could be easily integrated within the pullulan films [134–136]. Plasticizers such as polyethylene glycol (PEG) 300 and PEG 400 have also been used with pullulan in the formulation of fast-dissolving drugs [137]. Cetirizine hydrochloride coated with pullulan film has fast-dissolving property along with a shelf-life of 6 months at room temperature [138]. Pullulan-based enalapril maleate mucoadhesive buccal film was prepared for increased bioavailability and rapid therapeutic efficacy of the drug [139]. Nisoldipine sublingual tablets were prepared using solubility-enhancer pullulan and chitosan as adjuvant to decrease the flushing of saliva. Pullulan helps to increase the bioavailability and stability of drugs [140]. Orally disintegrating films using pullulan as a film former has also been developed for bitter levocetirizine dihydrochloride [141]. Similarly, fast-dissolving films containing pullulan were prepared for the easy dissolution of drugs like pilocarpine hydrochloride, lidocaine hydrochloride, dexamethasone, and lactoferin [142]. Matrix-assisted pulsed laser evaporation (MAPLE) technique was used to prepare high-quality pullulan films. These films were used in artificial organ lining, cell growth substrate, and immunology testing agents. Some of the different MAPLE-derived films were triacetate-pullulan, tosylate-pullulan, and cinnamate-pullulan [143, 144].

10.4.7 Plasma Expander

Plasma expander is a hypertonic and/or hyperoncotic colloidal or crystalloid fluid that expands the volume of circulating blood. The colloid osmotic pressure and viscosity of blood plasma could be comparable with the plasma expander. It should be non-toxic, non-pyrogenic, easily stored, and compatible with the body [145]. During any polytrauma or severe accident, if circulating blood volume is reduced, then blood transfusion is required. Modified pullulan could have been used as a plasma expander in order to replenish adequate blood volume and the osmotic pressure of blood vessel. Pullulan is non-toxic, non-immunogenic, and biocompatible and does not create any adverse cross-reaction or tissue reaction [146–148]. It could be easily metabolized and removed after producing the desired therapeutic effects [145, 149]. Low molecular weight γ-irradiated pullulan has been used as a plasma expander [150, 151]. Gamma-irradiated pullulan conjugated with cationic and anionic surfactants could produce different low molecular weight fractions. These fractions were also used as a blood volume expander [148]. The modified pullulan with a molecular weight range of 30,000 to 90,000 could have been used in the treatment and prevention of hemorrhage [145]. Methanol, ethanol, isopropanol, and acetone have also been used for the fractionation of modified pullulan.

10.4.8 Molecular Chaperones

Molecular chaperones are the group of proteins that help in the proper folding and assembly of unfolded or misfolded protein. Protein, after synthesis, should be folded into three-dimensional structures to gain its biological activity. Most of the proteins do not fold spontaneously and are at the risk of aberrant folding and aggregation. To overcome these dangers, networks of molecular chaperones ensure dynamic protein folding and maintain protein homeostasis [152]. The nanogel formulation of hydrophobized cholesterol-bearing pullulan has been widely used as molecular chaperones. It can be used in the refolding of citrate synthase enzyme [153, 154].

CHP assists in the refolding of chymotrypsin by increasing its thermal stability. Rhodanes protein was also efficiently folded by the CHP nanogel chaperone system. Cholesterol-bearing pullulan nanogel, upon the addition of cyclodextrin, could prevent the aggregation tendency of the proteins. Cyclodextrin binds to CHP and releases it in the form of folded protein [155].

The nanogel formulation of CHP has also been found efficient in the refolding and thermal stabilization of heat-denatured carbonic anhydrase B enzyme with cyclodextrin [149]. Cyclodextrin has the capacity to control the binding of CHP nanogels with protein [156]. Thermal stabilization of horseradish peroxidase (HRPO) could be increased by aggregation with CHP nanogels, and further addition of cyclodextrin promotes the recovery of active HRPO by up to 80% [157]. Heat-induced denaturation and lyophilization-induced lipase aggregation were also inhibited by the action of CHP nanogels [158]. Spiropyrane-induced pullulan nanogels were also used as molecular chaperones *via* photostimulation [159]. The polymerization of cholesterol-pullulan with 2-methacryloyloxyethyl phosphorylcholine and cross-linking of haluronan could have been used as molecular chaperones in the refolding of insulin, carbonic anhydrase B and glucagon-like peptide-1, insulin, and erythropoietin, respectively [160, 162, 163].

10.4.9 Insulinotropic Activity

Insulin-secreting cells with biomaterials have been developed in order to synthesize insulin from artificial pancreas, but the immune system of the body would not be properly compatible with artificial pancreas. That creates improper functioning of the artificial structure. Kim *et al.* had resolved the issues by using a sulphonourea–pullulan (SUP) conjugate [164]. SUP has insulinotropic effects on rat pancreatic islet microcapsules. The insulin secretion level has increased in the microcapsules of islets with SUP. These pullulan conjugate microcapsules can maintain their morphology in intact manner and have superior ability in responding to glucose changes.

10.5 Conclusion

The presence of distinct physico-chemical properties has made pullulan play an important role in biomedical research. In the recent years, owing to its biocompatibility and hemocompatibility, it has proved to be a promising carrier in gene delivery systems and tissue engineering. In the field of drug delivery, pullulan has become inevitable due to its contribution in improving drug release. Various derivatives of pullulan, for example, carboxymethyl pullulan, CHP, pullulan succinylate, pullulan acetate, and pullulan amine, are widely used for drug release. It also has a significant role in the vaccination and folding of proteins as molecular chaperones. As a result of these significant applications and effectiveness, pullulan is expected to be more useful in the delivery of drugs, diagnosis, and therapeutics, thus improving the healthcare system more and more in the near future.

References

1. Das, S.K. and Chakraborty, S., Functional characteristics of pullulan: Its physicochemical and physiological properties, in: *Pullulan- Processing, Properties, and Applications*, S. Ahmed and A. Soundararajan (Eds.), pp. 133–148, Jenny Stanford Publishing, New York, 2020.
2. Singh, R.S. and Kaur, N., Microbial biopolymers for edible film and coating applications, in: *Advances in Industrial Biotechnology*, N. Nawani, M. Khetmalas, P. Razdan, A. Pandey (Eds.), pp. 185–216, IK International Publishing House Pvt. Ltd, New Delhi, 2015.
3. Leathers, T.D., Biotechnological production and applications of pullulan. *Appl. Microbiol. Biotechnol.*, 62, 468, 2003.
4. Singh, R.S., Kaur, N., Singh, D., Purewal, S.S., Kennedy, J.F., Pullulan in pharmaceutical and cosmeceutical formulations: A review. *Int. J. Biol. Macromol.*, 231, 123353, 2023.
5. Singh, R. and Saini, G., Biosynthesis of pullulan and its applications in food and pharmaceutical industry, in: *Microorganisms in Sustainable Agriculture and Biotechnology*, T. Satyanarayana, B. Johri, A. Prakash (Eds.), pp. 509–553, Springer, Dordrecht, 2012.
6. Duan, X., Chi, Z., Wang, L., Wang, X., Influence of different sugars on pullulan production and activities of α-phosphoglucose mutase, UDPG-pyrophosphorylase and glucosyltransferase involved in pullulan synthesis in Aureobasidium pullulans Y68. *Carbohydr. Polym.*, 73, 587, 2008.
7. Soni, S.R., Kumari, N., Bhunia, B.K., Sarkar, B., Mandal, B.B., Ghosh, A., Synthesis and characterization of a non-cytotoxic and biocompatible acrylamide grafted pullulan -application in pH responsive controlled drug delivery. *Int. J. Biol. Macromol.*, 120, 753, 2018.
8. Petrov, P.T., Shingel, K.I., Scripko, A.D., Tsarenkov, V.M., Biosynthesis of pullulan by Aureobasidium pullulans strain BMP-97. *Biotechnologia*, 1, 36, 2002.
9. Trovatti, E., Fernandes, S.C., Rubatat, L., da Silva Perez, D., Freire, C.S., Silvestre, A.J., Neto, C.P., Pullulan-nanofibrillated cellulose composite films

with improved thermal and mechanical properties. *Compos. Sci. Technol.*, 72, 1556, 2012.
10. Zhang, C., Gao, D., Ma, Y., Zhao, X., Effect of gelatin addition on properties of pullulan films. *J. Food Sci.*, 78, C805, 2013.
11. Singh, R.S., Kaur, N., Hassan, M., Kennedy, J.F., Pullulan in biomedical research and development - A review. *Int. J. Biol. Macromol.*, 166, 694, 2021.
12. Rekha, M.R. and Sharma, C.P., Blood compatibility and *in vitro* transfection studies on cationically modified pullulan for liver cell targeted gene delivery. *Biomaterials*, 30, 6655, 2009.
13. Kabil, M.F., Dena, A.S.A., El-Sherbiny, I.M., Pullulan-based nanocarriers for therapeutic applications, in: *Polymeric Nanosystems*, M.S. Hasnain, A.K. Nayak, T.M. Aminabhavi (Eds.), pp. 309–331, Academic Press, Elsevier, 2023.
14. Cooke, W.B., An ecological life history of *Aureobasidium pullulans* (de Bary) Arnaud. *Mycopathologia*, 12, 1, 1959.
15. Kumar, D., Saini, N., Pandit, V., Ali, S., An insight to pullulan: A biopolymer in pharmaceutical approaches. *Int. J. Basic Appl. Sci.*, 1, 202, 2012.
16. Oguzhan, P. and Yangilar, F., Pullulan: Production and usage in food industry. *Afr. J. Food Sci.*, 4, 2141, 2013.
17. Singh, R.S., Saini, G.K., Kennedy, J.F., Pullulan: Microbial sources, production and applications. *Carbohydr. Polym.*, 73, 515, 2008.
18. Catley, B.J., Utilization of carbon sources by *Pullularia pullulans* for the elaboration of extracellular polysaccharides. *Appl. Microbiol.*, 22, 641, 1971.
19. Ono, K., Yasuda, N., Ueda, S., Effect of pH on pullulan elaboration by *Aureobasidium pullulans* S-1. *Agric. Biol. Chem.*, 41, 2113, 1977.
20. Seihr, D.J., Melanin biosynthesis in *Aureobasidium pullulans*. *J. Coat. Technol.*, 53, 23, 1981.
21. Singh, R.S., Saini, G.K., Kennedy, J.F., Downstream processing and characterization of pullulan from a novel colour variant strain of *Aureobasidium pullulans* FB-1. *Carbohydr. Polym.*, 78, 89, 2009.
22. Bulmer, M.A., Catley, B.J., Kelly, P.J., The effect of ammonium ions and pH on the elaboration of the fungal extracellular polysaccharide, pullulan, by Aureobasidium pullulans. *Appl. Microbiol. Biotechnol.*, 25, 362, 1987.
23. Yurlova, N.A. and De Hoog, G.S., A new variety of Aureobasidium pullulans characterized by exopolysaccharide structure, nutritional physiology and molecular features. *Antonie van Leeuwenhoek*, 72, 141, 1997.
24. Jung, M.Y., Kim, J.W., Kim, K.Y., Choi, S.H., Ku, S.K., Polycan, a beta-glucan from *Aureobasidium pullulans* SM-2001, mitigates ovariectomy-induced osteoporosis in rats. *Exp. Ther. Med.*, 12, 1251, 2016.
25. Leathers, T.D., Substrate regulation and specificity of amylases from *Aureobasidium* strain NRRLY-12,974. *FEMS Microbiol. Lett.*, 110, 217, 1993.
26. Chi, Z., Liu, N.N., Jiang, H., Wang, Q.Q., Chen, J.T., Liu, G.L., Hu, Z., Chi, Z.M., Relationship between β-D-fructofuranosidase activity,

fructooligosaccharides and pullulan biosynthesis in Aureobasidium melanogenum P16. *Int. J. Biol. Macromol.*, 125, 1103, 2019.
27. Oliva, E.M., Cirelli, A.F., De Lederkremer, R.M., Characterization of a pullulan in Cyttaria darwinii. *Carbohydr. Res.*, 158, 262, 1986.
28. Waksman, N., De Lederkremer, R.M., Cerezo, A.S., The structure of an a-D-glucan from *Cyttaria harioti* Fischer. *Carbohydr. Res.*, 59, 505, 1977.
29. Delben, F., Forabosco, A., Guerrini, M., Liut, G., Torri, G., Pullulans produced by strains of *Cryphonectria parasitica*-II. Nuclear magnetic resonance evidence. *Carbohydr. Polym.*, 63, 545, 2006.
30. Fraser, C.G. and Jennings, H.J., A glucan from Tremella mesenterica NRRL-Y6158. Can. *J. Chem.*, 49, 1804, 1971.
31. Reis, R.A., Tischer, C.A., Gorin, P.A., Iacomini, M., A new pullulan and a branched (1→ 3)-,(1→ 6)-linked β-glucan from the lichenised ascomycete Teloschistes flavicans. *FEMS Microbiol. Lett.*, 210, 1, 2002.
32. Corsaro, M.M., De Castro, C., Evidente, A., Lanzetta, R., Molinaro, A., Parrilli, M., Sparapano, L., Phytotoxic extracellular polysaccharide fractions from Cryphonectria parasitica (Murr.) Barr strains. *Carbohydr. Polym.*, 37, 167, 1998.
33. Gaur, R., Singh, R., Tiwari, S., Yadav, S.K., Daramwal, N.S., Optimization of physico-chemical and nutritional parameters for a novel pullulan-producing fungus, Eurotium chevalieri. *J. Appl. Microbiol.*, 109, 1035, 2010.
34. Chi, Z. and Zhao, S., Optimization of medium and cultivation conditions for pullulan production by a new pullulan-producing yeast strain. *Enzyme Microb. Technol.*, 33, 206, 2003.
35. Mishra, B. and Suneetha, V., Biosynthesis and hyper production of pullulan by a newly isolated strain of Aspergillus japonicus-VIT-SB1. *World J. Microbiol. Biotechnol.*, 30, 2045, 2014.
36. Singh, R.S. and Kaur, N., Biochemical and molecular characterization of a new pullulan producer Rhodosporidium paludigenum PUPY-06. *J. Appl. Biol. Biotechnol.*, 6, 28, 2018.
37. Mehta, A., Prasad, G.S., Choudhury, A.R., Cost effective production of pullulan from agri-industrial residues using response surface methodology. *Int. J. Biol. Macromol.*, 64, 252, 2014.
38. Sugumaran, K.R., Jothi, P., Ponnusami, V., Bioconversion of industrial solid wastecassava bagasse for pullulan production in solid state fermentation. *Carbohydr. Polym.*, 99, 22, 2014.
39. Sugumaran, K.R., Shobana, P., Balaji, P.M., Ponnusam, V., Gowdhaman, D., Statistical optimization of pullulan production from Asian palm kernel and evaluation of its properties. *Int. J. Biol. Macromol.*, 66, 229, 2014.
40. Sugumaran, K.R., Sindhu, R.V., Sukanya, S., Aiswarya, N., Ponnusami, V., Statistical studies on high molecular weight pullulan production in solid state fermentation using jack fruit seed. *Carbohydr. Polym.*, 98, 854, 2013.

41. Singh, R.S., Singh, H., Saini, G.K., Response surface optimization of the critical medium components for pullulan production by *Aureobasidium pullulans* FB-1. *Appl. Biochem. Biotechnol.*, 152, 42, 2009.
42. LeDuy, A. and Boa, J.M., Pullulan production from peat hydrolyzate. *Can. J. Microbiol.*, 29, 143, 1983.
43. Wang, D., Ju, X., Zhou, D., Wei, G., Efficient production of pullulan using rice hull hydrolysate by adaptive laboratory evolution of *Aureobasidium pullulans*. *Bioresour. Technol.*, 164, 12, 2014.
44. Youssef, F., Roukas, T., Biliaderis, C.G., Pullulan production by a nonpigmented strain of *Aureobasidium pullulans* using batch and fed-batch culture. *Process Biochem.*, 34, 355, 1999.
45. Ghosh, T., Priyadarshi, R., de Souza, C.K., Angioletti, B.L., Rhim, J.W., Advances in pullulan utilization for sustainable applications in food packaging and preservation: A mini-review. *Trends Food Sci. Technol.*, 125, 43, 2022.
46. Gniewosz, M. and Duszkiewicz-Reinhard, W., Comparative studies on pullulan synthesis, melanin synthesis and morphology of white mutant *Aureobasidium pullulans* B-1 and parent strain A.p.-3. *Carbohydr. Polym.*, 72, 431, 2008.
47. Farris, S., Introzzi, L., Fuentes-Alventosa, J.M., Santo, N., Rocca, R., Piergiovanni, L., Self-assembled pullulan–silica oxygen barrier hybrid coatings for food packaging applications. *J. Agric. Food Chem.*, 60, 782, 2012.
48. Shingel, K.I., Current knowledge on biosynthesis, biological activity, and chemical modification of the exopolysaccharide, pullulan. *Carbohydr. Res.*, 339, 447, 2004.
49. Muthusamy, S., Anandharaj, S.J., Kumar, P.S., Meganathan, Y., Vo, D.V.N., Vaidyanathan, V.K., Muthusamy, S., Microbial pullulan for food, biomedicine, cosmetic, and water treatment: A review. *Environ. Chem. Lett.*, 20, 3199, 2022.
50. Agrawal, S., Budhwani, D., Gurjar, P., Telange, D., Lambole, V., Pullulan based derivatives: Synthesis, enhanced physicochemical properties, and applications. *Drug Deliv.*, 29, 3328, 2022.
51. Mocanu, G., Mihai, D., Le Clerf, D., Picton, L., Muller, G., Synthesis of new associative gel microspheres from carboxymethyl pullulan and their interactions with lysozyme. *Eur. Polym. J.*, 40, 283, 2004.
52. Lu, D., Wen, X., Liang, J., Gu, Z., Zhang, X., Fan, Y., A pH-sensitive nano drug delivery system derived from pullulan/ doxorubicin conjugate. *J. Biomed. Mater. Res. Part B: Appl. Biomater.*, 89, 177, 2009.
53. Scomparina, A., Salmasoa, S., Bersania, S., Satchi-Fainarob, R., Caliceti, P., Novel folated and non-folated pullulan bioconjugates for anticancer drug delivery. *Eur. J. Pharm. Sci.*, 42, 547, 2011.
54. Hirakura, T., Yasugi, K., Nemoto, T., Sato, M., Shimoboji, T., Aso, Y., Morimoto, N., Akiyoshi, K., Hybrid hyaluronan hydrogel encapsulating nanogel as a protein nanocarrier: New system for sustained delivery of protein with a chaperone-like function. *J. Control. Release*, 142, 483, 2010.

55. Li, H., Bian, S., Huang, Y., Liang, J., Fan, Y., Zhang, X., High drug loading pH-sensitive pullulan-DOX conjugate nanoparticles for hepatic targeting. *J. Biomed. Mater. Res. Part A*, 102A, 150, 2014.
56. Satoh, K., Chen, F., Aoyama, A., Date, H., Akiyoshi, K., Nanoparticle of cholesterol-bearing pullulan as a carrier of anticancer drugs. *Eur. J. Cancer Suppl.*, 9, 139, 2008.
57. Yuan, L., Guo, B., Zhong, W., Nie, Y., Yao, X., Peng, X., Wang, R., Yu, H., Yang, S., He, C., Tao, X., Zhang Q., Interaction of mitoxantrone-loaded cholesterol modified pullulan nanoparticles with human serum albumin and effect on drug release. *J. Nanomater.*, 2019, Article ID 8036863, 13 pages, 2019, https://doi.org/10.1155/2019/8036863.
58. Shen, S., Li, H., Yang, W., The preliminary evaluation on cholesterol-modified pullulan as a drug nanocarrier. *Drug Deliv.*, 21, 501, 2014.
59. Jung, S.W., Jeong, Y.L., Kim, Y.H., Kim, S.H., Self-assembled polymeric nanoparticles of poly (ethylene glycol) grafted pullulan acetate as a novel drug carrier. *Arch. Pharm. Res.*, 27, 562, 2004.
60. Na, K., Lee, T.B., Park, K.H., Shin, E.K., Lee, Y.B., Choi, H.K., Self-assembled nanoparticles of hydrophobically-modified polysaccharide bearing vitamin H as a targeted anti-cancer drug delivery system. *Eur. J. Pharm. Sci.*, 18, 165, 2003.
61. Bishwambhar, M. and Suneetha, V., Release study of naproxen, a modern drug from pH sensitive pullulan acetate microsphere. *Int. J. Drug Dev. Res.*, 4, 184, 2012.
62. Juan, A.S., Hlawaty, H., Chaubet, F., Letourneur, D., Feldman, L.J., Cationized pullulan 3D matrices as new materials for gene transfer. *J. Biomed. Mater. Res. Part A*, 82, 354, 2007.
63. Constantin, M., Fundueanu, G., Bortolotti, F., Cortesi, R., Ascenzi, P., Menegatti, E., A novel multicompartimental system based on aminated poly (vinyl alcohol) microspheres/succinoylated pullulan microspheres for oral delivery of anionic drugs. *Int. J. Pharm.*, 330, 129, 2007.
64. Fundueanu, G., Constantin, M., Ascenzi, P., Preparation and characterization of pH-and temperature-sensitive pullulan microspheres for controlled release of drugs. *Biomaterials*, 29, 2767, 2008.
65. Nogusa, H., Yamamoto, K., Yano, T., Kajiki, M., Hamana, H., Okuno, S., Distribution characteristics of carboxymethylpullulan-peptide-doxorubicin conjugates in tumor-bearing rats: Different sequence of peptide spacers and doxorubicin contents. *Biol. Pharm. Bull.*, 23, 621, 2000.
66. Zheng, Y., Lv, X., Xu, Y., Cheng, X., Wang, X., Tang, R., pH-sensitive and pluronic-modified pullulan nanogels for greatly improved antitumor *in vivo*. *Int. J. Biol. Macromol.*, 139, 277, 2019.
67. Lee, J., Jeong, D., Seo, S., Na, K., Biodegradable nanogel based on all-trans retinoic acid/pullulan conjugate for anti-cancer drug delivery. *J. Pharm. Investig.*, 43, 63, 2013.

68. Cho, J.K., Park, W., Na, K., Self-organized nanogels from pullulan-g-poly (l-lactide) synthesized by one-pot method: Physicochemical characterization and *in vitro* doxorubicin release. *J. Appl. Polym. Sci.*, 113, 2209, 2009.
69. Li, F., Zhang, H., Gu, C., Fan, L., Qiao, Y., Tao, Y., Cheng, C., Wu, H., Yi, J., Self-assembled nanoparticles from folate-decorated maleilated pullulan–doxorubicin conjugate for improved drug delivery to cancer cells. *Polym. Int.*, 62, 165, 2013.
70. Balasso, A., Salmaso, S., Pontisso, P., Rosato, A., Quarta, S., Malfanti, A., Mastrotto, F., Caliceti, P., Re-programming pullulan for targeting and controlled release of doxorubicin to the hepatocellular carcinoma cells. *Eur. J. Pharm. Sci.*, 103, 104, 2017.
71. Chassot, J.M., Ferreira, L.M., Gomes, F.P., Da Silva, C.D., Tasso, L., Cruz, L., Pullulan as a stabilizer agent of polymeric nanocapsules for drug delivery. *Braz. J. Pharm. Sci.*, 52, 735, 2016.
72. Xu, E., Jiang, W., Xu, Y., Cai, Y., Li, H., Pullulan-modified spray-dried powders for pulmonary drug delivery. *J. Control. Release*, 213, e8, 2015.
73. Liang, Y., Zhao, X., Ma, P.X., Guo, B., Du, Y., Han, X., pH-responsive injectable hydrogels with mucosal adhesiveness based on chitosan-grafted-dihydrocaffeic acid and oxidized pullulan for localized drug delivery. *J. Colloid Interface Sci.*, 536, 224, 2019.
74. Shah, A., Ashames, A.A., Buabeid, M.A., Murtaza, G., Synthesis, *in vitro* characterization and antibacterial efficacy of moxifloxacin-loaded chitosan-pullulan-silver-nanocomposite films. *J. Drug Deliv. Sci. Technol.*, 55, 101366, 2020.
75. Huang, L., Wang, Y., Ling, X., Chaurasiya, B., Yang, C., Du, Y., Tu, J., Xiong, Y., Sun, C., Efficient delivery of paclitaxel into ASGPR over-expressed cancer cells using reversibly stabilized multifunctional pullulan nanoparticles. *Carbohydr. Polym.*, 159, 178, 2017.
76. FDA, US food and drug administration, center for food safety and applied nutrition, office of food safety, agency response letter, GRAS Notice No. GRN 000099, 2002, http://www.cfsan.fda.gov/~rdb/opa-g099.html.
77. Anderson, W., Human gene therapy. *Science*, 256, 808, 1992.
78. Hemapriya, J., Ravi, A. *et al.*, Biomedical applications of pullulan, in: *Pullulan-Processing, Properties, and Applications*, S. Ahmed and A. Soundararajan (Eds.), pp. 257–276, Jenny Stanford Publishing, New York, 2020.
79. Lee, I. and Akiyoshi, K., Single molecular mechanics of a cholesterol-bearing pullulan nanogel at the hydrophobic interfaces. *Biomaterials*, 25, 2911, 2004.
80. Singh, R.S., Kaur, N., Kennedy, J.F., Pullulan and pullulan derivatives as promising biomolecules for drug and gene targeting. *Carbohydr. Polym.*, 123, 190, 2015.
81. San Juan, A., Ducrocq, G., Hlawaty, H., Bataille, I., Guénin, E., Letourneur, D., Feldman, L.J., Tubular cationized pullulan hydrogels as local reservoirs for plasmid DNA. *J. Biomed. Mater. Res. Part A*, 83, 819, 2007.

82. Wang, J., Dou, B., Bao, Y., Efficient targeted pDNA/siRNA delivery with folate–low-molecular-weight polyethyleneimine–modified pullulan as non-viral carrier. *Mater. Sci. Eng. C*, 34, 98, 2014.
83. Nagane, K., Kitada, M., Wakao, S., Dezawa, M., Tabata, Y., Practical induction system for dopamine-producing cells from bone marrow stromal cells using spermine-pullulan-mediated reverse transfection method. *Tissue Eng. Part A*, 15, 1655, 2009.
84. Thomsen, L.B., Lichota, J., Kim, K.S., Moos, T., Gene delivery by pullulan derivatives in brain capillary endothelial cells for protein secretion. *J. Control. Release*, 151, 45, 2011.
85. Ueda, M., Jo, J., Gao, J.Q., Tabata, Y., Effect of lipopolysaccharide addition on the gene transfection of spermine-introduced pullulan-plasmid DNA complexes for human mesenchymal stem cells. *J. Biomater. Sci.*, 30, 1542, 2019.
86. Zhou, J., Wali, A.R.M., Ma, S., He, Y., Yue, D., Tang, J.Z., Gu, Z., Tailoring the supramolecular structure of guanidinylated pullulan toward enhanced genetic photodynamic therapy. *Biomacromolecules*, 19, 2214, 2018.
87. Juan, A.S., Hlawaty, H., Chaubet, F., Letourneur, D., Feldman, L.J., Cationized pullulan 3D matrices as new materials for gene transfer. *J. Biomed. Mater. Res. Part A*, 82, 354, 2007.
88. Kang, J.H., Tachibana, Y., Kamata, W., Mahara, A., Harada-Shiba, M., Yamaoka, T., Liver-targeted siRNA delivery by polyethylenimine (PEI)-pullulan carrier. *Bioorg. Med. Chem.*, 18, 3946, 2010.
89. Ambattu, L.A. and Rekha, M.R., Collagen synthesis promoting pullulan–PEI–ascorbic acid conjugate as an efficient anti-cancer gene delivery vector. *Carbohydr. Polym.*, 126, 52, 2015.
90. Ikuta, Y., Katayama, N., Wang, L., Okugawa, T., Takahashi, Y., Schmitt, M., Gu, X., Watanabe, M., Akiyoshi, K., Nakamura, H., Kuribayashi, K., Presentation of a major histocompatibility complex class 1–binding peptide by monocyte-derived dendritic cells incorporating hydrophobized polysaccharide–truncated HER2 protein complex: Implications for a polyvalent immuno-cell therapy. *Blood*, 99, 3717, 2002.
91. Kim, H. and Na, K., Evaluation of succinylated pullulan for long-term protein delivery in poly (lactide-co-glycolide) microspheres. *Macromol. Res.*, 18, 812, 2010.
92. Park, J.S., Park, J.K., Nam, J.P., Kim, W.S., Choi, C., Kim, M.Y., Jang, M.K., Nah, J.W., Preparation of pullulan-g-poly (L-lysine) and it's evaluation as a gene carrier. *Macromol. Res.*, 20, 667, 2012.
93. Liu, Y., Wang, Y., Zhang, C., Zhou, P., Liu, Y., An, T., Sun, D., Zhang, N., Wang, Y., Core–shell nanoparticles based on pullulan and poly (β-amino) ester for hepatoma-targeted codelivery of gene and chemotherapy agent. *ACS Appl. Mater. Interfaces*, 6, 18712, 2014.
94. Bhatia, N., Singh, A., Sharma, R., Singh, A., Soni, V., Singh, G., Bajaj, J., Dhawan, R., Singh, B., Evaluation of burn wound healing potential of

aqueous extract of Morus alba based cream in rats. *J. Phytopharmacol.*, 3, 378, 2014.
95. Tan, W.S., Arulselvan, P., Ng, S.F., Taib, C.N.M., Sarian, M.N., Fakurazi, S., Improvement of diabetic wound healing by topical application of Vicenin-2 hydrocolloid film on Sprague Dawley rats. *BMC Complement Altern. Med.*, 19, 1, 2019.
96. Elangwe, C.N., Morozkina, S.N., Olekhnovich, R.O., Polyakova, V.O., Krasichkov, A., Yablonskiy, P.K., Uspenskaya, M.V., Pullulan-based hydrogels in wound healing and skin tissue engineering applications: A review. *Int. J. Mol. Sci.*, 24, 4962, 2023.
97. Tabata, Y., Tissue regeneration based on drug delivery technology, in: *Topics in Tissue Engineering*, N. Ashammakhi and P. Ferretti (Eds.), pp. 1–32, University of Oulu, Oulu, 2003.
98. Singh, R.S., Kaur, N., Rana, V., Kennedy, J.F., Recent insights on applications of pullulan in tissue engineering. *Carbohydr. Polym.*, 153, 455, 2016.
99. Kuckling, D., Vo, C.D., Wohlrab, S.E., Preparation of nanogels with temperature-responsive core and pH-responsive arms by photo-cross-linking. *Langmuir*, 18, 4263, 2002.
100. Li, H., Yang, J., Hu, X., Liang, J., Fan, Y., Zhang, X., Superabsorbent polysaccharide hydrogels based on pullulan derivate as antibacterial release wound dressing. *J. Biomed. Mater. Res. A*, 98, 31, 2011.
101. Popescu, R.A., Magyari, K., Taulescu, M., Vulpoi, A., Berce, C., Bogdan, S., Lelescu, C., Dreancă, A., Tudoran, O., Papuc, I., Baia, L., New alginate-pullulan-bioactive glasses composites with copper oxide for bone tissue regeneration trials. *J. Tissue Eng. Regen. Med.*, 12, 2112, 2018.
102. Suguna, L., Wound healing potential of a biodegradable film from pullulan in rats. *J. Clin. Exp. Dermatol. Res.*, 5, 63, 2014.
103. Wong, V.W., Rustad, K.C., Galvez, M.G., Neofytou, E., Glotzbach, J.P., Januszyk, M., Major, M.R., Sorkin, M., Longaker, M.T., Rajadas, J., Gurtner, G.C., Engineered pullulan-collagen composite dermal hydrogels improve early cutaneous wound healing. *Tissue Eng. Part A*, 17, 631, 2011.
104. Iswariya, S., Bhanukeerthi, A.V., Velswamya, P., Uma, T.S., Perumal, P.T., Design and development of a piscine collagen blended pullulan hydrogel for skin tissue engineering. *RSC Adv.*, 6, 57863, 2013.
105. Karagianni, A., Tsierkezos, N.G., Prato, M., Terrones, M., Kordatos, K.V., Application of carbon-based quantum dots in photodynamic therapy. *Carbon*, 203, 273, 2023.
106. Arya, H., Kaul, Z., Wadhwa, R., Taira, K., Hirano, T., Kaul, S.C., Quantum dots in bioimaging: Revolution by the small. *Biochem. Biophys. Res. Commun.*, 329, 1173, 2005.
107. Michalet, X., Pinaud, F.F., Bentolila, L.A., Tsay, J.M., Doose, S.J.J.L., Li, J.J., Sundaresan, G., Wu, A.M., Gambhir, S.S., Weiss, S., Quantum dots for live cells, *in vivo* imaging, and diagnostics. *Science*, 307, 538, 2005.

108. Wagner, A.M., Knipe, J.M., Orive, G., Peppas, N.A., Quantum dots in biomedical applications. *Acta Biomater.*, 94, 44, 2019.
109. Cotta Mônica, A., Quantum dots and their applications: What lies ahead? *ACS Appl. Nano Mater.*, 3, 4920, 2020.
110. Rekha, M.R. and Sharma, C.P., Pullulan as a promising biomaterial for biomedical applications: A perspective. *Trends Biomater. Artif. Organs*, 20, 116, 2007.
111. Derfus, A.M., Chan, W.C.W., Bhatia, S.N., Intracellular delivery of quantum dots for live cell labeling and organelle tracking. *Adv. Mater.*, 16, 961, 2004.
112. Voura, E.B., Jaiswal, J.K., Mattoussi, H., Simon, S.M., Tracking metastatic tumor cell extravasation with quantum dot nanocrystals and fluorescence emission-scanning microscopy. *Nat. Med.*, 10, 993, 2004.
113. Hasegawa, U., Shin-ichiro, M.N., Kaul, S.C., Hirano, T., Akiyoshi, K., Nanogel-quantum dot hybrid nanoparticles for live cell imaging. *Biochem. Biophys. Res. Commun.*, 331, 917, 2005.
114. Chen, S., Xu, X.L., Zhou, B., Tian, J., Luo, B.M., Zhang, L.M., Acidic pH-activated gas-generating nanoparticles with pullulan decorating for hepatoma-targeted ultrasound imaging. *ACS Appl. Mater. Interfaces*, 11, 22194, 2019.
115. Biliuta, G., Sacarescu, L., Socoliuc, V., Iacob, M., Gheorghe, L., Negru, D., Coseri, S., Carboxylated polysaccharides decorated with ultrasmall magnetic nanoparticles with antibacterial and MRI properties. *Macromol. Chem. Phys.*, 218, 1700062, 2017.
116. Jo, J.I., Aoki, I., Tabata, Y., Design of iron oxide nanoparticles with different sizes and surface charges for simple and efficient labeling of mesenchymal stem cells. *J. Control. Release*, 142, 465, 2010.
117. Nimi, N., Saraswathy, A., Nazeer, S.S., Francis, N., Shenoy, S.J., Jayasree, R.S., Multifunctional hybrid nanoconstruct of zerovalent iron and carbon dots for magnetic resonance angiography and optical imaging: An *in vivo* study. *Biomaterials*, 171, 46, 2018.
118. Yim, H., Yang, S.G., Jeon, Y.S., Park, I.S., Kim, M., Lee, D.H., Bae, Y.H., Na, K., The performance of gadolinium diethylene triamine pentaacetate-pullulan hepatocyte-specific T1 contrast agent for MRI. *Biomaterials*, 32, 5187, 2011.
119. Huang, Y., Hu, H., Li, R.Q., Yu, B., Xu, F.J., Versatile types of MRI-visible cationic nanoparticles involving pullulan polysaccharides for multifunctional gene carriers. *ACS Appl. Mater. Interfaces*, 8, 3919, 2016.
120. Desbrée, A., Bonnard, T., Blanchardon, E., Petiet, A., Franck, D., Chauvierre, C., Le Visage, C., Evaluation of functionalized polysaccharide microparticles dosimetry for SPECT imaging based on biodistribution data of rats. *Mol. Imaging Biol.*, 17, 504, 2015.
121. Kong, S.H., Noh, Y.W., Suh, Y.S., Park, H.S., Lee, H.J., Kang, K.W., Kim, H.C., Lim, Y.T., Yang, H.K., Evaluation of the novel near-infrared fluorescence tracers pullulan polymer nanogel and indocyanine green/γ-glutamic acid

complex for sentinel lymph node navigation surgery in large animal models. *Gastric Cancer*, 18, 55, 2015.
122. Raychaudhuri, R., Naik, S., Shreya, A.B., Kandpal, N., Pandey, A., Kalthur, G., Mutalik, S., Pullulan based stimuli responsive and sub cellular targeted nanoplatforms for biomedical application: Synthesis, nanoformulations and toxicological perspective. *Int. J. Biol. Macromol.*, 161, 1189, 2020.
123. Cevher, E., Salomon, S.K., Makrakis, A., Li, X.W., Brocchini, S., Alpar, H.O., Development of chitosan–pullulan composite nanoparticles for nasal delivery of vaccines: Optimisation and cellular studies. *J. Microencapsul.*, 32, 755, 2015.
124. Nakahashi-Ouchida, R., Yuki, Y., Kiyono, H., Cationic pullulan nanogel as a safe and effective nasal vaccine delivery system for respiratory infectious diseases. *Hum. Vaccin. Immunother.*, 14, 2189, 2018.
125. Nochi, T., Yuki, Y., Takahashi, H., Sawada, S.I., Mejima, M., Kohda, T., Harada, N., Kong, I.G., Sato, A., Kataoka, N., Tokuhara, D., Nanogel antigenic protein-delivery system for adjuvant-free intranasal vaccines. *Nat. Mater.*, 9, 572, 2010.
126. Azegami, T., Yuki, Y., Sawada, S., Mejima, M., Ishige, K., Akiyoshi, K., Itoh, H., Kiyono, H., Nanogel-based nasal ghrelin vaccine prevents obesity. *Mucosal Immunol.*, 10, 1351, 2017.
127. Kong, I.G., Sato, A., Yuki, Y., Nochi, T., Takahashi, H., Sawada, S., Mejima, M., Kurokawa, S., Okada, K., Sato, S., Briles, D.E., Kunisawa, J., Inoue, Y., Yamamoto, M., Akiyoshi, K., Kiyono, H., Nanogel-based PspA intranasal vaccine prevents invasive disease and nasal colonization by Streptococcus pneumoniae. *Infect. Immun.*, 81, 1625, 2013.
128. Fukuyama, Y., Yuki, Y., Katakai, Y., Harada, N., Takahashi, H., Takeda, S., Mejima, M., Joo, S., Kurokawa, S., Sawada, S., Shibata, H., Nanogel-based pneumococcal surface protein A nasal vaccine induces microRNA-associated Th17 cell responses with neutralizing antibodies against Streptococcus pneumoniae in macaques. *Mucosal Immunol.*, 8, 1144, 2015.
129. Nagatomo, D., Taniai, M., Ariyasu, H., Taniguchi, M., Aga, M., Ariyasu, T., Ohta, T., Fukuda, S., Cholesteryl pullulan encapsulated TNF-α nanoparticles are an effective mucosal vaccine adjuvant against influenza virus. *BioMed. Res. Int.*, 2015, Article ID 471468, 15 pages, 2015, https://doi.org/10.1155/2015/471468.
130. Childers, R.F., Oren, P.L., Seidler, W.M.K., Film coating formulations, US Patent 5015480A, assigned to Eli Lilly and Co., USA, 1991.
131. Miyamoto, Y., Goto, H., Sato, H., Okano, H., Iijima, M., Process for sugar-coating solid preparation, US Patent 4610891A, assigned to Zeria-Shinyaku Kogyo KK, USA, 1986.
132. Scott, R., Cade, D., He, X., Pullulan film compositions, US Patent 6887307B1, assigned to Warner Lambert Co. LLC, USA, 2005.
133. Krull, S.M., Ma, Z., Li, M., Davé, R.N., Bilgili, E., Preparation and characterization of fast dissolving pullulan films containing BCS class II drug

nanoparticles for bioavailability enhancement. *Drug Dev. Ind. Pharm.*, 42, 1073, 2016.

134. Leung, S.H.S., Loene, R.S., Kumar, L.D., Kulkarni, N., Sorg, A.F., Fast dissolving orally consumable films, US Patent 7025983B2, assigned to Johnson and Johnson Consumer Inc., USA, 2006.
135. Leung, S.H.S., Loene, R.S., Kumar, L.D., Kulkarni, N., Sorg, A.F., Fast dissolving orally consumable films, US Patent 7407669B2, assigned to McNeil PPC Inc. & Johnson and Johnson Consumer Inc., USA, 2008.
136. Bess, W.S., Kulkarni, N., Ambike, S.H., Rampay, M.P., Fast dissolving orally consumable films containing a taste masking agent, US patent 7648712B2, assigned to McNeil PPC Inc. & Johnson and Johnson Consumer Inc., USA, 2010.
137. Bala, R. and Sharma, S., Formulation optimization and evaluation of fast dissolving film of aprepitant by using design of experiment. *Bull. Fac. Pharm. Cairo Univ.*, 56, 159, 2018.
138. Mishra, R. and Amin, A., Formulation and characterization of rapidly dissolving films of cetirizine hydrochloride using pullulan as a film forming agent. *Indian J. Pharm. Educ. Res.*, 45, 71, 2011.
139. Gherman, S., Zavastin, D., Ochiuz, L., Biliuta, G., Coseri, S., Enalapril maleate loaded pullulan film for mucoadhesive buccal drug delivery applications. *Cellul. Chem. Technol.*, 50, 593, 2016.
140. Chatap, V.K., Maurya, A.R., Deshmukh, P.K., Zawar, L.R., Formulation and evaluation of nisoldipne sublingual tablets using pullulan & chitosan for rapid oromucosal absorption. *Adv. Pharmacol. Pharm.*, 1, 18, 2013.
141. Kathpalia, H. and Patil, A., Formulation and evaluation of orally disintegrating films of levocetirizine dihydrochloride. *Indian J. Pharm. Sci.*, 79, 204, 2017.
142. Murata, Y., Isobe, T., Kofuji, K., Nishida, N., Kamaguchi, R., Preparation of fast dissolving films for oral dosage from natural polysaccharides. *Materials*, 3, 4291, 2010.
143. Cristescu, R., Dorcioman, G., Ristoscu, C., Axente, E., Grigorescu, S., Moldovan, A., Mihailescu, I.N., Kocourek, T., Jelinek, M., Albulescu, M., Buruiana, T., Matrix assisted pulsed laser evaporation processing of triacetate-pullulan polysaccharide thin films for drug delivery systems. *Appl. Surf. Sci.*, 252, 4647, 2006.
144. Jelinek, M., Cristescu, R., Axente, E., Kocourek, T., Dybal, J., Remsa, J., Plestil, J., Mihaiescu, D., Albulescu, M., Buruiana, T., Stamatin, I., Matrix assisted pulsed laser evaporation of cinnamate-pullulan and tosylate-pullulan polysaccharide derivative thin films for pharmaceutical applications. *Appl. Surf. Sci., 253*, 7755, 2007.
145. Igarashi, T., Nomura, K., Naito, K., Yoshida, M., Plasma extenders, US Patent 4370472A, assigned to Hayashibara Seibutsu Kagaku Kenkyujo KK, USA, 1983.
146. Cheng, K.C., Demirci, A., Catchmark, J.M., Pullulan: Biosynthesis, production, and applications. *Appl. Microbiol. Biotechnol.*, 92, 29, 2011.

147. Kulicke, W.M. and Heinze, T., Improvements in polysaccharides for use as blood plasma expanders. *Macromol. Symp.*, 231, 47, 2006.
148. Shingel, K.I. and Petrov, P.T., Behavior of γ-ray-irradiated pullulan in aqueous solutions of cationic (cetyltrimethylammonium hydroxide) and anionic (sodium dodecyl sulfate) surfactants. *Colloid Polym. Sci.*, 280, 176, 2002.
149. Akiyoshi, K., Sasaki, Y., Sunamoto, J., Molecular chaperone-like activity of hydrogel nanoparticles of hydrophobized pullulan: Thermal stabilization with refolding of carbonic anhydrase B. *Bioconjug. Chem.*, 10, 321, 1999.
150. Gapanovich, V.N., Petrov, P.T., Ivanov, E.P., Tsarenkov, V.M., Lapkovski, M.P., Tyurin, V.I., Blood plasma substituting preparation, the corrector of hemodynamical disfunctions–Neorondex, Russian Federation Patent Application 5059206, assigned to Production Association "Belmedpreparaty", Belarus, 1992.
151. Xie, C., Satake-Ozawa, M., Rashed, F., Khan, M., Ikeda, M., Hayashi, S., Sawada, S., Sasaki, Y., Ikeda, T., Mori, Y., Akiyoshi, K., Perforated hydrogels consisting of cholesterol-bearing pullulan (CHP) nanogels: A newly designed scaffold for bone regeneration induced by RANKL-binding peptides and BMP-2. *Int. J. Mol. Sci.*, 23, 7768, 2022.
152. Hartl, F.U., Bracher, A., Hayer-Hartl, M., Molecular chaperones in protein folding and proteostasis. *Nature*, 475, 324, 2011.
153. Kuroda, K., Fujimoto, K., Sunamoto, J., Akiyoshi, K., Hierarchical self-assembly of hydrophobically modified pullulan in water: Gelation by networks of nanoparticles. *Langmuir*, 18, 3780, 2002.
154. Sawada, S., Nomura, Y., Aoyama, Y., Akiyoshi, K., Heat shock protein-like activity of a nanogel artificial chaperone for citrate synthase. *J. Bioact. Compat. Polym.*, 21, 487, 2006.
155. Sasaki, Y., Asayama, W., Niwa, T., Sawada, S.I., Ueda, T., Taguchi, H., Akiyoshi, K., Amphiphilic polysaccharide nanogels as artificial chaperones in cell-free protein synthesis. *Macromol. Biosci.*, 11, 814, 2011.
156. Nomura, Y., Sasaki, Y., Takagi, M., Narita, T., Aoyama, Y., Akiyoshi, K., Thermoresponsive controlled association of protein with a dynamic nanogel of hydrophobized polysaccharide and cyclodextrin: Heat shock protein-like activity of artificial molecular chaperone. *Biomacromolecules*, 6, 447, 2005.
157. Sawada, S.I., Sasaki, Y., Nomura, Y., Akiyoshi, K., Cyclodextrin-responsive nanogel as an artificial chaperone for horseradish peroxidase. *Colloid Polym. Sci.*, 289, 685, 2011.
158. Sawada, S.I. and Akiyoshi, K., Nano-encapsulation of lipase by self-assembled nanogels: Induction of high enzyme activity and thermal stabilization. *Macromol. Biosci.*, 10, 353, 2010.
159. Hirakura, T., Nomura, Y., Aoyama, Y., Akiyoshi, K., Photoresponsive nanogels formed by the self-assembly of spiropyrane-bearing pullulan that act as artificial molecular chaperones. *Biomacromolecules*, 5, 1804, 2004.

160. Morimoto, N., Endo, T., Iwasaki, Y., Akiyoshi, K., Design of hybrid hydrogels with self-assembled nanogels as cross-linkers: Interaction with proteins and chaperone-like activity. *Biomacromolecules*, 6, 1829, 2005.
161. Aisverya, S., Annu, Ali, A., Sudha, P.N., 28 - Pullulan-based bionanocomposites in tissue engineering and regenerative medicine, in: *Bionanocomposites in Tissue Engineering and Regenerative Medicine*, S. Ahmed and Annu (Eds.), pp. 533–547, Woodhead Publishing, Sawston,Cambridge, 2021.
162. Morimoto, N., Endo, T., Ohtomi, M., Iwasaki, Y., Akiyoshi, K., Hybrid nanogels with physical and chemical cross-linking structures as nanocarriers. *Macromol. Biosci.*, 5, 710, 2005.
163. Hirakura, T., Yasugi, K., Nemoto, T., Sato, M., Shimoboji, T., Aso, Y., Morimoto, N., Akiyoshi, K., Hybrid hyaluronan hydrogel encapsulating nanogel as a protein nanocarrier: New system for sustained delivery of protein with a chaperone-like function. *J. Control. Release*, 142, 483, 2010.
164. Kim, S., Chae, S.Y., Na, K., Kim, S.W., Bae, Y.H., Insulinotropic activity of sulfonylurea/pullulan conjugate in rat islet microcapsule. *Biomaterials*, 24, 4843, 2003.

11

Biomedical Applications of Collagen/Gelatin

Jyoti Sarwan, Jasjeet Narang, Anushka Kumari, Monika Devi, Nazim Uddin, Savita Sharma and Jagadeesh Chandra Bose K*

University Institute of Biotechnology, Chandigarh University Gharuan, Punjab, India

Abstract
Collagen has been defined as a polymer that is found in abundance on planet Earth with a large number of applications. Like other molecules, collagen also has complex and simpler forms. There are different types of structures and modifications found in collagen molecules. Although there are several molecules present in nature to replace toxic chemical components, collagen is on the top of them. Numerous applications have been developing day by day to prove the vital role of collagen. In this chapter, we concluded some major applications of collagen and gelatine in medical sciences.

***Keywords*:** Fibrillogenesis, peptide, hydrogels, collagen, film, sheaths, elastin, gelatin

11.1 Introduction

Biomaterial research and development have achieved great success in recent decades. Many patients have been cured by these bio-material, and the quality of life has also improved. However, with the recent development in the field of technology and science, the demand for biomaterials has also increased due to the increasing number of injuries and illnesses—for instance, these types of biomaterials are widely used to heal wounds. Until now, to fix broken bones, metallic materials are in use, and to repair dentition defects, horsehairs and cotton fibers are also in use as stylolite.

**Corresponding author*: jcboseuibtlab@gmail.com

Annu (ed.) Biopolymers for Biomedical Applications, (287–314) © 2024 Scrivener Publishing LLC

Significant development has been made in metal biomaterials and polymer. However, due to material erosion, allergies, reoperations, and other deficiencies, development has been slow in recent years. Therefore, researchers are looking for new substitute materials. Just as a new material, an essential role is played by biodegradable materials.

Among biodegradable materials, collagen is most commonly in use [1]. Earlier, collagen has been used in food and pharmaceutical applications, etc. In the last decades, the use of gelatin, collagen, and its hydrolysates has expanded in the photography industry, medical cosmetology, and cosmetics. In biomedical applications, the use of collagen has grown fast because of its excellent biocompatibility and safety, and it is widely in use in the biotechnology field. However, along with that and due to their poor mechanical supplies, the difficulty of assuring adequate supplies, and ineffectiveness in managing infected sites, some disadvantages of collagen supplies emerged [84]. Improvement in biological, chemical, and physical properties will address some drawbacks in a collagen-based application. Since the 1980s, collagen has also been utilized as scaffolds in tissue engineering. Collagen scaffolds are in use basically for artificial skin, artificial bone, cartilage, tendon, etc. [35]. In cases like cancer studies, three-dimensional collagens are quite helpful for modulating calcium-concentrated channels [14]. For drug and growth factors, collagen and gelatin have been used in the form of gel, sponge, membrane, injection, and many more. The most significant constitutive protein is collagen, in comparison to other natural biodegradable polymers like chitosan, elastin, silk, glycosaminoglycan, fibroin, etc., and some synthetic polymers like polyethylene glycol, poly(lactic-co-glycolic acid) (PLGA), polylactic acid (PLA), etc. In the fundamental building blocks, the formation of connective tissues, and with many tissue engineering and physiological processes of biological adaptation, its metabolism is associated. Along with that, collagen has superiority (biocompatibility, absorbability in the body, hydrophilicity, abundant source, easy processing, low antigenicity, etc.); however, it also has poor physicochemical properties like resistance to an enzyme, mechanical strength, and thermostability [71].

In an aqueous solution, collagen can be extracted and molded into different forms of a delivery system. Shields in ophthalmology [54] and sponges for wounds get formulated in combination with liposome for sustained drug delivery [26], tablets, and mini-pallets for protein delivery [48], which are controlled material for transdermal delivery [74] and for gene delivery nanomaterials [63] and some of the main applications of collagen in drug delivery. Other than that, it is also used in hemostatic agents [11], surgical suture [54], and tissue engineering, which includes fundamental matrices for cell culture systems, and for artificial blood vessels and valves [1, 16], replacement/substitutes were reported earlier.

Other than natural polymers like albumin and gelatin, collagen exhibits weak antigenicity [49], biodegradability, and super biocompatibility. Self-aggregation and crosslinking is the main reason why collagen is helpful in biomedical application. In the drug delivery system of collagen, cross-linking agents controls the *in vivo* absorption of collagen—for example, chromium tanning [10], glutaraldehyde [4], formaldehyde ayleazide [58], carbodiimides [53], poly-epoxy compounds, and hexamethylenediisocyanate [16].

11.2 Structure of Collagen

The structure of a protein explains its function and properties. When a protein has a tertiary structure and above, it shows its biological function. Among all total identified collagen of 28 types, most of them are known to be type 1 collagen. The structure of collagen is shown in Figure 11.1.

From the perspective of *in vivo* biosynthesis of collagen, there is type 1 collagen which is demonstrated as hierarchical, natural, multi-layered structured polymer. A wide number of applications of collagen in cell biology lead to a deeper study of the biosynthetic process of collagen *in vivo* and constitute a huge number of extracellular and intracellular physiological steps.

The alpha chain of procollagen is first synthesized from the mRNA that carries the information which is necessary to encode the gene of collagen on the ribosome and genetic information and translate the amino acid sequence corresponding to the polypeptide chain of collagen in the ribosome. The signal peptide of the collagen polypeptide enters the endoplasmic reticulum; it would be localized and removed by a biological enzyme. The alpha chain of procollagen consists of five regions (N-terminal propeptide, N-terminal peptide, triple helix region, C-terminal front, and C-terminal peptide).

After that, the alpha chain would be glycosylated and hydroxylated by the action of hydroxylase, glucose invertase, and galactose invertase, resulting in a large amount of hydroxyproline in the alpha chain. After that, the alpha chain binds at the spherical C-terminal polypeptide sites and creates a disulfide bond to form a triple helix core. Finally, with the triple helix extending from the C-terminus to the N-terminus, there is a formation of heterophasic trimmer procollagen [24]. The secretion of a procollagen molecule is outside of cells which transform into tropocollagen, and there

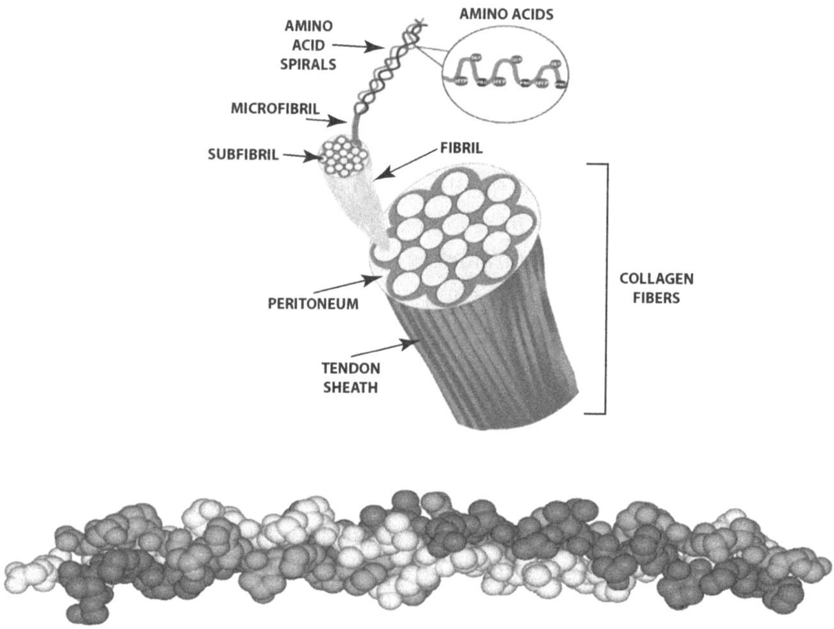

Figure 11.1 Structure of collagen.

is restrictive cleavage that takes place by biological enzymes [28]. However, the molecule form of tropocollagen in the body is not the final form of collagen. In physiological conditions, procollagen molecules continue to aggregate and assemble into supramolecular collagen. It is recognized by a 65–67-nm axial periodicity based on the so-called "1/4 staggered permutation rules". The natural collagen basic molecular conformational structure includes an amino acid triplet, i.e., primary structure; alpha helix, i.e., secondary structure; triple helix, i.e., tertiary structure; fibrils, i.e., quaternary structure; and a supermolecular aggregation structure.

11.2.1 Primary Structure

The molecule of collagen type 1 constitutes three alpha chains, and each chain is made up of 1,014 amino acids. Along the entire chain is the repeated permutation of amino acid, which is represented by -(Gly-X-Y-) n tripeptide sequence. Glycine (Gly) forms around 30% of the total collagen content of amino acids and X, Y are other amino acids except for glycine. The most commonly found amino acids are where X is proline and Y is hydroxyproline. The general structure of collagen is depicted in Figure 11.2.

Compared with other amino acids, hydroxyproline is most abundant (around 50%) in collagen protein. A characteristic of collagen's amino acid hydroxylase is that it undergoes hydroxylation, which changes lysine to proline to hydroxylysine and hydroxyproline. The structure of the amino acid is maintained by Vander waal interaction, covalent interactions between polypeptides, and hydrogen bond. Collagen molecule's triple helix structure is maintained by these molecular forces. With nuclear magnetic resonance (NMR) study, it was observed that to facilitate charge to charge interactions between neighboring molecules of collagen, a fibrillary structure of residues like Lys and Arg adopted a side-chain conformation and to minimize the hydrophobic area the hydrophobic residues like Leu and Ile fold along the molecular axis of collagen. Molecular modeling work and detailed NMR showed a conformational metastable proline which leads to the high flexibility of the molecule that was observed, i.e., triplets of an amino acid (Gly-Pro-Hyp) in triple helix collagen. This alignment of triplets within the fibril structure gives flexibility to fibril as well. These fibrillar bands show a high correlation with sites of collagen binding, which leads to the hypothesis that this fibril alignment is important to prevent ligand–collagen binding against tissue external stress [65].

11.2.2 Secondary Structure

Due to electrostatic repulsion between amino acid molecules that are abundant in collagen, X-proline and Y-hydroxyproline form a left-handed alpha-helical structure. The stability of the alpha-helical structure is maintained by the hydrogen bonds between residues. At the end of the alpha chain, two non-helical domains are present, which are C and N terminus. Triple helix formation initiates by C-terminus, and in the regulation of primary fibril diameters, N-terminus is involved. Additionally, to maintain helical structure stability of the chain of collagen on the alpha side chain, residues of free amino acids extend perpendicularly outward to a spiral axis and in a spiral chain formation of a large number of hydrogen bonds that take place. The motifs forming the structure of naturally occurring proteins (β-sheets, α-helices, and β-turns) are usually exploited to arrange the peptides, forming fibrils. The folding of peptides into their secondary structures is guided by the amino acid sequence, and the peptide conformation decides their packaging into fibrils—for instance, alpha helices generally assemble in the form of coiled coils which, in turn, forms fibril, leading to either hydrophobic interactions or hydrogen bonding in between particular residues of amino acids located in the exterior of the coiled coils. Being modular is the main advantage of nanofibrillar hydrogels that are peptide-based. Thus, their characteristics can be modified by altering the amino acid sequence in peptide molecules [3].

11.2.3 Tertiary Structure

In type 1, there is the formation of three left-handed alpha helix collagen along with two alpha 1 and one alpha 2 chains. However, in the structure of the triple helix, the third amino acid with a narrow alpha chain is located around a central axis. The third amino acid, glycine, is acceptable because of its minimal volume. A stabilizing interaction is there between the gly N–H group and adjacent O=C(X). The molecular weight of tropocollagen is approximately 300 kDa. Its approximate length and diameter are approximately 300 and 1.5 nm, respectively [47]. The triple-stranded conformation stabilizes the physiological and biomechanical functions which mainly depend upon links between the polypeptide chains of alpha and aldehyde amine condensation as well as aldol histidine between intramolecular and intermolecular crosslinking [22]—for instance, alpha helix assembles into fibrils and coiled for the interactions to hydrophobic or the hydrogen bonding in the outermost surface of specific amino acids.

11.2.4 Quaternary Structure

The end-to-end linkage of tropocollagens in the linear or parallel forms leads to the formation of stable microfibrils having D-periodic band spaces of approximately either 64 or 67 nm. Insoluble fibers are generally formed by inter- or intramolecular crosslinking present between the molecules of collagen. Moreover, there is the presence of covalent crosslinking between lysine and the C-terminus of histidine or N-terminus. Schmidt *et al.* showed the presence of a cavity with an area of approximately 0.54 D in between the tail and the head of any two molecules of anterior collagen placed on the same axis. A gap of approximately 0.46 is present in between the bottom and head of the two molecules of procollagen on the adjacent axis.

11.2.5 Super Molecular Aggregation Structure

Earlier observations of tissues that are collagen abundant showed that the nature of collagen is not as simple as a single microfibril of collagen. Microfibrils of collagens are collected in recent times as stated in the "1/4 staggered permutation rules". On the other side, axially elongated microfibrils would be in a non-helical simple manner along with an overlaying

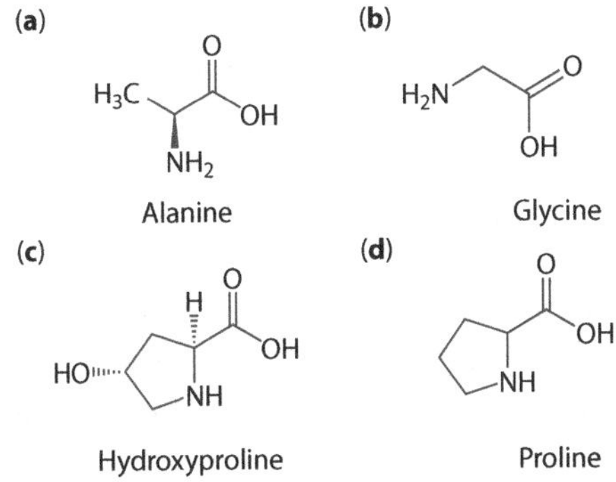

Figure 11.2 General structure of collagen: (a) alanine, (b) glycine, (c) hydroxyproline, and (d) proline.

lateral practice. Moreover, these microfibrils of collagens are collected in a crystallographic structure [9]. Bouquets of these fibrils having either the alike one or similar adaptation are bundled together, thus forming bundles of collagen fibers. These bundles intertwine or permeate among themselves, forming a basic organizational structure, with a 3D network, of the organism. Nowadays, the ordered pattern of arrangement of the fibrils made up of protofibrils generally includes quasi-hexagonal and positive pentagonal modes. It showed that the fibrils of collagen are composed of five protofibrils of collagen in the form of a regular pentagon. However, after the detailed study of skin collagen by Hulmes *et al.*, closely packed primary fibrils of collagen were found in the skin. For getting an inclined angle, primary fibril was mounted along with 5.2° maintained fiber axis; the protofibrils need to be arranged in the form of a quasi-hexagon. The fundamental properties such as thermal stability along with mechanical properties of various associative planning of the materials show the remarkable difference between a collagen-based molecule and mature tissues [64]. Thus, in the general structure of collagen, different materials are permitted for the interference interaction of the serially designed features with their particular application.

11.3 Modification of Collagen

In biomedical applications, there is the broad utilization of collagen-based biomaterials. In contrast, it has some unfavorable physicochemical properties like mechanical strength, degradation of enzyme that gets resistant, and thermostability—that is why before using collagen, modification is a must. Thus, some methods of crosslinking of collagens are changes in properties like chemical, physical, and biochemical. One effective method for the modification of collagen is the mixing of collagen with other biomaterials. All crosslinking methods have their pros and cons simultaneously. The general methods used for physical crosslinking are ultraviolet radiation, gamma radiation, and DHT, i.e., hydrothermal, etc. Compared with chemical crosslinking, these methods play a crucial role in the improvement of mechanical and physical properties without any toxic effect. By prolonged biological treatment, exposure degradation of collagen takes place. On the other hand, uniformity in crosslinking degree and high strength of collagen material are not attained by physical treatment. Therefore, from the synergistic effects among different crosslinking methods and physical modification, the collagen-based material properties might be beneficial. For the efficient modification of collagen-based biomaterials, some chemical crosslinking agents are used, like some aldehydes including glutaraldehyde, isocyanates including hexamethylene-diisocyanate, poly-epoxy compounds, and photoreactive agents. Some extracts of plants are also in use like tannic acid and genipin, and some inorganic crosslinkers include 1-(3-dimethylaminopropyl)-3-ethyl carbodiimide hydrochloride (EDC), etc. Among these widely used crosslinkers for crosslinking materials of amino acids in biomedical applications are glutaraldehyde and genipin [43]. The characteristics of materials based on collagen could be the collective result between physical moderation and other methods of crosslinking [36]. On the other side, the features and characteristics of collagen materials could be enhanced by proper chemical modification. In crosslinking, the agents involved are isocyanates (e.g., hexamethylenediisocyanate), aldehydes (e.g., glutaraldehyde, GA), photo reactive agents, plant extract (e.g., genipin), inorganic crosslinker EDC, tannic acid [46], and so on. The most common crosslinker for an amino acid is genipin and GA. Previously, the study has confirmed that a collagen-based scaffold could be promoted by synthetic crosslinkers correlated with GA that is easily compacted. Now, different kinds of crosslinkers that are chemical-based are found there, the non-harmful and suitable. Similar with biological and EDC, these crosslinking agents are seen as zero length. However, in the case of effective crosslinking, then it is limited because of limited groups which are ready to react with it. Therefore, recently, for their biocompatibility, natural crosslinkers out of cogitation were further placed upon it [5].

In the present scenario, a kind of moderately virulent chemical crosslinking agents found on traditional biomass have been arrived in the application of collagen moderation, specifically biomass with aldehyde classification after oxidation modification—for instance, dialdehydecarboxymethyl cellulose, [77] oxidized starch, dialdehyde alginate [37], etc. The biologic crosslinking representative could give a capability of functional modification for collagen substances. The possible interlinkage allying collagen molecules and alginate dialdehyde has appeared distinctly [31]. However, additional *in vitro* and *in vivo* studies shall be paid heed to the stability and durability of the crosslinking substance as an outcome of storage and chronic implant [79]. Moreover, collagen-based combined with synthetic polymers or natural polymer could also enhance the effects of collagen material to meet the contrasting application necessary for the material used in several fields. The utilization of synthetic polymers includes polyvinyl alcohol, polyurethane, PLA [82], PLGA, etc. Natural polymer materials have superior biological compatibility and biodegradable properties, which have captivated much attention in the present year. To date, these natural polymer materials, including natural carbohydrate polymer material such as cellulose, chitosan [38], and glycosaminoglycans (chondroitin sulfate, hyaluronic acid (HA), heparin, etc), have been appealed in collagen-based scaffolds for tissue engineering.

Gelatin is a pure protein food ingredient of thermal denaturation of collagen which is one of the most common proteins of an animal. The synthesis and structure of gelatin is depicted in Figure 11.3. It is a water-soluble

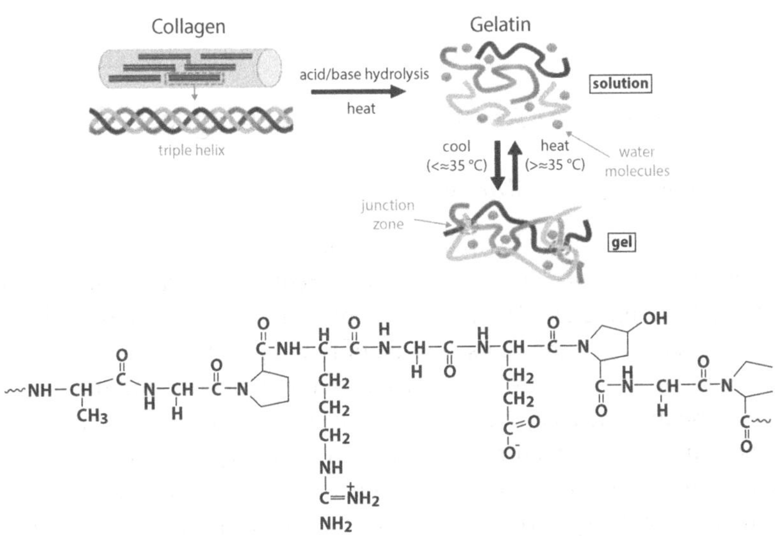

Figure 11.3 Synthesis and structure of gelatin.

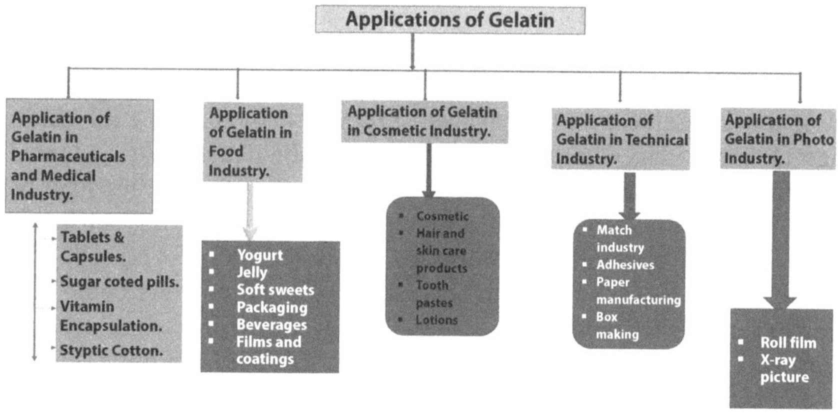

Figure 11.4 Applications of gelatin.

proteinaceous substance made by the destruction of tertiary, secondary, and, to some extent, primary structure of collagen, more specifically by the partial hydrolysis of collagen derived from white connective tissue, from the skin and bone of animals. It is a polypeptide with high molecular weight and an important hydrocolloid that has proved popular. Because of its gelling and thickening property, it is widely used in food products. It is a digestible protein with all essential amino acids except tryptophan. Recently, the primary commercial source of gelatin is insects, hides, cattle bones, pigskin, and fish. The application of gelatin in different fields of science is illustrated in Figure 11.4.

11.4 Biomedical Applications of Collagen/Gelatin

11.4.1 Collagen as Film or Membranes

A barrier membrane is the major application of collagen films [12]. An approximately 0.01–0.5-mm-thick film is required to the makeup with biodegradable material. The barrier membrane made of telopeptide that is free from collagen shows itself to be of low capacity to release in a drug that has incorporation of some bonds. Different types of bonding that take place in the drug delivery system by collagen are covalent bonding, hydrogen bonding, and only entrapment also. Another significance of collagen is that they are resistant to make changes and easy to sterilize. They are also liable to hydroxylation reactions. Collagen has the ability to hydrolyze when reacting with the eye within 5 to 6 h of treatment. There is a piece of evidence from previous studies that collagen and its derivates are suitable in many applications like acting as an ointment and liquid vehicle. In tissue infections and treatments, collagen also plays an important role. This collagen is very helpful in repairing injuries like liver cancer, corneal tissue, etc. Usually, during the treatment of corneal infections, there is ophthalmic insertion that takes place in the form of thin-film or water-soluble material that is required for repairing damaged corneal tissues. In this treatment therapy of corneal tissues, there are several antibiotics involved in higher amounts like gentamicin, tetracycline, and many more. During the treatment, a higher concentration of antibiotics is used in the incorporation of drugs. An evidence-based study shows that these drugs had been persisted in rabbits after 8 days of treatment. Therefore, collagen can act as a local carrier of delivery for the treatment of cancer and others. One of the common applications of collagen is to deliver a drug to the target site as in the case of cancer and liver dysfunction corrections. Different biomedical applications of collagen-based biomaterials and collagen fiber are represented in Figure 11.5 and Figure 11.6, respectively. Most of the time, formaldehyde and chromium tanning is used to make cross-linking with collagen in implantation recoveries. In a delivery system, the presence of nifedipine made up of chitosan in a device system of transdermal delivery shows its higher efficacy [45].

Collagen films or sheets are best suited for drug delivery systems as reported from past studies. With several applications, collagen has proven that it is a crucial polymer of nature. In the transplantation recoveries, collagen plays a beneficiary role by making them complex to simple drug delivery systems because they are attractive to the environment as well as to the economy and helpful for allogenic and even xenogenic patients. Therefore, with the help of collagen fibers, it has been possible to recover from several troubles, and by therapeutics treatments, various non-genetic diseases can also be treated. Thus, there

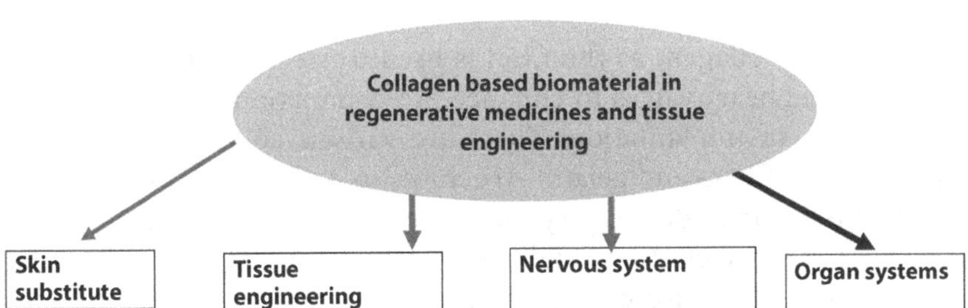

Figure 11.5 Biomedical applications of collagen-based biomaterials.

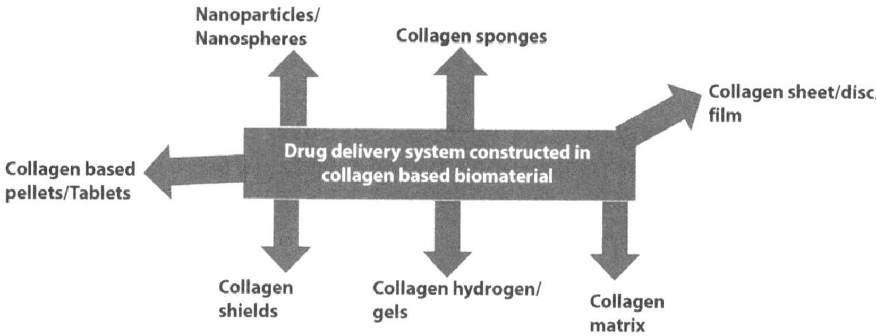

Figure 11.6 Biomedical applications of collagen fibers.

are many applications of collagen fibers that served as films or sheaths of collagen.

11.4.2 Collagen as a Matrix

In implantation treatments, the main issue that comes with biomaterials is their calcification in implanted structures. Therefore, in implanted structures, to avoid this calcification, there is a need to introduce new bio-based materials. In a bio-based material, calcification is the major issue, and implantation efficiency can be improved by using a cheap and suitable matrix. The issue related to the tissue by its calcification depends upon the nature of the biomaterial. Elastin and collagen both play a major role in connective tissues in which there are structures present with various collagen fibers associated with elastin fibers in a complex network to maintain the bonding of the polymer. Even in plastic surgeries and several medical applications, the combination of elastin and collagen showed its importance in numerous applications. In a laboratory for cardiovascular treatment, elastin and collagen combinations have been made to evaluate the results for analyzing the applications and calcification of tissues in regard to heart troubles [45].

Therefore, collagen, as the film, is used to treat heart diseases like bioprosthetic heart valves to enhance the stimulation of processing of calcification in implantation. There are various combinations found with collagen fibers and gelatin structures to make several suitable for treatments in drug delivery systems. Compared with normal tissues, collagen has more elasticity and calcification properties to make them a best-suited matrix for collagen as sheaths or matrix in drug delivery systems.

11.4.3 Collagen as Shields

Collagen shields were designed for making contact lenses in earlier days. These collagen contact lenses also have numerous applications in which one is to get easily dissolved within the cells of the cornea [45]. Originally, the idea had come to mind of the researcher to the treatment of drug delivery systems like ophthalmic drug deliveries. There is a shield of hydrogel lens that is required in the drug delivery system. Collagen is used as shields of ophthalmic drug delivery systems because they have the ability to self-administration in ocular treatments and can be used on ocular surfaces. The major quality of the collagen shields is that they can protect the eyelids and try to heal the corneal epithelium by blinking. The one good thing in collagen is it also promotes healing of the cornea after transplantation. Even collagen-based implants for fibroblast injuries or repair have been used since ages [34]. Collagen acts as a protective sheath to the cornea, and it also provides a biologically compatible layer to the cells of the eyes. Another good part of collagen is as a shield to remove tears and prevent eye blinking. The water-soluble capacity of the collagen sheath dissolves with the corneal cells and reduces the rubbing of the eyelids and protects the eyelids with lubricants. Therefore, collagen shields are proven as compatible support to the eye cells and in drug delivery systems.

11.4.4 Collagens as Sponges

Collagen is a very useful component to heal tissue wounds and burns. Collagen can treat issues like pressure sores, decubitus ulcers, donor sites, leg ulcers, and many more. Collagen is not only used in internal tissue repairs but is also able to cure *in vitro* healings. The biological dressing had a major role in collagen fibers since 1930 [45]. Human collagen membranes can be treated by collagen sponges as collagen is compatible with biological dressings. Collagen, as sponges, protect tissues from mechanical injuries by providing shields against wet wounds and bacterial infections. Collagen can be obtained from bovine serum and by combining collagen sponges and components like fibronectin, elastin, and glycosaminoglycans to provide tissue flexibility and speedy healing without coming in contact with moisture and to reduce microbial infections.

11.4.5 Collagen as Hydrogels

Collagen can be made as hydrogels or collagen as liposomes are used in applications of drug delivery systems. Due to larger applications, collagen applications in drug delivery systems can be used as collagen hydrogels. In drug delivery systems, various combinations of collagen can be included for medical treatments [60]. Collagen hydrogels have applications in cancer therapies. Collagen hydrogels are successfully replaced by synthetic fibers, and collagen hydrogels are combined with films and human growth hormones to treat cancer patients [45]. Collagen is used in drug delivery systems because of its use in injectable medicine for aqueous solutions. Collagen gels can cure or repair dermatological wounds and defects and sometimes urinary discomforts also [75].

11.4.6 Collagen as a Pellet

One more application of collagen in drug delivery systems is collagen used as a pellet in various processes of the medical field. Instant graft replacement is possible with collagen pellet in mice and human grafting also. Therefore, in the gene transfer mechanism, collagen pellet plays a major role [45].

11.4.7 Nanoparticles/Microspheres/Nanospheres

In the folding configuration, the mechanism of collagen suspension of crystallites leads to a multiple-chain system in gel aggregates. This formulation property gives an idea as a carrier for the delivery of drugs [39]. The formation of nanospheres is with a combination of electronics and electrostatic forces along with a liquefying agent that is sodium sulfate which facilitates more charge–charge interactions among collagen and plasmid DNA [50]. Based on molecular weight, temperature, pH, and non-covalent interactions (for molecular weight), collagen nanoparticles maintain stability [17]. Nanoparticles and nanospheres enhanced the uptake of exogenous compounds in cells like macrophages. They are based on biodegradable collagen, which makes it advantageous for carriers for drug delivery [62, 63]. These get sterilized easily and are thermally stable—for instance, steroidal drugs and cytotoxic drugs (campthocin, etc.), through nanoparticles of collagen, get delivered easily in systematic circulation. Due to its smaller size, large surface area, great ability to absorb, and diffusion capacity in water for the formation of colloid, such collagen-based nanoparticles are in use [70]—for example, enhancement of retinol's dermal delivery through collagen nanoparticles. Through the skin, faster transportation of the drug that is imported is possible due to the stable retinol drug delivery of collagen nanoparticles [23].

11.4.7.1 Collagen Nanoparticle Fabrication Techniques

To manufacture protein-based nanoparticles, four methods are usually defined; these are emulsification, coacervation, desolvation, and spray-drying. A few more methods include fluidization, solvent precipitation methods, interfacial polymerization, and jet milling technique, etc.

a. Emulsification: An aqueous phase of collagen constitutes water and hydrophilic surfactant and an organic phase which constitutes oil, water-miscible solvent, and lipophilic surfactant were all combined with rapid agitation at room temperature by a mechanical homogenizer, thus forming a homogenous emulsion. Then, this homogenous emulsion is mixed with preheated oil in a dropwise manner, which results in collagen nanoparticle formation.
b. Desolvation: This process was firstly done by Marty and co-workers. In this process, to alter the tertiary structure of collagen, natural salt or alcohol is added to a solution of collagen as a desolvation factor. When it attains desolvation critical level and collagen mass forms at that point as cross-linking material, glutaraldehyde is added for the final formation of collagen nanoparticles.
c. Coacervation: This process works similarly to the desolvation method. The only certain parameter that makes a difference are temperature, pH, homogenizer speed, cross-linker concentration, rate of addition of solvent, protein and organic solvent molar ratio, etc.
d. Spray-drying: By this process, the fabrication of spherical collagen nanoparticles is done. Using appropriate, temperature collagen's dilute solution spraying leads to hollow sphere formation. Collagen's triple helical structures get denatured at high temperatures; so, to prevent denaturation collagen solution spraying should be done in liquid nitrogen. After this step, fabricated nanospheres are successively cross-linked, sterilized, lyophilized, and frozen [13].

Collagen has good biological properties which play an essential role in current biomedical applications in tissue engineering such as in the substitution of skin, repair of bone, tendon, cartilage, and neural, and many more is under research.

11.4.8 Collagen in Substitution of Skin

Recent development in the biomedical use of bioengineered skin layers has led to more applications such as replacement of skin as done while repairing burn wounds [6, 47]. For clinical use, there are few types of products with collagen that are commercially available, for instance:

i. Dermo-epidermal or composite skin substitutes such as Orcel, Alpigraf, etc.
ii. Epidermal and dermal substitutes like Biobrane, Biobrane-L, TransCyte, Integra Dermal Regeneration template.

Although there is no permanent replacement of skin by such products, as these products take long durations to apply in therapy, along with their operating procedures being too complex [56]. Many researchers are currently working to achieve permanent achievement in the replacement of full-thickness skin.

According to [45], sponge implants are excellent for the job of skin recovery. Hence, artificial skin is often developed in the form of a sponge. For skin wounds and replacement of skin, collagen lattice is used in the development of cultured skin substitutes.

- To increase skin wound healing and control microbial contamination, there is the addition of certain selected antimicrobial drugs; one of them is amikacin (to bovine skin implantable collagen) [7].
- Some drawbacks are reported such as *in vitro* deficiency in barrier functions, and after grafting, it results in delay in keratinization if compared with native skin auto-grafts [72].
- To overcome such limitations, combinations of collagen are done with certain proteins such as glycosaminoglycan, biotin, and fibrin.
- Combination of collagen and glycosaminoglycan
- Dermal skin substitutes: To culture human epidermal keratinocytes, these act as suitable substrates [8].
- Cultured skin substitutes: To reduce the biodegradation rate and in *in vitro* low-humidity condition, from collagen, there is a restoration of the functional epidermis [32].

Table 11.1 Collagen fibers as a skin substitute.

Collagen-based implants	Features	Used for	References
Reconstituted type I Collagen	Prominent mechanical strength; biocompatibility	Skin replacement; burn wounds	[74]
Allogenic cultured skin substitute	Prepared from cryopreserved skin cells *As support for epithelial growth and differentiation, to replace pathological skin as cultured skin substitutes there is the use of contracted collagen lattice	Curing of chronic wounds (from diabetes)	[6, 81]
Allogenic cultured dermal substitute	Prepared by plating fibroblasts on collagen sponge matrix	Provides environment for the recovery of epithelial cells (by subsequent freeze-drying from 1% aqueous solution of atelocollagen)	[80]
Collagen sponges	Corneal cells' exhibition as normal cell phenotype; when culture on sponge matrix of collagen	Substrate for human corneal cells	[55]
Fibrillar collagen with gelatin	(i) Stabilization of collagen-based sponge metabolically and physically by dehydrothermal crosslinks (ii) Biocompatible; no evidence of immune response/lymphatic infiltration in osteochondral defect	(i) Artificial skin (ii) Carried matrix for human mesenchymal stem cells in cartilage regeneration therapy (iii) Implanted in osteochondral defect in rabbit femoral condyle	[42, 59]

Table 11.1 describes various applications of collagen in medical treatments like skin graftin, artificial skin, diabeters wound healing, cartilage regeneration therapy, implants in rabbits, grafting of corneal and epithelial cells, etc.

- Combination of collagen and fibrin: Direct grafting on the wound bed by cultured cells by using a thin layer of collagen in combination.
- Combination of collagen and biotin (covalently): By a process known as biotinylation of bovine skin, peptide growth factors' activity is retained and acts as a modulator in response to treat wound [69]. Many researchers demonstrate the *in vitro* condition; that in adult mammalian skin, multipotent stem cells play a vital role in the regeneration of hair follicles and sebaceous glands, and with epidermal stem cells along with the use of a modified collagen–chitin membrane, there is successful regeneration of full-thickness skin which constitutes all skin appendages and functions [47, 68].

11.4.9 Collagen in Bone Tissue Engineering

In our body, collagen is one of the essential building blocks of bone and vitamin C, helping the body make a protein named collagen which helps in the healing of a fracture to the bone.

If any injury occurs in the bone, so it has an intrinsic property to heal, but sometimes it is unable to repair fully. Around 5–10% of patient cases suffer from bone fracture, tissue loss, infection, or any other issue due to bone damages [61].

To repair critical defects of bone, in orthopedics, there is mostly the use of autografts which have a very less probability of immune rejections or infection transmission. However, autografts have certain drawbacks in the limitation of the material used in grafting. Due to this, researchers nowadays focus more on regenerative medicine and bone tissue engineering [20].

For bone tissue engineering, a large number of polymeric scaffolds are used, which include some natural polymers like fibrin, fibroin, chitosan, etc., and synthetic polymers like polylactic acid and polyglycolic acid and copolymers like polylactic co-glycolic acid, etc. Collagen-based biomaterials are most widely used in bone tissue engineering [47].

Type 1 collagen is usually produced by osteoblasts, which controls hydroxyapatite formation. It gives physical support to cells to attach and grow on, and along with that it also stimulates behavior and the fate of the cell through receptor-mediated interactions [21]. A wide variety of collagen modifications are used in bone tissue engineering due to its properties like fibrillogenesis, polymerization, biocompatibility, formation of fibers, etc. [22]. Recent fabrication for bone repair was done with collagen-1 type biomaterial in the form of cross-linked sponge-like scaffolds in which dehydrothermal and chemical glutaraldehyde were used in combination [41]. Additionally, the scaffold's surface also plays an essential role in the recognition of cells, repairmen, and regeneration. With the use of biomimetic proteins which modifies biomaterial surface and that leads to regeneration of tissue successfully. And for improvement of structural stability over months, cell proliferation and *in vivo* formation of bone in defects of calvarial over 6 weeks, etc, there is the use of hybrid biomimetic matrix which constitutes a fusion of osteocalcin and fibronectin protein which forms a network with matrices of fibrillar collagen protein [47].

11.4.10 Collagen in Tendon Repair

In severe cases, injury of a tendon leads to surgery with allografts, autografts, xenografts, or with certain devices, but it has many drawbacks such as the rejection of tendon tissue, the failure rates are high, and the probability of an injury occurring again or the recovery time which is very long or limited [76]. To overcome these limitations, tendon tissue engineering comes into a field that makes use of a combination of scaffolds, cells, bioactive molecules, and materials that boost the healing of defects and proper functioning of a tendon. In tendon tissue engineering, the cells that are used commonly are dermal fibroblasts. Mesenchymal stem cells and tenocytes also refer to tendon fibroblasts [44]. During the development of the tendon, the synthesis of some constituents of the extracellular matrix are glycoprotein, collagen, and proteoglycans, and these have a great impact on collagen fiber formation [27]. The ECMs of a tendon is composed of type 1 collagen, and collagen derivative scaffolds show high compatibility. Compared with polyester materials, derivatives of collagen show better support to cells—like cell adhesion, the proliferation of cells, and the bio-functionality of cells that also improves. By considering a rabbit as a model, gel of collagen is used to treat a tendon defect, and in collagen gel seeding of mesenchymal stem cells was performed to improve the mechanical and histological properties [2, 40, 73]. To enhance water resistance and mechanical strength, modification of collagen-derived scaffolds is done by co-fabrication or cross-linking [25].

11.4.11 Collagen in Cartilage Repair

In cartilage tissue engineering, scaffolds of collagen lack biomaterial fabrication, mechanical strength for the proper functioning of cells, cell adhesion, proliferation of the cell, and synthesis matrix of cartilage. So as to overcome all these challenges, cross-linking of scaffolds of collagen takes place. Some growth factors and seed cells are also used in the regeneration of cartilage injury [19].

By different fabrication methods, the scaffold properties and structure can be changed by electrospinning of polycaprolactone fibers. A five-layer cartilage-like tissue is fabricated, which is now alternated with fibrin-collagen hydro gel suspended on rabbit elastic chondrocytes with inkjet printing [79]. These alternated chondrocytes show more than 80% proliferation, viability, and maintenance of basic biological properties due to type 2 collagen and glycosaminoglycan deposition. Good mechanical properties were shown by fibrin–collagen gels or printed alginate compared with printed hybrid scaffolds [18]. Natural polymers like chitosan, fibrin glue, and chondroitin sulfate and some synthetic polymers like PLA and PLGA are used to improve the properties of collagen scaffolds [33, 83].

11.4.12 Collagen in Neural Repair

Neurodegenerative disease or neural injury leads to the improper functioning of nerves, which results in severe neural functional issues. Type 1 collagen shows attachment, *in vitro* growth, and primary nerve culture cell differentiation. To a large extent, enhancement in cell growth and differentiation is shown by type 1 collagen [78]. From a study by [52], to induce neurite morphogenesis and outgrowth, an evaluation of four types of collagen treatments was done.

Determination of primary nerve cell attachment by microscopy and by MTT colorimetric assay confirmation was done. After treatment with various non-hydrolyzed collagen (including collagen suspension and two-dimensional collagen matrix), there was increment in several cell attachments and a viable cell number. In hydrolyzed collagen, cells have neurite outgrowth and more viability. Due to uptake of collagen during proliferation of the cell, there is generation of differences between hydrolyzed and non-hydrolyzed collagen. For cell growth, small-sized collagen polymers are more preferable to non-hydrolyzed collagen [30]. Additionally, some cells have different responses toward natural and deformed collagen. Deformed collagen results from the damaged extracellular matrix, which persuades effective change in morphology and proliferation of the cell.

An *in vitro* model of damaged extracellular matrix is that from cell culture media with the treatment of hydrolyzed collagen [15]. Small pieces of collagen were disabled to form networks, due to which attachment of cell cannot take place properly. Hydrolyzed collagen mimics damaged extracellular matrix, which leads to its application in therapeutic research. As cellulose [66] is a major component of cell wall in plants, collagen peptide [67] has similarly almost the same moiety in animal tissues.

Improvement of attachment of cells by non-hydrolyzed collagen was observed by [52] on matrices with two- and three-dimensional attachment of a primary culture of nerve cell. Furthermore, for improved rate of cell viability and growth, collagen proved to be an essential factor for the interaction of cells [29]. Type 1 collagen constitutes a natural extracellular matrix that shows high suitability for the repair of nerves.

11.5 Conclusion

It can be concluded that collagen is an important biomaterial for the medical field. Collagen has been declared a commendable performance for various treatments. As has been concluded, collagen can be used in contact lenses, nanoparticles, bone replacements, tissue engineering, and so on. In this chapter, we tried to conclude some of the major applications of collagen as a matrix, gel, thread, bead, and many more, but we are not ensuring to include every application of the collagen. Hence, it has been proved that, without collagen, medical sciences are not able to treat diseases and injuries like cancer, ulcers, bone replacement, and many more.

References

1. Armentano, I., Puglia, D., Luzi, F., Arciola, C.R., Morena, F., Martino, S., Torre, L., Nanocomposites based on biodegradable polymers. *Materials*, 11, 5, 795, 2018.
2. Awad, H.A., Boivin, G.P., Dressler, M.R., Smith, F.N., Young, R.G., Butler, D.L., Repair of patellar tendon injuries using a cell–collagen composite. *J. Orthop. Res.*, 21, 3, 420–431, 2003.
3. Banwell, E.F., Abelardo, E.S., Adams, D.J., Birchall, M.A., Corrigan, A., Donald, A.M., Woolfson, D.N., Rational design and application of responsive α-helical peptide hydrogels. *Nat. Mater.*, 8, 7, 596–600, 2009.

4. Barbani, N., Giusti, P., Lazzeri, L., Polacco, G., Pizzirani, G., Bioartificial materials based on collagen: 1. Collagen cross-linking with gaseous glutaraldehyde. *J. Biomater. Sci. Polym. Ed.*, 7, 461–469, 1995.
5. Bhattacharyya, S.K., Dule, M., Paul, R., Dash, J., Anas, M., Mandal, T.K., Banerjee, S., Carbon dot cross-linked gelatin nanocomposite hydrogel for pH-sensing and pH-responsive drug delivery. *ACS Biomater. Sci. Eng.*, 6, 10, 5662–5674, 2020.
6. Boyce, S.T., Skin substitutes from cultured cells and collagen-GAG polymers. *Med. Biol. Eng. Comput.*, 36, 6, 791–800, 1998.
7. Boyce, S.T., Christianson, D.J., Hansbrough, J.F., Structure of a collagen-GAG dermal skin substitute optimized for cultured human epidermal keratinocytes. *J. Biomed. Mater. Res.*, 22, 10, 939–957, 1988.
8. Boyce, S.T., Supp, A.P., Warden, G.D., Holder, I.A., Attachment of an aminoglycoside, amikacin, to implantable collagen for local delivery in wounds. *Antimicrob. Agents Chemother.*, 37, 9, 1890–1895, 1993.
9. Bozec, L., van der Heijden, G., Horton, M., Collagen fibrils: Nanoscale ropes. *Biophys. J.*, 92, 1, 70–75, 2007.
10. Bradley, W.G. and Wilkes, G.L., Some mechanical property considerations of reconstituted collagen for drug release support. *Biomater. Med. Devices Artif. Organs*, 5, 159–175, 1977.
11. Cano Salazar, L.F., Claudio Rizo, J.A., Flores Guía, T.E., Cabrera Munguía, D.A., Nanocomposites comprise of collagen and acrylate-derived polymers for biomedical applications, in: *Green-Based Nanocomposite Materials and Applications*, pp. 173–189, Springer International Publishing, Cham, 2023.
12. Cao, S., Li, H., Li, K., Lu, J., Zhang, L., A dense and strong bonding collagen film for carbon/carbon composites. *Appl. Surf. Sci.*, 347, 307–314, 2015.
13. Chak, V., Kumar, D., Visht, S., A review on collagen-based drug delivery systems. *Int. J. Pharm. Teach Pract.*, 4, 4, 811–820, 2013.
14. Chen, C.N., Chen, Y.T., Yang, T.L., Application of three-dimensional collagen scaffolds to recapitulate and monitor the dynamics of epithelial-mesenchymal transition during tumor satellite formation of head and neck cancer. *Biomaterials*, 154, 134–146, 2018.
15. Chen, L., Cheng, W.L., Ke, Y.N., Wang, Y., Zhou, T.L., Pan, L., Guo, J., Aortic artery elastic lamina degradation, collagen remodeling, oxides stress, and inflammation in the apolipoprotein E deficient mice with or without aortic banding. *Zhonghua Xin Xue Guan Bing Za Zhi*, 40, 6, 505–510, 2012.
16. Coester, C.J., Langer, K., Von Briesen, H., Kreuter, J., Gelatin nanoparticles by two-step desolvation a new preparation method, surface modifications, and cell uptake. *J. Microencapsulation*, 17, 2, 187–193, 2000.
17. Cruz-Delgado, V.J., Rodríguez-Fuentes, N., Pat-Cetina, V.C., Cervantes-Uc, J.M., Nanocomposites based on biodegradable polymers for biomedical applications, in: *Green-Based Nanocomposite Materials and Applications*, pp. 317–337, Springer International Publishing, Cham, 2023.

18. Demircan, M., Cicek, T., Yetis, M.I., Preliminary results in single-step wound closure procedure of full-thickness facial burns in children by using the collagen–elastin matrix and review of pediatric facial burns. *Burns*, 41, 6, 1268–1274, 2015.
19. Deponti, D., Giancamillo, A.D., Gervaso, F., Domenicucci, M., Domeneghini, C., Sannino, A., Peretti, G.M., Collagen scaffold for cartilage tissue engineering: The benefit of fibrin glue and the proper culture time in an infant cartilage model. *Tissue Eng. Part A*, 20, 5-6, 1113–1126, 2014.
20. Dimitriou, R., Jones, E., McGonagle, D., Giannoudis, P.V., Bone regeneration: Current concepts and future directions. *BMC Med.*, 9, 1, 1–10, 2011.
21. Domingos, M., Intranuovo, F., Gloria, A., Gristina, R., Ambrosio, L., Bártolo, P.J., Favia, P., Improved osteoblast cell affinity on plasma-modified 3-D extruded PCL scaffolds. *Acta Biomater.*, 9, 4, 5997–6005, 2013.
22. Dou, L., Li, B., Zhang, K., Chu, X., Hou, H., Physical properties and antioxidant activity of gelatin-sodium alginate edible films with tea polyphenols. *Int. J. Biol. Macromol.*, 118, 1377–1383, 2018.
23. El Blidi, O., El Omari, N., Balahbib, A., Ghchime, R., El Menyiy, N., Ibrahimi, A., Barkiyou, M., Extraction methods, characterization and biomedical applications of collagen: A review. *Biointerface Res. Appl. Chem.*, *11*, 13587–13613, 2021.
24. Erdmann, R.S. and Wennemers, H., Conformational stability of collagen triple helices functionalized in the Yaa position by click chemistry. *Org. Biomol. Chem.*, 10, 10, 1982–1986, 2012.
25. Fessel, G., Gerber, C., Snedeker, J.G., Potential of collagen cross-linking therapies to mediate tendon mechanical properties. *J. Shoulder Elbow Surg.*, 21, 2, 209–217, 2012.
26. Fonseca, M.J., Alsina, M.A., Reig, F., Coating liposomes with collagen (Mr 50000) increases uptake into the liver. *Biochim. Biophys. Acta*, 1279, 2, 259–265, 1996.
27. Franchi, M., Quaranta, M., De Pasquale, V., Macciocca, M., Orsini, E., Trirè, A., Ruggeri, A., Tendon crimps and peritendinous tissues responding to tensional forces. *Eur. J. Histochem.*, 51, 9, 2007.
28. Gelse, K., Pöschl, E., Aigner, T., Collagens—Structure, function, and biosynthesis. *Adv. Drug Delivery Rev.*, 55, 12, 1531–1546, 2003.
29. Gloria, A., Causa, F., Russo, T., Battista, E., Della Moglie, R., Zeppetelli, S., Ambrosio, L., Three-dimensional poly (ε-caprolactone) bioactive scaffolds with controlled structural and surface properties. *Biomacromolecules*, 13, 11, 3510–3521, 2012.
30. Gowen, B.B., Borg, T.K., Ghaffar, A., Mayer, E.P., Selective adhesion of macrophages to denatured forms of type I collagen is mediated by scavenger receptors. *Matrix Biol.*, 19, 1, 61–71, 2000.
31. Gu, Z., Xie, H., Huang, C., Li, L., Yu, X., Preparation of chitosan/silk fibroin blending membrane fixed with alginate dialdehyde for wound dressing. *Int. J. Biol. Macromol.*, 58, 121–126, 2013.

32. Haaparanta, A.M., Järvinen, E., Cengiz, I.F., Ellä, V., Kokkonen, H.T., Kiviranta, I., Kellomäki, M., Preparation and characterization of collagen/PLA, chitosan/PLA, and collagen/chitosan/PLA hybrid scaffolds for cartilage tissue engineering. *J. Mater. Sci.: Mater. Med.*, 25, 4, 1129–1136, 2014.
33. Hardy, J.G., Lin, P., Schmidt, C.E., Biodegradable hydrogels composed of oxime crosslinked poly (ethylene glycol), hyaluronic acid, and collagen: A tunable platform for soft tissue engineering. *J. Biomater. Sci. Polym. Ed.*, 26, 3, 143–161, 2015.
34. Harriger, M.D., Supp, A.P., Warden, G.D., Boyce, S.T., Glutaraldehyde crosslinking of collagen substrates inhibits degradation in skin substitutes grafted to athymic mice. *J. Biomed. Mater. Res.: An Official Journal of The Society for Biomaterials and The Japanese Society for Biomaterials*, 35, 2, 137–145, 1997.
35. Hu, Y., Dan, W., Xiong, S., Kang, Y., Dhinakar, A., Wu, J., Gu, Z., Development of collagen/polydopaminecomplexed matrix as mechanically enhanced and highly biocompatible semi-natural tissue engineering scaffold. *Acta Biomater.*, 47, 135–148, 2017.
36. Hu, Y., Liu, L., Dan, W., Dan, N., Gu, Z., Yu, X., Synergistic effect of carbodiimide and dehydrothermal crosslinking on acellular dermal matrix. *Int. J. Biol. Macromol.*, 55, 221–230, 2013.
37. Hu, Y., Liu, L., Gu, Z., Dan, W., Dan, N., Yu, X., Modification of collagen with a naturally derived cross-linker, alginate dialdehyde. *Carbohydr. Polym.*, 102, 324–332, 2014.
38. Huang, R., Li, W., Lv, X., Lei, Z., Bian, Y., Deng, H., Li, X., Biomimetic LBL structured nanofibrous matrices assembled by chitosan/collagen for promoting wound healing. *Biomaterials*, 53, 58–75, 2015.
39. Jerome, A. and Ramshaw, J.A., Editorial: Collagen-based. *Biomaterials*, 9, 137–138, 1992.
40. Juncosa-Melvin, N., Shearn, J.T., Boivin, G.P., Gooch, C., Galloway, M.T., West, J.R., Butler, D.L., Effects of mechanical stimulation on the biomechanics and histology of stem cell–collagen sponge constructs for rabbit patellar tendon repair. *Tissue Eng.*, 12, 8, 2291–2300, 2006.
41. Klimek, K. and Ginalska, G., Proteins and peptides as important modifiers of the polymer scaffolds for tissue engineering applications—A review. *Polymers*, 12, 4, 844, 2020.
42. Kong, B., Liu, R., Guo, J., Lu, L., Zhou, Q., Zhao, Y., Tailoring micro/nanofibers for biomedical applications. *Bioact. Mater.*, *19*, 328–347, 2023.
43. Kozłowska, J. and Sionkowska, A., Effects of different crosslinking methods on the properties of collagen–calcium phosphate composite materials. *Int. J. Biol. Macromol.*, 74, 397–403, 2015.
44. Kuo, C.K., Marturano, J.E., Tuan, R.S., Novel strategies in tendon and ligament tissue engineering: Advanced biomaterials and regeneration motifs. *BMC Sports Sci. Med. Rehabil.*, 2, 1, 1–14, 2010.
45. Lee, C.H., Singla, A., Lee, Y., Biomedical applications of collagen. *Int. J. Pharm.*, 221, 1-2, 1–22, 2001.

46. Lee, J., Yeo, M., Kim, W., Koo, Y., Kim, G.H., Development of a tannic acid cross-linking process for obtaining 3D porous cell-laden collagen structure. *Int. J. Biol. Macromol.*, 110, 497–503, 2018.
47. Liu, X., Zheng, C., Luo, X., Wang, X., Jiang, H., Recent advances of collagen-based biomaterials: Multi-hierarchical structure, modification, and biomedical applications. *Mater. Sci. Eng.: C*, 99, 1509–1522, 2019.
48. Lucas, P.A., Syftestad, G.T., Goldberg, V.M., Caplan, A.I., Ectopic induction of cartilage and bone by water-soluble proteins from the bovine bone using a collagenous delivery vehicle. *J. Biomed. Mater. Res.*, 23, 23–39, 1989.
49. Maeda, M., Tani, S., Sano, A., Fujioka, K., Microstructure and release characteristics of the mini pellet, a collagen-based drug delivery system for controlled release of protein drugs. *J. Control. Release*, 62, 313–324, 1999.
50. Marty J.J., Oppenheim R.C., Speiser P.. Nanoparticles--a new colloidal drug delivery system. *Pharmaceutica Acta Helvetiae*, 53, 1, 17–23, 1978 Jan 1.
51. Miller, J.M., Zoll, D.R., Brown, E.O., Clinical observation on the use of an extrude collagen suture. *Arch. Surg.*, 88, 167–174, 1964.
52. Najafi, M.F., Zahri, S., Vahedi, F., Toosi, L.E., Ariaee, N., Which form of collagen is suitable for nerve cell culture? *Neural Regen. Res.*, 8, 23, 2165, 2013.
53. Nimni, M.E. and Harkness, R.D., Molecular structures and functions of collagen, in: *Collagen-Biochemistry*, vol. I, M.E. Nimni, (Ed.), pp. 1–79, CRC Press, Boca Raton, FL, 1988.
54. Olszta, M.J., Cheng, X., Jee, S.S., Kumar, R., Kim, Y.Y., Kaufman, M.J., Gower, L.B., Bone structure and formation: A new perspective. *Mater. Sci. Eng.: R: Rep.*, 58, 3-5, 77–116, 2007.
55. Orwin, E.J. and Hubel, A., *In vitro* culture characteristics of corneal epithelial, endothelial, and keratocyte cells in a native collagen matrix. *Tissue Eng.*, 6, 4, 307–319, 2000.
56. Ozgonul, C., Grisolia, A.B.D., Demirci, H., The use of Integra® dermal regeneration template for the orbital exenteration socket: A novel technique. *Ophthalmic Plast. Reconstr. Surg.*, 34, 1, 64–67, 2018.
57. Pashuck, E.T., Cui, H., Stupp, S.I., Tuning supramolecular rigidity of peptide fibers through the molecular structure. *J. Am. Chem. Soc.*, 132, 17, 6041–6046, 2010. Perspectives. *Eur. J. Orthop. Surg. Traumatol.*, 28, 3, 351.
58. Petite, H., Rault, I., Huc, A., Menasche, P.H., Herbage, D., Use of the acyl azide method for crosslinking collagen-rich tissues such as pericardium. *J. Biomed. Mater. Res.*, 24, 179–187, 1990.
59. Ponticiello, M.S., Schinagl, R.M., Kadiyala, S., Barry, F.P., Gelatin-based resorbable sponge as a carrier matrix for human mesenchymal stem cells in cartilage regeneration therapy. *J. Biomed. Mater. Res.: An Official Journal of The Society for Biomaterials, The Japanese Society for Biomaterials, and The Australian Society for Biomaterials and the Korean Society for Biomaterials*, 52, 2, 246–255, 2000.
60. Rao, K.P., Recent developments of collagen-based materials for medical applications and drug delivery systems. *J. Biomater. Sci.*, 7, 7, 623–645, 1995.

61. Roddy, E., DeBaun, M.R., Daoud-Gray, A., Yang, Y.P., Gardner, M.J.. Treatment of critical-sized bone defects: clinical and tissue engineering perspectives. *Eur. J. Orthop. Surg. Traumatol.*, 28, 351–362, 2018 Apr.
62. Rössler, B., Kreuter, J., Ross, G., Effect of collagen microparticles on the stability of retinol and its absorption into hairless mouse skin *in vitro*. *Pharmazie*, 49, 2-3, 175–179, 1994.
63. Rössler, B., Kreuter, J., Scherer, D., Collagen microparticles: Preparation and properties. *J. Microencapsul.*, 12, 1, 49–57, 1995.
64. Sam, S., Joseph, B., Thomas, S., Exploring the antimicrobial features of biomaterials for biomedical applications. *Results Eng.*, 100979, 2023.
65. Sanders, J.E. and Goldstein, B.S., Collagen fibril diameters increase and fibril densities decrease in skin subjected to repetitive compressive and shear stresses. *J. Biomech.*, 34, 12, 1581–1587, 2001.
66. Sarwan, J. and Bose, J.C., Importance of microbial cellulases and their industrial applications. *Ann. Rom. Soc. Cell Biol.*, 3568–3575, 2021.
67. Sarwan, J., Pandey, P., Bose, J.C., Proteases immobilization and pharmacological aspects of matrix metalloproteinases. *Ann. Rom. Soc. Cell Biol.*, 3604–3615, 2021.
68. Shen, Y., Dai, L., Li, X., Liang, R., Guan, G., Zhang, Z., Chen, H., Epidermal stem cells cultured on collagen-modified chitin membrane induce *in situ* tissue regeneration of full-thickness skin defects in mice. *PLoS One*, 9, 2, e87557, 2014.
69. Sivaperumal, P., Kamala, K., Dhanraj, G., Biomacromolecules from marine organisms and their biomedical application, in: *Marine Biochemistry*, pp. 299–313, CRC Press, Taylor and Francis, 2023.
70. Soji, S., Das, P., George, M., Anjali, C.S., Collagen shields-a novel drug delivery system, 2020.
71. Sousa, I., Mendes, A., Pereira, R.F., Bártolo, P.J., Collagen surface-modified poly (ε-caprolactone) scaffolds with improved hydrophilicity and cell adhesion properties. *Mater. Lett.*, 134, 263–267, 2014.
72. Supp, A.P., Wickett, R.R., Swope, V.B., Harriger, M.D., Hoath, S.B., Boyce, S.T., Incubation of cultured skin substitutes in reduced humidity promotes cornification *in vitro* and stable engraftment in athymic mice. *Wound Repair Regener.*, 7, 4, 226–237, 1999.
73. Takitoh, T., Bessho, M., Hirose, M., Ohgushi, H., Mori, H., Hara, M., Gamma-cross-linked nonfibrillar collagen gel as a scaffold for osteogenic differentiation of mesenchymal stem cells. *J. Biosci. Bioeng.*, 119, 2, 217–225, 2015.
74. Thacharodi, D. and Rao, K.P., Rate-controlling biopolymer membranes as transdermal delivery systems for nifedipine: Development and *in vitro* evaluations. *Biomaterials*, 17, 1307–1311, 1996.

75. Vermonden, T., Censi, R., Hennink, W.E., Hydrogels for protein delivery. *Chem. Rev.*, 112, 5, 2853–2888, 2012.
76. Voleti, P.B., Buckley, M.R., Soslowsky, L.J., Tendon healing: Repair and regeneration. *Annu. Rev. Biomed. Eng.*, 14, 47–71, 2012.
77. Wang, Q., Yang, X., Ren, M., Hu, Y., Chen, Q., Xing, L., Liu, T., Effect of chitosan/type I collagen/gelatin composites in biocompatibility and nerve repair. *Neural Regener. Res.*, 7, 15, 1179, 2012.
78. Wang, X., Wang, Y., Li, L., Gu, Z., Yu, X., Feasibility study of the naturally occurring dialdehydecarboxymethyl cellulose for biological tissue fixation. *Carbohydr. Polym.*, 115, 54–61, 2015.
79. Xu, Y., Li, L., Yu, X., Gu, Z., Zhang, X., Feasibility study of a novel crosslinking reagent (alginate dialdehyde) for biological tissue fixation. *Carbohydr. Polym.*, 87, 2, 1589–1595, 2012.
80. Yamada, N., Uchinuma, E., Kuroyanagi, Y., Clinical evaluation of an allogeneic cultured dermal substitute composed of fibroblasts within a spongy collagen matrix. *Scand. J. Plast. Reconstr. Surg. Handb. Surg.*, 33, 2, 147–154, 1999.
81. Yannas, I.V., Lee, E., Orgill, D.P., Skrabut, E.M., Murphy, G.F., Synthesis and characterization of a model extracellular matrix that induces partial regeneration ofmadult mammalian skin. *Proc. Natl. Acad. Sci.*, 86, 3, 933–937, 1989.
82. Yin, D., Wu, H., Liu, C., Zhang, J., Zhou, T., Wu, J., Wan, Y., Fabrication of composition-graded collagen/chitosan–polylactide scaffolds with gradient architecture and properties. *React. Funct. Polym.*, 83, 98–106, 2014.
83. Zhou, X., Wang, J., Fang, W., Tao, Y., Zhao, T., Xia, K., Chen, Q., Genipin cross-linked type II collagen/chondroitin sulfate composite hydrogel-like cell delivery system induces differentiation of adipose-derived stem cells and regenerates degenerated nucleus pulposus. *Acta Biomater.*, 71, 496–509, 2018.
84. Zhu, S., Yuan, Q., Yin, T., You, J., Gu, Z., Xiong, S., Hu, Y., Self-assembly of collagen-based biomaterials: Preparation, characterizations, and biomedical applications. *J. Mater. Chem. B*, 6, 18, 2650–2676, 2018.

12

Biomedical Applications of Pectin

Heenu Sharma[1], Jyoti Sarwan[2], Mariyam Chaudhary[2], Riya Chugh[3] and Jagadeesh Chandra Bose K[2*]

[1]*Department of Nutrigenetics, Rayat Bahara University, Mohali, Punjab, India*
[2]*Department of Biotechnology, Chandigarh University, Mohali, Punjab, India*
[3]*Department of Chemistry, UIS, Chandigarh University, Mohali, Punjab, India*

Abstract
Nowadays, macromolecules are seeking the attraction of researchers with their characteristics like biocompatible nature, biodegradable capabilities, and low toxicity. Pectin is among the polysaccharides which have many biomedical applications like drug delivery and other biomedical activities. Due its various applications and easy availability, pectin is a major demand of research and medical science. In this chapter, we have summarized the sources of pectin and the various biomedical applications of pectin, followed by drug conveyance, wound healing, tissue designing, gene modulations, and many more.

Keywords: Pectin, polysaccharide, tissue engineering, drug delivery, polyplexes, ocular, wound

12.1 Introduction

Biopolymers are majorly used in the biomedical, food, and drug industries because of their high sustainability and minimal production expense [1, 2]. It is estimated that the worldwide biopolymers market is relied upon to reach $27.9 billion by 2025 [3], among which pectin (a water-dissolvable anionic biopolymer) stands for the most popularized biopolymers which are assessed at $1 billion as of 2019 and is relied upon to reach $1.5 billion by 2025 [4]. It can be defined as a polymer of carbon that has molecules of a huge weight and forms the polysaccharide integrity of the plant cell wall, which is convertible into a movable complex of repeating units and

**Corresponding author*: jcboseuibtlab@gmail.com

Annu (ed.) Biopolymers for Biomedical Applications, (315–332) © 2024 Scrivener Publishing LLC

formation of hydrogel [5]. Pectin is extracted commercially from fruits such as apples and oranges (citrus fruits are majorly used for its extraction) and/or from several industrial by-products. Obtaining pectin as a mechanical by-product introduces itself a green choice for the concept of bio-economy [6]. It is a plant-inferred anionic biopolymer that addresses a few qualities. It is removed from the cell mass of plant sources to an extensive variety which is 0.1–30 % reliant upon the technique for extraction [17, 20, 65].

This macro-biopolymer is composed of homo-galacturonan, rhamno-galacturonan I, rhamano-glacturonan II, and xylo-galacturonan structure and composition of pectin is mentioned in Figure 12.1 and Table 12.1 [7]. The complex chain of biopolymers formed by pectin is highly flexible, expandable, and hydrophobic in nature. The physical gels of pectin are formed due to coil entanglements resulting from low pH and a highly concentrated pectin solution. The formation of hydrophobic pectin gels can

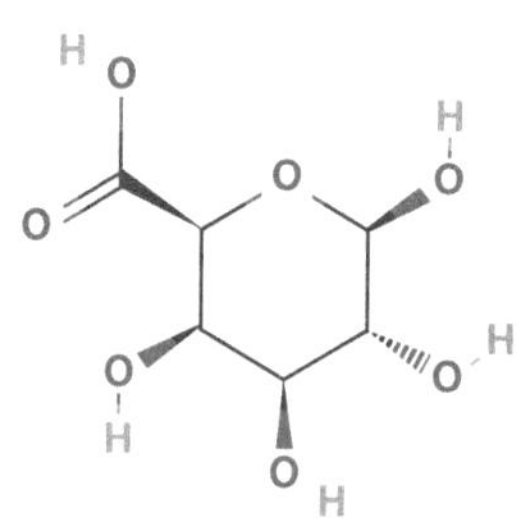

Figure 12.1 Chemical structure of pectin [54, 55].

Table 12.1 Composition of the pectin molecule.

Sub-domain title	Structural composition	Amount	References
Rhamnogalacturonan I (RGI)	Repeated disaccharides made up of rhamnose (Rha) and GalA (galacturonic acid) residues	20–35	[56, 57]
Xilogalacturonan (XG)	Highly complex branched structure	<10	[58, 59]
Homogalacturonan (HG)	Linear homopolymer of GalA	65	[60, 61]
Rhamnogalacturonan (RG II)	HG backbone composed of GalA (7–9 units)	<10	[63, 64]

also be acquired by using divalent or trivalent cations. These hydrogels are of scientific interest and are used in pharmaceuticals, tissue and genetic engineering [53], and wound treatment processes. However, biopolymers show assorted structural changes depending on the source and extraction procedure and are vulnerable to physical, compound, and enzymatic changes [8, 66]. Hence, while designing a particular gel for biomedical use, the structural formation of pectin, viz., the content of mono-saccharides, branching, and the cross-linkers are considered carefully [9, 11]. Other factors like molecular weight, extent of esterification, extraction source, method of extraction, etc., are measured carefully while designing a pectin gel for any specific use, as the various functional groups present in pectin are capable of stimulating various functional capabilities, with small modifications, thereby broadening the applications of pectin in various fields [5, 62].

12.2 Biomedical Applications of Pectin

Pectin is widely used not only because of its structural flexibility but also due to its non-toxic nature, making it safe and effective. Besides this, the production cost of pectin is very low with high productivity; hence, the demand and the availability of pectin, respectively, is always high [10]. Some of the applications of pectin in the biomedical field are listed in the following discussion mentioned in Table 12.2.

Table 12.2 Commercial sources and biomedical applications of pectin [44].

Commercial source	Biomedical application	References
CP Kelko GENU (type LM104 AS LM101 both AS-FS) Herbstreith & Fox KG GENU (type LM104 AS-FS) and CP Kelko, Otto Kemi, Mumbai, India East-Asiatic Company, type 7210 Sigma Aldrich Herbstreith & Fox KG Herbstreith & Fox KG, type CU70 Extracted pectin from cactus N.A. Cesap, Cesapectin® LM 32	Oral drug delivery Gene delivery Tissue engineering Cancer-targeted drug delivery Ocular drug delivery	[44–51]
Pectin type E440 Pectin type FPA	Nasal drug delivery	[52]

12.2.1 Pectin in Drug Administration

12.2.1.1 Pectin for Drug Delivery

Pectin is used in drug transfer systems to enhance the efficiency and safety of the pharma-drugs. It is used to provide muco-adhesiveness which is essential in controlling and targeting drug delivery in the case of nasal and/or gastric environment. It also provides ease of dissolution in an alkaline environment [84]. As the colon is resistant to protease and amylase, pectin can successfully release drugs into the colon. Additionally, pectin is used for the construction of gels in an acidic environment, thereby increasing the duration of the medicines for gastric or ocular treatments [18]. Pectin can also be used in cancer pathologies, as it can identify cancer-causing gelatin substances by targeting the onco-cells in various stages of chemotherapies. Pectin-based formulations offer several advantages for nasal drug delivery:

- Mucoadhesive properties: Pectin exhibits mucoadhesive properties, which means that it can adhere to the nasal mucosa and prolong the contact time between the drug and the nasal tissue. This prolonged residence time enhances drug absorption and bioavailability [78].
- Gel formation: Pectin can form gels when exposed to the nasal environment, particularly in the presence of calcium ions. These gels can act as drug carriers, providing sustained release and improved drug retention within the nasal cavity [79, 80].
- Biocompatibility: Pectin is derived from natural sources and is generally considered safe and biocompatible. It has been widely used in food and pharmaceutical industries, which adds to its appeal for nasal drug delivery applications [81, 82].
- Permeation enhancement: Pectin has been shown to enhance the permeation of drugs across the nasal mucosa. It can open the tight junctions between epithelial cells, allowing for improved drug absorption [83].

Overall, pectin shows promise as a material for nasal drug delivery due to its mucoadhesive properties, gel-forming capabilities, biocompatibility, and potential for controlled drug release. Ongoing research aims to harness these properties to improve drug delivery efficiency and therapeutic outcomes *via* the nasal route.

Figure 12.2 shows pure and drug-loaded pectin SEM images. Kocaaga *et al.*, in their study, revealed the controlled drug release ability of pectin-based

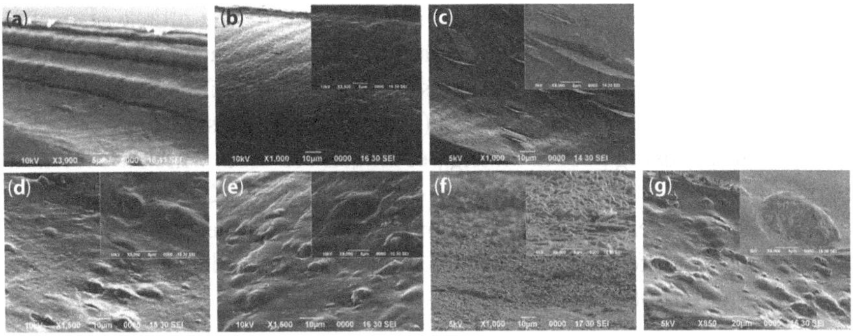

Figure 12.2 SEM images of hydrogels: (a) pure LM pectin, (b) LM pectin hydrogel, (c) drug-loaded LM pectin hydrogel, (d) A7.30Zn0, (e) A7.50Zn0, (f) A7.30Zn30, and (g) A7.50Zn30. Wiley [85].

pectin–zeolite hydrogel. The diffusion of theophylline drug into the channels of pectin hydrogel has been observed as pores in morphological studies. The controlled drug release from the pectin hydrogel can be addressed to wound healing application [85].

12.2.1.2 Nasal Drug Delivery

It is used to cure common nasal diseases, viz., allergy, congestion, and infection in the nasal cavity. Administration of medicines through the nasal passage is ideal for patients having a difficulty in the oral administration of drugs, for which drugs in the form of drops or sprays are used, which are directed in the nose [19]. Pectin is one of the best biopolymer used in nasal drug delivery due to its rapid and constant drug transportation—for instance, the nasal spray combination of an opioid painkiller fentanyl and pectin is proven to relieve pain due to cancer and is effective in cancer treatment by rapidly transporting the required drug into the body [21]. The combination of pectin with chlorpromazine hydrochloride is effective in treating psychological disorders [22]. Pectin is also used in association with nicotine for smoking cessation, where pectin helps the immediate supply and absorption of nicotine in the system [23].

12.2.1.3 Oral Drug Delivery

Polypeptide or protein-based drugs are not suitable for oral administration, as the different environmental conditions of the gut would limit the effectivity of the incorporated medicine due to the attack of protease in the small intestine, acidic surroundings, decreased absorbability, and movement of the drug particles across the intestinal barriers [24]. Hence, these drugs are encapsulated (using nano-encapsulation or micro-encapsulation) using pectin. Strong pectin gels are used, which are formed by combining the biopolymer with divalent cations (Ca^{2+} and Zn^{2+}) or polymers (ethyl-cellulose and chitosan) so that a complete dosage of drug is administered into the colon [4, 25, 26] and the degradation of pectin by micro-flora present in the intestine is prevented.

12.2.1.4 Ocular Drug Delivery

The bioavailability of ocular drugs is less due to sensitive eye mechanisms. As pectin is non-toxic and a safe biopolymer, pectin-based hydrogels can be used in ocular drug delivery systems. Pectin in ophthalmic drugs would improve the drug efficiency by increasing the duration of drug contact period with the eye and drug penetration [27]. An advanced approach of pectin-based gelation system is now being used, in which pectin is applied in the form of liquid, and the pH and electrolytes of the eye tears initiate the gelation of pectin, making a thin film over the eye. This is done to elevate the time period of drug contact with the eye [25]. Fundamental investigations were evaluated on pectin microspheres as a visual conveyance framework to perioxicam (i.e., not a steroid-mitigating chemical) and also, potentially, for different medications. Therefore, the framework ensures guaranteeing with higher bioavailability. It has displayed a decrease for a few detriments of other ophthalmic conveyance frameworks, for example, vision obscuring, need for inclusion and evacuation, and soundness issues.

12.2.2 Cancer-Designated Drug Conveyance

Pectin emerges to prevent the spread of cancer cells to one or more parts of the body and the development of primary cyst in different organisms and animals [28]. The inhibitory impact was proposed because of the identification of the galectin constituents of gelatin (Gal3). This Gal3 is associated with different phases of cancer, metastasis, ontogenesis, and physical cell reaction to cytotoxic drugs [29] and also disclosed in an assortment of metastatic cancer cells [30]. The citrus gelatin, which is achieved from the pulp and peels of citrus products and then changed through high temperature and pH conditions [31], has been additionally used to target Gal3, and this citrus gelatin effectively restrained its metastatic capacities. Hence, the altered gelatin prone to be stacked with antineoplastic to instigate the neoplastic cell apoptosis holds the possibility of expanding the proficiency to regulate therapy with chemicals adequately [46].

12.2.3 Pectin Polyplexes for Gene Conveyance

The gene therapy treatment is a potential methodology for the treatment of hereditary issues. The major aim in gene therapy is substituting the deficient and defected genes, dislodging the missing genes or suppressing the undesirable gene to expel the main cause of infection. Different studies have been carried out to focus upon the transmittance of hereditary materials to thee designated tissues, in association with some transfection vectors (viral or non-viral). The ability of the viral transfection vectors is far greater than that of non-viral ones; thus, the adverse cell-interceded immunogenic reaction is accomplished by utilizing viral vectors, and this is intentionally to make the non-viral quality transporters the vectors of choice. Amidst the multitude of non-viral vectors, the polycationic stretched polyethylenimine polymer (b-PEI) or chitosan [32] has displayed great transfection productivity both *in vitro* as well as *in vivo*. However, a drawback is that the positive charge on the poly-cationic polymers may lead to increased cytotoxicity, which can cause precariousness inside the area of the serum and/or extracellular matrix/grid parts. To vanquish the above-mentioned weakness or obstruction, various kinds of anionic polymers have been used to lower the charge of the external layer of the poly-cation and poly-anion complexes [32]. To cover the b-PEI polyplexes, the gelatin was observed to be more reasonable to form gelatin-covered polyplexes having a diminished surface charge, which thus prompted the diminished transfection going with low cytotoxicity and high stability of the complex [33]. Katav *et al.* proposed a substance that must be altered gelatin to create the anionic polymer suitable for DNA movement using some exceptional approach: The organization of gelatin was adjusted with the amine gatherings, and a short time later from the changed substances the edifices are framed with DNA plasmid with high stability. Distinctive investigators have announced the development of gelatin nanoparticles with a number of various cations to get the DNA for transfection [34].

12.2.4 Pectin Gels for Tissue Designing

There are different types of natural polymers that have been utilized as a stage to invigorate the particular cell capacities and to guide cell-to-cell collaboration inserts that are at first without cell, perhaps stacked with development components to advance tissue recovery [35] and to insert to help in cell development. These substances have been planned in two to three estimations to create multi-cellular tissue models which possibly can recover sound tissues in a restorative site. Some investigations have reported the utilization of gelatin for tissue recuperation, probably considering the non-adherence of this cellulose for cells. Therefore, the gelatin colloidal gels abide to have incredible possibilities for osseous tissue planning applications, as they advance the nucleation of a mineral stage at whatever point immersed in acceptable physiological arrangements [36], with the improvement of biomimetic fabricates reflecting the typical design of the bone that has been noticed [37]. Concerning the engineered changes, gelatin oxidation was evaluated to hasten the debasement of gels to extend the appearance of immobilized cells; hydrolytic medications and medicines were shown to convey homo-galacturonan (some portion of the gelatins), making bristly regions, which are exhibited to start a decent coordinated effort with variable cell types [38], and an arginine–glycine–aspartate destructive (RGD) organization was united on pectin spine to further develop cell adherence, in like manner fortifying the development of cell and extension. Additionally, the composite hydrogels framed with gelatin and polyvinyl liquor were surveyed as prosthetic nucleus pulpous substitutes, showing the incredible biocompatibility with phagocytes and no proof of uprooting or debasement subsequent to installation in the creature model [39].

One of the advantages of using pectin gels is their biocompatibility and biodegradability. Pectin is derived from natural sources and is generally considered safe for biological applications. It can be enzymatically degraded by various enzymes, including pectinases, which allows for controlled degradation of the gel over time as new tissue forms [67].

Pectin gels can also be modified to incorporate bioactive molecules such as growth factors, peptides, or drugs. These modifications can enhance the functionality of the gel and promote specific cellular responses, such as cell adhesion, proliferation, or differentiation. Additionally, pectin gels have been investigated for their immunomodulatory properties [74, 75]. They can potentially regulate the immune response in tissue engineering applications, minimizing inflammation and promoting tissue regeneration. While pectin gels show promise in tissue engineering, further research is needed to optimize their properties and understand their interactions with different cell types [76]. Scientists are actively exploring various strategies to improve the mechanical strength, stability, and bioactivity of pectin-based gels for tissue engineering applications.

Overall, pectin gels have the potential to serve as versatile biomaterials for tissue designing, offering a combination of biocompatibility, tunable properties, and controllable degradation, making them suitable for various tissue engineering applications [77].

12.2.5 Pectin in Wound-Mending Patches

Sometimes the utilization of hydro-colloidal layers upon injuries, even on ulcers meant to forestall a microbial disease, assist in the surgical intervention and keeping a damp healing climate. Besides this, an additional proof proposed the acceptability of hydrogels as restricted medication or development factor conveyance systems, hindrances to forestall post-careful adhesions and grids for skin recuperation. Thus, hydrogel movies ought to take part effectively in the injury recuperation process rather than just being inactive hindrances against parching. The natural properties of pectin offer a few benefits to the injury dressings, like aquaphilicity, which allows the evacuation of transude, the maintenance of an acidic climate, which might go about as a boundary against microbes and the capacity for binding the dynamic particles as medications or development promoters that help to provide relief in injuries. As an injury recovers, the surrounding cells are invigorated by development factors to multiply and develop into the injury [40]. The aquaphilic particles of pectin react with the injury liquid inside the dressing to shape a fragile gel onto the injury bed, thus killing the exudates in injury. The acidic climate acquired with pectin solubilization further goes about as a bacterial or viral block. The gelatin hydrogels likewise give an additionally created process for stacking and conveying drugs, for example, anti-toxins, anti-infection agents, pain killers, and, additionally, tissue-healing factors at the site of action.

Pectin-based wound mending patches or dressings have gained attention in recent years due to their potential benefits in wound healing. These patches utilize the unique properties of pectin to create an environment conducive to wound healing. Here are some key aspects of pectin-based wound mending patches:

- Moist wound healing: Pectin-based patches promote moist wound healing, which is known to enhance the healing process. These dressings have the ability to absorb and retain moisture from the wound, maintaining a moist environment that facilitates cell migration, angiogenesis (formation of new blood vessels), and the formation of granulation tissue [67, 68].

- Exudate absorption: Wounds often produce exudate, which can hinder the healing process if not managed properly. Pectin-based dressings have a high absorptive capacity, allowing them to effectively manage the exudate by absorbing excess fluid from the wound while maintaining the desired moisture level.
- Gel formation: Pectin has the unique ability to form a gel-like structure in the presence of water. This gel formation provides a protective barrier over the wound, shielding it from external contaminants while allowing gas exchange to occur. The gel also helps to maintain the moist environment required for optimal wound healing [68, 69].
- Biocompatibility and biodegradability: Pectin is derived from natural sources, primarily fruits, and is generally considered biocompatible and biodegradable. These properties make pectin-based dressings safe for use on wounds and reduce the need for frequent dressing changes.
- Anti-inflammatory and antimicrobial properties: Pectin has been reported to possess anti-inflammatory and antimicrobial properties, which can aid in the wound healing process. The anti-inflammatory effects help reduce inflammation at the wound site, while the antimicrobial properties can inhibit the growth of bacteria and prevent wound infections [70–72].
- Sustained release of bioactive agents: Pectin-based dressings can be modified to incorporate bioactive agents, such as growth factors or antimicrobial agents, which can be released gradually over time. This sustained release capability allows for the targeted delivery of therapeutic agents to the wound site, promoting healing and preventing infections [73].

Pectin-based wound mending patches are still under active research. While they show promise in facilitating wound healing, further studies and clinical trials are necessary to fully understand their effectiveness, optimize their properties, and establish their clinical utility in different types of wounds.

12.2.6 Other Formulations

Shaikh *et al.* created cross-linked pectin, gelation-based wafer systems for progressive buccal delivery [41]. They created three different arrangements of medication from stacked cross-linked pectin wafers by utilizing a model. Water-soluble diphenhydramine and antihistamine were compared with non-cross-linked wafers. These wafers were cross-linked with $BaCl_2$, $ZnSO_4$, and $CaCl_2$. They recommended that medication discharge execution was reliant upon the created wafer crosslinking arrangement. The authors also recommended sub-atomic system reenactment authenticated with the exploratory information and established that Ba^{++} having the biggest nuclear radii formed various ionic bridges delivering wafers of great porosity and with extra effect on drug discharge [42]. They arranged chitosan/gelatin polyelectrolyte buildings at pH 5 with various polycation and polyanion proportions and then lyophilized in little insects in the presence of a chemical, i.e., chlorpromazine hydrochloride. They proposed that the increase in content of pectin will permit communication with the chemical chlorpromazine hydrochloride which induced the arrangement of less hydra table insects, in this way restricting medication delivery and penetration. The creators additionally recommended that these nasal creepy crawlies are prepared for achieving antipsychotic transport in nasal depression [43]. Nano-emulsions were framed by simple assimilation to turn away the high tension conditions. They reasoned that pectin, with its undeniable degree of esterification, will offer incredible emulsion properties due to its large number of aquaphobic atoms. Moreover, they proposed that the acquired nano-emulsions might be created as a self-emulsifying drug conveyance system.

The properties of pectin, such as biocompatibility, biodegradability, and gel-forming ability, make it an attractive material for various drug delivery applications. However, the specific formulation and optimization of pectin-based drug delivery systems depend on the characteristics of the drug, desired release profile, and the targeted route of administration. Ongoing research continues to explore the potential of pectin in drug delivery and to develop innovative strategies to enhance its effectiveness.

12.2.7 Conclusion

In this chapter, the biomedical applications and certain non-ordinary domains of use of extraordinary polysaccharide, viz., pectin, are discussed. The structural and chemical features of pectin have been depicted to include which of them may impact the gels' advancement and, explicitly, which are concerned with the different biomedical uses. The tremendous collection of applications of pectin in various fields of industrial and biomedical horizons is recommendable. The increasing quality studies on pectin hydrogels in various fields endorse the ability of this polysaccharide as a novel and versatile biomaterial which will be substantially more significant in the future.

References

1. DiahSetiowati, A., Rwigamba, A., Van der Meeren, P., The influence of degree of methoxylation on the emulsifying and heat stabilizing activity of whey protein-pectin conjugates. *Food Hydrocolloids*, *96*, 54–64, 2019.
2. González-Henríquez, C.M., Sarabia-Vallejos, M.A., Rodriguez-Hernandez, J., Polymers for additive manufacturing and 4D-printing: Materials, methodologies, and biomedical applications. *Prog. Polym. Sci.*, *94*, 57–116, 2019.
3. Gong, J., Chen, X., Tang, T., Recent progress in controlled carbonization of (waste) polymers. *Prog. Polym. Sci.*, *94*, 1–32, 2019.
4. Martău, G.A., Mihai, M., Vodnar, D.C., The use of chitosan, alginate, and pectin in the biomedical and food sector—Biocompatibility, bioadhesiveness, and biodegradability. *Polymers*, *11*, 11, 1837, 2019.
5. Rodsamran, P. and Sothornvit, R., Lime peel pectin integrated with coconut water and lime peel extract as a new bioactive film sachet to retard soybean oil oxidation. *Food Hydrocolloids*, *97*, 105173, 2019.
6. Yang, J.S., Mu, T.H., Ma, M.M., Extraction, structure, and emulsifying properties of pectin from potato pulp. *Food Chem.*, *244*, 197–205, 2018.
7. Wicker, L., Kim, Y., Kim, M.J., Thirkield, B., Lin, Z., Jung, J., Pectin as a bioactive polysaccharide–Extracting tailored function from less. *Food Hydrocolloids*, *42*, 251–259, 2014.
8. Wang, W., Chen, W., Zou, M., Lv, R., Wang, D., Hou, F., Liu, D., Applications of power ultrasound in oriented modification and degradation of pectin: A review. *J. Food Eng.*, *234*, 98–107, 2018.
9. Mishra, R.K., Banthia, A.K., Majeed, A.B.A., Pectin based formulations for biomedical applications: A review. *Asian J. Pharm. Clin. Res.*, *5*, 4, 1–7, 2012.
10. Naqash, F., Masoodi, F.A., Rather, S.A., Wani, S.M., Gani, A., Emerging concepts in the nutraceutical and functional properties of pectin—A review. *Carbohydr. Polym.*, *168*, 227–239, 2017.
11. Munarin, F., Petrini, P., Fare, S., Tanzi, M.C., Structural properties of polysaccharide-based microcapsules for soft tissue regeneration. *J. Mater. Sci.: Mater. Med.*, *21*, 1, 365–375, 2010.
12. Munarin, F., Petrini, P., Tanzi, M.C., Barbosa, M.A., Granja, P.L., Biofunctional chemically modified pectin for cell delivery. *Soft Matter*, *8*, 17, 4731–4739, 2012.
13. Tan, H. and Marra, K.G., Injectable, biodegradable hydrogels for tissue engineering applications. *Materials*, *3*, 3, 1746–1767, 2010.

14. Munarin, F.A.B.I.O.L.A., Tanzi, M.C., Petrini, P.A.O.L.A., Advances in biomedical applications of pectin gels. *Int. J. Biol. Macromol.*, *51*, 4, 681–689, 2012.
15. Ravi, V., Pramod Kumar, T.M., S., Novel colon targeted drug delivery system using natural polymers. *Indian J. Pharm. Sci.*, *70*, 1, 111, 2008.
16. Ludwig, A., The use of mucoadhesive polymers in ocular drug delivery. *Adv. Drug Delivery Rev.*, *57*, 11, 1595–1639, 2005.
17. Glinsky, V.V. and Raz, A., Modified citrus pectin anti-metastatic properties: One bullet, multiple targets. *Carbohydr. Res.*, *344*, 14, 1788–1791, 2009.
18. Garna, H., Mabon, N., Robert, C., Cornet, C., Nott, K., Legros, H., Paquot, M., Effect of extraction conditions on the yield and purity of apple pomace pectin precipitated but not washed by alcohol. *J. Food Sci.*, *72*, 1, C001–C009, 2007.
19. Marttin, E., Schipper, N.G., Verhoef, J.C., Merkus, F.W., Nasal mucociliary clearance as a factor in nasal drug delivery. *Adv. Drug Delivery Rev.*, *29*, 1-2, 13–38, 1998.
20. Sriamornsak, P., Wattanakorn, N., Takeuchi, H., Study on the mucoadhesion mechanism of pectin by atomic force microscopy and mucin-particle method. *Carbohydr. Polym.*, *79*, 1, 54–59, 2010.
21. Fisher, A., Watling, M., Smith, A., Knight, A., Pharmacokinetics and relative bioavailability of fentanyl pectin nasal spray 100-800 µg in healthy volunteers. *Int. J. Clin. Pharmacol. Ther.*, *48*, 12, 860–867, 2010.
22. Luppi, B., Bigucci, F., Abruzzo, A., Corace, G., Cerchiara, T., Zecchi, V., Freeze-dried chitosan/pectin nasal inserts for antipsychotic drug delivery. *Eur. J. Pharm. Biopharm.*, *75*, 3, 381–387, 2010.
23. L. Illum, Nasal drug delivery composition containing nicotine. US Patent no. 5935604, 1999.
24. Liu, L., Fishman, M.L., Hicks, K.B., Pectin in controlled drug delivery–a review. *Cellulose*, *14*, 1, 15–24, 2007.
25. Paharia, A., Yadav, A.K., Rai, G., Jain, S.K., Pancholi, S.S., Agrawal, G.P., Eudragit-coated pectin microspheres of 5-fluorouracil for colon targeting. *AAPS PharmSciTech*, *8*, 1, E87–E93, 2007.
26. Jain, A., Khare, P., Agrawal, R.K., Jain, S.K., Metronidazole loaded pectin microspheres for colon targeting. *J. Pharm. Sci.*, *98*, 11, 4229–4236, 2009.
27. Ludwig, A., The use of mucoadhesive polymers in ocular drug delivery. *Adv. Drug Delivery Rev.*, *57*, 11, 1595–1639, 2005.
28. Pienta, K.J., Nailk, H., Akhtar, A., Yamazaki, K., Replogle, T.S., Lehr, J., Raz, A., Inhibition of spontaneous metastasis in a rat prostate cancer model by oral administration of modified citrus pectin. *JNCI: J. Natl. Cancer Inst.*, *87*, 5, 348–353, 1995.
29. Nangia-Makker, P., Honjo, Y., Sarvis, R., Akahani, S., Hogan, V., Pienta, K.J., Raz, A., Galectin-3 induces endothelial cell morphogenesis and angiogenesis. *Am. J. Pathol.*, *156*, 3, 899–909, 2000.

30. Baldus, S.E., Zirbes, T.K., Weingarten, M., Fromm, S., Glossmann, J., Hanisch, F.G., Dienes, H.P., Increased galectin-3 expression in gastric cancer: Correlations with histopathological subtypes, galactosylated antigens and tumor cell proliferation. *Tumor Biol.*, *21*, 5, 258–266, 2000.
31. Platt, D. and Raz, A., Modulation of the lung colonization of B16-F1 melanoma cells by citrus pectin. *Journal of the National Cancer Institute (JNCI)*, 84, 6, 438–442, 1992.
32. Guo, W. and Lee, R.J., Efficient gene delivery using anionic liposome-complexedpolyplexes (LPDII). *Biosci. Rep.*, *20*, 5, 419–432, 2000.
33. Munarin, F.A.B.I.O.L.A., Tanzi, M.C., Petrini, P.A.O.L.A., Advances in biomedical applications of pectin gels. *Int. J. Biol. Macromol.*, *51*, 4, 681-689, 2012.
34. Katav, T., Liu, L., Traitel, T., Goldbart, R., Wolfson, M., Kost, J., Modified pectin-based carrier for gene delivery: Cellular barriers in gene delivery course. *J. Controlled Release*, *130*, 2, 183–191, 2008.
35. Tabata, Y., The importance of drug delivery systems in tissue engineering. *Pharm. Sci. Technol. Today*, *3*, 3, 80–89, 2008.
36. Munarin, F., Giuliano, L., Bozzini, S., Tanzi, M.C., Petrini, P., Mineral phase deposition on pectin microspheres. *Mater. Sci. Eng.: C*, *30*, 3, 491–496, 2010.
37. Munarin, F., Petrini, P., Tanzi, M.C., Barbosa, M.A., Granja, P.L., Biofunctional chemically modified pectin for cell delivery. *Soft Matter*, *8*, 17, 4731–4739, 2012.
38. Kokkonen, H.E., Ilvesaro, J.M., Morra, M., Schols, H.A., Tuukkanen, J., Effect of modified pectin molecules on the growth of bone cells. *Biomacromolecules*, *8*, 2, 509–515, 2007.
39. Yao, N.Z., Huang, C., Jin, D.D., Evaluation of biocompatibility of a pectin/polyvinyl alcohol composite hydrogel as a new nucleus material. *Orthop. Surg.*, *1*, 3, 231–237, 2009.
40. R. Harris and W. Nasi, Gels for use in wound management. Patent WO2008/015475, February, 7, 2008.
41. Shaikh, R.P., Pillay, V., Choonara, Y.E., Toit, L.C.D., Ndesendo, V.M., Kumar, P., Khan, R.A., The application of a crosslinked pectin-based wafer matrix for gradual buccal drug delivery. *J. Biomed. Mater. Res. Part B: Appl. Biomater.*, *100*, 4, 1029–1043, 2012.
42. Luppi, B., Bigucci, F., Abruzzo, A., Corace, G., Cerchiara, T., Zecchi, V., Freeze-dried chitosan/pectin nasal inserts for antipsychotic drug delivery. *Eur. J. Pharm. Biopharm.*, *75*, 3, 381–387, 2009.
43. Burapapadh, K., Kumpugdee-Vollrath, M., Chantasart, D., Sriamornsak, P., Fabrication of pectin-based nanoemulsions loaded with itraconazole for pharmaceutical application. *Carbohydr. Polym.*, *82*, 2, 384–393, 2010.
44. Dhalleine, C., Assifaoui, A., Moulari, B., Pellequer, Y., Cayot, P., Lamprecht, A., Chambin, O., Zinc-pectinate beads as an *in vivo* self-assembling system for pulsatile drug delivery. *Int. J. Pharm.*, *414*, 1-2, 28–34, 2011.

45. Sachan, N.K., Pushkar, S., Ghosh, S.K., Evaluation of assam bora rice as a natural mucoadhesivematrixing agent for controlled drug delivery, in: *Chemistry of Phytopotentials: Health, Energy and Environmental Perspectives*, pp. 165–170, Springer, Berlin, Heidelberg, 2012.
46. Sharma, H. and Bajwa, J., Potential role and mechanism of probiotics. *Ann. Rom. Soc. Cell Biol.*, 25, 1, 3616–3624, 2021.
47. Asnaghi, M.A., Candiani, G., Farè, S., Fiore, G.B., Petrini, P., Raimondi, M.T., Mantero, S., Trends in biomedical engineering: Focus on regenerative medicine. *J. Appl. Biomater. Biomech.*, *9*, 2, 73–86, 2011.
48. Coimbra, P., Ferreira, P., De Sousa, H.C., Batista, P., Rodrigues, M.A., Correia, I.J., Gil, M.H., Preparation and chemical and biological characterization of a pectin/chitosan polyelectrolyte complex scaffold for possible bone tissue engineering applications. *Int. J. Biol. Macromol.*, *48*, 1, 112–118, 2011.
49. Glinsky, V.V. and Raz, A., Modified citrus pectin anti-metastatic properties: One bullet, multiple targets. *Carbohydr. Res.*, *344*, 14, 1788–1791, 2009.
50. Pienta, K.J., Nailk, H., Akhtar, A., Yamazaki, K., Replogle, T.S., Lehr, J., Raz, A., Inhibition of spontaneous metastasis in a rat prostate cancer model by oral administration of modified citrus pectin. *JNCI: J. Natl. Cancer Inst.*, *87*, 5, 348–353, 1995.
51. Sharma, H. and Bajwa, J., Isolation and characterization of lactic acid bacteria isolated from fermented food of North-West Himalayas. *J. Pure Appl. Microbiol.*, 15, 3, 1567–1572, 2021, doi: 10.22207/JPAM.15.3.50.
52. Ni, Y. and Yates, K.M., U.S. Patent and Trademark Office, Washington, DC, 2004.
53. Nangia-Makker, P., Honjo, Y., Sarvis, R., Akahani, S., Hogan, V., Pienta, K.J., Raz, A., Galectin-3 induces endothelial cell morphogenesis and angiogenesis. *Am. J. Pathol.*, *156*, 3, 899–909, 2000.
54. https://pubchem.ncbi.nlm.nih.gov/compound/Pectin#section=2D-Structure.
55. Maxwell, E.G., Belshaw, N.J., Waldron, K.W., Morris, V.J., Pectin an emerging new bioactive food polysaccharide. *Trends Food Sci. Technol.*, 24, 64–73, 2012.
56. Wicker, L., Kim, Y., Kim, M.J., Thirkield, B., Lin, Z., Jung, J., Pectin as a bioactive polysaccharide extracting tailored function from less. *Food Hydrocolloids*, 42, 251–259, 2014.
57. Ngouémazong, E.D., Christiaens, S., Shpigelman, A., Van Loey, A., Hendrickx, M., The emulsifying and emulsion-stabilizing properties of pectin: A review. *Compr. Rev. Food Sci. Food Saf.*, 14, 705–718, 2015.
58. Mohnen, D., Pectin structure and biosynthesis. *Curr. Opin. Plant Biol.*, 11, 266–277, 2008.
59. Marenda, F.R.B., Mattioda, F., Demiate, I.M., de Francisco, A., de Oliveira Petkowicz, C.L., Canteri, M.H.G., de Mello Castanho Amboni, R.D., Advances in studies using vegetable wastes to obtain pectic substances: A review. *J. Polym. Environ.*, 27, 549–560, 2019.

60. Yang, J.S., Mu, T.H., Ma, M.M., Extraction, structure, and emulsifying properties of pectin from potato pulp. *Food Chem.*, 244, 197–205, 2018.
61. Freitas, C.M.P., Costa, A.R., Rodrigues, F.Á., Júnior, M.M.J., Dias, M.M., Sousa, R.C.S., Optimization of pectin extraction from passion fruit (Passiflora edulis flavicarpa) using the response surface method. *Braz. J. Dev.*, 6, 25609–25625, 2020.
62. Freitas, C.M.P., Sousa, R.C.S., Dias, M.M., Coimbra, J.S., Extraction of pectin from passion fruit peel. *Food Eng. Rev.*, 12, 460–472, 2020.
63. Badaró, A.T., Garcia-Martin, J.F., del C. López-Barrera, M., Barbin, D.F., Alvarez-Mateos, P., Determination of pectin content in orange peels by near infrared hyperspectral imaging. *Food Chem.*, 323, 126861, 2020.
64. Grassino, A.N., Halambek, J., Djaković, S., Rimac Brnčić, S., Dent, M., Grabarić, Z., Utilization of tomato peel waste from canning factory as a potential source for pectin production and application as tin corrosion inhibitor. *Food Hydrocolloids*, 52, 265–274, 2016.
65. Moslemi, M., Reviewing the recent advances in application of pectin for technical and health promotion purposes: From laboratory to market. *Carbohydr. Polym.*, *254*, 117324, 2021.
66. Eivazzadeh-Keihan, R., Noruzi, E.B., Aliabadi, H.A.M., Sheikhaleslami, S., Akbarzadeh, A.R., Hashemi, S.M., Gorab, M.G., Maleki, A., Cohan, R.A., Mahdavi, M., Poodat, R., Recent advances on biomedical applications of pectin-containing biomaterials. *Int. J. Biol. Macromol.*, 217, 1–18, 2022.
67. Bostancı, N.S., Büyüksungur, S., Hasirci, N., Tezcaner, A., Potential of pectin for biomedical applications: A comprehensive review. *J. Biomater. Sci. Polym. Ed.*, *33*, 14, 1866–1900, 2022.
68. Zhang, S., Waterhouse, G.I., Xu, F., He, Z., Du, Y., Lian, Y., Wu, P., Sun-Waterhouse, D., Recent advances in utilization of pectins in biomedical applications: A review focusing on molecular structure-directing health-promoting properties. *Crit. Rev. Food Sci. Nutr.*, 1–34, 2021.
69. Michailidou, G., Terzopoulou, Z., Kehagia, A., Michopoulou, A., Bikiaris, D.N., Preliminary evaluation of 3D printed chitosan/pectin constructs for biomedical applications. *Mar. Drugs*, *19*, 1, 36, 2021.
70. Kim, M.S., Chandika, P., Jung, W.K., Recent advances of pectin-based biomedical application: Potential of marine pectin. *J. Mar. Biosci. Biotechnol.*, *13*, 1, 28–47, 2021.
71. Vaikundam, M., Shanmugam, S., Aldawood, S., Arunkumar, P., Santhanam, A., *et al.* Preparation of biopolymer pectin fascinate hydroxyapatite nanocomposite for biomedical applications. *Appl. Nanosci.*, 13, 6039–6051, 2023.
72. Freitas, C.M.P., Coimbra, J.S.R., Souza, V.G.L., Sousa, R.C.S., Structure and applications of pectin in food, biomedical, and pharmaceutical industry: A review. *Coatings*, *11*, 8, 922, 2021.
73. Diao, J., Bai, F., Wang, Y., Han, Q., Xu, X., Zhang, H., Luo, Q., Wang, Y., Engineering of pectin-dopamine nano-conjugates for carrying ruthenium

complex: A potential tool for biomedical applications. *J. Inorg. Biochem.*, *191*, 135–142, 2019.

74. Li, D.Q., Li, J., Dong, H.L., Li, X., Zhang, J.Q., Ramaswamy, S., Xu, F., Pectin in biomedical and drug delivery applications: A review. *Int. J. Biol. Macromol.*, *185*, 49–65, 2021.
75. Robledo, V.R. and Vázquez, L.I.C., Pectin-Extraction, purification, characterization and applications, in: *Pectins-Extraction, Purification, Characterization and Applications*, pp. 1–19, Intechopen, London, UK, 2019.
76. Morello, G., De Iaco, G., Gigli, G., Polini, A., Gervaso, F., Chitosan and pectin hydrogels for tissue engineering and *in vitro* modeling. *Gels*, *9*, 2, 132, 2023.
77. Oh, G.W., Nam, S.Y., Heo, S.J., Kang, D.H., Jung, W.K., Characterization of ionic cross-linked composite foams with different blend ratios of alginate/pectin on the synergistic effects for wound dressing application. *Int. J. Biol. Macromol.*, *156*, 1565–1573, 2020.
78. Rascón-Chu, A., Gomez-Rodriguez, G.H., Carvajal-Millan, E., Campa-Mada, A.C., Pectin in drug delivery applications, in: *Natural Polysaccharides in Drug Delivery and Biomedical Applications*, pp. 249–262, 2019.
79. Hernandez, G., J., Pazmino, M., Barrios-Rodriguez, Y.F., Tonda Turo, C., Wills, C., Cucinotta, F., Benlloch-Tinoco, M., Gentile, P., The influence of organic acids coupled with ultrasound on pectin extracted from apple pomace for biomedical applications. *Heliyon*, 9, 7, 2023.
80. Demir, D., Ceylan, S., Göktürk, D., Bölgen, N., Extraction of pectin from albedo of lemon peels for preparation of tissue engineering scaffolds. *Polym. Bull.*, *78*, 2211–2226, 2021.
81. Sood, A., Arora, V., Kumari, S., Sarkar, A., Kumaran, S.S., Chaturvedi, S., Jain, T.K., Agrawal, G., Imaging application and radiosensitivity enhancement of pectin decorated multifunctional magnetic nanoparticles in cancer therapy. *Int. J. Biol. Macromol.*, *189*, 443–454, 2021.
82. Tortorella, S., Inzalaco, G., Dapporto, F., Maturi, M., Sambri, L., Buratti, V.V., Chiariello, M., Franchini, M.C., Locatelli, E., Biocompatible pectin-based hybrid hydrogels for tissue engineering applications. *New J. Chem.*, *45*, 47, 22386–22395, 2021.
83. Kedir, W.M., Deresa, E.M., Diriba, T.F., Pharmaceutical and drug delivery applications of pectin and its modified nanocomposites. *Heliyon*, e10654, 2022.
84. Induru, J., Pectin-based nanomaterials in drug delivery applications, in: *Biopolymer-Based Nanomaterials in Drug Delivery and Biomedical Applications*, pp. 87–117, Academic Press, 2021, https://doi.org/10.1016/B978-0-12-820874-8.00011-7.
85. Kocaaga, B., Kurkcuoglu, O., Tatlier, M., Batirel, S., Guner, F.S., Low-methoxyl pectin–zeolite hydrogels controlling drug release promote *in vitro* wound healing. *J. Appl. Polym. Sci.*, 136, 24, 47640, 2019.

13

Biomedical Applications of Lignin Derived from Bio-Waste Materials

Nancy George[1*], Anshika Adlakha[1], Mridula[1], Pragati Gupta[1] and Abhrajit Debroy[1,2]

[1]*University Institute of Biotechnology, Chandigarh University, Mohali, Punjab, India*
[2]*Centre for Nanobiotechnology, School of Bio Sciences & Technology, Vellore Institute of Technology (VIT), Vellore, India*

Abstract
Lignin is a complex organic polymeric substance constituting the key biochemical and structural component in plant support tissues, thus providing rigidity and stability to plant cells. It is the second most abundant organic polymer on earth, accounting for nearly 30% of carbon sequestered in the ecosystem. Owing to its abundance, there is increased attention towards valorization of lignin in terms of developing value-added chemicals, fuel alternatives, and other commercial applications. In this regard, lignin extracted from biological waste generated from agriculture, households, and industries represents a sustainable approach in lignin valorization and waste management. Various properties of lignin, *viz.*, its renewability, ability to imitate the extracellular environment, biocompatibility, ease of integration with nanotechnology, and antioxidant and antimicrobial ability, make it the center of attraction for its utilization in the biomedical industry and, presently, an exciting field of research. Appropriate chemical modifications of lignin extracted from biomass pave the way for its application in a wide range of biomedical technologies. Utilization of bio-waste-derived lignin for such applications will boost the bio-economy and also serve as an eco-friendly approach for waste management. This chapter summarizes the structure, sources, extraction, and characteristics of lignin polymer with a major focus on its biomedical applications.

***Keywords*:** Lignin, biopolymer, extraction, biocompatible, drug delivery, wound, hydrogel, tissue

**Corresponding author*: nancy.george@cumail.in

Annu (ed.) Biopolymers for Biomedical Applications, (333–372) © 2024 Scrivener Publishing LLC

13.1 Introduction

Biopolymers are natural polymers synthesized by living systems and comprise monomeric units which covalently bind to build up large molecules. Natural polymers play a vital role and diverse functions based on their indigenous setting. In this regard, lignin is an organic polymer and primary constituent of plant dry matter, popularly known as lignocellulosic biomass. Lignocellulosic biomass is one of the most encouraging renewable carbon compounds on earth. Based on the source and type of biological mass, lignin comprises about 20%–35% of lignocellulosic biomass [1, 2]. The composition and structure of this polymer mainly depend on its source of origin and the methods used for its extraction. Once extracted, lignin can be altered and improved by various chemical reactions and used for numerous commercial applications.

Subsequently, lignin is one of the principal crude biological resources which can be utilized for the production of eco-friendly bio-based material. The high ratio of aromatic compounds present in it offers a great potential of replacing industrially produced synthetic polymers and harmful chemicals. Currently, the largest amount of lignin is synthesized and utilized by pulp and biorefinery industries which utilize it to generate power by combustion [3]. However, lignin can be utilized for various other commercial applications, *viz.*, production of biofuels, production of synthetic polymers and biodegradable plastics, and application in therapeutics. In recent times, significant research has been done for the use of lignin in biomedical applications. The properties of lignin like high carbon content and thermostability, adhesiveness, biocompatibility, and bio-activity make it a suitable candidate for biomedical purposes. Lignin-based biomaterials are popular in biomedical applications, including drug and gene delivery carriers, tissue engineering scaffolds, cosmetics, wound dressing materials, and therapeutics [4]. In addition, lignin can enhance the mechanical stability and strength of other processed biomaterials, too.

Though terrestrial plants like shrubs, bushes, trees, and grasses serve as a good source of lignin, waste biomass generated as a byproduct of diverse industrial activities, *viz.*, agriculture (wheat bran, wheat stalks, straws, corn stover, etc.) forestry, food industry, and household waste, is an attractive approach for lignin valorization and waste management as well. Taking into account the huge amount of lignin-based waste biomass generated per year, it is indeed an attractive and sustainable approach of waste management. Lignin derived from bio-waste reduces the overall cost involved in extraction and application [5, 6].

In this chapter, we have primarily focused on lignin derived from waste biomass and reviewed the application of lignin in the area of biomedical technology. Although this field still remains unexplored and is mostly based on theoretical facts, it still opens the door to many possibilities that are not only beneficial to the developing area but also benign on the environment.

13.2 Structure of Lignin

Lignin is a phenolic biopolymer that has an irregular three-dimensional structure with many branches and has an amorphous structure. It is a complex biopolymer having phenylpropanoids units cross-linked randomly and does not have any optical activity [7]. Different species of plants have different compositions of lignin, which lead to substantial diversity in overall structure and chemical nature.

Since the chemical structure of lignin is highly intricate, it is believed to be synthesized through certain irregular biosynthesis processes in which p-hydroxyphenyl (H), guaiacyl (G), and syringyl (S) play the role of building units [8, 9]. Sinapyl alcohol, coumaryl alcohol, and guaiacyl alcohol are the three monolignols (Figure 13.1) which form these building blocks. The major component of lignin in softwoods is G, whereas the G–S units are prevalent in hardwoods; herbaceous plants, on the other hand, have G–S–H units [10]. Ecological factors like climate, nutrition, and photoperiod are also known to affect lignin's chemical structure [11]. The linkages through which the building blocks of lignin are interlinked are majorly ether bonds, namely, phenyl ether, aryl ether, and carbon–carbon bonds, such as biphenyls, diphenyl ethane, or pinoresinol.

Thus, the combination of monolignols in different amounts and with different distribution and substitution patterns results in the variations and intricacies that are found in lignin present in various plants [12]. Systematic monolignol linkage formation was discussed in one of the research conducted by Gou *et al.* in 2018 [13]. Monolignol units have certain methylation extensions in the phenolic rings at 3′ or 3′–5′ [8]. As the site of polymerization is the cell wall of the plant, the monolignol precursors are transferred to the cell wall where the final macromolecule of lignin is synthesized.

The amount of H/G/S units present in lignin affects its structural features, and based on these structural features, there are nine interlinkages which, in turn, contribute to the formation of lignin's 3D structure. The ideal site for coupling reaction is the β-position in these linkages, and as a result, spirodienones (SD), diphenylethane dimers (β-1′), arylglycerol-β-aryl

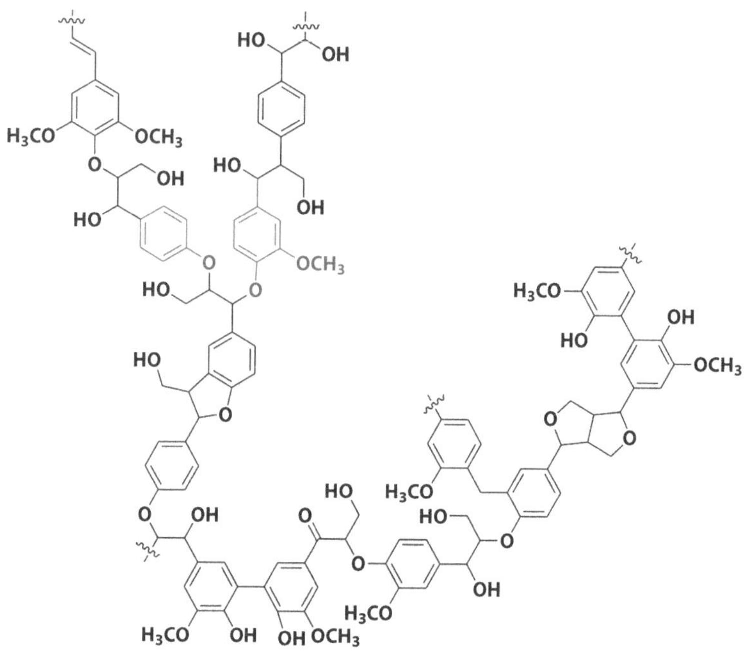

Figure 13.1 Schematic representation of the generalized structure of lignin (important monolignol units are colored as follows: guaiacyl alcohol—red, sinapyl alcohol—blue, and p-coumaryl alcohol—pink).

ethers (β-O-4), phenylcoumarans (β-5′), and pinoresinols (β-β′) are formed. In a study, it was shown that diphenylethane dimers (β-1′) constitute up to 5%–10% of lignin [14, 15]. The functional groups present in lignin are phenolic hydroxyl, noncyclic benzyl ether, carbonyl groups, aliphatic hydroxyl, benzyl alcohol, and methoxyl groups, and the amount of these functional groups affects the activity of lignin. The important functional groups of lignin are discussed below:

a) Hydroxyl Groups: The phenolic and aliphatic hydroxyl groups are present in much higher content in lignin. In the synthesis, modification, and degradation of lignin, the hydroxyl functional groups play a significant role [16, 17]. Due to the presence of the phenolic hydroxyl group, the cleavage of the β-O-4 bond increases during the oxidative and base-catalyzed depolymerization of lignin.

Wood lignin has a greater amount of aliphatic hydroxyl group than industrial lignin, whereas the phenolic hydroxyl group content is more in the industrial lignin [10, 18].

b) Benzyl Ether and Benzyl Alcohol Groups: These two groups are very crucial in the chemical reactions in which lignin takes part. Benzyl ether and benzyl alcohol attack the benzyl position and break the β-O-4 bond. This is the basic role of these two groups. Non-cyclic benzyl aryl ethers, cyclic benzyl aryl ethers, for example, phenylcoumarans and pinoresinols, and benzyl alkyl ethers are the different types of benzyl ethers. To understand the structural and chemical characters of lignin, information about these two groups is very important [19, 20].

c) Carbonyl Groups: Aldehydes, ketones, and carboxyl groups are the different carbonyl groups that are present in lignin. Carbonyl groups can be conjugated and non-conjugated. The conjugated carbonyl groups are of CαO and cinnamaldehyde (CγHO) types, and the non-conjugated carbonyls are of CβO type. Some other carbonyl groups can also be found on aromatic rings like quinones. The non-conjugated carbonyl groups can be found in the form of aldehyde, whereas the conjugated aldehyde is present at the γ carbon and the conjugated ketone is attached to the α-carbon of the carbonyl group [21].

The isolation and identification process of lignin has become difficult due to its molecular complexity, and there are many unexplored questions regarding the structure of lignin. Lignin is an exception to the other natural polymers such as proteins and polysaccharides because it is degraded at a very slow rate by living organisms, and they cannot even use it as a sole source of carbon.

13.3 Sources of Lignin

Lignin is the material that makes plants woody. For commercial purposes, lignin can be derived from natural sources (terrestrial plants or energy crops) and organic wastes of biological origin. Various sources of lignin are discussed in detail bin the following subsections.

13.3.1 Lignin Derived from Natural Sources

Lignin is one of the most plentiful polymers in nature and is found in all vascular plants. It is renewable due to its formation *via* the photosynthetic reaction, and thus its estimated annual production on earth is found to be in the range of 5–36 × 10^8 tons. In softwoods of gymnosperms, e.g., spruces (*Picea* spp.) and pines (*Pinus* spp.), lignin content is found to be in the range of 25% to 35% and constitutes only guaiacyl units, a low fraction of p-hydroxyphenyl units (HG-lignin) along with methoxyl content in the range of 15%–16% [22]. In hardwoods from angiosperms (*viz.*, *Eucalyptus globulus, Betula pendula, Cynara cardunculus*, and *Acacia mangium*), lignin is found in the range of 15%–28%, comprising mainly guaiacyl and syringyl units (GS-lignin), and a methoxyl content of approximately 21% [22–24]. Monocotyledons belonging to families Arecaceae (essentially palms), Poaceae (*viz.*, grasses, sugarcane, and bamboo), and Musaceae (bananas) are very economically significant. The lignin content found in monocotyledon species varies structurally from the one found in softwoods and hardwoods. It is an HGS-lignin with a higher content comprising p-hydroxyphenyl units, a major concentration of hydroxycinnamate esters along with some acylation by p-coumarates in the S units, and constitutes a range of 5% to 12% [25].

The major difference between the three types of lignin is that gymnosperm lignin is found to have more G units, angiosperm lignin is observed to be a combination of G and S units, whereas lignin in grasses is a blend of all three units. Based on the presence of the type of monolignols, lignin from softwood, hardwood, and grass are classified as G type, G–S type, and H–G–S type, respectively. The higher levels of methoxyl groups in lignin make it less condensed and friendlier to chemical conversion.

In addition to the above-mentioned plant groups, energy crops can also be used for the extraction of lignin. This category of crops is cultivated exclusively with the objective of energy production by combustion and not for utilization as food crops. Some extensively used energy crops include maize or sweet sorghum and herbaceous and woody plants—for example, kenaf (*Hibiscus cannabinus*), reed canary grass (*Phalaris arundinacea*), (*Miscanthus giganteus*), switchgrass (*Panicum virgatum*), or giant reed (*Arundo donax*) and (*Eucalyptus globulus*). The accessibility of these energy crops for energy production and their being simultaneously harmless to the environment are factors that made them acquire exceptional interest worldwide [26]. They can be harnessed for lignin production, too.

13.3.2 Lignin Derived from Bio-Waste

Lignin, in the objective of its extraction from biowaste, was initially observed to have a higher significance in the paper and pulp industry where lignin is extracted from wood to produce fibrous products rich in cellulose [27]. Currently, lignocellulosic woody materials and wastes are highly utilized for the manufacture of green chemicals, bio-based products, and energy utilization by biorefineries [15]. The concentration of lignocellulose in agricultural plant residues is found to be in the range of 50%–90% of the total organic matter. Grasses have a relatively less lignocellulosic content, and wastes like straws, stovers, wood, and tea waste are observed to have a comparatively higher concentration of lignin [28].

Several studies show varying amounts of acid-insoluble lignin in different plants, namely, fern [29], hazelnut shells [30], loblolly pine [31], and banana pseudo-stems [32], and in sludge produced by paper and pulp factories, which contain softwood and hardwood [33]. The paper industry generates the highest amount of lignin, and future lignocellulosic biorefineries are predicted to increase the production even more. In urban and sub-urban locations, where lignin-based materials are found to be less than 15%, grass branches, leaves, and solid trash act as a source of minimal lignin [34]. In addition, huge amounts of lignin can be derived from agricultural and food industry sources.

13.3.2.1 Paper and Pulp Industry

Hardwood and softwood species as well as certain types of annual plants exist as the most preferred source of commercial interest for cellulose fibers for manufacturing paper and board goods. Huge quantities of lignin are produced as lignosulfonates and alkali lignin and in a typical process of fiber liberation like alkaline or sulfite pulping [35]. An estimated amount of 70 million tons of lignin is accessible annually from the paper-making business, out of which only 2% of it is processed for lignin production, while most of it is utilized to make fuels. Industrial lignin is the type of lignin that is available commercially through the process of industrial delignification. The major source of the paper-making business is industrial lignin, with two of its most commercially significant types: kraft lignin and lignosulfonates. It is noted that kraft lignin accounts for around 85% of total lignin production worldwide; however, lignosulfonates are the major form of lignin available commercially. The industry sector has benefited from current technology that aims towards improving the recovery of lignin from black liquor using the method of CO_2 acidification, resulting in a readily available dry lignin powder stream [36].

13.3.2.2 Forest Residue/Wood Waste

Organic waste from forest residue is the widest source of cheap and abundant lignocellulosic material rich in lignin. Forest residues mostly include broken branches, roots, and leaves. Moreover, cutting stands for timber and lands cleared for construction or other uses also yield waste branches that are usable for the extraction of lignin. Wood waste is mainly produced from wood processing industries like wooden panels, sawmills, plywood, and other wood products, which produce significant amounts of byproducts [37]. The main types of waste include sawdust, off-cuts, and wood cuttings and trims that can be later converted into pellets [38].

13.3.2.3 Agricultural Waste

Agricultural waste includes corn stover, rice straw, rice husk, and wheat straw, mainly the wastes left over on the fields after harvests and mostly utilized as fodder and landfill materials or burnt in many places [28]. Using agricultural wastes for lignin extraction has three distinct advantages: cost-effective, environmentally safe, and higher significance of use commercially. Agricultural pulps, unlike wood pulps, provide ease of being processed and are bleached in a more environmentally friendly manner. The majority of the pulp in the wood pulping process is subjected to bleaching with chlorine or through other chlorine-based chemicals, whereas agricultural wastes can be handled with minimum chlorine-free chemicals, resulting in a safer technology. Furthermore, agricultural leftovers are found to have a lower lignin content and more porous structure than woody plants, which makes the process of pulping comparatively easier. Rice straw is one of the major agricultural biomass produced with a turnover of 525 million tons over the world each year. It contains about 30%–35% cellulose, 25%–30% hemicellulose, 15%–28% lignin, and 4%–7% ash. In addition to rice straw, 529 million tons of wheat straw (*Triticum aestivum*) is also produced each year. Wheat straw may be generated on a larger scale when extensively farmed. Wheat straw contains lignin content in the range of 5% to 17%, which, when compared to hardwoods, varies with the geographical location of the plants [39].

13.3.2.4 Household and Food Industry Waste

Lignin is found in fruit peels, nutshells, and vegetable peels. Fruit peels contain lignin, cellulose, hemicellulose as well as pectin [40]. Lignin can also be found in citrus peel waste [41]. The lignin in walnut and almond shells is SG, while the lignin in pine nutshells is G. The lignin content of almond shells is around 29% [42, 43]. Other lignocellulosic sources generated in food industries and household include cereal straws (e.g., wheat, rice, barley), corn stover, fruits and vegetable peels, or fruit leaf fibers, which contain 7%–38% lignin [44]. Bagasse is the fiber left over after the sugars were removed from sugarcane. Sugarcane bagasse (*Saccharum officinarum*) is a rich source of lignin derived through food industry residues. It contains around 25%–32% lignin and is a byproduct of the sugarcane industry. Compared to softwood and hardwood lignins, sugarcane bagasse lignin contains a higher content of H-type lignin, hydroxyphenyl, and a relatively lower methoxy concentration. Annually, a total of 54 million tons of bagasse is produced, and each ton of sugarcane produces approximately 250–280 kg of bagasse [45].

Thus, wastes of biological origin are a rich resource of lignin and are a great field for research and exploration to develop cost-effective and commercially viable technologies for lignin application.

13.4 Extraction of Lignin

Native lignin is a branched biopolymer that is formed by the arbitrary covalent crosslinkages between cellulose and hemicelluloses [46], due to which a highly stable lignin carbohydrate complex that hinders lignin extraction is formed [47, 48].

For the efficient and complete usage of the molecule, various strategies have been employed to isolate the compound while sustaining the purity of the same (Figure 13.2). Lignin extracted after the execution of such processes on the biomass available is then called technical lignin [49]. The major source of technical lignin is paper and pulp which yield cheap byproducts that are rich in lignocellulosic materials [50]. The effective isolation of lignin from plant tissue is a cumbersome task due to the unavailability of lignases or enzymatic proteins that could rupture the bonds in the stable complex formed as a result of interaction between carbohydrate moieties and lignin itself, but it is an important step for the valorization of lignin [51]. Nowadays, lignin production at the pilot scale is extensively carried out by the pulp and papermaking industry which executes the

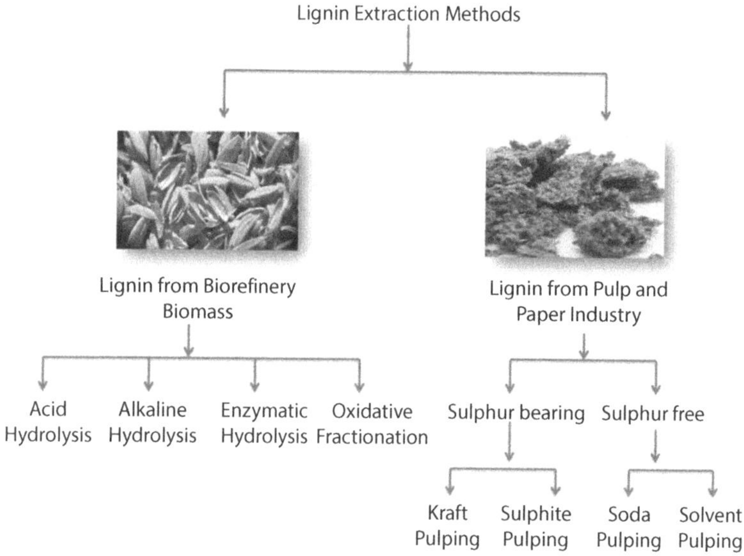

Figure 13.2 Schematic representation of the various methods employed for lignin extraction.

same by the application of biorefinery methodologies that include various pulping strategies and sets of enzymes; some are based on the differences in the pH of the solvents being used as well [52].

13.4.1 Lignin Extraction from Biorefinery Biomass

This industry endeavors to utilize biomass as an inexhaustible and replenishable resource with an aim to substitute the fossil fuels to generate energy. The constituents of the biomass matrix have different characteristics based on the functional groups it possesses and physical properties such as density and melting and boiling points, which altogether determine its reactivity. Therefore, by utilizing these facets, different isolation strategies have been employed to separate the compound of interest [53]. Completion is often achieved by subjecting the biomass to pretreatment and fractionation stages which is the initial step of biorefining. While dealing with the biomass stock, it has been observed that even the same matrix produces lignin with varying purity levels and alternate structures when treated with different fractionation processes [54]. Some biorefinery processes will be discussed in the following subsections [48].

13.4.1.1 *Acid Hydrolysis of Biomass*

This process makes both dilute (0–5 wt% HCl, H_2SO_4 or H_3PO_4, 120°C–240°C) and concentrated acids (concentrated HCl, H_2SO_4 or HF, 18°C–25°C) for depolymerization to take place. Lignin is collected in dissolved form, forming soluble fragments which are highly reactive in nature [55]. Three types of lignin forms have been known to be isolated during this process:

(a) Lignin Droplet: It is produced as a result of the difference in the temperature conditions, i.e., the pretreatment temperature (kept raised) and the phase transition temperature. The polymer gets liquefied, and *via* traveling, cell wall matrix gets accumulated on the surface of biomass.
(b) Pseudolignin: formed when the various lignin intermediates react with carbohydrates; hence, repolymerization reaction occurs between the two.
(c) Lignin Forming: A highly condensed form of lignin is produced, called Klason lignin, by acid hydrolysis carried out with the concentrated acids (72 wt% H_2SO_4 hydrolysis; at 25°C–27°C). This form is highly condensed, which makes it difficult to use as a fuel/valuable chemical [56].

13.4.1.2 *Alkaline Hydrolysis*

This extraction procedure is alkali dependent, hence the name alkaline hydrolysis. The pulverized polymer is dissolved in alkali such as hydroxides of sodium, calcium, or ammonia to bring about the faster depolymerization of heteropolymer. As soon as the bond between crosslinkages in the polymer weakens, constituents of the biomass matrix in their native and altered forms get separated, i.e., monomers, dimers, and oligomers. However, the success rate of alkaline hydrolysis is not that high, specifically when induced with strong alkalis due to their reactive nature and strong binding affinity. However, when carried out with ammonia, the process was benefited due its high volatility which eases the separation phase. The nucleophilic property of ammonia aids in bond breakage to recover lignin from the biomass matrix [52].

13.4.1.3 *Enzymatic Hydrolysis of Biomass*

To bring about efficient bond cleavage in the polymer, enzymatic proteins, cellulases, and hemicellulases have been extensively used; this leaves lignin-enriched solid residues at the end of the process. Before the actual process is carried out, the biomass is made to undergo some pretreatment strategies which are preferably including ball milling, hot water pretreatment, and mechanical refining. It is much needed because the molecule is firm and obdurate enough due to the presence of lignocellulose. Furthermore, some post-treatment techniques can be applied as enzymatic hydrolysis does not interfere much with the native structure of the compound. Therefore, either mild acid hydrolysis or extraction of organic solvent might be used to bring about the process. The changes occurring as a result of it are tolerable at significant levels. This method is commonly supported by the biorefining industries and gaining popularity to carry out the valorization [52].

13.4.1.4 Reductive Catalyst Fractionation of Biomass

To forge ahead the breakage in the ether linkages of the biopolymer, the use of reductive reagents or oxidants has been devised. This mainly includes the hydrogen-donating molecules or oxygen-rich species such as molecular oxygen, ozone, or hydrogen peroxide. The series of events in the reaction is further aided by the redox catalysts [57]. The lignin products obtained at the end have highly complicated and complex structures [52].

The fractionation of biomass feedstock is worked out with the organic solvents (for example: alcohols, ethers, ketones, esters, and acids); both concentrated and diluted forms are used, keeping into account the biomass source and its properties. Considering the fact that lignin solubility is higher in natural solvents than in water, therefore the polymer isolation proficiency is much higher, thus retaining the original nature of the compound [58].

13.4.2 Extraction of Lignin from Pulp and Paper Industry Biomass

13.4.2.1 Extraction of Sulphur-Bearing Lignin

Sulphur lignin is composed of two categories of lignin: kraft and lignosulphonate biopolymer, mainly recovered from the pulp and paper industries. This extraction methodology focuses on lignin extraction from another abundant plant polymer, i.e., cellulose.

a) Kraft Pulping: It is named after the peculiar solvent used, i.e., sulfide solution of sodium, also known as white liquor, dissolved in 1-L solution of sodium hydroxide of concentration 1 N. The chipped wood along with the white liquor [4:1 (w:w) liquid/solid] is packed in the containers of steel, and they are heated constantly rotated while being heated in an oil bath for the better homogenization of the content and equal distribution of heat. The heating temperatures differ for the hard and soft woods, being 155°C for the former and 170°C–180°C for the latter ones. The process altogether takes 1 to 2 h. Thereafter, washing is done with distilled water and freeze-drying [59].

b) Sulphite Pulping: The process is executed by treating the wood with aqueous sodium sulphite. The pH is dependent on the wood type; hence, it can be acidic, neutral, or alkaline. The process formulation differs depending upon the cooking medium preferred—for instance, in neutral medium conditions, heating of crushed wood is carried out at 175°C for 90 min in sulphite liquor [60].

13.4.2.2 Sulphur-Free Process

As soon as the fractionation procedure comes to an end, it leashes out the low fragments of lignin which are sulphur-free and therefore have low molecular weight. They resemble the native lignin very closely in the structural aspect. Sulphur-free lignin is grouped under two heads: organosolv lignin obtained by solvent pulping and soda lignin produced as a result of alkaline pulping.

a) Solvent Pulping

The technique is so called because of the high solubility of lignin in organic solvents. They are the purest forms of lignin [31]. They are produced by precipitation reaction at optimal reaction conditions. The most generalized solvent pulping method is based on the use of ethanol or water-blended ethanol (for example, Alcell) and further pulping with acetic acid, consisting of strong acids such as hydrochloric acid in trace amounts [61]. Moreover, another extraction strategy is achieved by using a mixture of acetic acid, formic acid, and water. The formulations were devised by CIMV Company (France). The obtained bio-compound was called bio-lignin, and its properties are still under investigation [62].

b) Soda Pulping

Soda lignin is a product of soda pulping process and specifically soda-anthraquinone technique. It targets at utilizing annual plant species that could yield good amounts of straw, flax, and bagasse along with some selective hardwood species [63]. The success rate of this process is much higher when measured in terms of the purity of the product. Although the principle of lignin isolation rests on the bond cleavage and rupturing of the crosslinkages peculiarly ether bonds, yet, comparatively, a decent amount of unmodified polymer is recovered.

The major issue is preserving the product purity levels as the compound shows significant alterations in structure configuration when subjected to various treatment procedures. Therefore, numerous strategies have been proposed, such as the use of ionic liquids, enzymatic proteins, hydrothermals, etc., but when tested for industrial scale, all efforts were in vain due to the failure of such techniques. Researchers are still working to emerge with successful approaches while learning from the shortcoming [62].

13.5 Properties of Lignin for Biomedical Applications

As stated, the lignin structure from various plant species differs from each other and so do its physicochemical properties. Lignin, being a versatile biomass, is known to possess varied desirable characteristics that make it a favorable material for the biomedical industry. The extracted lignin has to be modified through certain specific modifications in its chemical structure to enhance the development of lignin biomaterials. Some of the important properties of lignin polymer are discussed in the following subsections.

13.5.1 Molecular Weight

It is not possible to obtain the accurate molecular weight of lignin as it may degrade during isolation, separation, or identification technique. Gel permeation, chromatography size exclusion chromatography, osmotic method, etc., are some of the techniques used for molecular weight determination, and it ranges from 10^3–10^5 Da in isolated lignin [64].

13.5.2 Thermal Characterization

Lignin under high temperature has slight friability and cannot form film in solution and is treated as an amorphous thermoplastic polymer. The glassy transfer properties of lignin are described using the glass transition temperature (T_g) which also depends on various factors just like molecular weight [65, 66].

13.5.3 Solubility

The lignin obtained from biomass that is natural and unmodified is usually insoluble in common solvents [35]. The solubility of lignin can be improved by degrading lignin into low molecular weight fragments. There is a wide range of polar functional groups in modified lignin which are able to bind with the solvent through hydrogen bonds. Lignin gets dissolved well under alkaline conditions due to carboxyl and phenolic hydroxyl groups, but this is not in the case of acidic or enzymatic conditions.

13.5.4 Relative Density and Color

The techniques used to determine the relative density and the liquid used are the factors which influence the relative density, and it varies from 1.35 to 1.50. The natural lignin is colorless or white whereas the color of synthesized lignin depends upon the separation and preparation techniques.

13.5.5 Biocompatibility

The polymers like protein and polysaccharides have been used by researchers comprehensively for biomedical applications like for drug delivery and as nanocarriers due to their biocompatible nature. The ability of the biomaterials to co-exist with the human body and demonstrate a suitable response when required is the main distinguishing factor that differentiates them from the other materials. Biocompatibility and cytotoxicity are closely related. Low cytotoxicity is a prerequisite for the interaction of biomaterials with the cells. Lignin is a polymer with high biocompatibility [67]. In the two-cell line test in which four lignins were studied, it was found that Kraft lignin was the most cytotoxic and lignosulfonate was the least cytotoxic [68].

13.5.6 Biodegradability

Although lignin is the toughest constituent of the plant cell structure, certain microorganisms have the ability to degrade it. The powerful extracellular oxidative system of the white-rot fungi is capable of lignin biodegradation. Different fungal species have this system, and its components are Mn-peroxidase and laccase. The oxidative system of fungi has low substrate specificity [69]. Even though certain enzymes and bacterial and fungal species are capable of degrading lignin, the biodegradation of lignin in the context of its application in the biomedical industry is in terms of its degradation inside the human body which can cause its breakdown.

13.5.7 Antioxidant and Antimicrobial Properties

In order to replace the currently used harmful antioxidants, there is a great need for natural antioxidants with high biocompatibility. The antioxidant properties of lignin have been studied by various scientists. The high antioxidant ability of lignin was determined by Faustino and co-workers (2010) [70] and Azadfar and co-workers (2015) [71]. It is known that the antioxidant property of lignin corresponds to that of commercially used antioxidants such as guaiacol, and butylated hydroxytoluene [71]. The high amount of phenolic group and the other functional groups which are present in lignin contribute to its antioxidant property [72]. These functional groups react with the free radicals and terminate the oxidative chain reaction [73]. Lignin is able to scavenge free radicals by hydrogen transfer and single electron transfer with the aid of the phenolic hydroxy groups (Ar-OH) present in it [74]. Similarly, the intensive antibacterial properties have been studied in lignin over years [75]. The double bond Cα = Cβ of the phenolic group induces the antibacterial activity of lignin. When the bacteria come in contact with the phenolic group of lignin it breaks the cell membrane causing the lysis of bacteria as depicted in Figure 13.3. The bacterial species and the type of lignin used are the factors that affect the antibacterial properties [76, 77].

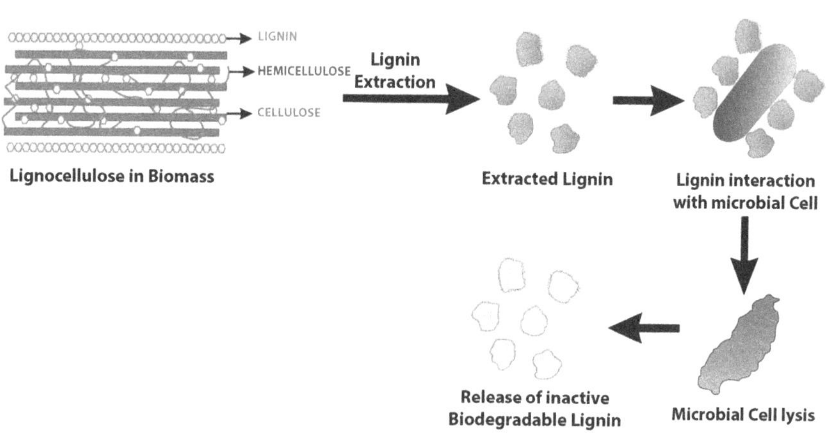

Figure 13.3 Schematic representation of the antimicrobial effect of lignin polymer.

13.5.8 Anti-Ultraviolet Properties

Lignin helps in protecting the cellulose fibers of plants which are exposed to UV radiations due to its good ultraviolet-absorbing properties [78]. The functional groups of lignin that have the UV-absorbing property are phenols, ketones, and other chromophores [79]. Using a mixture of lignin and other synthetic anti-UV materials provides increased UV protection than using just synthetic materials [80].

The different extraction processes alter the properties of lignin like molecular weight and thereby its solubility, exposure of functional groups, abundance of aromatic groups, and presence of impurities added during extraction. This leads to a significant difference in native and extracted lignin. During the process of extraction, high temperature and harsh chemical conditions are used, and as a result, there is a reduction in the ether bond linkages, especially β-O-4 bonds that cause the technical lignin to be less active in comparison to native lignin, and the highly condensed extracted lignin has irreversible degradation. Therefore, for different lignins, some crucial studies are conducted to understand its biocompatibility and biodegradability to a greater extent.

13.6 Biomedical Applications of Lignin

Lignin is a potential polymer and crude material for biomedical applications. This heteropolymer exhibits a unique set of properties like higher thermal stability, biocompatibility, adhesive, antimicrobial, and antioxidant properties as well as anti-ultraviolet properties, which make lignin a preferable choice for its use in biomedical applications. In addition, lignin also gains benefits of cost-effective processing as a by-product of paper and pulp industries and in agriculture, too. Various preliminary studies have shown that lignin has its biomedical applications in the development of drug delivery vehicles, biomedical hydrogels, biomaterials for tissue engineering and wound healing, etc. [81]. The various biomedical applications of lignin and recent advancements in this field are discussed in the following subsections.

13.6.1 Lignin in Hydrogel Development

Hydrogels are polymers with three-dimensional structures of crosslinked materials which have the ability to absorb and retain water and other biological liquids within their structural arrangement [82]. Hydrogels can be obtained from both natural and synthetic biopolymers [83]. The synthetic hydrogels that are used commercially are made of polyacrylic acids, polyvinyl liquor (PVA), and polyvinyl pyrrolidones. However, the development of hydrogel from natural polymer over synthetic polymer is preferred for medical applications. These are preferred due to their characteristics like less toxicity, biocompatibility, propensity to degradation by enzymes, and eco-friendly properties [84]. In this regard, the application of lignin for developing hydrogels is noteworthy. Recently, lignin-based hydrogels have been widely favored for various biomedical applications, like gene and drug delivery, wound healing, and tissue engineering [81]. The main reason behind lignin being favored for the synthesis of hydrogels is owing to its biocompatibility, low toxicity, and promising thermal properties [85].

In one of the reports, lignin-based hydrogels were utilized for the controlled delivery of curcumin for as long as 4 days [86]. In a study with a similar objective by Raschip *et al.* (2015) [87], lignin-based hydrogels were synthesized from curcumin by mixing it with xanthan for the controlled delivery of bisoprolol for the purpose of using it as a successful treatment for high blood pressure and heart failure. Lignin has also been effectively experimented and combined with various different polymers (*viz.*, chitosan, PVA, alginate, or cellulose) to create hydrogels for the application of wound healing [88] and tissue engineering [89]. The inclusion of lignin in the preparation of hydrogel systems was found to prevent infections and enhance the wound healing effect.

Hydrogels prepared from lignin are synthesized through various techniques such as hydrothermal, cross-linking, and ultrasonication processes. Such materials are presently being used for various applications in biomedical areas, including being preferred as a therapeutic agent, adhesive agents, and adsorbing agents and in other miscellaneous applications, namely, in the production of conductive carbon microfibers and supercapacitors and for preparing biosensors. The utilization of lignin-derived hydrogels from biomass and its application are still being studied extensively, and the role of lignin in hydrogels has turned out as an environment-friendly option over harsh chemical-based counterparts.

13.6.2 Lignin in Wound Healing

Healing of wound is an intricate process of action for re-establishing and maintaining any particular damaged area of the skin [90]. A suitable environment provided by the wound dress material is a major requirement for proper wound healing. The major characteristics of an ideal wound dressing includes promoting a clean area at the wound surface, allowing proper gaseous exchange, protecting the site from possible external contamination, capable enough of absorbing exudates from the wound area, and able to decrease wound pain. It is also expected to be non-allergic, comfortable, non-adherent, non-toxic, and sterile [91].

To achieve all these objectives in a wound dressing material, extensive research is being carried out for the successful development of products that have all the required properties required for healing the wounds efficiently. It is observed that the use of conventional wound dressings like bandages, gauze-woven cotton, and cotton wool is not enough to provide suitable conditions for a speedy and complete curing of the wound. Some of the modern wound dressing materials that are widely preferred and available in the market include hydrogels, hydrocolloid, medicated and semi-permeable foam, semipermeable film, and composite and bioactive wound dressings [92]. Polymers of biological origin like alginate, gelatin, cellulose, chitosan, etc., are regarded as some of the promising candidates for the preparation of modern wound dressing material.

Among other biopolymers, lignin is observed to be the most preferred source for the development of wound dressings. This is due to its antioxidant, antimicrobial, and high water retention properties [93]. Various studies were performed on lignin to check its efficiency and viability as a potential wound healing material. One such study by Ravishankar *et al.* (2019) [88] proposed the idea of integrating the benefits of alkali lignin with chitosan to develop novel materials for regenerative medical applications. The cytotoxicity of chitosan–alkali lignin hydrogels was examined by utilizing *in vitro* 3-(4,5-dimethylthiazol-2-yl)-2,5-diphenyl tetrazolium bromide assay, and the *in vivo* toxicity of the gel was tested on a zebrafish (*Danio rerio*) embryo model to check its complete suitability and efficiency for regeneration. Furthermore, the healing capacity of the hydrogel was assured by using a scratch assay with mouse fibroblast cell lines. The results proved the biocompatibility of chitosan–alkali lignin hydrogels in every aspect and the suitability to support cell attachment and growth with a conductive surface [88]. Jaganathan *et al.* (2018) [94] extracted lignin from the peels of a tropical fruit (*Artocarpus heterophyllus*) to develop biocomposites made of lignin and chitosan. The porosity of these composites has a huge impact on the contamination of wound tissue and also on the growth of cells. The composites were found to be of potentially wound healing materials, as they were found to be biocompatible and showed a higher antibacterial activity against *Escherichia coli* and *Klebsiella* sp. [94].

In one of the research, towards similar objectives, curcumin-loaded lignin nanoparticles (LNPs) were used and studied to explore their wound curing activity. Toxicity studies on cells showed that LNPs loaded with curcumin were highly biocompatible with keratinocytes of the skin and have no cytotoxic effects, confirming that they do not interfere with cell proliferation during wound healing. In addition, it displayed powerful *in vitro* antibacterial action against gram-positive bacteria, especially against the most common bacterium found on wounds, *Staphylococcus aureus* [94]. Thus, lignin holds a great scope in developing an efficient and viable wound healing material.

13.6.3 Lignin in Drug Delivery

A drug delivery system is an engineered technique which aims at achieving higher therapeutic effects of drugs, on the precise site which is diseased, with minimal toxicological effect [95]. These days, natural polymers are preferred for this purpose over synthetic polymers. Being an abundant polymer with various useful characteristics, *viz.*, antimicrobial and antioxidant behavior, biocompatibility, low toxicity, and presence of functional groups, researchers prefer lignin for potential future drug delivery endeavors [96]. Lignin is highly preferred for the preparation of composite films for drug delivery systems. In a work with a similar objective, lignin was extracted from the residues of hazelnut shells and was used on cornstarch films to enhance its mechanical properties. The biofilms obtained showed a speedy discharge of ciprofloxacin within 60 min, and afterward the rate of medication discharge was found to decelerate and achieved a stationary state by the end of the 10th day. Therefore, it was proved through this study that the ciprofloxacin/starch/lignin biofilm might be considered as a promising carrier for drug delivery [97].

The current phase of development has witnessed a significant interest of the researchers in the field of utilization of nanoparticles for various applications within food, polymers, cosmetics, drugs, and similar applications due to their nanoscopic and environmentally friendly properties. Lignin-based nano- and microcarriers are observed to be efficient biodegradable carriers for fungicide delivery in agricultural crops. Lignin nanotubes synthesized from the cell walls of aromatic plants have potential applications as carriers for drug delivery into human cells. In one of the studies, lignin-based nanocapsules were synthesized for the controlled arrival of hydrophobic particles. These complex micelles have application in the controlled delivery of hydrophobic oral medications, namely, aspirin, indomethacin, and nifedipine [96].

Some lignin/complexed lignin nanoparticles were explored for expected biomedical applications. In one of the studies, Fe_3O_4 lignin nanoparticle complex showed superparamagnetic conduct, which proved it to be a promising agent for the application of magnetic targeting and magnetic resonance imaging. In a study by Uraki *et al.* (2016) [98], ethylene glycol diglycidyl ether was cross-connected with hydroxypropyl cellulose–lignin gel, which, at body temperature, shrinks and releases the content outside the gel network. Hence, an engineered technique on drug delivery system utilizing lignin can help to achieve the better release of a drug at the exact site of the disease with minimal toxicological effects [98].

13.6.4 Lignin in Gene Delivery

The process of transfer of external genetic material, i.e., DNA or RNA, to the host cell is called gene delivery. To induce gene expression, it is necessary to make sure that gene delivery must reach the genome of the host cell. Lignin polymer has a great prospective to be used as a carrier for gene delivery. A promising non-viral lignin-based graft copolymer was developed by Liu *et al.* [99]. They selected atom transfer radical polymerization method for co-polymerization of lignin with 2-(dimethylamino) ethyl methacrylate. Hyper-branched cationic amphiphilic lignin-based graft copolymer was constructed as it has multiple cationic hydrophilic arms of poly-(2-dimethylaminoethylmethacrylate) and a hydrophobic backbone of lignin. This positively charged lignin-based graft copolymer was effectively condensed with anionic plasmid DNA, and nanoparticles of size ranging from 100 to 200 nm were formed. Effective gene transfer efficiency was shown by these nanoparticles in Hela, Cos-7, and MDA-MB-231 cell lines [99].

Ten and coworkers [100], in their research, showed the physical association of DNA with lignin nanotubes. These nanotubes were able to penetrate the cytosol and even the nucleus in some cases. It was observed that cells were able to tolerate concentrations of up to 90 mg/mL of lignin nanotubes, which is 10 times higher than that observed for carbon nanotubes. This minimal cytotoxicity of lignin nanotubes enhanced the delivery of DNA because of the ease of penetration of the nuclei. Thus, an optimum always persists between cytotoxicity and transfection efficiency [100]. Some of the clinical trials in this regard have failed due to inefficiency in tissue penetration and lower blood circulation time. The efficiency of lignin-based gene delivery systems can be improved by a combination of other polymers and systems. In this regard, lignin-based nanoparticles can be beneficial in terms of targeted and controlled delivery and maintenance of the therapeutic dosage, and this field has lot of research scope.

13.6.5 Lignin in Tissue Engineering

Tissue engineering is an interdisciplinary branch which employs the knowledge and application of biology, chemistry, and engineering for the regeneration of damaged tissue or organ. For this biomedical purpose, fabrication of natural biomaterials is necessary [101]. The application of lignin biopolymer in tissue engineering is an exciting area of research. In this regard, much has been investigated in recent years by developing lignin into nanofibers and hydrogels, and this field is rapidly growing [102, 103].

Electrospun fibers have a high surface area–volume ratio, and these nanofibers having a porous and fibrous framework have been expectedly explored in tissue engineering applications [102, 104]. Using this electrospinning technique for tissue engineering, Saudi and co-workers prepared an electrospun PVA, poly glycerol sebacate (PGS), and PVA–PGS with different amounts of lignin. The addition of lignin enhanced the uniformity and smoothness of the fabricated fibers for application in tissue engineering [105].

Natural polymers like lignin are known to have great contributions in the development of hydrogels for tissue engineering [101]. Lignin hydrogels have a three-dimensional network structure and high water-absorbing capacity similar to living tissues, which makes them promising in this field. In one of the studies, physically crosslinked hydrogel was fabricated from a mixture of cellulose, xylan, and kraft lignin by Kalinowski *et al.* with the aid of ionic liquids [82]. Lignin-based scaffolds are yet other biomaterials that may fulfill several requirements of tissue engineering. Cell adhesion is the first step in the cell scaffold interaction. Surface polarity, roughness, and topography are the factors which strongly influence the interactions between the cell and the scaffold [106, 107]. Lignin hybrid polymers can be used as scaffolds for the regeneration of bones and other stem cell-based therapies, too.

13.7 Therapeutic Applications of Lignin

13.7.1 Treatment of Diabetes

Diabetes is a chronic metabolic disease in which the blood sugar level increases. A diabetic person's body either cannot produce a sufficient amount of insulin or cannot use the synthesized insulin effectively. Thus, insulin hormone is taken externally to transport blood sugar to the cell where it can be stored or used for energy. If high blood sugar is left untreated, it damages the kidneys, eyes, and other organs. Products of lignin and lignin-derived compounds including polyphenols are known to have potential effects in controlling diabetes. The bio-modified alkali lignin which is extracted from *Acacia nilotica*, a deciduous plant, has a significant inhibitory action on α-amylase activity *in vitro*, thus having potential properties of anti-hyperglycemic activity. The structural and functional modifications of alkali lignin changed its binding efficiency with glucose molecule, which further affected the cellular movement of glucose and limited the postprandial glucose absorption. Thus, by inhibiting theα-amylase activity and affecting glucose movement, lignin provided a potential therapeutic role in the treatment of diabetes [108]. In another study, Hasegawa *et al.* (2015) [109] suggested that lignosulfonic acid, when taken with glucose, could potentially cause a delay in glucose uptake by intestine to treat diabetes. This effect was studied on intestinal glucose absorption. As lignosulfonic acid is a non-competitive inhibitor of α-glucosidase, it can therefore easily inhibit the activity of α-glucosidase and thereby delay glucose absorption in the intestine [109].

Lignophenols are known to reduce vascular inflammation and oxidative stress by inhibiting the activity of nicotinamide adenine dinucleotide (phosphate) oxidase, hence providing a cure for certain ailments such as vascular impairment of diabetes [110]. In a study done by Chen *et al.*, lignin was found to have a potential for inhibition on the enzyme α-glucosidase [111]. Thus, lignin has a potential role in diabetes treatment, and this field has an immense research scope.

13.7.2 Cancer Treatment

13.7.2.1 Antioxidant Properties

Reactive Oxygen Species (ROS), primarily contained in hydrogen peroxide, hydroxyl, peroxy radical, and superoxide, are unceasingly generated, reworked, and used by the cells throughout the housekeeping reactions in the human body [112]. Accumulations of ROSs in excessive amounts can cause depletion of the molecular machinery in the body which may potentially become the cause of abnormal functioning of tissues and organs, substantially leading to the occurrence of dreadful diseases [113].

Therefore, this raises the urge of the usage of antioxidants whether available artificially or naturally to counterbalance the appalling effects of ROSs. Antioxidants are molecular species that hunt free radicals present in living systems and edible items by scavenging them, hence refraining the occurrence of life-threatening chemical reactions [114]. Applications of antioxidants has dug its way far enough, reaching several industries—primarily pharmaceuticals, followed by cosmetics, foods, plastics, and fuel-generating ones [115].

13.7.2.2 *Cytotoxicity*

There is an inevitable relation between the inhibiting potential of the heteropolymer with cytotoxicity, stating that the stronger the inhibiting ligand, the more is the extent of cytotoxicity, and the compounds are antioxidants at non-cytotoxic concentrations. The major acknowledged issue with the inhibiting potential of the polymer is that a fully recognized pathway with facts and figures has been not elucidated yet. However, a group of researchers, while investigating the pathway, annotated three attainable possibilities for the inhibitory reaction mechanism to take place. These include lignin acting as follows:

(1) A chelating agent as it hunts for the metal ions of Fenton's reagent
(2) A suicide molecule, scavenging hydroxyl radicals that may potentially have a halting effect on alternative target molecules
(3) An inhibitor, by blocking the enzymatic proteins which pave the way for the metabolic pathways generating free radicals [116]. Several examinations throw light on the possibility of lignin scavenging free radicals. This quality of it acting as an anti-oxidant is attributed to the facts that it is enriched with the presence of oxygen-containing species such as phenolic resin structures [117].

In these commodities, the antioxidants usually used contain artificial compounds like butylated hydroxyanisole, butylated hydroxytoluene, tert-butylhydroquinone, and propyl gallate [114]. Synthetic antioxidants pose both relieving and dreadful biological effects at the physiological level in living systems [118]. However, malignant neoplastic disease effects are determined for artificial antioxidants once administered in excessive amounts; their maximum levels are counseled and governed by numerous lawmaking authorities like the FDA or/and European Food Safety Authority reckoning on the ultimate use.

A pair of carbohydrate–lignin complex was obtained from water-soluble fractions, and a detailed study was carried out to examine their antiproliferative action. Chemical and spectral analyses demonstrated that each of the functional compounds of lignin contained macromolecular (carbohydrate) moieties which made the hydrophobic polymer extremely soluble. Reports for toxicity analysis exhibited that these polymer derivatives evoked programmed cell death, which was mostly independent of cell cycle. Moreover, each complex of lignin–carbohydrate restrained the activation of transcription issue in cancerous cells.

13.7.3 Antimicrobial and Antiviral Activity

Lignin is one of the dominant natural antimicrobial compounds. The microbe-inhibiting trait of the biopolymer is a factor in choosing the plant species and the isolation method to be employed [102]—for instance, Klapisezewski *et al.* (2015) [119], while experimenting with 10 dissimilar forms of the heteropolymer on five species of microorganisms, determined that a variation in the selection of the extraction methodology results in the production of different forms of lignin. As a result of which the effectiveness of each form varies when used during the treatment processes [119].

Microorganisms like *E. coli*, *Saccharomyces cerevisiae*, *Bacillus licheniformis*, *Enterics enteric*, *Staphyllococcus aureus*, *Bacillus mycoids*, fungus lypolytica, and *Aspergillus niger* are restrained by fragments of polymer. The inhibitory trait of the biopolymer is attributed to the double bond present in the Cα=Cβ site of the side chain and an alkyl (–CH3) group in the seventh position. Generally, the polyphenolic species is represented by a carbon–carbon double bond at the alpha and beta positions of the side chain and is also characterized by the presence of a methyl functional group, which is more reactive compared to the phenols consisting of the oxygen-containing functional groups (OH, CO, and COOH) in the side chain [74]. In general, polyphenolic compounds deplete the plasma membrane and rupture the bacterium by bursting out the cell contents [76]. Wang *et al.* (2019) [120] also presented a novel approach to decrease the structural heterogeneity of lignin, after which it was observed that there was a considerable increase by many folds in the antimicrobial activity of the compound.

Dumitriu *et al.* (2013) [121] formulated a study stating the relationship of phenolic compounds and their bactericidal properties. He demonstrated that cinnamaldehyde (monolignol/one of the monomers of lignin) deprives the cells of adenosine ATP by invading the bacteria and lowering down the pH. Black liquors obtained from bagasse and cotton stalks restricted the growth of gram-positive bacterial species. However, gram-negative microorganisms (*Escherichia coli*) and the thready fungus (*Aspergillus niger*) remained unaffected. The chemical compound p-hydroxycinnamaldehyde is featured as a broad-spectrum antibacterial drug not solely effective against gram-negative but additionally against gram-positive microorganisms, too. The antibacterial drug activities of hemp (*Cannabis sativa*) and different fibrous plants like jute (*Corchorus capsularis*), flax (*Linum usitatissimum*), kenaf (*Hibiscus cannabinus*), sisal (*Agave sisalana*), and bamboo (subfamily Bambusoideae) were shortly reviewed, which showed that the antibacterial drug quality of these fibers was related to the presence of cannabinoids, alkaloids, and synthetic resin molecules contained in polymer. The current analysis has showed that the antibacterial drug property of bamboo (*Phyllostachys pubescens*) is due to the presence of lignin and not the other available chemical elements in the biomass matrix. Synthetic resin compounds like isoeugenol, group p-hydroxybenzoate, p-coumaric acid, and ferulic acid impart the antimicrobial activity in kraft pulping by subjecting the compound under chemical modifications. Furthermore, the antimicrobial activity of the polymer isolated from the corn fodder residue in bio-ethanol production was analyzed [74]. Silver (Ag) nanoparticles lodged with lignin have been discovered to have disinfectant properties. The ions, when coupled with lignin, react with the proteins (specifically interacting with the thiol group) present in the bacterial cells, thus puncturing the structure of the microbe [122].

Since the fatal outbreak of COVID-19, the world has been in alarming need of antiviral drugs. A number of viral inhibitors have been formulated after the first virus inhibitory compound, idoxuridine, has been introduced in 1963 [123]. It was found that the microwave solvolysis of biowaste–sugarcane bagasse produced a viral inhibitor that was derived from lignin. The purified extract of the compound FR-200 was obtained and was effective against encephalomyocarditis virus (EMCV) in the cell line L929. Deciphering the post-introduction effect in the cell line, it was found that there are no cytotoxic effects of the lignin-derived antiviral compound. The process included heating of bagasse (byproduct of sugarcane) at 200°C temperature in aq. glycerol containing sulfuric acid [124]. The same method has been used for the degradation of beech wood and further conversion of the same into a novel class of virus inhibitory substances, i.e., lignin–carbohydrate complexes. It was also efficient in suppressing the effect of EMCV. This provides evidence for the eco-friendly approach to fight the war against viruses [125]. Gords *et al.* (2015) [126] demonstrated that a lignin derivative, lignosulfonic acid, has the capacity to inhibit HIV and HSV by interacting with the surface of glycoproteins present on the viral envelope. The derivative showcased a broad spectrum antiviral activity by restricting virus entry into vulnerable target cells. This potent virus inhibitor has further various microbicidal applications and is inexpensive, too [127]. Another successful inhibitory effect was shown against influenza A virus, whose RNA synthesis was suppressed by the heavy compound obtained from the pinecones of *Pinus parviflora*. The highlighted result of such data was that it is the polymerized phenolic structure of lignin that inhibited the entry of influenza virus into the susceptible target cells [128]. Another set of studies also formulated that lignin–carbohydrate protein structure complexes extracted from *Pimpinella anisum* exhibited restricting effects on HSV 1 and 2. The suppression is attributed to the fact that there was enough production of nitric oxide and mRNA expression of IL- beta.

13.7.4 Anticoagulant Activity

Lignin is known to exhibit anticoagulant activity when present in sulfonated form [129]. In modern times, technologists have come up with new anticoagulants consisting of a couple of properties, primarily hydrophobicity, followed by anionic nature. Lignin-derived sulfated molecules of low molar mass served as the practical mimetics of heparin [130]. Sulfated lignin represents altogether a novel category of anticoagulants that induces the allosteric inhibition of a specific class of enzymes called proteases by binding at the heparin binding site. Though different ligands were examined to inhibit the functionality of proteases by targeting the heparin binding sites, none exhibited an immediate anticoagulant impact as effectively as the sulfated lignins. Sulfated β-O-lignin are a newly discovered class of anticoagulants. The success rate, while analyzing the restrictive effects on certain principles of enzymatic proteins (blood serine proteases, factor Xa, and thrombin), responsible to bring about the coagulation process was fair enough to capture the interests of scientists as the study proposed the idea of lignin being able to cure ailments (e.g., respiratory illness, pulmonary emphysema) [131].

13.7.5 Lignin in Cosmetics

Lignin derived from bio-wastes such as walnut and hazelnut shell has potential to be used as a bioactive component of cosmetic products [132–134]. The antioxidant and anti-ultraviolet properties of lignin make it a suitable ingredient for the cosmetic industry. The non-toxic and versatile nature of lignin has increased its application in the cosmetic industry [135]. The phenolic and other UV-absorbing functional groups and the fact that lignin-based sunscreens have rapid sun protection factor lignin warrant that it can be used as a component in sunscreens and other moisturizers [136]. In the lignin structure, chromophores like carbon–carbon double bonds, carbonyl groups, and aromatic groups are the reason for its UV-absorbing properties. Moreover, the auxochromes present in lignin such as the hydroxyl groups and ethers exhibit the sunblock action of lignin [137, 138]. Being safe for the eyes and skin, lignin is surely beneficial for products at this level. Studies which depict its high antioxidant ability and its safety when applied on the eyes and skin open new outlook in their potential use in topical formulations and cosmetics [139]. Lignin, to date, has not been used in cosmetics, but its unique properties and versatile nature make it a promising candidate.

13.8 Feasibilities, Challenges, and Prospects

Lignin, as a natural polymer, is available at high amounts in terrestrial plants and waste biomass generated from pulping and bio-refinery industries [140]. Out of the 50 to 70 million tons of lignin extracted annually, only 20% of it is used for various applications [141]. Lignin valorization can help in the enhancement of economic viability of modern biorefineries [142–144] and promotes the execution of bioeconomy [145]. The unique property of lignin and its easy availability make it a potential feedstock for advancement in chemical and biomedical applications [52].

Despite the prospects and feasibilities that make lignin the most favorable material for biomedical applications, its full potential is still not explored. For biomedical applications, lignin-based products need to be approved by various regulatory bodies which itself is a long process and needs financial investment. This limits the number of upcoming materials and proves to be a negative factor in the development of drug products in the future [146]. The chemical structure of lignin varies among species and depends on the method of extraction, which makes the application of lignin challenging [147]. Different techniques such as solvent fractionation [148], ultra-filtration [149], or a combination of these techniques is applied to ensure regularity in molecular size, reactivity, and functionality and to reduce the heterogeneity of isolated lignin. Similar to other biopolymers such as chitosan, lignin can also be explored for biomedical applications [150–152]. The challenges of developing lignin-based biomaterials demand a deeper and comprehensive study of the structure as well as its possible alternate forms, explicating a vivid picture of all the possible chemical and physical properties and, along with it, knowledge about the mechanism behind isolation methods, chemical modifications, and transformation to nano/micro-materials for various applications.

Hence, purposeful chemical changes are needed to stretch the horizons of lignin applications. Since lignin is related to multifunctional teams, varied structural changes within the chemical compound could occur throughout the pre- and post-treatments of the biomass. This must be studied in an elaborate manner for the proper regulation of the polymer structure. Determination of lignin molar mass is a crucial parameter determining polymer reactivity and performance. This is highly challenging and a key disadvantage in understanding the polymer potential.

An excellent valorization of lignin polymer desires further experimentation to disclose variations within the chemical properties and its implications on applications. In addition, biomedical application studies related to toxicity and other interactions with human cellular environment need to be studied in depth with clinical implications. Researchers in this field must consider these limitations and find out possible remedies to take lignin-based biomaterials towards a promising future.

13.9 Conclusion

This chapter presented a detailed study of lignin and its activity potential and furthered the expanding horizons of lignin polymer in the field of biomedical technology. The polymer is present in abundant amounts in nature and cheap, too. This chapter presents a vivid idea of the utilization of bio-waste for polymer extraction and application. Since polymer exhibits a high level of heterogeneity, extensive studies have therefore been made on the alternative forms and intermediate compounds. Its ability to show biodegradability, biocompatibility, and low toxicity has been a boon for its usage in various biomedical applications. The inclusion of nanobiotechnology with this field has made perfect use of lignin by determining its polymer reactivity and performance. Lignin has the potential to become an indispensable part of gene delivery and tissue engineering systems. Its therapeutic ability can inhibit the occurrence of certain ailments like diabetes, cancer, and microbial infections, and it can even act as an anti-coagulant agent. This natural compound has indeed emerged out successfully in the disciplines of biomedical technology. It is a much-needed step by mankind to revert back to their roots because there is still a lot much concealed in nature that could vastly benefit life.

References

1. Doherty, W.O., Mousavioun, P., Fellows, C.M., Value-adding to cellulosic ethanol: Lignin polymers. *Ind. Crops Prod.*, *33*, 2, 259–276, 2011.
2. Bruijnincx, P.C. and Weckhuysen, B.M., Lignin up for break-down. *Nat. Chem.*, 6, 12, 1035–1036, 2014.

3. Domínguez-Robles, J., Cárcamo-Martínez, Á., Stewart, S.A., Donnelly, R.F., Larrañeta, E., Borrega, M., Lignin for pharmaceutical and biomedical applications–Could this become a reality? *Sustain. Chem. Pharm.*, 18, 100320, 2020.
4. Figueiredo, P., Lintinen, K., Hirvonen, J.T., Kostiainen, M.A., Santos, H.A., Properties and chemical modifications of lignin: Towards lignin-based nanomaterials for biomedical applications. *Prog. Mater. Sci.*, 93, 233–269, 2018.
5. Ramakrishna, S., Mayer, J., Wintermantel, E., Leong, K.W., Biomedical applications of polymer-composite materials: A review. *Compos. Sci. Technol.*, 61, 9, 1189–1224, 2001.
6. Thakur, V.K., Thakur, M.K., Gupta, R.K., Rapid synthesis of graft copolymers from natural cellulose fibers. *Carbohydr. Polym.*, *98*, 1, 820–828, 2013.
7. Leisola, M., Pastinen, O., Axe, D.D., Lignin—Designed randomness. *BIO-Complexity*, 1, 11, 2012.
8. Dence, C.W., The determination of lignin, in: *Methods in Lignin Chemistry*, *Springer Sci. Rev.*, pp. 33–61, Berlin, Heidelberg, 1992.
9. Lu, F. and Ralph, J., The DFRC method for lignin analysis. 2. Monomers from isolated lignins. *J. Agric. Food. Chem.*, 46, 547, 1998.
10. Melro, E., Filipe, A., Sousa, D., Medronho, B., Romano, A., Revisiting lignin: A tour through its structural features, characterization methods and applications. *New J. Chem.*, 45, 16, 6986–7013, 2021.
11. Hatfield, R. and Vermerris, W., Lignin formation in plants. The dilemma of linkage specificity. *Plant Physiol.*, 126, 1351–7, 2001.
12. Mann, D.G.J., Labbé, N., Sykes, R.W. *et al.*, Rapid assessment of lignin content and structure in switchgrass (*Panicum virgatum* L.) grown under different environmental conditions. *Bioenergy Res.*, 2, 246–256, 2009.
13. Gou, M., Ran, X., Martin, D.W., Liu, C.J., The scaffold proteins of lignin biosynthetic cytochrome P450 enzymes. *Nat. Plants*, 4, 299–310, 2018.
14. Azadi, P., Inderwildi, O.R., Farnood, R., King, D.A., Liquid fuels, hydrogen and chemicals from lignin: A critical review. *Renew. Sustain. Energy Rev.*, 21, 506–523, 2013.
15. Zakzeski, J., Bruijnincx, P.C., Jongerius, A.L., Weckhuysen, B.M., The catalytic valorization of lignin for the production of renewable chemicals. *Chem. Rev.*, 110, 3552–3599, 2010.
16. Chen, C.L. and Connors, W.J., New carbonyl compounds from the alkaline ferricyanide dehydrogenation of p-cresol. *J. Org. Chem.*, 39, 3877–3880, 1974.
17. Gierer, J., Chemistry of delignification. *Wood Sci. Technol.*, 20, 1–33, 1986.
18. Lai, Y.-Z., Methods in lignin chemistry, *Springer Sci. Rev.*, pp. 423–434, Berlin, Heidelberg, 1992.
19. Adler, E. and Gierer, J., The alkylation of lignin with alcoholic hydrochloric acid. *Acta Chem. Scand.*, 9, 84–93, 1955.
20. Katahira, R., Elder, T.J., Beckham, G.T., A brief introduction to lignin structure, in: *Lignin Valorization: Emerging Approaches*, pp. 1–20, 2018.

21. Chen, C.-L., Methods in lignin chemistry, *Springer Sci. Rev.*, pp. 446–457, Berlin, Heidelberg, 1992.
22. Rowell, R.M., *Handbook of wood chemistry and wood composites*, CRC Press, 2015.
23. Christiernin, M., Ohlsson, A.B., Berglund, T., Henriksson, G., Lignin isolated from primary walls of hybrid aspen cell cultures indicates significant differences in lignin structure between primary and secondary cell wall. *Plant Physiol. Biochem.*, 43, 8, 777–785, 2005.
24. Abreu, H.S., Latorraca, J.V., Pereira, R.P., Monteiro, M.B.O., Abreu, F.A., Amparado, K.F., A supramolecular proposal of lignin structure and its relation with the wood properties. *An. Acad. Bras. Cienc.*, 81, 137–142, 2009.
25. Boerjan, W., Ralph, J., Baucher, M., Lignin biosynthesis. *Annu. Rev. Plant Biol.*, 54, 1, 519–546, 2003.
26. Blanco-Canqui, H., Energy crops and their implications on soil and environment. *J. Agron.*, 102, 2, 403–419, 2010.
27. Rahimi, A., Ulbrich, A., Coon, J.J., Stahl, S.S., Formic-acid-induced depolymerization of oxidized lignin to aromatics. *Nature*, 515, 7526, 249–252, 2014.
28. Bernal, M.P., Sommer, S.G., Chadwick, D., Qing, C., Guoxue, L., Michel Jr., F.C., Current approaches and future trends in compost quality criteria for agronomic, environmental, and human health benefits, in: *Adv. Agron*, vol. 144, pp. 143–233, Academic Press, 2017.
29. Carrier, M., Loppinet-Serani, A., Denux, D., Lasnier, J.M., Ham-Pichavant, F., Cansell, F., Aymonier, C., Thermogravimetric analysis as a new method to determine the lignocellulosic composition of biomass. *Biomass Bioenergy*, 35, 1, 298–307, 2011.
30. Haykiri-Acma, H., Yaman, S., Kucukbayrak, S., Comparison of the thermal reactivities of isolated lignin and holocellulose during pyrolysis. *Fuel Process. Technol.*, 91, 7, 759–764, 2010.
31. Sannigrahi, P., Ragauskas, A.J., Miller, S.J., Effects of two-stage dilute acid pretreatment on the structure and composition of lignin and cellulose in loblolly pine. *Bioenergy Res.*, 1, 3, 205–214, 2008.
32. Cordeiro, N., Belgacem, M.N., Torres, I.C., Moura, J.C.V.P., Chemical composition and pulping of banana pseudo-stems. *Ind. Crops Prod.*, 19, 2, 147–154, 2004.
33. Jackson, M.J. and Line, M.A., Organic composition of a pulp and paper mill sludge determined by FTIR, 13C CP MAS NMR, and chemical extraction techniques. *J. Agric. Food Chem.*, 45, 6, 2354–2358, 1997.
34. Haghdan, S., Renneckar, S., Smith, G.D., Sources of lignin. *Polym. Polym. Compos.*, 10, 1–11, 2016.
35. Sainsbury, P.D., *Biocatalytic valorisation of lignin via genetic or chemical intervention of bacterial aromatic degradation pathways*, Doctoral dissertation, University of Warwick, England, 2013.

36. Öhman, F., Theliander, H., Tomani, P., Axegard, P., U.S. Patent 8486224. U.S. Patent and Trademark Office, Washington, DC, 2013.
37. Milner, H.R. and Woodard, A.C., Sustainability of engineered wood products, in: *Sustainability of Construction Materials*, pp. 159–180, Woodhead Publishing, 2016.
38. El Bassam, N., Energy resources, global contribution, and applications, in: *Distributed Renewable Energies for Off-Grid Communities*, pp. 165–211, Elsevier, 2021.
39. Buranov, A.U. and Mazza, G., Lignin in straw of herbaceous crops. *Ind. Crops Prod.*, *28*, 237–259, 2008.
40. Brijwani, K., Oberoi, H.S., Vadlani, P.V., Production of a cellulolytic enzyme system in mixed-culture solid-state fermentation of soybean hulls supplemented with wheat bran. *Process Biochem.*, 45, 1, 120–128, 2010.
41. Mamma, D. and Christakopoulos, P., Citrus peels: An excellent raw material for the bioconversion into value-added products. *WORLD*, *94793*, 59041.4, 19224–9, 2008.
42. Luo, M., Lin, H., Li, B., Dong, Y., He, Y., Wang, L., A novel modification of lignin on corncob-based biochar to enhance removal of cadmium from water. *Bioresour. Technol.*, 259, 312–318, 2018.
43. Allerdings, E., Ralph, J., Schatz, P.F., Gniechwitz, D., Steinhart, H., Bunzel, M., Isolation and structural identification of diarabinosyl 8-O-4-dehydrodiferulate from maize bran insoluble fibre. *Phytochemistry*, 66, 1, 113–124, 2005.
44. Shoda, M. and Sugano, Y., Recent advances in bacterial cellulose production. *Biotechnol. Bioprocess Eng.*, 10, 1, 1, 2005.
45. Singh, R., Singh, S., Trimukhe, K.D., Pandare, K.V., Bastawade, K.B., Gokhale, D.V., Varma, A.J., Lignin–carbohydrate complexes from sugarcane bagasse: Preparation, purification, and characterization. *Carbohydr. Polym.*, 62, 1, 57–66, 2005.
46. Funaoka, M., Matsubara, M., Seki, N., Fukatsu, S., Conversion of native lignin to a highly phenolic functional polymer and its separation from lignocellulosics. *Biotechnol. Bioeng.*, *46*, 6, 545–552, 1995.
47. Ragauskas, A.J., Beckham, G.T., Biddy, M.J., Chandra, R., Chen, F., Davis, M.F., Davison, B.H., Dixon, R.A., Gilna, P., Keller, M., Langan, P., Naskar, A.K., Saddler, J.N., Tschaplinski, T.J., Tuskan, G.A., Wyman, C.E., Lignin valorization: Improving lignin processing in the biorefinery. *Science*, 344, 1246843, 2014.
48. Schutyser, W., Renders, T., Van den Bosch, S., Koelewijn, S.F., Beckham, G., Sels, B., Chemicals from lignin: An interplay of lignocellulose fractionation, depolymerisation, and upgrading. *Chem. Soc. Rev.*, 47, 852–908, 2018.
49. Osbert, K., Lignin to materials: A focused review on recent novel lignin applications. *Appl. Sci.*, 13, 2020, 4626, 2020.
50. Al Arni, S., Extraction and isolation methods for lignin separation from sugarcane bagasse: A review. *Ind. Crops Prod.*, Elsevier, *115*, 330–339, 2018.

51. Garedew, M., Lin, F., Song, B., DeWinter, T.M., Jackson, J.E., Saffron, C.M., Anastas, P.T., Greener routes to biomass waste valorization: Lignin transformation through electrocatalysis for renewable chemicals and fuels production. *ChemSusChem*, *13*, 17, 4214–4237, 2020.
52. Wang, H., Pu, Y., Ragauskas, A., Yang, B., From lignin to valuable products–strategies, challenges, and prospects. *Bioresour. Technol.*, *271*, 449–461, 2019.
53. Kim, K.H., Dutta, T., Sun, J., Simmons, B., Singh, S., Biomass pretreatment using deep eutectic solvents from lignin derived phenols. *Green Chem.*, *20*, 4, 809–815, 2018.
54. Li, M., Pu, Y., Ragauskas, A.J., Current understanding of the correlation of lignin structure with biomass recalcitrance. *Front. Chem.*, *4*, 45, 2016.
55. Selig, M.J., Viamajala, S., Decker, S.R., Tucker, M.P., Himmel, M.E., Vinzant, T.B., Deposition of lignin droplets produced during dilute acid pretreatment of maize stems retards enzymatic hydrolysis of cellulose. *Biotechnol. Prog.*, *23*, 6, 1333–1339, 2007.
56. Ibáñez, A.B. and Bauer, S., Downscaled method using glass microfiber filters for the determination of Klason lignin and structural carbohydrates. *Biomass Bioenergy*, *68*, 75–81, 2014.
57. Galkin, M.V. and Samec, J.S., Lignin valorization through catalytic lignocellulose fractionation: A fundamental platform for the future biorefinery. *ChemSusChem*, *9*, 13, 1544–1558, 2016.
58. Lancefield, C.S., Panovic, I., Deuss, P.J., Barta, K., Westwood, N.J., Pretreatment of lignocellulosic feedstocks using biorenewable alcohols: Towards complete biomass valorisation. *Green Chem.*, *19*, 1, 202–214, 2017.
59. Lundquist, K. and Kirk, T.K., Fractionation-purification of an industrial kraft lignin. *OSTI.GOV*, *63*, 1, 1980.
60. Radotić, K. and Mićić, M., Methods for extraction and purification of lignin and cellulose from plant tissues, in: *Sample Preparation Techniques for Soil, Plant, and Animal Samples*, pp. 365–376, Humana Press, New York, NY, 2016.
61. Shukry, N., Fadel, S.M., Agblevor, F.A., El-Kalyoubi, S.F., Some physical properties of acetosolvlignins from bagasse. *J. Appl. Polym. Sci.*, *109*, 1, 434–444, 2008.
62. Mandlekar, N., Cayla, A., Rault, F. *et al.*, An overview on the use of lignin and its derivatives in fire retardant polymer systems, in: *Lignin-Trends Applications*, M. Poletto (Ed.), pp. 207–231, 2018.
63. Rodríguez, A., Sánchez, R., Requejo, A., Ferrer, A., Feasibility of rice straw as a raw material for the production of soda cellulose pulp. *J. Cleaner Prod.*, *18*, 10-11, 1084–1091, 2010.
64. Calvo-Flores, F.G., Dobado, J.A., Isac-García, J., Isac-García, J., Martín-Martínez, F.J., *Lignin and lignans as renewable raw materials: Chemistry, technology, and applications*, Wiley, 2015.

65. Chen, H., Chemical composition and structure of natural lignocellulose, in: *Biotechnology of Lignocellulose, Springer Sci. Rev.*, pp. 25–71, Netherlands 2017.
66. Upton, B.M. and Kasko, A.M., Strategies for the conversion of lignin to high-value polymeric materials: Review and perspective. *Chem. Rev.*, 116, 2275–2306, 2015.
67. Yu, O. and Kim, K.H., Lignin to materials: A focused review on recent novel lignin applications. *Appl. Sci.*, *10*, 13, 4626, 2020.
68. Ugartondo, V., Mitjans, M., Vinardell, M.P., Comparative antioxidant and cytotoxic effects of lignins from different sources. *Bioresour. Technol.*, 99, 14, 6683–6687, 2008.
69. Reid, I.D., Biodegradation of lignin. *Can. J. Bot.*, 111, 1011–1018, 1995.
70. Faustino, H., Gil, N., Baptista, C., Duarte, A.P., Antioxidant activity of lignin phenolic compounds extracted from kraft and sulphite black liquors. *Molecules*, 15, 12, 9308–9322, 2010.
71. Azadfar, M., Gao, A.H., Bule, M.V., Chen, S., Structural characterization of lignin: A potential source of antioxidants guaiacol and 4-vinylguaiacol. *Int. J. Biol. Macromol.*, 75, 58–66, 2015.
72. Ogan, A., Yüce-Dursun, B., Abdullah, D., Beyler-Çiğil, A., Kahraman, M.V., Çağlayan, P., Birbir, M., Mutlu, Ö., Gülsoy, N., Preparation and antimicrobial properties of LL-37 peptide immobilized lignin/caprolactone polymer film. *Polym. Adv. Technol.*, 31, 10, 2222–2228, 2020.
73. Majira, A., Godon, B., Foulon, L., van der Putten, J.C., Cezard, L., Thierry, M., Pion, F., Bado-Nilles, A., Pandard, P., Jayabalan, T., Aguie-Beghin, V., Ducrot, P.H., Lapierre, C., Marlair, G., Gosselink, R.J.A., Baumberger, S., Cottyn, B., Enhancing the antioxidant activity of technical lignins by combining solvent fractionation and ionic-liquid treatment. *ChemSusChem*, 12, 21, 4799–4809, 2019.
74. Alzagameem, A., Klein, S.E., Bergs, M., Do, X.T., Korte, I., Dohlen, S., Huwe, C., Kreyenschmidt, J., Kamm, B., Larkins, M., Schulze, M., Antimicrobial activity of lignin and lignin-derived cellulose and chitosan composites against selected pathogenic and spoilage microorganisms. *Polymers*, 11, 4, 1–18, 2019.
75. Nada, A.M.A., El-Diwany, A.I., Elshafei, A.M., Infrared and antimicrobial studies on different lignins. *Acta Biotechnol.*, 9, 3, 295–298, 2004.
76. Espinoza-Acosta, J.L., Torres-Chávez, P.I., Ramírez-Wong, B., López-Saiz, C.M., Montaño-Leyva, B., Antioxidant, antimicrobial, and antimutagenic properties of technical lignins and their applications. *BioResources*, 11, 2, 5452–5481, 2016.
77. Lee, E., Song, Y., Lee, S., Crosslinking of lignin/poly(vinyl alcohol) nanocomposite fiber webs and their antimicrobial and ultraviolet-protective properties. *Text. Res. J.*, 89, 1, 3–12, 2017.

78. Kim, Y., Suhr, J., Seo, H.W., Sun, H., Kim, S., Park, I.K., Kim, S.-H., Lee, Y., Kim, K.J., Nam, J.D., All biomass and UV protective composite composed of compatibilized lignin and poly (lactic-acid). *Sci. Rep.*, 7, 1, 1–11, 2017.
79. Mehta, M.J. and Kumar, A., Ionic liquid stabilized gelatin-lignin films: A potential UV-shielding material with excellent mechanical and antimicrobial properties. *Chemistry*, 25, 5, 1269–1274, 2019.
80. Sadeghifar, H. and Ragauskas, A., Lignin as a UV light blocker-a review. *Polymers*, 12, 5, 1–10, 2020.
81. Mohammadinejad, R., Maleki, H., Larrañeta, E., Fajardo, A.R., Nik, A.B., Shavandi, A., Sheikhi, A., Ghorbanpour, M., Farokhi, M., Govindh, P., Cabane, E., Status and future scope of plant-based green hydrogels in biomedical engineering. *Appl. Mater. Today*, 16, 213–246, 2019.
82. Garcıa, D.M., Escobar, J.L., Bada, N., Casquero, J., Hernáez, E., Katime, I., Synthesis and characterization of poly (methacrylic acid) hydrogels for metoclopramide delivery. *Eur. Polym. J.*, 40, 8, 1637–1643, 2004.
83. Raschip, I.E., Hitruc, E.G., Vasile, C., Semi-interpenetrating polymer networks containing polysaccharides. II. Xanthan/lignin networks: A spectral and thermal characterization. *High Perform. Polym.*, 23, 3, 219–229, 2011.
84. Feng, Q., Cheng, H., Xie, Y., Chen, F., Development of sulfomethylated lignin-based adsorbent for methylene blue removal. *J. Biobased Mater. Bioenergy*, 11, 6, 655–661, 2017.
85. Raschip, I.E., Hitruc, G.E., Vasile, C., Popescu, M.C., Effect of the lignin type on the morphology and thermal properties of the xanthan/lignin hydrogels. *Int. J. Biol. Macromol.*, 54, 230–237, 2013.
86. Larrañeta, E., Imízcoz, M., Toh, J.X., Irwin, N.J., Ripolin, A., Perminova, A., Domínguez-Robles, J., Rodríguez, A., Donnelly, R.F., Synthesis and characterization of lignin hydrogels for potential applications as drug eluting antimicrobial coatings for medical materials. *ACS Sustain. Chem. Eng.*, 6, 7, 9037–9046, 2018.
87. Răschip, I.E., Panainte, A.D., Pamfil, D., Profire, L., Vasile, C., *In vitro* testing of xanthan/lignin hydrogels as carriers for controlled delivery of bisoprolol fumarare. *Rev. Med. Chir. Soc. Med. Nat. Iasi*, 119, 1189–1194, 2015.
88. Ravishankar, K., Venkatesan, M., Desingh, R.P., Mahalingam, A., Sadhasivam, B., Subramaniyam, R., Dhamodharan, R., Biocompatible hydrogels of chitosan-alkali lignin for potential wound healing applications. *Mater. Sci. Eng. C*, 102, 447–457, 2019.
89. Berglund, L., Forsberg, F., Jonoobi, M., Oksman, K., Promoted hydrogel formation of lignin-containing arabinoxylan aerogel using cellulose nanofibers as a functional biomaterial. *RSC Adv.*, 8, 67, 38219–38228, 2018.
90. Sezer, A.D. and Cevher, E., Biopolymers as wound healing materials: Challenges and new strategies, in: *Biomaterials Applications for Nanomedicine*, pp. 383–414, 2011.

91. Mutlu, G., Calamak, S., Ulubayram, K., Guven, E., Curcumin-loaded electrospun PHBV nanofibers as potential wound-dressing material. *J. Drug Deliv. Sci. Technol.*, 43, 185–193, 2018.
92. Dhivya, S., Padma, V.V., Santhini, E., Wound dressings–A review. *BioMedicine*, 5, 4, 1–5, 2015.
93. Reesi, F., Minaiyan, M., Taheri, A., A novel lignin-based nanofibrous dressing containing arginine for wound-healing applications. *Drug Deliv. Transl.*, 8, 1, 111–122, 2018.
94. Jaganathan, G., Manivannan, K., Lakshmanan, S., Sithique, M.A., Fabrication and characterization of Artocarpus heterophyllus waste derived lignin added chitosan biocomposites for wound dressing application. *Sustain. Chem. Pharm.*, 10, 27–32, 2018.
95. Jacob, J., Haponiuk, J.T., Thomas, S., Gopi, S., Biopolymer based nanomaterials in drug delivery systems: A review. *Mater. Today Chem.*, 9, 43–55, 2018.
96. Zhou, Y., Han, Y., Li, G., Yang, S., Xiong, F., Chu, F., Preparation of targeted lignin–based hollow nanoparticles for the delivery of doxorubicin. *Nanomaterials*, 9, 2, 188, 2019.
97. Çalgeris, İ., Çakmakçı, E., Ogan, A., Kahraman, M.V., Kayaman-Apohan, N., Preparation and drug release properties of lignin–starch biodegradable films. *Starke*, 64, 5, 399–407, 2012.
98. Uraki, Y., Imura, T., Kishimoto, T., Ubukata, M., Body temperature-responsive gels derived from hydroxypropylcellulose bearing lignin II: Adsorption and release behavior. *Cellulose*, 13, 3, 225–234, 2016.
99. Liu, X., Yin, H., Zhang, Z., Diao, B., Li, J., Functionalization of lignin through ATRP grafting of poly(2-dimethylaminoethyl methacrylate) for gene delivery. *Colloids Surf. B Biointerfaces*, 125, 230–237, 2015.
100. Ten, E., Ling, C., Wang, Y., Srivastava, A., Dempere, L.A., Vermerris, W., Lignin nanotubes as vehicles for gene delivery into human cells. *Biomac*, 15, 1, 327–338, 2014.
101. Morganti, P., Danti, S., Coltelli, M.B., Chitin and lignin to produce biocompatible tissues. *Res. Clin. Dermatol.*, 01, 01, 5–11, 2018.
102. Kai, D., Tan, M.J., Chee, P.L., Chua, Y.K., Yap, Y.L., Loh, X.J., Towards lignin-based functional materials in a sustainable world. *Green Chem.*, 18, 5, 1175–1200, 2016.
103. Moraes, M.A., Silva, C.F., Vieira, R.S. (Eds.), *Biopolymer Membranes and Films*, Elsevier, 2020.
104. Kumar, R., Aadil, K.R., Ranjan, S., Kumar, V.B., Advances in nanotechnology and nanomaterials based strategies for neural tissue engineering. *J. Drug Deliv. Sci. Technol.*, 57, 101617, 2020.
105. Saudi, A., Amini, S., Amirpour, N., Promoting neural cell proliferation and differentiation by incorporating lignin into electrospun poly(vinyl alcohol) and poly(glycerol sebacate) fibers. *Mater. Sci. Eng. C*, 104, 110005, 2019.
106. Zippel, N., Tobiasch, E., Schulze, M., Biomaterials for stem cell differentiation. *Recent Pat. Biotechnol.*, 4, 1–22, 2020.

107. Jafari, M., Paknejad, Z., Rad, M.R., Motamedian, S.R., Eghbal, M.J., Nadjmi, N., Khojasteh, A., Polymeric scaffolds in tissue engineering: A literature review. *Biomed. Mater. Res.*, 105, 1552–4973, 2017.
108. Barapatre, A., Aadil, K.R., Tiwary, B.N., Jha, H., *In vitro* antioxidant and antidiabetic activities of biomodified ligninfrom Acacia nilotica wood. *Int. J. Biol. Macromol.*, 75, 81–89, 2015.
109. Hasegawa, Y., Kadota, Y., Hasegawa, C., Kawaminami, S., Lignosulfonic acid-induced inhibition of intestinal glucose absorption. *Nutr. Sci. Vitaminol.*, 61, 449–454, 2015.
110. Mukai, Y., Norikura, T., Fujita, S., Mikame, K., Funaoka, M., Sato, S., Effect of lignin-derived lignophenols on vascular oxidative stress and inflammation in streptozotocin-induced diabetic rats. *Mol. Cell. Biochem.*, 348, 117, 2011.
111. Chen, Y., Liu, Y., Li, X., Zhang, J., Li, G., Lignin interacting with α-glucosidase and its inhibitory effect on the enzymatic activity. *Food Biophys.*, 10, 264, 2015.
112. Xu, Q., He, C., Xiao, C., Chen, X., Reactive oxygen species (ROS) responsive polymers for biomedical applications. *Macromol. Biosci., 16*, 5, 635–646, 2016.
113. Albayrak, S. and Aksoy, A., Phenolic contents and biological activity of endc mic Origanum minutiflorum grown in Turkey. *Indian J. Pharm. Educ. Res.*, 53, 1, 160–170, 2019.
114. Gülçin, I., Antioxidant activity of food constituents: An overview. *Arch. Toxicol.*, 86, 3, 345–391, 2012.
115. Sindhi, V., Gupta, V., Sharma, K., Bhatnagar, S., Kumari, R., Dhaka, N., Potential applications of antioxidants–A review. *J. Pharm. Res.*, 7, 9, 828–835, 2013.
116. Perez-Perez, E. and Rodríguez-Malaver, A.J., Antioxidant capacity of Kraft black liquor from the pulp and paper industry. *J. Environ. Biol.*, 26, 3, 603–608, 2005.
117. Cazacu, G., Capraru, M., Popa, V.I., Advances concerning lignin utilization in new materials, in: *Advances in Natural Polymers*, pp. 255–312, Springer, Berlin, Heidelberg, 2013.
118. Carocho, M. and Ferreira, I.C., A review on antioxidants, prooxidants and related controversy: Natural and synthetic compounds, screening and analysis methodologies and future perspectives. *FCT, 51*, 15–25, 2013.
119. Klapiszewski, Ł., Rzemieniecki, T., Krawczyk, M., Malina, D., Norman, M., Zdarta, J., Majchrzak, I., Dobrowolska, A., Czaczyk, K., Jesionowski, T., Kraft lignin/silica–AgNPs as a functional material with antibacterial activity. *Colloids Surf. B*, 134, 220–228, 2015.
120. Wang, G., Pang, T., Xia, Y., Liu, X., Li, S., Parvez, A.M., Si, C., Subdivision of bamboo kraft lignin by one-step ethanol fractionation to enhance its water-solubility and antibacterial performance. *Int. J. Biol. Macromol., 133*, 156–164, 2019.

121. Dumitriu, S. (Ed.), *Polysaccharides: Structural Diversity and Functional Versatility*, CRC Press, Boca Raton, 2004.
122. Ndaba, B., Roopnarain, A., Daramola, M.O., Adeleke, R., Influence of extraction methods on antimicrobial activities of lignin-based materials: A review. *Sustain. Chem. Pharm.*, 18, 100342, 2020.
123. De Clercq, E. and Li, G., Approved antiviral drugs over the past 50 years. *Clin. Microbiol. Rev.*, *29*, 3, 695–747, 2016.
124. Kimura, C., Li, R., Ouda, R., Nishimura, H., Fujita, T., Watanabe, T., Production of antiviral substance from sugarcane bagasse by chemical alteration of its native lignin structure through microwave solvolysis. *ChemSusChem*, *13*, 17, 4519–4527, 2020.
125. Li, R., Ouda, R., Kimura, C., Narita, R., Nishimura, H., Fujita, T., Watanabe, T., Conversion of beech wood into antiviral lignin–carbohydrate complexes by microwave acidolysis. *ACS Sustain. Chem.*, 9, 28, 9248–9256, 2021.
126. Gordts, S.C., Férir, G., D'huys, T., Petrova, M.I., Lebeer, S., Snoeck, R., Schols, D., The low-cost compound lignosulfonic acid (LA) exhibits broad-spectrum anti-HIV and anti-HSV activity and has potential for microbicidal applications. *PLoS One*, *10*, 7, e013121, 2015.
127. Oeyen, M., Noppen, S., Vanhulle, E., Claes, S., Myrvold, B.O., Vermeire, K., Schols, D., A unique class of lignin derivatives displays broad anti-HIV activity by interacting with the viral envelope. *Virus Res.*, *274*, 197760, 2019.
128. Harada, H., Sakagami, H., Nagata, K., Oh-hara, T., Kawazoe, Y., Ishihama, A., Konno, K., Possible involvement of lignin structure in anti-influenza virus activity. *Antiviral. Res.*, 15, 1, 41–49, 1991.
129. Shukla, A.K. and Iravani, S. (Eds.), *Green Synthesis, Characterization and Applications of Nanoparticles*, Elsevier, 2018.
130. Mehta, A.Y., Mohammed, B.M., Martin, E.J., Brophy, D.F., Gailani, D., Desai, U.R., Allosterism-based simultaneous, dual anticoagulant and antiplatelet action: Allosteric inhibitor targeting the glycoprotein Ibα-binding and heparin-binding site of thrombin. *J. Thromb. Haemost.*, *14*, 4, 828–838, 2016.
131. Vinardell, M.P. and Mitjans, M., Lignins and their derivatives with beneficial effects on human health. *Int. J. Mol. Sci.*, *18*, 6, 1219, 2017.
132. Ariyanta, H.A., Santoso, E.B., Suryanegara, L., Arung, E.T., Kusuma, I.W., Taib, M.N.A.M., Hussin, M.H., Yanuar, Y., Batubara, I., Fatriasari, W., Recent progress on the development of lignin as future ingredient biobased cosmetics. *Sustain. Chem. Pharm.*, 32, 100966, 2023.
133. Antunes, F., Mota, I.F., Fangueiro, J.F., Lopes, G., Pintado, M., Costa, P.S., From sugarcane to skin: Lignin as a multifunctional ingredient for cosmetic application. *Int. J. Biol. Macromol.*, 234, 123592, 2023.
134. Gagosian, V.S., Claro, F.C., de AP Schwarzer, A.C., Cruz, J.V., Thá, E.L., Trindade, E.D.S., Magalhães, W.L., Pestana, C.B., Leme, D.M., The potential use of kraft lignins as natural ingredients for cosmetics: Evaluating their photoprotective activity and skin irritation potential. *Int. J. Biol. Macromol.*, 222, 2535–2544, 2022.

135. Reddy, N. and Yang, Y., Biofibers from agricultural byproducts for industrial applications. *Trends Biotechnol.*, 23, 22–27, 2005.
136. Diffey, B.L. and Robson, J., A new substrate to measure sunscreen protection factors throughout the ultraviolet spectrum. *J. Soc. Cosmet. Chem.*, 40, 127–133, 1989.
137. Qian, Y., Qiu, X., Zhu, S., Lignin: A nature-inspired sun blocker for broad-spectrum sunscreens. *Green Chem.*, 17, 320–324, 2015.
138. Gould, R.F. (Ed.), *Lignin Structure and Reactions*, American Chemical Society, Washington, DC, 1966.
139. Vinardell, M.P., Ugartondo, V., Mitjans, M., Potential applications of antioxidant lignins from different sources. *Ind. Crops Prod.*, 27, 2, 220–223, 2008.
140. Jääskeläinen, A.S., Liitiä, T., Mikkelson, A., Tamminen, T., Aqueous organic solvent fractionation as means to improve lignin homogeneity and purity. *Ind. Crops Prod.*, 103, 51–58, 2017.
141. Bajwa, D.S., Pourhashem, G., Ullah, A.H., Bajwa, S.G., A concise review of current lignin production, applications, products, and their environmental impact. *Ind. Crops Prod.*, 139, 111526, 1–11, 2019.
142. Wang, X., Leng, W., Nayanathara, R.O., Milsted, D., Eberhardt, T.L., Zhang, Z., Zhang, X., Recent advances in transforming agricultural biorefinery lignins into value-added products. *J. Agric. Food Res.*, 12, 100545, 2023.
143. Sharma, V., Tsai, M.L., Nargotra, P., Chen, C.W., Sun, P.P., Singhania, R.R., Patel, A.K., Dong, C.D., Journey of lignin from a roadblock to bridge for lignocellulose biorefineries: A comprehensive review. *Sci. Total Environ.*, 861, 160560, 2023.
144. Khaire, K.C., Moholkar, V.S., Goyal, A., A biorefinery approach for sequential extraction of commercial grade xylan and alkali lignin from alkali pretreated sugarcane tops hydrolysate. *Ind. Crops Prod.*, 187, 115545, 2022.
145. Graichen, F.H.M., Grigsby, W.J., Hill, S.J., Raymond, L.G., Sanglard, M., Smith, D.A., Thorlby, G.J., Torr, K.M., Warnes, J.M., Yes, we can make money out of lignin and other bio-based resources. *Ind. Crops Prod.*, 106, 74–85, 2017.
146. Challener, C.A., Exploring a new approval process for continued excipient innovation. *Pharm. Technol.*, 536, 2, 563–569, 2014.
147. Domínguez-Robles, J., Sánchez, R., Espinosa, E., Savy, D., Mazzei, P., Piccolo, A., Rodríguez, A., Isolation and characterization of gramineae and fabaceae soda lignins. *Int. J. Mol. Sci.*, 18, 327, 1-17, 2017.
148. D'Orsi, R., Lucejko, J.J., Babudri, F., Operamolla, A., Kumagawa and Soxhlet solvent fractionation of lignin: The impact on the chemical structure. *ACS Omega*, 7, 29, 25253–25264, 2022.
149. Zhang, J., Tian, Z., Ji, X.X., Zhang, F., Light-colored lignin extraction by ultrafiltration membrane fractionation for lignin nanoparticles preparation as UV-blocking sunscreen. *Int. J. Biol. Macromol.*, 231, 123244, 2023.

150. Annu, Ahmed, S., Nirala, R.K., Kumar, R., Ikram, S., Green synthesis of chitosan/nanosilver hybrid bionanocomposites with promising antimicrobial, antioxidant and anticervical cancer activity. *Polym. Polym. Compos.*, 29, 9_suppl, S199–S210, 2021.
151. Annu, Ahmed, S., Ikram, S., Perspectives of chitosan and alginate membranes for biomedical applications, in: *Natural Polymers: Derivatives, Blends and Composites*, vol. 2, pp. 157–182, Nova Science Publishers, 2017.
152. Nan, N., Hu, W., Wang, J., Lignin-based porous biomaterials for medical and pharmaceutical applications. *Biomedicines*, 10, 747, 2022.

14

Biomedical Applications of Polycaprolactone

Anindita Ray (Chakravarti)[1]* and Annu[2]

[1]*Department of Food & Nutrition, Maharani Kasiswari College, University of Calcutta, Kolkata, West Bengal, India*
[2]*School of Mechanical Engineering, Yeungnam University, Gyeongsan, Republic of Korea*

Abstract
There is an emerging demand for new and novel substances and techniques in the field of biomedical sciences with every passing minute, and this can only be met with the development and production of unique polymers which will possess special properties. One such highly popular polymer which has taken the scientific world by storm is polycaprolactone produced from ε-caprolactone and from fossil carbon. This polymer is being extensively used in the production of various sutures and scaffolds and is an important substrate for tissue engineering. It is an important vehicle for drug delivery in the physiological system as well. It is being used in the human body for various physiological purposes in its original form and also as a blended polymer with several other natural and artificial substances. It is not only a polymer with its own unique properties but one which is extremely flexible to modifications and requires changes, which makes it one of the most sought-after polymers in modern-day engineering and scientific research.

***Keywords*:** Polycaprolactone, biodegradable, tissue engineering, sutures, pharmaceuticals, scaffolds, drug delivery

14.1 Introduction

The modern world has created an increasing demand in the fields of biomedicine, biological science, and medical technology, and there has to be

**Corresponding author*: anindita.ray.chakravarti@gmail.com

Annu (ed.) Biopolymers for Biomedical Applications, (373–392) © 2024 Scrivener Publishing LLC

a steady supply of technological advances in these areas at every moment to keep up with the pace of this expanding new world. Polymer chemistry is such an important and novel field, where highly developed products are being manufactured, that is setting the ground for putting forth some answers to this new demand of the 21st century [63]. One such semicrystalline polymer is polycaprolactone, which is a synthetic polymer produced from Ɛ-caprolactone [65], commonly derived from fossil carbon and is being used extensively in the field of biomedical sciences, especially for controlled drug delivery systems, absorbable surgical sutures, nerve guides, 3D scaffolds, etc. [64].

14.2 Structure of PCL

Polycaprolactone or PCL is a synthetic biodegradable aliphatic polyester which is a new-generation biodegradable aliphatic polyester like many others and which not only have good processibility but also have a wide range of functions. As long back as in 1930s, PCL was studied by producing it from the cyclic monomer ε-caprolactone by the technique of ring opening polymerization. There are various ways in which this polymer can be generated—for example, anionic, cationic, coordination, and radical polymerizations are some of the mechanisms used. Each of these methods can affect the various characteristic features of the polymer like the final molecular weight, the molecular weight distribution, and the composition of the terminal group as well as the structure and chemical nature of the copolymers [1]. The average molecular weight of PCL samples generally varies from 3,000 to 8,000 g/mol, and the molecular weight can determine the degradative properties of the polymer [2]. Stannous octoate, as a catalyst, can generally be used for the process of polymerization. The molecular weight of the polymer, during synthesis, can also be controlled by using alcohols of low molecular weight [3].

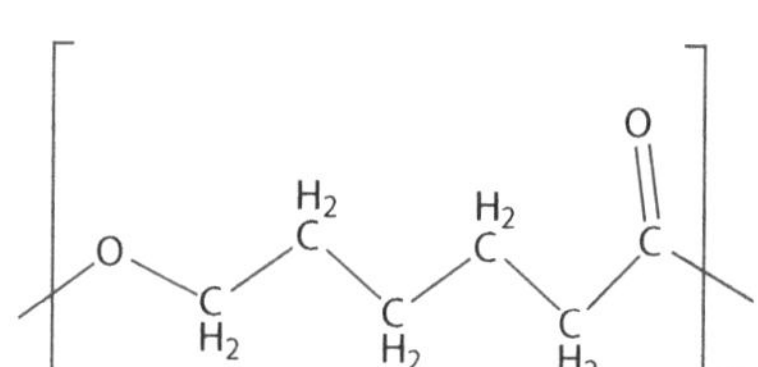

Figure 14.1 Structure of polycaprolactone.

As is evident from Figure 14.1, the polycarolactone is capable of aggregate formation through self-assembly behavior. These aggregates form various types of structures which can be used for many purposes, especially in the field of biomedical sciences [4].

14.3 Characteristic Properties of PCL

Figure 14.2 shows the physical and chemical characteristics of PCL which are used for applications in biomedical sciences.

- PCL has a melting point of 59–64°C and a glass-transition temperature of 60°C.
- It is soluble in chloroform, dichloromethane, carbon tetrachloride, benzene, toluene, cyclohexanone, and 2-nitropropane at room temperature. It has low solubility in acetone and is insoluble in alcohol, petroleum ether, and diethyl ether [5].
- It is an extremely versatile and popular polymer since it offers itself to various types of blending and formation of new and more adaptable copolymers. These copolymers not only have varied physical properties but also show novel chemical and mechanical properties as well.
- Under physiological conditions (such as those found in the human body), PCL is biodegradable and can be broken down by the hydrolysis of its ester bonds. When the highly crystalline PCL polymer is taken up because of its low molecular weight (less than 3,000) and fragmented structure, it typically undergoes a two-stage degradation process: first, the non-enzymatic hydrolytic cleavage of ester groups and, second, the intracellular degradation that occurs in macrophage phagosomes and within fibroblasts [6]. PCL has ester linkages which can be randomly hydrolytically cleaved and degraded, leading to a decrease in molecular weight. A complete degradation may take place in the homopolymer of PCL which has been observed to take 2 to 4 years approximately, but the rate of hydrolysis depends on the extent of copolymerization with other lactones or glycolides/lactides [7, 8].

In Figure 14.2, the characteristics of polycaprolactone which are of significance in the formation of polymers and utilization of these polymers in different fields where these characters are essential are shown.

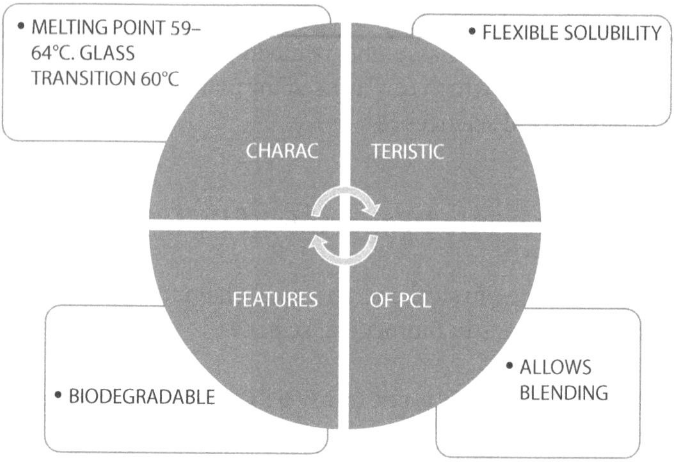

Figure 14.2 Characteristics of PCL polymer.

Low melting point, biodegradability, biocompatibility, good solubility, pliability, and an extremely high blend of compatibility of PCL has made it one of the most sought-after polymers in the biomedical field. The primary applications of PCL mainly include drug delivery and tissue engineering [9]. It is found to be compatible and conducive with different types and categories of drugs and so is the perfect choice as a drug delivery and distribution agent. Moreover, it is found to have a sustainable release mechanism that makes it possible for long-term release ranging up to several months [10]. For tissue engineering applications, such as scaffold for bone tissue engineering and other advanced 3D prototype mix composites for hard tissue engineering, PCL was widely explored [11].

PCL is also commercially used for the past three or more decades in monofilament sutures, made of a PCL-glycolide copolymer, and a contraceptive product, which can deliver a drug for over a year [12]. These applications will be discussed later in this chapter to highlight not only the importance of the polymer but also the role it will play in the field of biomedical sciences in the future.

14.4 Synthesis

Recently, PCL/polyethylene oxide (PEO)-blended electrospinned nanofibers with functional properties have been developed with enhanced plasma processing [62]. There may be several other methods though which can be economically used for the industrial and laboratory preparation of PCL. Figure 14.3 shows the method of polycondensation of a hydrocarboxylic acid, 6-hydrohexanoic acid, and another method of ring opening polymerization of lactone and epsilon–lactone. Both of these methods are used extensively nowadays to produce PCL polymers of high quality [62].

The Figure 14.4 shows a method of production of 3D polymers from PCL which is now being used commercially.

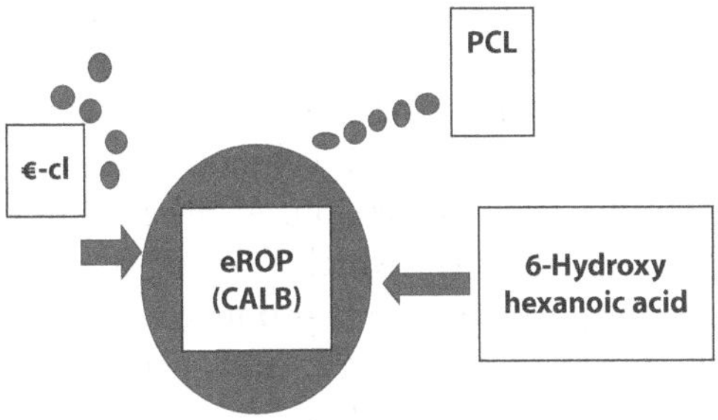

Figure 14.3 Synthesis of PCL through enzymatic action [66].

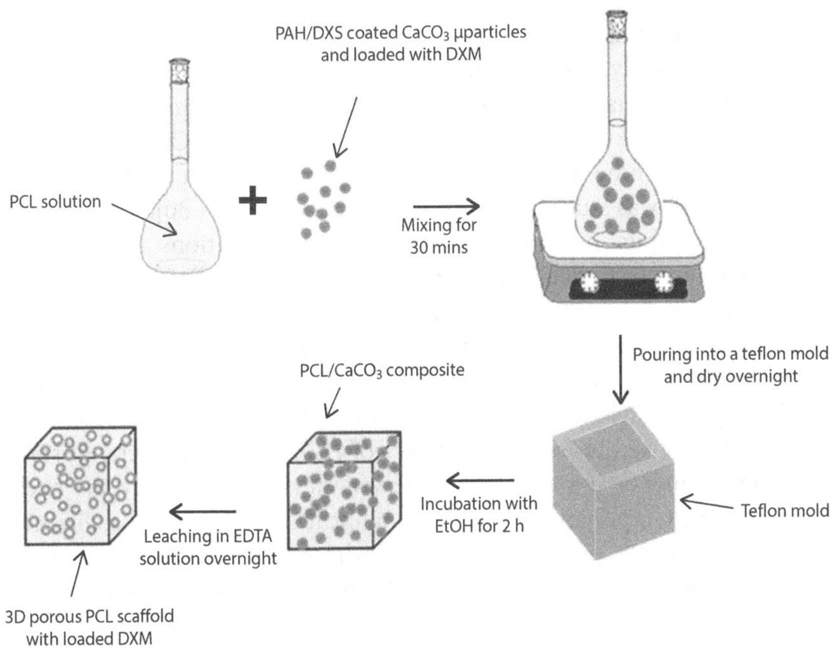

Figure 14.4 Method of production of 3D porous polymers from PCL [69].

14.5 Modifications of PCL

PCL can be used in its original form or its functionality can be altered using some other substances, thereby modifying the original properties as desired. The modification of PCL can be done to increase its cell binding capacity and also to enhance its compression and tensile strength. Modification can also help to increase the rate of degradation, which can help to modify the functions of this polymer. Alginate, gelatin, collagen, egg shell powder, and chitosan as well as some other natural fillers have been used to improve the cell compatibility and hydrophilicity of PCL [13–16, 26–28]—for example, the chitosan–PCL composite has been found to support neuron-like PC-12 cell adhesion and show a significantly higher β-tubulin gene expression. Another example is a composite of gelatin, chitosan, and PCL which is being used for cardiac tissue engineering. This composite, which is produced from the cardiac patch, shows significant mechanical strength and also allows the migration or pre-loading of cardiac cells in a biomimetic environment. For some types of wound healing applications and skin tissue engineering, PCL and PCL–gelatin may be coated by collagen type-I superficially to form a composite of great functionality for these applications in the human body science. Some examples of how PCL polymers have been modified for use are summarized and shown in Table 14.1.

In another study, a network of PCL and elastin has been devised, which has a unique partial-penetration matrix due to the polymeric mesh, and this is being used for some types of biomedical purposes. Cell proliferation and adhesion can be increased by the inclusion of elastin in the structure of PCL, which was studied *in vitro* in chondrocytes [29, 30]. These modifications have thus shown fabulous biomedical applications and are being used for a number of beneficial purposes as discussed.

The chemical modifications of PCL have altered the properties of the polymer as well—for example, the copolymerization of caprolactone with functionalized monomers like lactide [31], ethylene glycol [32–35], monomethoxy poly(ethylene glycol) [36], acryloxy [37], and propylene fumarate [38] has been used to form new classes of PCL-based polymers. In some other experiments, another technique was devised to modify the backbone of the PCL polymer and functionalize it by using various chemical groups like amines, peptides, carboxyl, hydroxyl, etc. [39–41]. In a particular study, Hu *et al.* could carry out a 10-fold increase in the fibroblast cell ingrowth, functionalizing the surface of PCL by coating it with biotin to

Table 14.1 Examples of certain modifications of the PCL polymer that are being used currently.

Polymer	Modifier	Concentration (wt%)	Enhanced properties/ application	References
PCL	Chitosan	25	Hydrophilicity and cell binding for tissue engineering	[13]
	Collagen	Coating	Tissue engineering	[14]
	Gelatin and Collagen	20% gelatin and 1.5% collagen	Skin tissue engineering	[15]
	Elastin	30	Composite scaffold	[16]
	Alginate	5	Bone tissue engineering	[17]
	Nanofiber PLA	10	Mechanical strength and porosity, bone tissue engineering	[18]
	Multi-walled carbon nanotubes	2	Cellular activities for tissue engineering	[19]
	Phlorotannin nanofibers	5	Bioactive properties, bone tissue engineering	[20]
	Silica	5.4	Degradation behavior and bioactivity, bone tissue engineering	[21]
	Bioactive glass (BG)	21 vol%	Bone tissue engineering	[22]
	Nanoscale BG	30	Bone tissue engineering	[23]
	Calcium phosphate	10		[24]
	Chitosan and gelatin hydrogel	50% gelatin, 50% chitosan	Cardiac tissue	[25]

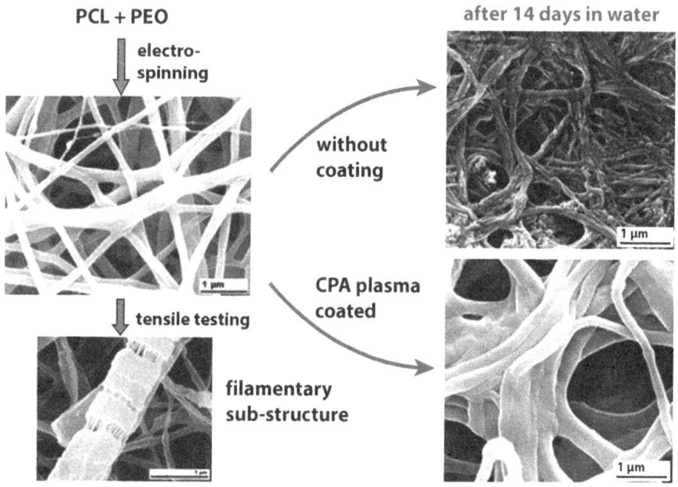

Figure 14.5 Conversion of the original polymer to a blended form to enhance the mechanical properties [62].

enhance superficial cellular proliferation [42]. The hydrolytic and mechanical properties of PCL blended with PEO are observed to be enhanced as shown in Figure 14.5 [62].

14.6 Biomedical Applications of PCL

Some of the most common applications of PCL in the field of biomedical sciences are discussed in the following section of this chapter. Some of the biomedical applications of PCL polymers are highlighted in Figure 14.6.

14.6.1 Drug Delivery

The development of the drug delivery system is a key to various treatments in the modern world. It was developed for regulating drug pharmacokinetics, including its uptake, assimilation, metabolism localization, and especially targeting of the drugs in terms of right time, period, dose, and place. The discovery, development, and wide-scale production of polymers have added to the technology further by imparting their own characters to the drugs. A revolutionary occurrence in 1970 made it possible to utilize PCLs for drug delivery with the advent of the techniques for its use in controlled narcotic release that had started during this time. Since PCL has a very high permeability to many compounds, especially drugs, and owing

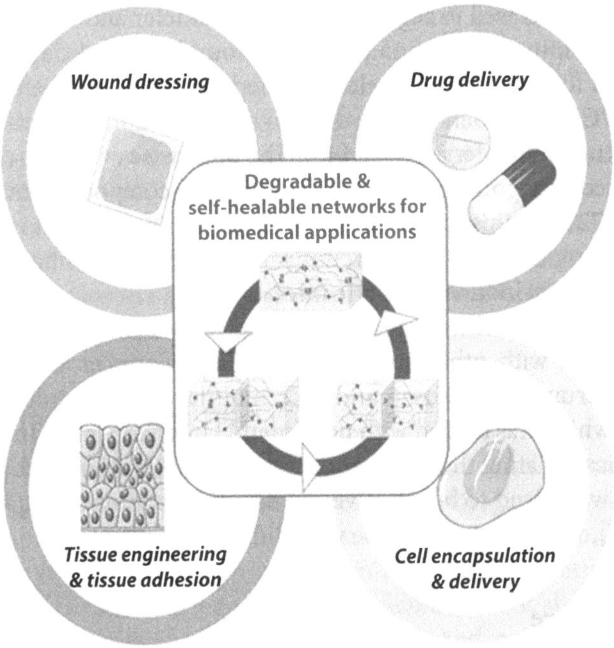

Figure 14.6 A total overview of the biomedical applications of PCL polymers [67].

to its high biocompatibility and property of being excreted from the body after resorption, it is being used as a potential candidate for a drug delivery system. The biodegradation rate of PCL is much slower in comparison to other similar polymers, and so it is used for a more sustained or long-term release of the drugs in the system, for as long as almost a year or so. Compatible blends with other polymers, as will be discussed later, can also influence the kinetics and also create desirable modifications.

Microspheres, nanospheres, nanocapsules, and microfibers are some of the vesicles that have been developed to control the release of drugs and/or protein. Williamson *et al.* [43] devised a lipophilic drug (progesterone) as well as a hydrophilic macromolecule (ovalbumin). They did so by using dispersions of particulate size along with solutions composed of PCL and steroid and used the method of gravity spinning. Gravity spinning has been found to be a very good method for the programmed delivery of bioactive molecules, especially using PCL-fiber based platforms. These are proving to be extremely beneficial in tissue engineering and drug delivery in the present day. Gentamicin sulphate infusion in PCL fibers has also shown positive effects in the local delivery of antibiotics from wound closure materials and tissue substitutes [44]. Heparin can be used in electrospun

These are being used to develop certain types of tissues which can form meshes which have a very high permeability and thus can be used to deliver nutrients and necessary substances to target cells. Like in the textile industry, fibers from different sources can be extracted and extruded, collected on spools, and then converted to fabrics. Van Lieshout *et al.* [54] also produced scaffolds made of PCL which could be used as valves of the aorta. He covered the PCL scaffolds using knitted multifilament fibrins which exhibited greater durability, proper opening, and also contraction of tissues upon closure. On the other hand, it showed greater leakage compared to other valves like porcine valves.

A number of research studies have been carried out throughout the world to study the techniques and fabrication of PCL scaffolds in biomedical applications, especially tissue engineering. An example of such a biodegradable and bioactive 3D porous scaffold produced from PCL polymers is shown in Figure 14.7. It has the advantage of being both biodegradable as well as bioactive. This makes its use extremely suitable and beneficial for current biomedical practices.

Some of the common experiments using PCL polymers in biomedical science are listed in Table 14.2.

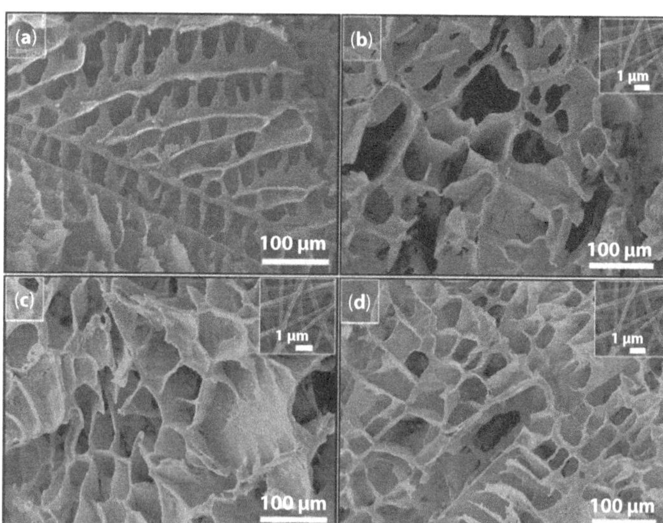

Figure 14.7 A biodegradable and bioactive PCL 3D porous scaffold [68]. (a) PCL/PLA/GNF, (b) PCL/PLA/GNF/Tau 0.1%, (c) PCL/PLA/GNF/Tau 1%, (d) PCL/PLA/GNF/Tau 10%. The figure shows SEM micrographic photos.

Table 14.2 Current research in the field of biomedical sciences and the contribution of PCL polymers.

Year	Outcome	References
2016	Development of self-healing coating with multiple action using acrylated polycaprolactone	[55]
2017	Used for the replacement of hard tissues due to its prolonged degradation time and suitable biological and chemical properties	[56]
2020	Development of porous $CaCO_3$ and Ag nano-particle-modified scaffolds which are PCL-based and using them as SERS platform with specific adsorption at the molecular level	[57]
2020	Using ditelluride-bridged PEG-PCL copolymers as nanoparticles for folic acid target and redox responsiveness needed for enhanced cancer therapy	[58]
2020	Development of bioresorbable stents for coronary artery disease using polycaprolactone, chitosan, and other polymers	[59]
2021	Using polycaprolactone in hydroxyapatite particles and regulating their density	[60]
2021	Coating of mineralized collagen/chitosan cast film with berberine-loaded PCL	[61]

14.6.4 Other Applications

These miraculous polymers are now being used to repair teeth and in root canal procedures. Its biodegradability makes it suitable for use and superior to other substances used for the same purpose. They are being used as dermal fillers as well similar to collagen and has almost comparable outcomes. 3D porous polymers are being used extensively in modern applications in the fields of biology and medicine. A total overview of the applications of PCL polymers is presented in Figure 14.6, which shows how these polymers are presently being used. It also gives an idea regarding the future uses that can be conceptualized using these polymers.

14.7 Current Challenges

PCLs are found to possess high viscoelastic properties but have very low melting points, which make them very suitable for use in the biomedical field. Although PCLs are easy to process and offer themselves with too much flexibility, there can be some issues regarding the use of these polymers. To mention a few of them, they have poor mechanical properties, poor cell adhesion in some cases, and very slow degradation, and these may pose challenges to biomedical scientists who will need to address and modify these according to the need of the application for which these are to be used.

14.8 Conclusion

It is evident from the discussion and examples cited in the chapter that polycaprolactones and its modified derivatives are extremely important in the present times and that they can be as well an answer to future research in biomedicine as well as in various fields. They are especially important in the development of sutures and various scaffolds and has a boom in the techniques of tissue engineering. They are also being used in implants, drug delivery, and other fields as well. This environment-friendly polymer is also being used in other industries as well and being used for substituting plastics, films, covers, etc. The polymer, be it in the original or in the blended form, can definitely be the answer to several problems that have posed themselves throughout the history of mankind and the answer which holds the key to human health.

References

1. Okada, M., Chemical syntheses of biodegradable polymers. *Prog. Polym. Sci.*, 27, 1, 87–133, Feb. 1, 2002.
2. Hayashi, T., Biodegradable polymers for biomedical uses. *Prog. Polym. Sci.*, 19, 4, 663–702, Jan. 1, 1994.
3. Storey, R.F. and Taylor, A.E., Effect of stannous octoate concentration on the ethylene glycol-initiated polymerization of epsilon-caprolactone, in: *Abstracts of Papers of the American Chemical Society*, vol. 211, p. 114, 1996.
4. Mandal, P. and Shunmugam, R., Polycaprolactone: A biodegradable polymer with its application in the field of self-assembly study. *J. Macromol. Sci. A*, 58, 2, 111–29, Oct. 16, 2020.

5. Woodruff, M.A. and Hutmacher, D.W., The return of a forgotten polymer—Polycaprolactone in the 21st century. *Prog. Polym. Sci.*, 35, 10, 1217–56, Oct. 1, 2010.
6. Woodward, S.C., Brewer, P.S., Moatamed, F., Schindler, A., Pitt, C.G., The intracellular degradation of poly (ε-caprolactone). *J. Biomed. Mater. Res.*, 19, 4, 437–44, Apr. 1985.
7. Gunatillake, P.A., Adhikari, R., Gadegaard, N., Biodegradable synthetic polymers for tissue engineering. *Eur. Cell. Mater.*, 5, 1, 1–6, May 20, 2003.
8. Middleton, J.C. and Tipton, A.J., Synthetic biodegradable polymers as orthopedic devices. *Biomaterials*, 21, 23, 2335–46, Dec. 1, 2000.
9. Dash, T.K. and Konkimalla, V.B., Poly-ϵ-caprolactone based formulations for drug delivery and tissue engineering: A review. *J. Control. Release*, 158, 1, 15–33, Feb. 28, 2012.
10. Sinha, V.R., Bansal, K., Kaushik, R., Kumria, R., Trehan, A., Poly-ϵ-caprolactone microspheres and nanospheres: An overview. *Int. J. Pharm.*, 278, 1, 1–23, Jun. 18, 2004.
11. Yoshimoto, H., Shin, Y.M., Terai, H., Vacanti, J.P., A biodegradable nanofiber scaffold by electrospinning and its potential for bone tissue engineering. *Biomaterials.*, 24, 12, 2077–82, May 1, 2003.
12. Ulery, B.D., Nair, L.S., Laurencin, C.T., Biomedical applications of biodegradable polymers. *J. Polym. Sci. B Polym. Phys.*, 49, 12, 832–64, Jun. 15, 2011.
13. Cooper, A., Bhattarai, N., Zhang, M., Fabrication and cellular compatibility of aligned chitosan–PCL fibers for nerve tissue regeneration. *Carbohydr. Polym.*, 85, 1, 149–56, Apr. 22, 2011.
14. Kim, M. and Kim, G.H., Electrohydrodynamic direct printing of PCL/collagen fibrous scaffolds with a core/shell structure for tissue engineering applications. *Chem. Eng. J.*, 279, 317–26, Nov. 1, 2015.
15. Gautam, S., Chou, C.F., Dinda, A.K., Potdar, P.D., Mishra, N.C., Surface modification of nanofibrous polycaprolactone/gelatin composite scaffold by collagen type I grafting for skin tissue engineering. *Mater. Sci. Eng. C*, 34, 402–9, Jan. 1, 2014.
16. Kim, M.S. and Kim, G., Three-dimensional electrospun polycaprolactone (PCL)/alginate hybrid composite scaffolds. *Carbohydr. Polym.*, 114, 213–21, Dec. 19, 2014.
17. Guarino, V., Causa, F., Taddei, P., Di Foggia, M., Ciapetti, G., Martini, D., Fagnano, C., Baldini, N., Ambrosio, L., Polylactic acid fibre-reinforced polycaprolactone scaffolds for bone tissue engineering. *Biomaterials*, 29, 27, 3662–70, Sep. 1, 2008.
18. Pan, L., Pei, X., He, R., Wan, Q., Wang, J., Multiwall carbon nanotubes/polycaprolactone composites for bone tissue engineering application. *Colloids Surf. B Biointerfaces*, 93, 226–34, May 1, 2012.
19. Kim, M. and Kim, G., Electrospun PCL/phlorotannin nanofibres for tissue engineering: Physical properties and cellular activities. *Carbohydr. Polym.*, 90, 1, 592–601, Sep. 1, 2012.

20. Lee, H., Hwang, H., Kim, Y., Jeon, H., Kim, G., Physical and bioactive properties of multi-layered PCL/silica composite scaffolds for bone tissue regeneration. *Chem. Eng. J.*, 250, 399–408, Aug. 15, 2014.
21. Dziadek, M., Menaszek, E., Zagrajczuk, B., Pawlik, J., Cholewa-Kowalska, K., New generation poly (ε-caprolactone)/gel-derived bioactive glass composites for bone tissue engineering: Part I. Material properties. *Mater. Sci. Eng. C*, 56, 9–21, Nov. 1, 2015.
22. Mohammadkhah, A., Marquardt, L.M., Sakiyama-Elbert, S.E., Day, D.E., Harkins, A.B., Fabrication and characterization of poly-(ε)-caprolactone and bioactive glass composites for tissue engineering applications. *Mater. Sci. Eng. C*, 49, 632–9, Apr. 1, 2015.
23. Lei, B., Shin, K.H., Noh, D.Y., Jo, I.H., Koh, Y.H., Kim, H.E., Kim, S.E., Sol–gel derived nanoscale bioactive glass (NBG) particles reinforced poly (ε-caprolactone) composites for bone tissue engineering. *Mater. Sci. Eng. C*, 33, 3, 1102–8, Apr. 1, 2013.
24. Zhao, X., Lui, Y.S., Choo, C.K., Sow, W.T., Huang, C.L., Ng, K.W., Tan, L.P., Loo, J.S., Calcium phosphate coated Keratin–PCL scaffolds for potential bone tissue regeneration. *Mater. Sci. Eng. C*, 49, 746–53, Apr. 1, 2015.
25. Pok, S., Myers, J.D., Madihally, S.V., Jacot, J.G., A multilayered scaffold of a chitosan and gelatin hydrogel supported by a PCL core for cardiac tissue engineering. *Acta Biomater.*, 9, 3, 5630–42, Mar. 1, 2013.
26. Biscaia, S.I., Viana, T.F., Almeida, H.A., Bártolo, P.J., Production and characterisation of PCL/ES scaffolds for bone tissue engineering. *Mater. Today Proc.*, 2, 1, 208–16, Jan. 1, 2015.
27. Gautam, S., Dinda, A.K., Mishra, N.C., Fabrication and characterization of PCL/gelatin composite nanofibrous scaffold for tissue engineering applications by electrospinning method. *Mater. Sci. Eng. C*, 33, 3, 1228–35, Apr. 1, 2013.
28. Annabi, N., Fathi, A., Mithieux, S.M., Weiss, A.S., Dehghani, F., Fabrication of porous PCL/elastin composite scaffolds for tissue engineering applications. *J. Supercrit. Fluids*, 59, 157–67, Nov. 1, 2011.
29. Annabi, N., Fathi, A., Mithieux, S.M., Martens, P., Weiss, A.S., Dehghani, F., The effect of elastin on chondrocyte adhesion and proliferation on poly (ε-caprolactone)/elastin composites. *Biomaterials*, 32, 6, 1517–25, Feb. 1, 2011.
30. Patricio, T., Domingos, M., Gloria, A., Bártolo, P., Characterisation of PCL and PCL/PLA scaffolds for tissue engineering. *Proc. CIRP*, 5, 110–4, Jan. 1, 2013.
31. Shim, W.S., Yoo, J.S., Bae, Y.H., Lee, D.S., Novel injectable pH and temperature sensitive block copolymer hydrogel. *Biomacromolecules*, 6, 6, 2930–4, Nov. 14, 2005.
32. Fu, S., Ni, P., Wang, B., Chu, B., Zheng, L., Luo, F., Luo, J., Qian, Z., Injectable and thermo-sensitive PEG-PCL-PEG copolymer/collagen/n-HA hydrogel

composite for guided bone regeneration. *Biomaterials*, 33, 19, 4801–9, Jun. 1, 2012.
33. Hwang, M.J., Joo, M.K., Choi, B.G., Park, M.H., Hamley, I.W., Jeong, B., Multiple sol-gel transitions of PEG_PCL_PEG triblock copolymer aqueous solution. *Macromol. Rapid Commun.*, 31, 23, 2064–9, Dec. 1, 2010.
34. Hwang, M.J., Suh, J.M., Bae, Y.H., Kim, S.W., Jeong, B., Caprolactonic poloxamer analog: PEG-PCL-PEG. *Biomacromolecules*, 6, 2, 885–90, Mar. 14, 2005.
35. Miao, B., Song, C., Ma, G., Injectable thermosensitive hydrogels for intra-articular delivery of methotrexate. *J. Appl. Polym. Sci.*, 122, 3, 2139–45, Nov. 5, 2011.
36. Li, Z. and Tan, B.H., Towards the development of polycaprolactone based amphiphilic block copolymers: molecular design, self-assembly and biomedical applications. *Mater. Sci. Eng. C*, 45, 620–34, Dec. 1, 2014.
37. Cai, L. and Wang, S., Poly (ε-caprolactone) acrylates synthesized using a facile method for fabricating networks to achieve controllable physicochemical properties and tunable cell responses. *Polymer*, 51, 1, 164–77, Jan. 6, 2010.
38. Wang, S., Lu, L., Gruetzmacher, J.A., Currier, B.L., Yaszemski, M.J., A biodegradable and cross-linkable multiblock copolymer consisting of poly (propylene fumarate) and poly (ε-caprolactone): Synthesis, characterization, and physical properties. *Macromolecules*, 38, 17, 7358–70, Aug. 23, 2005.
39. Hu, W.W., Elkasabi, Y., Chen, H.Y., Zhang, Y., Lahann, J., Hollister, S.J., Krebsbach, P.H., The use of reactive polymer coatings to facilitate gene delivery from poly (ε-caprolactone) scaffolds. *Biomaterials*, 30, 29, 5785–92, Oct. 1, 2009.
40. Riva, R., Lenoir, S., Jérôme, R., Lecomte, P., Functionalization of poly (ε-caprolactone) by pendant hydroxyl, carboxylic acid and epoxide groups by atom transfer radical addition. *Polymer.*, 46, 19, 8511–8, Sep. 8, 2005.
41. Hu, X., Chen, X., Xie, Z., Cheng, H., Jing, X., Aliphatic poly (ester-carbonate) s bearing amino groups and its RGD peptide grafting. *J. Polym. Sci. A Polym. Chem.*, 46, 21, 7022–32, Nov. 1, 2008.
42. Kasper, F.K., Tanahashi, K., Fisher, J.P., Mikos, A.G., Synthesis of poly (propylene fumarate). *Nat. Protoc.*, 4, 4, 518–25, Apr. 2009.
43. Williamson, M.R., Chang, H.I., Coombes, A.G., Gravity spun polycaprolactone fibres: Controlling release of a hydrophilic macromolecule (ovalbumin) and a lipophilic drug (progesterone). *Biomaterials*, 25, 20, 5053–60, Sep. 1, 2004.
44. Chang, H.I., Lau, Y.C., Yan, C., Coombes, A.G., Controlled release of an antibiotic, gentamicin sulphate, from gravity spun polycaprolactone fibers. *J. Biomed. Mater. Res. A*, 84, 1, 230–7, Jan. 2008.
45. Luong-Van, E., Grøndahl, L., Chua, K.N., Leong, K.W., Nurcombe, V., Cool, S.M., Controlled release of heparin from poly (ε-caprolactone) electrospun fibers. *Biomaterials*, 27, 9, 2042–50, Mar. 1, 2006.

46. Merrell, J.G., McLaughlin, S.W., Tie, L., Laurencin, C.T., Chen, A.F., Nair, L.S., Curcumin loaded poly (ε-caprolactone) nanofibers: diabetic wound dressing with antioxidant and anti-inflammatory properties. *Clin. Exp. Pharmacol. Physiol.*, 36, 12, 1149, Dec. 2009.
47. Soares, I., Faria, J., Marques, A., Ribeiro, I.A., Baleizão, C., Bettencourt, A., Ferreira, I.M., Baptista, A.C., Drug delivery from PCL/chitosan multilayer coatings for metallic implants. *ACS Omega*, 7, 27, 23096–106, Jun. 28, 2022.
48. Pillai, C.K. and Sharma, C.P., Absorbable polymeric surgical sutures: Chemistry, production, properties, biodegradability, and performance. *J. Biomater. Appl.*, 25, 4, 291–366, Nov. 2010.
49. Hu, J., Song, Y., Zhang, C., Huang, W., Chen, A., He, H., Zhang, S., Chen, Y., Tu, C., Liu, J., Xuan, X., Highly aligned electrospun collagen/polycaprolactone surgical sutures with sustained release of growth factors for wound regeneration. *ACS Appl. Bio Mater.*, 3, 2, 965–76, Jan. 20, 2020.
50. Soufdoost, R.S., Mosaddad, S.A., Salari, Y., Yazdanian, M., Tebyanian, H., Tahmasebi, E., Yazdanian, A., Karami, A., Barkhordari, A., Surgical suture assembled with tadalafil/polycaprolactone drug-delivery for vascular stimulation around wound: Validated in a preclinical model. *Biointerface Res. Appl. Chem.*, 10, 5, 6317–27, 2020.
51. Soufdoost, R.S., Mosaddad, S.A., Salari, Y., Yazdanian, M., Tebyanian, H., Tahmasebi, E., Yazdanian, A., Karami, A., Barkhordari, A., Surgical suture assembled with tadalafil/polycaprolactone drug-delivery for vascular stimulation around wound: Validated in a preclinical model. *Biointerface Res. Appl. Chem.*, 10, 5, 6317–27, 2020.
52. Arif, Z.U., Khalid, M.Y., Noroozi, R., Sadeghianmaryan, A., Jalalvand, M., Hossain, M., Recent advances in 3D-printed polylactide and polycaprolactone-based biomaterials for tissue engineering applications. *Int. J. Biol. Macromol.*, 218, 930–68, Oct. 1, 2022.
53. Hutmacher, D.W., Scaffolds in tissue engineering bone and cartilage. *Biomaterials*, 21, 24, 2529–43, Dec. 15, 2000.
54. Van Lieshout, M.I., Vaz, C.M., Rutten, M.C., Peters, G.W., Baaijens, F.P., Electrospinning versus knitting: two scaffolds for tissue engineering of the aortic valve. *J. Biomater. Sci. Polym. Ed.*, 17, 1-2, 77–89, Jan. 1, 2006.
55. Lutz, A., Van den Berg, O., Wielant, J., De Graeve, I., Terryn, H., A multiple-action self-healing coating. *Front. Mater.*, 2, 73, Jan. 7, 2016.
56. Malikmammadov, E., Tanir, T.E., Kiziltay, A., Hasirci, V., Hasirci, N., PCL and PCL-based materials in biomedical applications. *J. Biomater. Sci. Polym. Ed.*, 29, 7-9, 863–93, Jun. 13, 2018.
57. Weng, M. and Yu, X., Electrochemical oxidation of para-aminophenol with rare earth doped lead dioxide electrodes: Kinetics modeling and mechanism. *Front. Chem.*, 7, 382, Jun. 11, 2019.
58. Pang, Z., Zhou, J., Sun, C., Ditelluride-bridged PEG-PCL copolymer as folic acid-targeted and redox-responsive nanoparticles for enhanced cancer therapy. *Front. Chem.*, 8, 156, Feb. 28, 2020.

59. Qiu, T., Jiang, W., Yan, P., Jiao, L., Wang, X., Development of 3D-printed sulfated chitosan modified bioresorbable stents for coronary artery disease. *Front. Bioeng. Biotechnol.*, 8, 462, May 19, 2020.
60. Juhl IV, O.J., Merife, A.B., Zhang, Y., Lemmon, C.A., Donahue, H.J., Hydroxyapatite particle density regulates osteoblastic differentiation through β-catenin translocation. *Front. Bioeng. Biotechnol.*, 8, 591084, Jan. 8, 2021.
61. Zhang, Y., Wang, T., Li, J., Cui, X., Jiang, M., Zhang, M., Wang, X., Zhang, W., Liu, Z., Bilayer membrane composed of mineralized collagen and chitosan cast film coated with berberine-loaded PCL/PVP electrospun nanofiber promotes bone regeneration. *Front. Bioeng. Biotechnol.*, 199, 684335, 2021.
62. Kupka, V., Dvořáková, E., Manakhov, A., Michlíček, M., Petruš, J., Vojtová, L., Zajíčková, L., Well-blended PCL/PEO electrospun nanofibers with functional properties enhanced by plasma processing. *Polymers*, 12, 6, 1403, 2020.
63. González, S.C., Galindo, A.S., Herrera, R.R., Badillo, C.M., Palomo-Ligas, L., Fernandez, I.M., Valencia, V.D., Health applications of biodegradable polymers, in: *Biodegradable Polymers*, pp. 207–217, CRC Press, Boca Raton, Mar. 16, 2023.
64. Bruder, S.P. and Roy, A., *Orthobiologics: Scientific and Clinical Solutions for Orthopaedic Surgeons*, Lippincott Williams & Wilkins, Germany, Jan. 6, 2023.
65. Saudi, S., Jun, S., Fialkova, S., Surendran, V., Chandrasekaran, A., Bhattarai, S.R., Sankar, J., Bhattarai, N., Incorporating nanoconfined chitin-fibrils in poly (ε-caprolactone) membrane scaffolds improves mechanical and chemical properties for biomedical application. *J. Biomed. Mater. Res. A*, 111, 3, 1–15, Jan. 28, 2023.
66. Posada, N.C., Sierra, C.A., Perez, L.D., An evaluation of the effect of reaction conditions in the enzymatically catalyzed synthesis of poly (ϵ-caprolactone). *ChemistrySelect*, 7, e202202372, Oct. 2022.
67. Grosjean, M., Gangolphe, L., Nottelet, B., Degradable self-healable networks for use in biomedical applications. *Adv. Funct. Mater.*, 33, 2205315, 2023.
68. Samadian, H., Farzamfar, S., Vaez, A., Ehterami, A., Bit, A., Alam, M., Goodarzi, A., Darya, G., Salehi, M., A tailored polylactic acid/polycaprolactone biodegradable and bioactive 3D porous scaffold containing gelatin nanofibers and Taurine for bone regeneration. *Sci. Rep.*, 10, 13366, 2020.
69. Palamà, I.E., Arcadio, V., D'Amone, S., Biasiucci, M., Gigli, G., Cortese, B., Therapeutic PCL scaffold for reparation of resected osteosarcoma defect. *Sci. Rep.*, 7, 12672, 2017.

15

Biopolymers for Wound Healing Applications

Hitesh Chopra

Department of Biosciences, Saveetha School of Engineering, Saveetha Institute of Medical and Technical Sciences, Chennai, Tamil Nadu, India

Abstract
Wound healing is a complex phenomenon involving various growth factors and mediators. Wound healing can be aided by natural extracts, synthetic molecules, and growth factors. The wound healing dressings act as mediators for delivering the active substances for accelerated wound healing. The biopolymeric wound dressings act as a barrier for the wound, thus protecting it from shock and micro-environment. Biopolymers such as chitosan, hyaluronic acid, gelatin, etc., have shown prominent results in the past decade. The conjugation of biopolymers with other additives supplies an excellent environment for accelerated wound healing. This article provides a bird's eye view of the latest advances in biopolymers used in wound healing.

***Keywords*:** Wound healing, biopolymers, chitosan, hyaluronic acid, natural polymers, alginate, electrospinning, gelatin

15.1 Introduction

A wound is any harm to the integrity of biological tissues or organs [1]. Wounds can be caused by various types of trauma, and it is a basic necessity to keep them clean so that the spread of infection could be lowered [2–6]. Based on cleanliness and condition, a wound can be classified into four different classes as follows:

1.1. Class 1: The wound is clean, uninfected, and non-inflammatory. They are generally superficially present on the

Email: chopraontheride@gmail.com

Annu (ed.) Biopolymers for Biomedical Applications, (393–416) © 2024 Scrivener Publishing LLC

skin and are absent in respiratory, alimentary, genital, and urinary tracts.

1.2. Class II: They are clean-contaminated and show the absence of unusual contamination. Class II wound enters in respiratory, alimentary, genital, or urinary tracts.

1.3. Class III: These wounds are infected and open owing to inadequate sterile procedures or GI tract leakages to wounds.

1.4. Class IV: These kinds of wounds are poorly cared for traumatic injuries. They possess devitalized tissue and have the presence of microorganisms due to flawed operative procedures [7].

15.2 Wound Healing Stages

The wound healing process consists of three major phases [8] as shown in Figure 15.1.

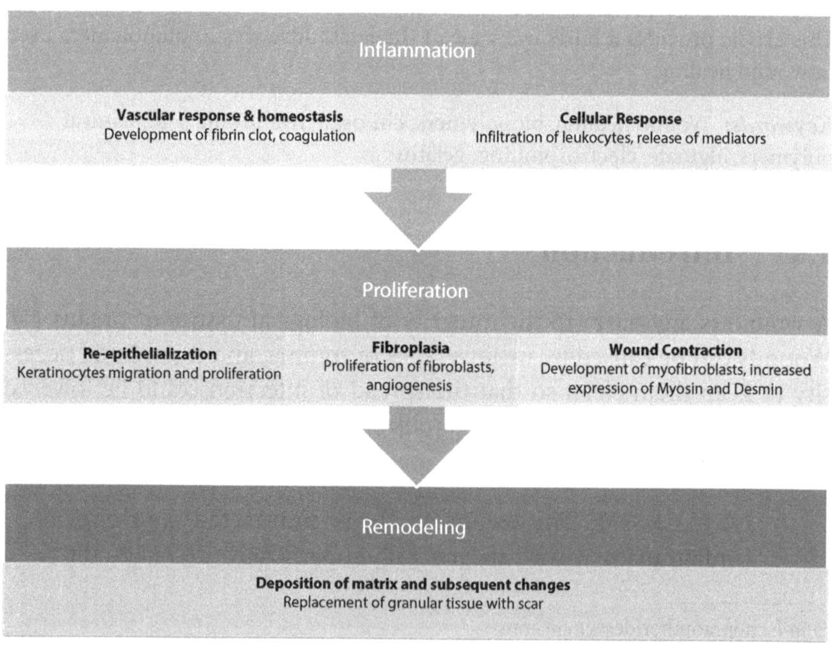

Figure 15.1 Flow chart representing the stages of wound healing.

15.2.1 Inflammation

This phase sub-consisted of hemostasis and inflammation. Immediate injury results in activation of clotting cascade involving the formation of fibrin clot plug at the injury site [9, 10]. Vasoconstriction stops future bleeding and preserves the wound. The fibrin plug attracts leukocytes, keratinocytes, and fibroblasts to the wound site. Injury-induced sub-endothelium collagen and tissue factor promote platelet aggregation and degranulation. Chemotactic agents also end hemostasis and initiate inflammation [9] (Figure 15.1).

In the inflammation phase, the neutrophils are recruited in the wound area within 1 day and stay back for around 5 days [10]. Phagocytosis is induced, followed by macrophages. ROS-releasing phagocyte kills germs and debrides necrotic tissues. As a result, neutrophils attract other cells and produce pro-inflammatory cytokines [11].

15.2.2 Proliferation

Proliferation is characterized by granulation tissue development and vascular network repair and takes 3–10 days to complete [12]. TGF-beta 1, TGF-beta 2, interleukins, and angiogenesis factors are implicated. The proliferation of fibroblasts and endothelial cells needs sufficient blood supply [8]. Both angiogenesis and vasculogenesis produce new blood vessels [9]. Angiogenesis includes sprouting of new arteries, while vasculogenesis involves a *de novo* route where progenitor stem cells differentiate and create new vessels [13]. Endothelial progenitor cells are recruited after damage. Endothelial cell immobilization requires nitric oxide, VEGF, and MMP-9. The newly created vessel transports nutrients and gases.

Meanwhile, inflammatory cytokines and growth hormones stimulate epithelialization. Epithelialization involves keratinocytes, epithelial stem cells, and apocrine glands. Stem cells become keratinocytes and migrate to the wound edge to make contact. Forming granulation tissue is the last step. Fibroblasts travel to the wound site and multiply. Side processes produce collagen III, glycosaminoglycans, and fibronectin [14].

15.2.3 Remodeling

Restoration and remodeling are the last phases, which begin on day 21 and may continue for 1 year [14]. During this time, a perfect balance between new tissue synthesis and breakdown is required. Chronic wounds are formed as a result of the disruption. In addition, the development of granulation tissue comes to a stop, and the process of maturation starts. Collagen type I replaces collagen III, increasing tensile strength and extending collagen synthesis for 4 to 5 weeks [11]. Collagen production is controlled by hydroxylases, which need oxygen and ascorbic acid to function properly; thus, the quantity of these enzymes may impact the strength of a wound.

The wound contracts at the same time, and TGF-beta 1 promotes the formation of myoblasts in the fibroblasts [15]. Meanwhile, collagen types I, II, III, IV, and XVIII, as well as collagen type XVIII, have a role in wound contraction. The margins of the wound and the epithelialization of the wound are where the contraction starts. Irregular myofibroblast activity results in the death of the cells called myofibroblasts, and this results in the development of fibrosis and scar tissue. The process of apoptosis results in the formation of a mature wound that is acellular. As the angiogenic response comes to an end, blood flow decreases and acute metabolic activity comes to a stop. This results in the complete closure of the wound and the restoration of mechanical strength.

15.3 Biopolymers

Unlike synthetic polymers, biopolymers are generated or produced solely from natural resources such as plants and animals [16]. Bio is an abbreviation for polymeric biomolecules, which are made up of monomeric components that are linked together by a covalent connection. They are chemically varied and have a wide range of characteristics depending on where they are derived from. These biopolymers are inherently distinct in structure and have 3D forms and architectures, which allow them to function in the body [17, 18]. Because of their shape and structure, they are able to carry out extremely specialized tasks. Biopolymers have the ability to absorb large quantities of water while dry and release it when completely hydrated [19]. Medicines, antibiotics, cells, and other things have been impregnated into biopolymers to control their release [20, 21]. They have also been utilized in cell delivery and tissue engineering [21–23].

15.3.1 Chitosan

Chitosan is made from chitin, which is acetylated to become chitosan. Chitosan is a polysaccharide complex with a high molecular weight and a high level of molecular entropy that is stable at room temperature and solubilizes in water when acidic [24]. Technically, it is a polycationic derivative of poly-N-acetyl-d-glucosamine, which is found in crustacean bones [25]. The degree of acetylation for chitosan varies from 0 to 1. Protein acetylation and deacetylation levels have different values—for example, chitin has a protein acetylation level of 1, whereas a value of 0 is seen for fully deacetylated proteins. Biopolymers are composed of bi-acetylated monomers to de-acetylated monomers. The composition of biopolymers, in turn, affects their physical and biochemical characteristics. Solubility, viscosity, and biocompatibility are all influenced by the acetylation amount. Acetylation is proportional to molar mass and inversely proportional to biodegradability [26, 27].

Cai and colleagues created a freeze-drying method, a sponge-like dressing of hydroxybutyl chitosan in alkali/urea aqueous solutions [28]. The degree of substitutions ranged from 0.41 to 1.38, and the dressings were formed using the freeze-drying process. The pore size ranged from 14.8 to 18.4 μm, while the porosity ranged from 98% to 99%. It was shown that the compressive strength rose as the degree of acetylation decreased. The sponges were shown to be cytocompatible with normal human dermal fibroblasts and stimulate the fibroblasts' development in the laboratory. Additionally, the study results revealed that sponges had higher hemocompatibility and a lower blood clotting index.

Infections are an unavoidable component of wound healing, and they may represent a severe health risk if not treated promptly. Spray drying-based derivatized chitosan as tiny particles sensitive to lysozyme was the inspiration for the formulation proposed by the researchers [29]. Infected wound fluid and fake wound fluid were tested for infection using the particles and zeta potential, which were utilized to identify the presence of infection. Compared to control particles, the spray-dried N-acetylated chitosan particles demonstrated a fivefold increase in responsiveness to lysozyme compared to control particles. In a similar vein, Wu *et al.* developed non-leaching chitosan-based wound dressings *via* chemical modification with ampicillin [30]. *Staphylococcus aureus*, *Candida albicans*, and *E. coli* were all killed by the sponges, and the sponges were shown to be very biocompatible when tested against the HEK293 cell line. Wound healing materials were produced by Ray *et al.* by utilizing eggshell membrane, chitosan/polycaprolactone nanofibers, and eggshell membrane to construct bilayered scaffolds [31]. The XPS and FTIR analyses showed that the layers were crosslinked using a carbodiimide-based chemical linkage. The layers were also shown to be cytocompatible with human skin fibroblasts. Because of the presence of egg shell membranes, the scaffolds naturally attached to the wound bed. In contrast to natural egg shell, the nanofibers induced significant host cell migration and skin regeneration in the test subjects. In one study, Li *et al.* developed a chitosan wound dressing based on silver–gold nanoparticles [32]. In addition to having an antibacterial effect, the dressing released silver ions at a quicker pace and in a higher concentration than the other dressings. The L929 cells were not harmed by the dressing since it was non-cytotoxic.

Zou *et al.* recently reported the electrospinning-based synthesis of polyvinyl alcohol/chitosan nanofibers encapsulating the antimicrobial peptide OH-CATH30 [33]. The nanofibers demonstrated antibacterial activity against *E. coli* and *Staphylococcus aureus.* Khorasani and colleagues have also developed heparinized polyvinylalcohol/chitosan wound dressings and nano zinc-based wound dressings [34]. The water permeability, porosity, and swelling ratio were all near-optimal, supporting the candidate's claim to be the best wound dressing. The presence of nano zinc oxide resulted in an improvement in the mechanical characteristics and a reduction in the rate of heparin release. In the presence of polyvinyl alcohol/chitosan wound dressing, lignin has been shown to improve the mechanical characteristics of the wound dressing [35]. Hassan *et al.* developed a multifunctional polyelectrolyte wound dressing membrane based on chitosan and hyaluronan bonded with phosphatidylcholine dihydroquercetin, which was used in the treatment of burn victims [36]. The membrane demonstrated scavenging activity against free radicals and hemocompatibility, which resulted in improved blood coagulation. The inclusion of the quercitin component increased the antibacterial and anti-inflammatory activity of the molecule. Chopra *et al.* studied the chitosan–PVA hydrogels comprising honey as active molecule [37]. The *in vitro* and *in silico* studies showed that they can be used for wound healing purposes. Similarly, chitosan PVA-based curcumin was evaluated for wound healing purposes [38].

With the good biological qualities of chitosan (CS) and the higher oxygen permeability of polyvinyl alcohol (PVA), it was possible to make a wound healing patch that could also supply L-arginine (L-arg) and caffeine (Caff) [39]. Using the freezing and reheating method, 12 hydrogels were made with PVA/CS polymer ratios of 90:10, 75:25, and 60:40. Each hydrogel was loaded with L-arg, Caff, or a mixture of L-arg and Caff. After that, an artificial material called zeolite-Z was added to the best polymeric ratio (75:25) and loaded with active substances. In the presence of the inert material, the release profile of the active substances is delayed, and the *in vitro* absorption rates show that the steady state is not reached even after 4 h. The way the parts of the polymer network work together suggests that they could be used in medical uses, such as patches for wounds.

A star-shaped eight-arm cross-linker made of benzaldehyde-terminated polyethylene glycol (POSS-PEG-CHO) was made [40]. It was then cross-linked with hydroxypropyltrimethyl ammonium chloride chitosan using the Schiff base reaction to make chitosan-based POSS-PEG hybrid hydrogels. The hybrid hydrogels that were made had good mechanical strength, could be injected, self-healed well, did not harm cells, and were antibacterial. The hybrid hydrogels could also speed up cell movement and growth, which was predicted because they helped diabetes mice heal wounds very quickly. When the wounds were treated with the compound hydrogels, the epithelial tissue grew back faster, there were fewer inflammatory cells, more collagen was laid down, and the amount of VEGF production was higher. Because of this, chitosan-based POSS-PEG hybrid hydrogel has a lot of promise as a treatment to help diabetic wounds heal faster.

Metwally *et al.* want to find out how well chemically and naturally made chitosan nanoparticles (Ch-NPs) help wounds heal by using extract from *Lawsonia inermis* leaves. A total of 45 adult female albino rats were put into three groups [41]. The surgery cuts on the skin were treated directly with 0.9% normal saline (the control group), green Ch-NPs (the second group), and chemical Ch-NPs gels (the third group), in that order. On the 3rd, 7th, 14th, and 21st day after treatment, the wounds were looked at from a clinical and anatomical point of view. The X-ray diffraction research showed that the nanoparticles were very crystalline and pure. The nanoparticles studied increased the wound contraction percent and decreased the healing time and wound surface area. The histological results showed that the treated rats had better epithelialization, dermal differentiation, collagen deposition, and angiogenesis than control rat. Table 15.1 shows the various chitosan-based wound dressing studies.

Table 15.1 A literature review of the various chitosan-based wound dressings studied.

Name of polymer	Active ingredient	Main characteristic of the wound dressing	References
Chitosan, polyvinyl pyrrolidone	Silver oxide nanoparticles	The films showed higher antibacterial action and low cytotoxicity against the L929 cell line	[42]
Bacterial cellulose, low molecular weight chitosan	Ciprofloxacin	The films showed the release of ciprofloxacin decreased by 30% and antibacterial action against *P. aeruginosa* and *S. aureus*	[43]

(*Continued*)

Table 15.1 A literature review of the various chitosan-based wound dressings studied. (*Continued*)

Name of polymer	Active ingredient	Main characteristic of the wound dressing	References
Chitosan nanoparticles, poly(ε-caprolactone)/ collagen	Insulin	After 14 days of therapy, full wound closure was reported	[44]
Polycaprolactone (PCL), chitosan	Mupirocin and lidocaine hydrochloride	The fibers demonstrated enhanced antimicrobial action in wound sites containing *Staphylococcus aureus* and *Escherichia coli*	[45]
Polyethylene oxide (PEO)/ chitosan)/ poly(ε-caprolactone)	Olive oil	The fabrics exhibited greater antibacterial activity against the strains of *Staphylococcus aureus* and *Escherichia coli* found in wounds	[46]
Poly(ε-caprolactone)/ chitosan	Caffeic acid	Cell attachment and proliferation are higher on the PCL/CCA fibrous mat than on the PCL and PCL/CH fibrous mat, whereas the antibacterial effect is greater on the PCL/CCA fibrous mat	[47]
PCL/PVA/ chitosan	Eugenol	Additionally, *S. aureus* and *P. aeruginosa* antimicrobial activities were discovered, and *in vitro* cytotoxicity assays on normal human dermal fibroblasts revealed that these cells were alive for at least 7 days following direct contact with the electrospun fiber mats	[48]

(*Continued*)

Table 15.1 A literature review of the various chitosan-based wound dressings studied. (*Continued*)

Name of polymer	Active ingredient	Main characteristic of the wound dressing	References
Polycaprolactone, chitosan	Gamma oryzanol	Wound area and overall bacterial count have been significantly decreased	[49]
Polycaprolactone, chitosan	Curcumin	Antibacterial, antioxidant, and cell proliferation characteristics make it noteworthy for cell and tissue growth	[50]

15.3.2 Hyaluronic Acid

Hyaluronic acid (HA) is a chemical found in the body's eye and joint fluids. Hyaluronic acid is a naturally occurring molecule that has a hydrating effect. Concentrations of this substance are located in the most significant amounts in secretions from the eyes and joints. HA is a polysaccharide having a variety of physicochemical characteristics. Because of its viscoelastic and rheological characteristics, HA may be found in nature as glycosaminoglycan. To promote skin regeneration and enlargement, HA is a suitable option.

In a recent study, El-Aassar *et al.* used an electrospinning method to create silver nanoparticles–polygalaturonic acid and HA-based nanofibrous mats [51]. Silver nanoparticles are antioxidants and anti-inflammatory agents that help wounds heal faster. Hydrogels can be defined as three-dimensional structures which swell when they encounter water. Supramolecular hydrogels based on stimuli were created by researchers, and they have enhanced the structural and functional recapitulation of injured organs [52]. Supramolecular polysaccharide-based hydrogels with photoresponsive properties were created by using the interaction between azobenzene and beta-cyclodextrin coupled with HA as the host–guest. In this experiment, azobenzene is exposed to a supramolecular hydrogel under the influence of various wavelengths of photons. The hydrogel can quickly release EGF into the wound. Compared to the control, the EGF-loaded supramolecular hydrogel had superior granulation tissue formation, growth factor levels, and angiogenesis.

Makvandi *et al.* developed injectable hydrogels containing HA, corn silk extract, and nano-silver based on HA [53]. Because of their high mechanical strength and gelation temperatures, near-physiological temperatures may be applied locally to wound skins and have antibacterial activity against both Gram-positive and Gram-negative bacteria, among other things. It was discovered that arginine derivatives combined with dopamine-functionalized HA may be used to create scaffolds with antioxidant properties [54]. They demonstrated increased DPPH production and scavenging rate, improved reactivity to external oxidative stress, and faster wound healing after being implanted. A wound closure injectable hydrogel composed of collagen I and HA that mimicked the ECM was developed for wound closure [55]. The hydrogel was made by linking the phenol moieties of collagen I-hydroxybenzoic acid and hyaluronic acid–tyramine using horseradish peroxidase. By having porous structures, it was possible to achieve a gaseous exchange of gases and nutrients. However, the proliferation rate of the human microvascular endothelial cells and fibroblasts was quite high. Antimicrobial peptides are now being integrated by researchers to combat the issue of antibiotic medication resistance. According to Lin *et al.*, AMP Tet213 peptides were immobilized on a substrate comprising alginate, hyaluronic acid, and collagen to create a wound dressing [56]. The wound dressings demonstrated antibacterial efficacy against bacterial strains such as *E. coli*, MRSA, and *S. aureus* as well as the ability to promote the proliferation of NIH 3T3 fibroblasts.

Electrospinning was utilized to create PVA–HA–cellulose nanocrystals loaded with L-arginine–citric acid nanofibers and then used to promote wound healing [57]. According to the researchers, the cellulose nanocrystals served as nanofillers, while L-arginine functioned as a wound healing accelerator. The inclusion of nanofiller enhanced the mechanical and swelling characteristics of the material. On HFB-4 cells, the nanofibers demonstrated improved hemocompatibility, high protein adsorption, and good proliferative and adhesive potential, with a wound gap closure rate of 99.9% after 48 h of treatment. The formulation demonstrated the continuous release of arginine for more than 48 h, with a 90% release of amino acid. Hybrid hydrogels with fast hemostasis and long-lasting antibacterial activity are also being developed.

Diabetes wounds have been a big health problem because they stay inflamed for a long time and do not get enough blood flow. Ding *et al.* reported on new oil-in-water (o/w) nanoemulsions (NEs) containing Poria cocos triterpenes extract (PTE) to make hyaluronic acid hydrogels (PTE–NEs) to treat diabetic burns [58]. The structure of PTE–NE gel is found to be uniform and porous, and it has good elasticity. *In vitro* tests also show that PTE–NE gel keeps more than 85% of cells alive without killing them. Experiments with diabetes rats in the wild show that the PTE–NE gel can not only speed up wound healing, collagen formation, M2 macrophage polarization, and angiogenesis but can also stop inflammation. In conclusion, PTE helps wounds heal and has anti-inflammatory benefits, showing that it has great promise for treating wounds caused by diabetes.

Chemical cross-linking agents called 1-(3-dimethylaminopropyl) 3-ethylcarbondiimide and N-hydroxysuccinimide were used to make a hydrogel made of HA and carboxymethyl chitosan (CMCS) [59]. Immersion method was used to make the HA/CMCS hydrogel with *panax notoginseng* saponins (PNS). The swelling, sustained release, biocompatibility, anti-inflammatory, and anti-oxidant qualities of the hydrogel were looked into, and the hydrogel's ability to heal skin wounds in type II diabetic SD rats was tested by putting it on a cut on their skin. The results showed that HA/CMCS hydrogel had good swelling properties and could soak up water up to 46 times its own dry weight. The PNS was released into the artificial body fluid at a rate of 76% for 24 h. *In vitro* cytotoxicity, anti-oxidant, and anti-inflammatory experiments that stopped macrophages from releasing NO showed that adding low concentrations of PNS (5 mg/mL) helped the cells grow and made the material more compatible with living organisms. The rate at which HA/CMCS-PNS can get rid of ABTS free radicals can hit 99%. In tests on animals, HA/CMCS-PNS hydrogel helped SD rats heal wounds much faster, lowered the expression of TNF- and IL-6, quickly increased the expression of IL-10, and slowed down the body's response to inflammation. When a wound was healing, HA/CMCS-PNS hydrogel increased the production of VEGF and bFGF in the wound, sped up angiogenesis and epithelial cell regrowth, and helped the wound heal quickly.

15.3.3 Alginates

Alginates are biopolymers that arise naturally as anionic peptides produced from brown algae. Using alginates to make wound dressings creates a moist environment that reduces the likelihood of bacterial infections [60]. Because the ratio of guluronate to mannuronate varies depending on the source, its physical and chemical characteristics are determined by this factor [61]. Alginate-based dressings are manufactured *via* an ionic cross-linking process that uses divalent and trivalent cations to produce the gel. Through freezing and dehydrating techniques, this gel may be readily converted into foam or fibrous dressing [62, 63].

Ammonium salt-based alginate dressings have been developed by researchers who have used different counter-cations such as sodium, triethylammonium, tributyl ammonium, and dihexylammonium in their formulations [64]. The anti-hemolytic and antibacterial activities of tributyl ammonium alginate are superior to those of its rivals compared to those of the other two ingredients. The results were favorable when the dressings were evaluated for their potential to accelerate the re-epithelialization of diseased full-thickness lesions in a clinical setting. Okeil *et al.* developed a new topical bioactive HA/sodium alginate crosslinked with divalent ion films formulation [65]. These films were loaded with silver nanoparticles, and the film that was crosslinked with calcium ions had the greatest antibacterial activity of the three films tested. In another study, a formulator reported electrospinning of chitosan–alginate dressings with gentamicin as the entrapped drug [66]. As fibers, no bead was reported; however, as the concentration of gentamicin was increased, the fiber diameter decreased. The higher the amount of gentamicin, the greater the antibacterial activity also. These films were loaded with silver nanoparticles, and the film that

Table 15.2 Commercially available alginate-based dressings.

Marketed alginate-based formulations	Compositions	References
Algicell™	Sodium alginate, 1.4% silver	[70, 71]
AlgiSite M™	Calcium alginate	[72, 73]
Comfeel Plus™	Sodium carboxymethylcellulose and calcium alginate	[74, 75]
Kaltostat™	Sodium alginate	[76, 77]
Sorbsan™	Calcium alginate	[78, 79]
Tegagen™	Sodium alginate	[80]
Guardix-SG®	Sodium alginate and poloxamer	[81, 82]
SeaSorb®	Calcium alginate	[83, 84]
Algivon®	Calcium alginate and Manuka honey	[85]
Fibracol™Plus	Calcium alginate and collagen	[86]

was crosslinked with calcium ions had the greatest antibacterial activity of the three films tested.

Chalitangkoon *et al.* reported wound dressings made of hydroyethylacryl chitosan and sodium alginate with silver nanoparticles loaded using immersion technique [67]. As the silver loading was increased, the swelling degree also increased along with other mechanical properties. An increase in cross-linking density also retarded the release rate for silver ions and showed a linear release rate. Similarly, cefazolin loaded 0.5% films of sodium alginate, and pectin films showed antibacterial action [68].

Pires *et al.* evaluated the effect of poly(dimethylsiloxane) on the mechanical properties of chitosan–alginate polyelectrolyte complex as material for wound dressing applications [69]. It was reported that the highest effect was observed when 0.1 g of poly(dimethylsiloxane) was added per gram to polyelectrolyte complex. The formulation showed non-hemolytic behavior and was reported to induce thrombus formation and reduce the bleeding. Compounds such as thymol and beta-carotene showed increased wound healing ability. Table 15.2 shows the commercially available alginate-based dressing materials.

15.3.4 Collagen

A protein present in abundance in the human body, collagen is one of the most prevalent proteins. While collagen plays a vital role in wound healing, it does so *via* the attraction of fibroblasts and the stimulation of new collagen deposition at the wound bed. Angiogenesis, re-epithelialization, and autolytic debridement are all aided by collagen dressings in conjunction with other treatments. Collagen dressings bind with matrix metalloproteinases and inactivate them. Xie *et al.* prepared composite dressings composed of collagen, chitosan, and alginates, and it was attached to polyurethane to form composite dressing [87]. The composite cushion has excellent wound healing activity and has good water absorption ability. Treatment up to the 5th day showed a healing rate of 48.5% for composite dressing compared to 28.0% and 38.9% in gauze and chitosan alone. As already known, collagen attracted more fibroblasts and intact re-epithelialization in histological samples. The presence of growth factors such as EGF, bFGF, and TGF-β increased significantly. Ghorbani *et al.* studied the nanofibers fabricated with zein, polycaprolactone, and collagen using zinc oxide nanoparticles and aloe vera [88]. The water contact angle of nanofibers improved with the concentration of zein and polycaprolactone ratio. The nanofibers showed good potential as having antimicrobial activity against *S. aureus* and *E. coli*.

Tanha *et al.* investigated chitosan loaded with recombinant human granulocyte colony-stimulating factor and coated with collagen type 1 [89]. The scaffolds showed good proliferation and well-adherent morphology of stem cells. The skin regeneration was accelerated with scaffolds.

Researchers found that rats administered with PCS, ACS, BCE I, and BCE II exhibited more significant wound contraction, less inflammatory infiltration, and quicker epithelialization and healing [90]. PCS, ACS, BCE I, and BCE II enhanced the total protein and hydroxyproline levels in rats. Although not by much, PCS and ACS had less influence on fibroblast cell differentiation than BCE I and BCE II. They were able to regulate collagen deposition, which resulted in excellent wound healing alignment. Mei *et al.* investigated *Salmo salar* and *Tilapia nilotica* skin collagen peptides in mouse models [91]. The healing rate was relatively faster, and reduced visibility of TNF-, IL-6, and IL-8 levels, upregulated BD14, NOD2, and IL-10 levels, and increased VEGF and –FGF were reported to be linked with it.

15.3.5 Gelatin

Gelatin is an inexpensive, water-soluble polymer utilized in many industries for a millennia. It is made from animal collagen, usually epidermis or skeleton-type I collagen. These basic materials are used to make types A and B gelatins. The isoelectric values of the gelatins influence the net charge at physiological pH. In contrast, gelatin B has an isoelectric point of just 4.9. Understanding the pH of the environment may help choose the suitable gelatin that will become cationic or anionic when hydrated, affecting its function. A maximum of six extractions are required, the first at 55°C and the others increasing to just below boiling point. Initial extractions provide high molecular weight gelatins with greater viscosity and gel strength. The molecular weight is linked to the "bloom value" of gelatin hydrogels. Stronger gels come from higher bloom values.

Bioactive glasses (BGs) have been shown to aid in wound healing and hard tissue restoration [92]. Nano-scale BGs produced by the sol-gel technique performed better because of a higher specific surface area. BP-12 bioactive nanoparticles were combined with gelatin to create an easy-to-use hydrogel bandage for skin wounds. The BP-12/gelatin combination formed a hydrogel (BP-12/gel) at 25°C with good thixotropy and a low shear rate, making it suitable for wound cover. The BP-12/gelatin composite hydrogel was biocompatible with fibroblast cells *in vitro*. Wound healing experiments on rats *in vivo* revealed fast cutaneous tissue regeneration and tissue structure development within 7 days.

Prefabricated dressings of cellulose acetate and gelatin filled with berberine were tested in streptozotocin-induced diabetic rats [93]. The dressings had strong antibacterial action, according to the tests. These results indicated that adding berberine to the dressing did not alter its physical characteristics while it increased its biological activity. With cross-linked gelatin/chitosan biopolymers, novel biodegradable and antibacterial membranes encapsulated cinnamaldehyde (as antibacterial agent) [94]. The inhibition (%) increased significantly with increasing cinnamaldehyde concentration in the membrane matrix. Moreover, hemocompatibility, biodegradability, and cytotoxicity tests revealed that the biodegradable gelatin/chitosan/cinnamaldehyde membranes may be utilized as antibacterial wound dressings. The electrostatic complexation of gelatin and gellan creates a continuous hydrogel network [95]. The resulting mix of hydrogel is shear-thinning and self-recovering. Tannic acid was added to the mix hydrogel to kill microorganisms.

Moreover, tannic acid-loaded mix hydrogel accelerated wound healing in mice model considerably faster than the control groups. Due to their unique features, one-layer wound dressings cannot fulfil all therapeutic requirements [96], so bilayer wound dressings, which contain two layers with distinct characteristics, have gained popularity. Researchers used PCL/gel scaffold electrospun over a polyurethane/ethanolic extract of propolis (PU/EEP) thick membrane. The PU/EEP barrier protected the wound from external contamination and dehydration, while the PCL/gel scaffold promoted cell adhesion and proliferation. The inhibition zone test revealed antibacterial activity against *Staphylococcus aureus*, *Escherichia coli*, and *Staphylococcus epidermidis*. The biodegradability and biocompatibility of the bilayer wound dressing were high. It may speed up wound healing and collagen deposition in the Wistar rat skin wound model.

15.4 Conclusion and Future Directions

Even if the quantity of biopolymers is considerable, there are several problems associated with extracting these molecules from natural sources. Extracting the proper quantity of crude extracts would be a difficult challenge. The extraction methods need to be revamped to meet the growing demand. Wound dressings are difficult to prepare due to the technique of preparation. Manufacturing processes are inefficient and not reproducible. As a result, three-dimensional printing must also be taken into consideration. Because the wound dressings are three-dimensional, the active component is uniformly distributed over the wound [97]. Each dressing created has the same therapeutic effect.

Moreover, no rigid regulation mandates lab-prepared wound dressings to be evaluated to determine their effectiveness. A variety of animal models are being used for testing due to their simplicity. In other words, FDA should investigate this so that scale-up and repeatability are easy.

References

1. Chopra, H., Gandhi, S., Gautam, R.K., Kamal, M.A., Bacterial nanocellulose based wound dressings: Current and future prospects. *Curr. Pharm. Des.*, 27, 570–580, 2021.
2. Chopra, H., Islam, M.A., Sharun, K., Emran, T.B., Al-Tawfiq, J.A., Dhama, K., Recent advances in the treatment of biofilms induced surgical site infections. *Int. J. Surg.*, 109, 1, 65–67, 2023.
3. Chopra, H., Kumar, S., Safi, S.Z., Singh, I., Emran, T.B., Wound dressings : Recent updates. *Int. J. Surg.*, 104, 106793, July 2022.
4. Chopra, H., Gandhi, S., Gautam, R.K., Kamal, M.A., Bacterial nanocellulose based wound dressings: Current and future prospects. *Curr. Pharm. Des.*, 28, 7, 570–580, 2021.
5. Chopra, H., Kumar, S., Singh, I., Biopolymer-based scaffolds for tissue engineering applications. *Curr. Drug Targets*, 22, 3, 282–295, 2021.
6. Chopra, H., Kumar, S., Singh, I., Strategies and therapies for wound healing: A review. *Curr. Drug Targets*, 23, 1, 87–98, 2022.
7. Onyekwelu, I., Yakkanti, R., Protzer, L., Pinkston, C.M., Tucker, C., Seligson, D., Surgical wound classification and surgical site infections in the orthopaedic patient. *JAAOS Glob. Res. Rev.*, 1, 3, 1–10, 2017.
8. Broughton, G., Janis, J.E., Attinger, C.E., The basic science of wound healing. *Plast. Reconstr. Surg.*, 117, 7, 13s–34s, suppl., 2006.
9. Heng, M.C.Y., Wound healing in adult skin: Aiming for perfect regeneration. *Int. J. Dermatol.*, 50, 9, 1058–1066, 2011.
10. Reinke, J.M. and Sorg, H., Wound repair and regeneration. *Eur. Surg. Res.*, 49, 1, 35–43, 2012.
11. Gantwerker, E.A. and Hom, D.B., Skin: Histology and physiology of wound healing. *Clin. Plast. Surg.*, 39, 1, 441–453, 2012.
12. Sorg, H., Tilkorn, D.J., Hager, S., Hauser, J., Mirastschijski, U., Skin wound healing: An update on the current knowledge and concepts. *Eur. Surg. Res.*, 58, 1–2, 81–94, 2017.
13. Velazquez, O.C., Angiogenesis and vasculogenesis: Inducing the growth of new blood vessels and wound healing by stimulation of bone marrow-derived progenitor cell mobilization and homing. *J. Vasc. Surg.*, 45, 6, A39–A47, suppl., 2007.

14. Teller, P. and White, T.K., The physiology of wound healing: Injury through maturation. *Perioper. Nurs. Clin.*, 6, 2, 159–170, 2011.
15. Li, B. and Wang, J.H.C., Fibroblasts and myofibroblasts in wound healing: Force generation and measurement. *J. Tissue Viability*, 20, 4, 108–120, 2011.
16. Chopra, H., Kumar, S., Singh, I., Biopolymer-based scaffolds for tissue engineering applications. *Curr. Drug Targets*, 22, 3, 282–295, 2020.
17. Chopra, H., Kumar, S., Singh, I., Bioinks for 3D printing of artificial extracellular matrices, in: *Advanced 3D-Printed Systems and Nanosystems for Drug Delivery and Tissue Engineering*, pp. 1–37, Elsevier, Netherlands, 2020.
18. Shekh, M.I., Annu, Ahmed, S., Chapter 1-Biopolymers: An overview, in: *Advanced Applications of Biobased Materials*, S. Ahmed, and Annu (Eds.), pp. 3–21, Elsevier, Netherlands, 2023.
19. Sharma, A., Chopra, H., Singh, I., Emran, T.B., Physically and chemically crosslinked hydrogels for wound healing applications. *Int. J. Surg.*, 106, 1–2, 2022.
20. Chopra, H., Kumar, S., Singh, I., Strategies and therapies for wound healing: A review. *Curr. Drug Targets.*, 23, 1, 87–98, 2021.
21. Annu, Pandit, P., Maity, S., Bhattacharya, T., Shekh, M.I., Ahmed, S., Chapter 30-Chitosan biobased materials in textile industry, in: *Advanced Applications of Biobased Materials*, S. Ahmed, and Annu (Eds.), pp. 717–735, Elsevier, Netherlands, 2023.
22. Annu, and Ahmed, S., 1-Bionanocomposites: An overview, in: *Bionanocomposites in Tissue Engineering and Regenerative Medicine*, S. Ahmed, and Annu (Eds.), pp. 1–6, Woodhead Publishing, Netherlands, 2021.
23. Aisverya, S., Annu, Ali, A., Sudha, P.N., 28-Pullulan-based bionanocomposites in tissue engineering and regenerative medicine, in: *Bionanocomposites in Tissue Engineering and Regenerative Medicine*, S. Ahmed, and Annu (Eds.), pp. 533–547, Woodhead Publishing, Netherlands, 2021.
24. Chopra, H., Mishra, A.K., Singh, I., Mohanta, Y.K., Sharma, R., Emran, T.B., Bibi, S., Nano-chitosan: A novel material for glioblastoma treatment. *Int. J. Surg.*, 104, 1–4, May 2022.
25. Annu, Ahmed, S., Ahmed, S., Ikram, S., Chitin and chitosan: History, composition and properties, in: *Chitosan*, pp. 1–24, John Wiley & Sons, Ltd, United States, 2017.
26. Annu, Ahmed, S., Nirala, R.K., Kumar, R., Ikram, S., Green synthesis of chitosan/nanosilver hybrid bionanocomposites with promising antimicrobial, antioxidant and anticervical cancer activity. *Polym. Polym. Compos.*, 29, 9_suppl, S199–S210, 2021.
27. Annu, Manzoor, K., Ahmad, S., Soundarajan, A., Ikram, S., Ahmed, S., Chapter 30-Chitosan based nanomaterials for biomedical applications, in: *Handbook of Nanomaterials for Industrial Applications*, C. Mustansar Hussain, (Ed.), pp. 543–562, Elsevier, Netherlands, 2018.

28. Cai, Y., Zhong, Z., He, C., Xia, H., Hu, Q., Wang, Y., Ye, Q., Zhou, J., Homogeneously synthesized hydroxybutyl chitosans in alkali/urea aqueous solutions as potential wound dressings. *ACS Appl. Bio Mater.*, 2, 10, 4291–4302, 2019.
29. Tallian, C., Tegl, G., Quadlbauer, L., Vielnascher, R., Weinberger, S., Cremers, R., Pellis, A., Salari, J.W.O., Guebitz, G.M., Lysozyme-responsive spray-dried chitosan particles for early detection of wound infection. *ACS Appl. Bio Mater.*, 2, 3, 1331–1339, 2019.
30. Wu, J., Su, C., Jiang, L., Ye, S., Liu, X., Shao, W., Green and facile preparation of chitosan sponges as potential wound dressings. *ACS Sustain. Chem. Eng.*, 6, 7, 9145–9152, 2018.
31. Ray, P.G., Pal, P., Srivas, P.K., Basak, P., Roy, S., Dhara, S., Surface modification of eggshell membrane with electrospun chitosan/polycaprolactone nanofibers for enhanced dermal wound healing. *ACS Appl. Bio Mater.*, 1, 4, 985–988, 2018.
32. Li, Q., Lu, F., Zhou, G., Yu, K., Lu, B., Xiao, Y., Dai, F., Wu, D., Lan, G., Silver inlaid with gold nanoparticle/chitosan wound dressing enhances antibacterial activity and porosity, and promotes wound healing. *Biomacromolecules*, 18, 11, 3766–3775, 2017.
33. Zou, P., Lee, W.H., Gao, Z., Qin, D., Wang, Y., Liu, J., Sun, T., Gao, Y., Wound dressing from polyvinyl alcohol/chitosan electrospun fiber membrane loaded with OH-CATH30 nanoparticles. *Carbohydr. Polym.*, 232, 1–10, 2020.
34. Khorasani, M.T., Joorabloo, A., Moghaddam, A., Shamsi, H., MansooriMoghadam, Z., Incorporation of ZnO nanoparticles into heparinised polyvinyl alcohol/chitosan hydrogels for wound dressing application. *Int. J. Biol. Macromol.*, 114, 1203–1215, 2018.
35. Zhang, Y., Jiang, M., Zhang, Y., Cao, Q., Wang, X., Han, Y., Sun, G., Li, Y., Zhou, J., Novel lignin–chitosan–PVA composite hydrogel for wound dressing. *Mater. Sci. Eng. C*, 104, 1–8, 2019.
36. Hassan, M.A., Tamer, T.M., Valachová, K., Omer, A.M., El-Shafeey, M., Mohy Eldin, M.S., Šoltés, L., Antioxidant and antibacterial polyelectrolyte wound dressing based on chitosan/hyaluronan/phosphatidylcholine dihydroquercetin. *Int. J. Biol. Macromol.*, 166, 18–31, 2021.
37. Chopra, H., Bibi, S., Kumar, S., Khan, M.S., Kumar, P., Singh, I., Preparation and evaluation of chitosan/PVA based hydrogel films loaded with honey for wound healing application. *Gels*, 8, 2, 1–21, 2022.
38. Chopra, H., Bibi, S., Mohanta, Y.K., Kumar Mohanta, T., Kumar, S., Singh, I., Saad Khan, M., Ranjan Rauta, P., Alshammari, A., Alharbi, M., Alasmari, A.F., *In vitro* and *in silico* characterization of curcumin-loaded chitosan–pva hydrogels: Antimicrobial and potential wound healing activity. *Gels*, 9, 5, 1–21. 2023.
39. Gherman, S.P., Biliuță, G., Bele, A., Ipate, A.M., Baron, R.I., Ochiuz, L., Șpac, A.F., Zavastin, D.E., Biomaterials based on chitosan and polyvinyl alcohol as a drug delivery system with wound-healing effects. *Gels*, 9, 2, 1–18, 2023.

40. Li, C., Jiang, T., Zhou, C., Jiang, A., Lu, C., Yang, G., Nie, J., Wang, F., Yang, X., Chen, Z., Injectable self-healing chitosan-based POSS-PEG hybrid hydrogel as wound dressing to promote diabetic wound healing. *Carbohydr. Polym.*, 299, 1–15, 2023.
41. Metwally, A.A., Soliman, A.S., Abdel-Hady, A.-N.A.A., Ebnalwaled, K., Mohamedien, D., Abdelhameed, A.A., Saied, A.A., *In vivo* wound-healing effect of chemical and green synthesized chitosan nanoparticles using lawsonia inermis ethanolic extract. *Microsc. Microanal.*, ozad026, 29, 3, 1178–1189, 2023.
42. Archana, D., Singh, B.K., Dutta, J., Dutta, P.K., Chitosan-PVP-nano silver oxide wound dressing: *In vitro* and *in vivo* evaluation. *Int. J. Biol. Macromol.*, 73, 1, 49–57, 2015.
43. Cacicedo, M.L., Pacheco, G., Islan, G.A., Alvarez, V.A., Barud, H.S., Castro, G.R., Chitosan-bacterial cellulose patch of ciprofloxacin for wound dressing: Preparation and characterization studies. *Int. J. Biol. Macromol.*, 147, 1136–1145, 2020.
44. Ehterami, A., Salehi, M., Farzamfar, S., Vaez, A., Samadian, H., Sahrapeyma, H., Mirzaii, M., Ghorbani, S., Goodarzi, A., *In vitro* and *in vivo* study of PCL/COLL wound dressing loaded with insulin-chitosan nanoparticles on cutaneous wound healing in rats model. *Int. J. Biol. Macromol.*, 117, 601–609, 2018.
45. Li, X., Wang, C., Yang, S., Liu, P., Zhang, B., Electrospun PCL/mupirocin and chitosan/ lidocaine hydrochloride multifunctional double layer nanofibrous scaffolds for wound dressing applications. *Int. J. Nanomedicine*, 13, 5287–5299, 2018.
46. Zarghami, A., Irani, M., Mostafazadeh, A., Golpour, M., Heidarinasab, A., Haririan, I., Fabrication of PEO/chitosan/PCL/olive oil nanofibrous scaffolds for wound dressing applications. *Fibers Polym.*, 16, 6, 1201–1212, 2015.
47. Oh, G.W., Ko, S.C., Je, J.Y., Kim, Y.M., Oh, J.H., Jung, W.K., Fabrication, characterization and determination of biological activities of poly(ε-caprolactone)/chitosan-caffeic acid composite fibrous mat for wound dressing application. *Int. J. Biol. Macromol.*, 93, 1549–1558, 2016.
48. Mouro, C., Simões, M., Gouveia, I.C., Emulsion electrospun fiber mats of pcl/pva/chitosan and eugenol for wound dressing applications. *Adv. Polym. Technol.*, 2019, 1–11, 2019.
49. Hajilou, H., Farahpour, M.R., Hamishehkar, H., Polycaprolactone nanofiber coated with chitosan and Gamma oryzanol functionalized as a novel wound dressing for healing infected wounds. *Int. J. Biol. Macromol.*, 164, 2358–2369, 2020.
50. Fahimirad, S., Abtahi, H., Satei, P., Ghaznavi-Rad, E., Moslehi, M., Ganji, A., Wound healing performance of PCL/chitosan based electrospun nanofiber electrosprayed with curcumin loaded chitosan nanoparticles. *Carbohydr. Polym.*, 259, 1–13, 2021.

51. El-Aassar, M.R., Ibrahim, O.M., Fouda, M.M.G., El-Beheri, N.G., Agwa, M.M., Wound healing of nanofiber comprising Polygalacturonic/Hyaluronic acid embedded silver nanoparticles: *In-vitro* and *in-vivo* studies. *Carbohydr. Polym.*, 238, 1–11, 2020.
52. Zhao, W., Li, Y., Zhang, X., Zhang, R., Hu, Y., Boyer, C., Xu, F.J., Photo-responsive supramolecular hyaluronic acid hydrogels for accelerated wound healing. *J. Control. Release*, 323, 24–35, 2020.
53. Makvandi, P., Ali, G.W., Della Sala, F., Abdel-Fattah, W.I., Borzacchiello, A., Biosynthesis and characterization of antibacterial thermosensitive hydrogels based on corn silk extract, hyaluronic acid and nanosilver for potential wound healing. *Carbohydr. Polym.*, 223, 1–12, 2019.
54. Zhang, S., Hou, J., Yuan, Q., Xin, P., Cheng, H., Gu, Z., Wu, J., Arginine derivatives assist dopamine-hyaluronic acid hybrid hydrogels to have enhanced antioxidant activity for wound healing. *Chem. Eng. J.*, 392, 1–10, 2020.
55. Ying, H., Zhou, J., Wang, M., Su, D., Ma, Q., Lv, G., Chen, J., *In situ* formed collagen-hyaluronic acid hydrogel as biomimetic dressing for promoting spontaneous wound healing. *Mater. Sci. Eng. C*, 101, 487–498, 2019.
56. Lin, Z., Wu, T., Wang, W., Li, B., Wang, M., Chen, L., Xia, H., Zhang, T., Biofunctions of antimicrobial peptide-conjugated alginate/hyaluronic acid/collagen wound dressings promote wound healing of a mixed-bacteria-infected wound. *Int. J. Biol. Macromol.*, 140, 330–342, 2019.
57. Hussein, Y., El-Fakharany, E.M., Kamoun, E.A., Loutfy, S.A., Amin, R., Taha, T.H., Salim, S.A., Amer, M., Electrospun PVA/hyaluronic acid/L-arginine nanofibers for wound healing applications: Nanofibers optimization and *in vitro* bioevaluation. *Int. J. Biol. Macromol.*, 164, 667–676, 2020.
58. Ding, X., Li, S., Tian, M., Yang, P., Ding, Y., Wang, Y., Duan, G., Zhang, D., Chen, B., Tan, Q., Facile preparation of a novel nanoemulsion based hyaluronic acid hydrogel loading with Poria cocos triterpenoids extract for wound dressing. *Int. J. Biol. Macromol.*, 226, 1490–1499, 2023.
59. Zhang, L., Tan, W., Zhang, M., Ma, Z., Zhao, T., Zhang, Y., Preparation and characterization of Panax notoginseng saponins loaded hyaluronic acid/carboxymethyl chitosan hydrogel for type o diabetic wound healing. *Mater. Today Commun.*, 34, 105284, 2023.
60. Lee, K.Y. and Mooney, D.J., Alginate: Properties and biomedical applications. *Prog. Polym. Sci.*, 37, 1, 106–126, 2012.
61. Annu, Ahmed, S., Ikram, S., Perspectives of chitosan and alginate membranes for biomedical applications, in: *Natural Polymers: Derivatives, Blends and Composites*, vol. 2, pp. 157–182, Nova Science Publishers, Incorporated, United States, 2017.
62. Ching, S.H., Bansal, N., Bhandari, B., Alginate gel particles–A review of production techniques and physical properties. *Crit. Rev. Food Sci. Nutr.*, 57, 6, 1133–1152, 2017.
63. Topuz, F., Henke, A., Richtering, W., Groll, J., Magnesium ions and alginate do form hydrogels: A rheological study. *Soft Matter*, 8, 18, 4877–4881, 2012.

64. Zare-Gachi, M., Daemi, H., Mohammadi, J., Baei, P., Bazgir, F., Hosseini-Salekdeh, S., Baharvand, H., Improving anti-hemolytic, antibacterial and wound healing properties of alginate fibrous wound dressings by exchanging counter-cation for infected full-thickness skin wounds. *Mater. Sci. Eng. C*, 107, 110321, 2020.
65. Abou-Okeil, A., Fahmy, H.M., El-Bisi, M.K., Ahmed-Farid, O.A., Hyaluronic acid/Na-alginate films as topical bioactive wound dressings. *Eur. Polym. J.*, 109, 101–109, 2018.
66. Bakhsheshi-Rad, H.R., Hadisi, Z., Ismail, A.F., Aziz, M., Akbari, M., Berto, F., Chen, X.B., *In vitro* and *in vivo* evaluation of chitosan-alginate/gentamicin wound dressing nanofibrous with high antibacterial performance. *Polym. Test.*, 82, 106298, 2020.
67. Chalitangkoon, J., Wongkittisin, M., Monvisade, P., Silver loaded hydroxyethylacryl chitosan/sodium alginate hydrogel films for controlled drug release wound dressings. *Int. J. Biol. Macromol.*, 159, 194–203, 2020.
68. Shahzad, A., Khan, A., Afzal, Z., Umer, M.F., Khan, J., Khan, G.M., Formulation development and characterization of cefazolin nanoparticles-loaded cross-linked films of sodium alginate and pectin as wound dressings. *Int. J. Biol. Macromol.*, 124, 255–269, 2019.
69. Pires, A.L.R., de Azevedo Motta, L., Dias, A.M.A., de Sousa, H.C., Moraes, Â.M., Braga, M.E.M., Towards wound dressings with improved properties: Effects of poly(dimethylsiloxane) on chitosan-alginate films loaded with thymol and beta-carotene. *Mater. Sci. Eng. C*, 93, 595–605, 2018.
70. Boateng, J. and Catanzano, O., Advanced therapeutic dressings for effective wound healing–A review. *J. Pharm. Sci.*, 104, 11, 3653–3680, 2015.
71. Jacob, S., Nair, A.B., Shah, J., Sreeharsha, N., Gupta, S., Shinu, P., Emerging role of hydrogels in drug delivery systems, tissue engineering and wound management. *Pharmaceutics*, 13, 3, 1–36, 2021.
72. Kamoun, E.A., Kenawy, E.R.S., Chen, X., A review on polymeric hydrogel membranes for wound dressing applications: PVA-based hydrogel dressings. *J. Adv. Res.*, 8, 3, 217–233, 2017.
73. McBride, C.A., Kimble, R.M., Stockton, K.A., Prospective randomised controlled trial of Algisite™ M, Cuticerin™, and Sorbact® as donor site dressings in paediatric split-thickness skin grafts. *Burn. Trauma*, 6, 1–10, 2018.
74. Goodhead, A., Clinical efficacy of comfeel plus transparent dressing. *Br. J. Nurs.*, 11, 4, 286–287, 2002.
75. Ågren, M.S. and Everland, H., Two hydrocolloid dressings evaluated in experimental full-thickness wounds in the skin. *Acta Derm. Venereol.*, 77, 2, 127–131, 1997.
76. Williams, C., Kaltostat. *Br. J. Nurs.*, 3, 18, 965–967, 1994.
77. Ribeiro, D.M.L., Carvalho Jr., A.R., Vale de Macedo, G.H.R., Chagas, V.L., Dos Santos Silva, L., Da Silva Cutrim, B., Santos, D.M., Soares, B.L.L., Zagmignan, A., de Miranda, R.C.M., de Albuquerque, P.B.S., da Silva, L.C.N., Polysaccharide-based formulations for healing of skin-related wound

infections: Lessons from animal models and clinical trials. *Biomolecules*, 10, 1, 1–16, 2019.
78. Shigeki, S., Murakami, T., Tani, Y., Ikuta, Y., Use of Sorbsan, a calcium alginate fiber dressing, as a drug reservoir in iontophoretic transdermal delivery. *Plast. Reconstr. Surg.*, 102, 7, 2509–2511, 1998.
79. Gilchrist, T. and Martin, A.M., Wound treatment with Sorbsan–an alginate fibre dressing. *Biomaterials*, 4, 4, 317–320, 1983.
80. Williams, C., Tegagen alginate dressing for moderate to heavily exuding wounds. *Br. J. Nurs.*, 7, 9, 550–552, 1998.
81. Hong, J.H., Choe, J.W., Kwon, G.Y., Cho, D.Y., Sohn, D.S., Kim, S.W., Woo, Y.C., Lee, C.J., Kang, H., The Effects of barrier materials on reduction of pericardial adhesion formation in rabbits: A comparative study of a hyaluronan-based solution and a temperature sensitive poloxamer solution/gel material. *J. Surg. Res.*, 166, 2, 206–213, 2011.
82. Haryanto, Singh, D., Han, S.S., Son, J.H., Kim, S.C., Poly(ethylene glycol) dicarboxylate/poly(ethylene oxide) hydrogel film co-crosslinked by electron beam irradiation as an anti-adhesion barrier. *Mater. Sci. Eng. C*, 46, 195–201, 2015.
83. Leveriza-Oh, M. and Phillips, T.J., Dressings and postoperative care, in: *Lower Extremity Soft Tissue & Cutaneous Plastic Surgery*, 2nd Edition, pp. 471–488, 2012.
84. Wietlisbach, C.M., Wound care, in: *Cooper's Fundamentals of Hand Therapy*, pp. 154–166, 2020.
85. Szweada, P., Gorczyca, G., Tylingo, R., Comparison of antimicrobial activity of selected, commercially available wound dressing materials. *J. Wound Care*, 27, 5, 320–326, 2018.
86. Shi, L., Ermis, R., Kiedaisch, B., Carson, D., The effect of various wound dressings on the activity of debriding enzymes. *Adv. Skin Wound Care*, 23, 10, 456–462, 2010.
87. Xie, H., Chen, X., Shen, X., He, Y., Chen, W., Luo, Q., Ge, W., Yuan, W., Tang, X., Hou, D., Jiang, D., Wang, Q., Liu, Y., Liu, Q., Li, K., Preparation of chitosan-collagen-alginate composite dressing and its promoting effects on wound healing. *Int. J. Biol. Macromol.*, 107, PartA, 93–104, 2018.
88. Ghorbani, M., Nezhad-Mokhtari, P., Ramazani, S., Aloe vera-loaded nanofibrous scaffold based on Zein/Polycaprolactone/Collagen for wound healing. *Int. J. Biol. Macromol.*, 153, 921–930, 2020.
89. Tanha, S., Rafiee-Tehrani, M., Abdollahi, M., Vakilian, S., Esmaili, Z., Naraghi, Z.S., Seyedjafari, E., Javar, H.A., G-CSF loaded nanofiber/nanoparticle composite coated with collagen promotes wound healing *in vivo*. *J. Biomed. Mater. Res. A*, 105, 10, 2830–2842, 2017.
90. Chen, J., Gao, K., Liu, S., Wang, S., Elango, J., Bao, B., Dong, J., Liu, N., Wu, W., Fish collagen surgical compress repairing characteristics on wound healing process *in vivo*. *Mar. Drugs*, 17, 1, 1–12, 2019.

91. Mei, F., Liu, J., Wu, J., Duan, Z., Chen, M., Meng, K., Chen, S., Shen, X., Xia, G., Zhao, M., Collagen peptides isolated from salmo salar and tilapia nilotica skin accelerate wound healing by altering cutaneous microbiome colonization via upregulated NOD2 and BD14. *J. Agric. Food Chem.*, 68, 6, 1621–1633, 2020.
92. Wang, C., Zhu, F., Cui, Y., Ren, H., Xie, Y., Li, A., Ji, L., Qu, X., Qiu, D., Yang, Z., An easy-to-use wound dressing gelatin-bioactive nanoparticle gel and its preliminary *in vivo* study. *J. Mater. Sci. Mater. Med.*, 28, 1, 1–10, 2017.
93. Samadian, H., Zamiri, S., Ehterami, A., Farzamfar, S., Vaez, A., Khastar, H., Alam, M., Ai, A., Derakhshankhah, H., Allahyari, Z., Goodarzi, A., Salehi, M., Electrospun cellulose acetate/gelatin nanofibrous wound dressing containing berberine for diabetic foot ulcer healing: *In vitro* and *in vivo* studies. *Sci. Rep.*, 10, 1, 1–12, 2020.
94. Kenawy, E., Omer, A.M., Tamer, T.M., Elmeligy, M.A., Eldin, M.S.M., Fabrication of biodegradable gelatin/chitosan/cinnamaldehyde crosslinked membranes for antibacterial wound dressing applications. *Int. J. Biol. Macromol.*, 139, 440–448, 2019.
95. Zheng, Y., Liang, Y., Zhang, D., Sun, X., Liang, L., Li, J., Liu, Y.N., Gelatin-based hydrogels blended with gellan as an injectable wound dressing. *ACS Omega*, 3, 5, 4766–4775, 2018.
96. Eskandarinia, A., Kefayat, A., Agheb, M., Rafienia, M., Amini Baghbadorani, M., Navid, S., Ebrahimpour, K., Khodabakhshi, D., Ghahremani, F., A novel bilayer wound dressing composed of a dense polyurethane/propolis membrane and a biodegradable polycaprolactone/gelatin nanofibrous scaffold. *Sci. Rep.*, 10, 1, 1–15, 2020.
97. Chopra, H., Priyanka, Choudhary, O.P., Emran, T.B., Three dimensional printed wound dressings: recent progresses. *Int. J. Surg.*, 109, 3, 549–550, 2023.

16

Bioplastics in Medical Devices

Reshma Soman[1*], Divyanshi Mangla[2] and Annu[3†]

[1]*School of Biosciences, Mahatma Gandhi University, Kottayam, Kerala, India*
[2]*Bio/Polymers Research Laboratory, Department of Chemistry, Jamia Millia Islamia, New Delhi, India*
[3]*Thin Film Engineering and Materials Laboratory, School of Mechanical Engineering, Yeungnam University, Gyeongsan, Republic of Korea*

Abstract
Plastics are a catastrophe to nature and to the globe, making troubles to human beings. The global accumulation of plastics is increasing day by day as they have not been removed from nature due to their non-biodegradable nature. Therefore, the need for degradable plastics gives rise to the thought of bioplastics. Bioplastics are degraded by microorganisms or could be lowered to lesser toxic compounds so that further degradation will take place eventually. Bioplastics have a wider use in industries like toy manufacturing, construction works, and so on. One of the important applications of bioplastics is its use in medical applications. Bioplastics have far taken the role in tissue engineering, drug delivery, and dentistry besides wound treatment and so on. The use of plastics should be limited and pace towards bioplastics for the environment and living beings.

***Keywords*:** Plastics, biodegradable, bioplastics, environment, medical application, drug delivery, tissue engineering, devices

16.1 Introduction

Plastics are a global threat that imposes problems in the life and livelihood of living organisms. The need for the removal of plastics is increasing day by day, which makes scientists rack their brains for the development of

**Corresponding author*: reshmasoman50@gmail.com
†*Corresponding author*: annuchem92@gmail.com

Annu (ed.) Biopolymers for Biomedical Applications, (417–438) © 2024 Scrivener Publishing LLC

novel biodegradable plastics. Thus, the discovery of biodegradable plastics is slowly taking the era.

Bioplastics can be degraded by microorganisms to a lesser toxic substance, thereby imposing a lesser threat to human beings and living organisms. Bioplastics could be used in various fields including construction works, toy manufacturing, and industries along with medicine. It is used for drug delivery, tissue designing, and dentistry as well [1]. This chapter emphasizes plastics, bioplastics, and their uses in medical applications.

16.2 Plastics and Their Environmental Concern

Plastics are man-made organic polymers that have been around for slightly over a century. Plastics are lightweight, robust, resilient, and inexpensive compounds, which make them ideal for manufacturing a broad variety of items [2]. Most plastics, such as those used in bottles and textiles, are thermoplastics made up of unbranched polymer chains that can tolerate high temperatures and used for reshaping the polymers, while some monomers are polymerized *in vitro* to form thermosets that are sturdier than thermoplastic polymers and have a definite shape that is nonchalant to heat [3]. Due to its buoyancy, a growing amount of plastic garbage is being scattered across great distances in seas and oceans along with soil, and once deposited in sediments, it may remain for generations. These same attributes are precisely the reasons why plastics pose such a major threat to the environment [2].

Plastic manufacturing is already over 380 million tonnes per year, and it is growing at a pace of 4% annually; as a result, 6,300 million tonnes of plastic garbage have been created since 1950. The shift to a "circular plastic economy" (Figure 16.1) is motivated by a growing concern about the ecological effect of plastic surplus and the plastic-associated discharge of greenhouse gases (GHGs). In a circular economy, nonrenewable resources and waste creation are reduced, and material life cycles are dominated by reuse and recycling [3].

The plastic business has conventionally been established on undeviating lifespan (gray arrows) as shown in Figure 16.1: the splintered crude oil is cultured into monomeric and polymeric goods using fossil fuel, which is then either disposed of (80%) with a probable leakage into the surroundings, reduced to ashes (10%), or mechanically recycled as less significant landfill products. Plastic trash becomes unprocessed material for a salvaging procedure at the termination of its life cycle in a "circular plastic economy" (green arrows), and all the recycling as well as manufacturing

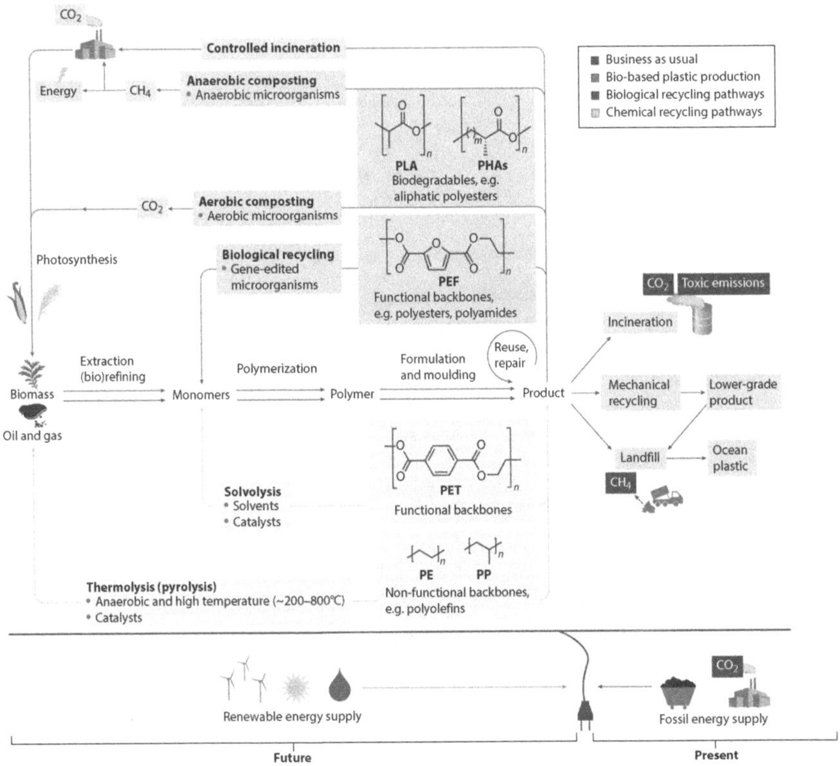

Figure 16.1 Representation of the "circular plastic economy" [3].

operations are powered by renewable energy. Polymer goods are made from natural assets (lignocellulosic biomass and pyrolysis oils), and they altogether have a specified spherical end-of-life conditions. CO_2 added to the carbon cycle by bioplastic ignition (blue arrow), composting in the presence of air, or burning of CH_4 by composting in the absence of air is a nil input since it is caught by photosynthesis and converted into new biomass. Upcycling of plastic trash is possible because polymers with functional backbones (polyamides or polyesters) may be depolymerized chemically or biologically. The resultant monomers are then polymerized into tailored virgin and high-quality goods. Polyolefins (including bio-based PE, polyethylene (PE), polystyrene, and polypropylene (PP)) with non-functional backbones are more suitable for thermolysis breaking into hydrocarbon oil and gas and subsequently follow a similar upcycling path. Polyethylene furanoate, polyethylene terephthalate, polyhydroxyalkanoate,

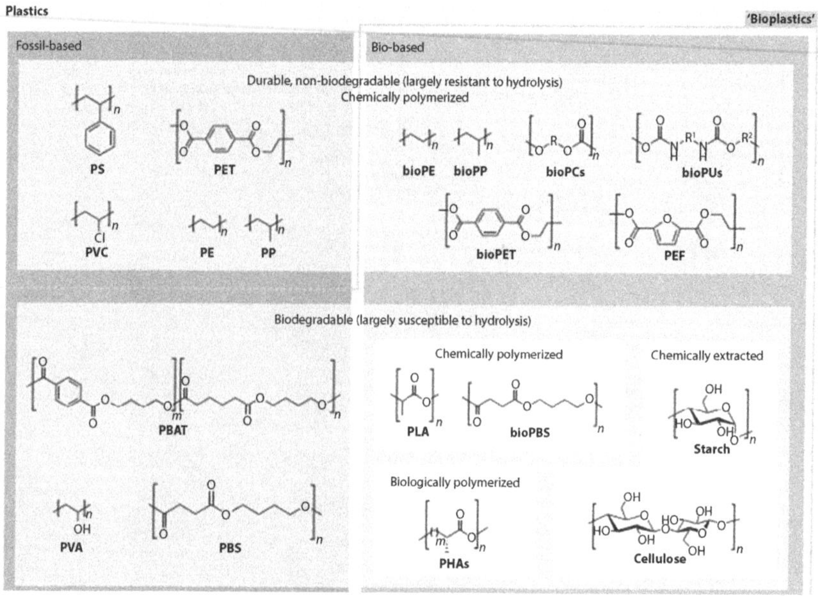

Figure 16.2 Various plastics and bioplastic-based derivatives [3].

and polylactic acid are represented as PEF, PET, PHA, and PLA, respectively, in Figure 16.2 [3].

16.3 Need for Bioplastics

The majority of the Earth's warming and rise in extreme weather are caused by human activities, which is based on the release of GHGs from the use of fossil resources, such as CO_2 and CH_4. Plastic manufacturing uses 5–7% of the world oil supply and emits >850 million tonnes of CO_2, accounting for 2% of global CO_2 emissions in 2019. Raw material extraction (61%) and polymer synthesis (30%) account for the vast bulk of plastic-related CO_2 emissions, whereas the EOL stage (mostly incineration) accounts for just 9%. Unreported open burning of plastic waste might increase these numbers by more than 1 Gt CO_2 equivalent. Simulations show that switching to sugarcane as a feedstock would reduce the greenhouse gas emissions by 25%; recycling all present plastics would have a similar effect. Plastic-related emissions would be reduced by 62% by switching to a renewable energy source because manufacturing has the largest carbon footprint of all polymers. The current levels would be reduced by 93% if these efforts were combined with halving consumption, implying that these four systematic adjustments must be implemented in tandem in order to successfully reduce emissions [3].

16.4 Bioplastics

Bioplastics are materials that decompose to their constituent monomers without emitting hazardous gases into the atmosphere and are biocompatible when utilized in biomedical applications. Based on their synthesis resources, bioplastics may be biologically degradable or biodegradable. Different plastics and bioplastic-based derivatives are shown in Figure 16.2.

Biodegradable plastics are created from renewable feedstocks that deteriorate physically and chemically before completely decomposing when exposed to fungus and bacteria that produce CO_2, CH_4, and water as well as other gases. Biodegradable polycaprolactone (PCL) and non-biodegradable (PCL) are available bio-derived polymers [4].

Biodegradable aliphatic polyesters such as poly lactic-co-glycolic acid (PLGA), PLA, PHA, and PCL, as well as their copolymers, are now used in biomedical applications in the human body. People use the term "degradable" in a broad sense and "biodegradable" for polymers that are biologically destroyed by enzymes, introduced *in vitro*, or created by surrounding live cells according to Li (2006) [5]. Many biodegradable polymers, such as esters, orthoesters, anhydrides, carbonates, amides, urea, and urethane, have hydrolyzable links in their backbones. Aliphatic polyesters offering ester linkages experience a wide variety of physical, chemical, and biological properties along with great biocompatibility. PHA is the most frequently employed aliphatic polyester in biomedical applications because of its excellent biocompatibility [6].

Because of their primary-chain structure, hydrophilicity, and crystallinity, biodegradable polyesters may decompose in the environment. According to recent studies, the balance of hydrophobicity as well as hydrophilicity of polyester molecules will become critical for enzyme–substrate binding and subsequent hydrolytic enzyme operations [7].

16.4.1 Polylactic Acid

PLA is the tiniest biological molecule created by animals, plants, and microbes in the environment that shows optical activity. Lactic acid was discovered and reported for the first time in 1780. In his review, Carothers described the ring-opening polymerization of lactic acid to lactide

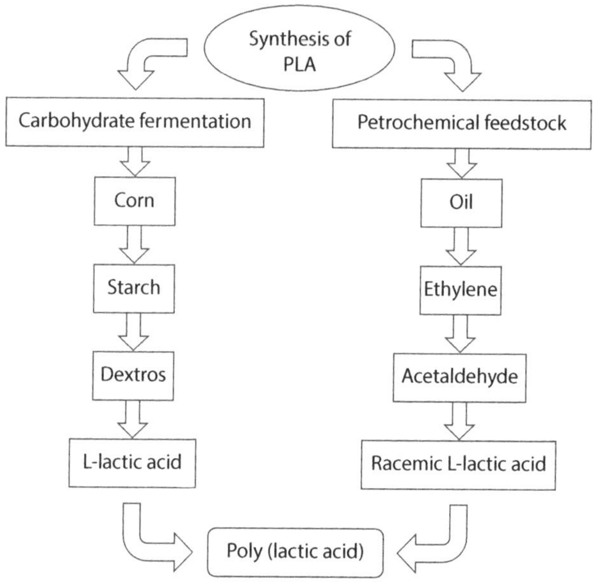

Figure 16.3 The synthesis route of polylactic acid [13].

dimerization. He also claimed that lactic acid will go through reversible polymerization, which is a six-atom cyclic ester characteristic. Linear polyesters are polymers made up of six-membered cyclic esters that have had their chains opened and substituted with hydroxyl (-OH) and carboxylic (-COOH) groups in some cases [8–10]. The synthesis route of PLA is described in Figure 16.3.

Both polymerization and depolymerization are accomplished by exchanging esters. The ease of degradation and innocuous nature of these polymers for biological applications was developed apparently in the 1960s. PLA has emerged as one of the most favorable polymers owing to its biocompatibility and biodegradability, and it has an extensive diversity of usages in biomedical research and biotechnology [11, 12].

16.5 Biomedical Applications of Polylactic Acid

16.5.1 PLA in Tissue Engineering

PLA is one of the most common biopolymers in medical applications due to its bio-dissolvability and biocompatibility in the human body and hydrolysis of the ester framework with innocuous and non-hazardous chemical production. The most common route of deterioration for PLA polymers utilized in medical applications is hydrolysis [14]. Figure 16.4 shows the biomedical applications of PLA.

Zhang *et al.* had created PLA/octa-decyl amine-functionalized nano-diamond (ND-ODA) composites for their application in tissue engineering. PLA was dissolved in chloroform while ND-ODA was dispersed in chloroform, and the mixtures were then sonicated together to create the composites. They obtained thin films of PLA/octa-decyl amine composites by evaporating the PLA (with chloroform) and ND-ODA dispersion solution. Additionally, murine osteoblasts are unaffected by ND-ODA and composites. PLA and its copolymers, such as PLA-p-dioxanone- polyethylene glycol (PLA-p-DPEG) block copolymer and PLA-polyethylene glycol (PLA-PEG) block copolymer, are also used as transporters for bone morphogenetic proteins (BMPs) [15].

BMPs are physiologically dynamic chemicals, which have the capacity to initiate fresh bone production. BMPs are utilized in therapeutic settings in conjunction with natural resources—for instance, bone graft substitutes to promote bone regeneration. The bone which forms as a result of PLA breakdown, on the other hand, was in extremely minute quantities. As a result, PLA copolymers were developed to address the issue of low molecular weight PLA. Chang *et al.* had developed a PLA scaffold and investigated

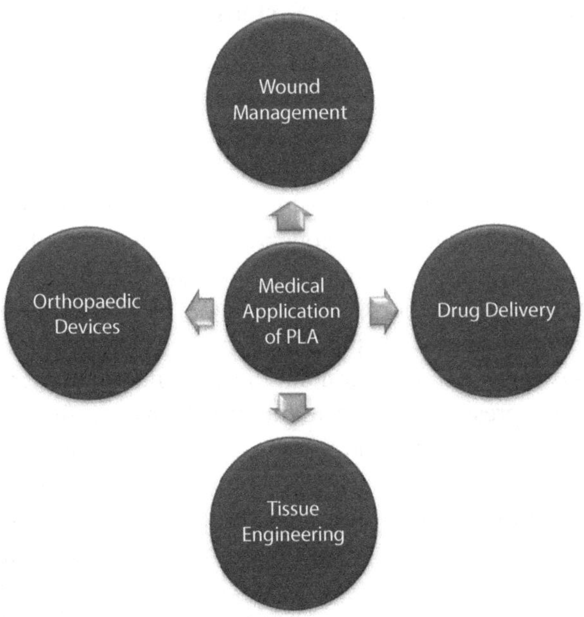

Figure 16.4 Biomedical applications of polylactic acid.

its capacity to function as a transporter for recombinant bone morphogenetic protein 2 (rhBMP2) [14, 15].

16.5.2 PLA in Wound Management

PLA and its copolymers are utilized in a diversity of wound treatment applications together with the creation of surgical sutures, the restoration of dental wounds, and the prevention of postoperative adhesions. The capacity and contingency of PLA ureteral stents used to treat ureteral injuries were investigated by Li *et al.* [16]. PLA stents are biodegradable and may be removed from the human body afterwards. As a result, PLA stents appear to have a bright future in dealing with ureteral damages. In their research, Qin *et al.* (2006) employed PLA polymer blends to avoid postoperative adhesions. Since the mechanical properties of pure PLA, for example, glass transition temperature, tensile strength, and Young's modulus, were greater than the PLA blends, blended PLA is much more flexible than pristine PLA [17].

Brekke noted that PLA might be used to improve dental wound treatment, and a surgical bandage composed of PLA could minimize the rate of mandibular third molar extraction wound botch [18].

16.5.3 PLA in Drug Delivery Systems

In medication delivery frameworks, the medication may be delivered over and over for as long as a year. PLA is utilized in drug delivery systems since it is completely decomposable, has a higher limit with respect for encapsulation, is biocompatible, and is less poisonous. Erosion, diffusion, and swelling are the three methods by which polymeric drugs are released. At the point where water goes into a biodegradable polymer containing monomers connected through ester linkages, the breakdown begins [19, 20]. The ester bonds are randomly broken by hydrolytic ester cleavage, resulting in device degradation. Erosion happens in two ways for degradable polymers: homogeneous/mass disintegration and heterogeneous/surface disintegration. Numerous prescriptions, including psychosis, restenosis, chemicals, oridonin, dermal treatment, and protein, were exemplified involving PLA and their copolymers as nanoparticles. Solvent evaporation, diffusion of emulsion solvent, salting out, and solvent displacement are few techniques for acquiring these nanoparticles. For loading the medication paclitaxel, Ling and Huang employed poly (lactic-co-glycolic) acid nanoparticles [21].

The practice of using PLA nanoparticles (PLA-NPs) coated with luminous dyes as exporters for trans-epidermal medication supply was examined by Rancan *et al.* PLA stands for polymer nanoparticles, which are made by evaporating a solvent. PLA was initially dissolved in acetone and afterward mixed with water while stirring continuously, and the solvent was then vaporized at room temperature under reduced pressure to generate PLA-NPs. To make fluorescent particles, acetone was used to dissolve fluorescent dye and PLA, and then the same process was used. PLA-NPs were shown to be good candidates for building drug delivery systems that possibly will aim active chemicals into hair follicles when tested on human skin [22].

16.5.4 PLA in Orthopedic Devices

In orthopedic applications, biodegradable polymers are utilized to prevent a second surgical operation to remove superfluous hardware. Biodegradable suture anchors, screws, and fixation pins all require PLA polymers [23]. These absorbable screws and pins are typically utilized in clinical uses wherever significant automated power is not mandatory. In certain circumstances, high PLA mechanical force was necessary; thus, strategies to increase PLA mechanical characteristics, especially influencing ductile strength and crack modulus in bone obsession, with pole, pins, and both biodegradable as well as metal plates, have restricted their applicability in bone break fixing [24]. PLA copolymers were biocompatible in the human body, according to Bostman *et al.* [25]. They also indicated that six out of 120 patients cured with needles made from PLA/PGA copolymers may develop a sterile cavity by the side of the imbedding spot, which is extremely rare and may be rectified with further modifications [26].

16.5.5 Poly Lactic-co-glycolic Acid

PLGA has received approval from both the US Food and Drug Administration (FDA) and the European Medicine Agency as one of the best synthetic biodegradable polymers used in the biomedical industry [27]. The biodegradation rate and the biocompatibility of PLGA, which is influenced by the polymer's molecular weight and copolymer ratio, have drawn interest as a key material for medical applications. According to the FDA, its usage is safe for the human body and improves biological material interaction by changing its surface qualities.

When compared to other biodegradable polymers, PLGA is a hydrophilic, crystalline polymer that possesses a rapid degradation rate. PLGA co-polymers are frequently favored over their component homopolymers for the integration of bone replacement structures because PLGA advises high-grade control over its deteriorating properties by adjusting its monomeric ratios. The PLGA has an inclusive variety of degradation rates, which are ruled by the combination of hydrophilic as well as hydrophobic chains and the polymer's crystalline structure [28]. To make PLGA more bionic and capable of intensifying bone reformation, it is generally combined with other materials such as biologically active glass and ceramics [29]. As a result, PLGA-based bone replacements have been divided into scaffolds, fibers, hydrogels, and microspheres depending on their kinds and applications [30].

16.6 Biomedical Applications of PLGA

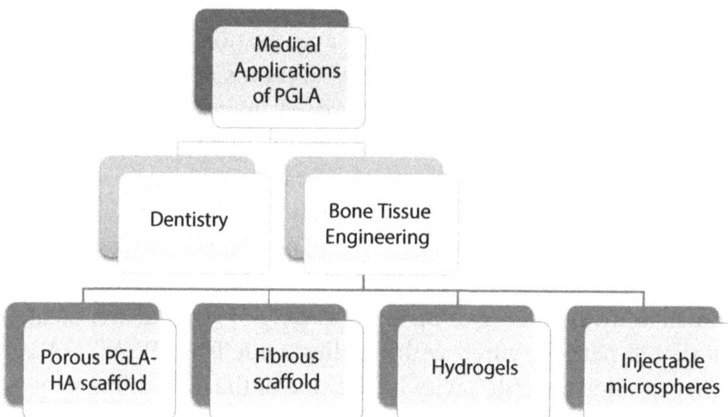

Figure 16.5 Various biomedical applications of poly lactic-co-glycolic acid (PLGA).

16.7 PLGA in Bone Tissue Engineering

PLGA is used in bone tissue engineering in various ways as depicted in Figure 16.5 and discussed in the following subsections.

16.7.1 PLGA-HA Scaffolds

Kim and colleagues (2006) described a unique approach for fabricating a polymeric/nano-HA composite scaffold devoid of the use of organic solvents [31] using gas-forming and particulate leaching (GF/PL). In contradiction with scaffolds made through the solvent casting/particulate leaching (SC/PL) process, scaffolds manufactured by means of gas-shaping other than particulate draining (GF/PL) have an exceptionally permeable design, worked on mechanical properties, with and remarkable cell development, basic phosphatase action, and *in vitro*-mineralized frameworks.

In 2011, Ebrahimian-Hosseinabadi [32] used the thermally induced phase separation approach to create a bionic skeleton based on PLGA and a nano-biphasic component (nBCP), which included crushed versions of HA and tri-calcium phosphate as strengthening ingredients at a temperature of 60°C. The scaffolds formed by composites of nBCP had maximum and optimal yield strength and Young's modulus of 20–30% (w/w), respectively [33].

16.7.2 Fibrous Scaffolds

Fibrous scaffolds possess remarkable bone tissue restructuring capacity. Numerous methods have been employed to create micro- and nano-fibrous composite scaffolds [34]. Morgan (2007) employed the wet-spinning approach to generate hollow fibers, which were then used as supports in conjunction with humanoid bone marrow stromal cells to aid in natural bone rejuvenation besides rebuilding [35]. Nano-fibrous composites, on the other hand, have a structure comparable to the natural bone extracellular matrix and could provide subordinate stimulation to artificial cells.

The electro-spinning method is a modest and adaptable approach for generating exceedingly thin non-interweaved fibers of diameters ranging between nanometers and micrometers. Furthermore, electrospun fibers are believed to have a role in bone regeneration by preserving the mechanical qualities, permitting bio-degradability, and performing as a true osteoconductive framework following accumulation or coating with ceramic constituents [36].

16.7.3 Hydrogels

Hydrogels are another type of scaffold that is frequently employed in tissue engineering. Hydrogels containing hyaluronic acid, fibrin, and pluronic F127 have revealed potential in the delivery of growth factors [37]. Dhillon *et al.* [38] stated that combining PLGA with a plasticizer, for example, PEG, enabled the fabrication of thermo-sensitive materials with a T_g of 37°C. In a mouse calvarial defect model [39], this scaffolded approach was recently revealed to aid bone healing *in vivo*. On the other hand, there are certain disadvantages of utilizing hydrogels for bone redevelopment, such as their poor mechanical strength that could make them unsuitable for use as individual bone substitutes [40].

16.7.4 Injectable Microspheres

As amorphous PLGA copolymers enable an additional homogenous distribution of the dynamic class in the polymer milieu [41], they are suited for biomedical applications. The PLGA microspheres were made using a traditional oil/water emulsification procedure, and then HA was added to make biomimetic injectable microspheres. To make a cohesive colloidal gel [42], adversely charged inorganic HA nanoparticles were blended in with decidedly charged PLGA microspheres scattered in deionized water. Electrostatic forces held this material together, which might be broken by extrusion, molding, or injection.

16.7.5 PLGA in Dentistry

In an extensive variety of dental applications, PLGA materials have proven to be beneficial. From designing screws for bone fixation [43–45] to treating periodontal infections and manufacturing buccal mucosa [46] to indirect pulp-capping treatments, they are applied in a variety of ways. As PLGA implants, discs, and dental layers, PLGA could be employed in periodontal therapy to improve local antibiotic administration and reduce the systemic adverse effects of general antibiotic delivery [47].

Furthermore, PLGA gel composite fabrics utilized in bone redevelopment [48], as well as extraordinarily degradable PLGA and SiO_2–CaO gel non-woven textiles, were subjected to simulated bodily fluid for 7 days, and a coating of apatite crystals was deposited on their exteriors. The granular composite of gatifloxacin-stacked b-tricalcium phosphate and PLGA is a nearby exchange mode for relieving osteomyelitis, as it conveyed gatifloxacin bit by bit and showed palatable bacterial movement *in vitro* against *Bacteroides fragilis* and *Streptococcus milleri*, the microbes that cause osteomyelitis. GFLX-loaded PLGA and TCP were also able to considerably decrease inflammation and promote osteoconduction and vascularization of the affected areas in the rabbit jaw at just 4 weeks after implantation. Furthermore, sterilized PLGA scaffolds appear to be a good constituent for flesh engineering buccal mucosa. PLGA mixtures containing bio-ceramics could also be employed in undeviating pulp capping *via* any means of integrating progress dynamics into PLGA microparticles or by directly capping mechanically exposed teeth with PLGA composites [49].

Regardless of the biocompatibility demonstrated in the cellular test, there was no stiff tissue in adjacent pulp capping with PLGA, and pulp necrosis was observed to be caused by the poor adhesion between PLGA and the pulp. As a result, PLGA composites containing bio-ceramics are still a superior alternative for pulp capping than PLGA alone, with greater tissue response than calcium hydroxide. The encouraging consequences of the PLGA constituents point to the requirement for more research, particularly in the areas of substance distribution to dental tissues and the pulp capping capacity of PLGA composites [49].

16.8 Biomedical Applications of Poly(ε-caprolactone)

Polycaprolactone emerged as a key polymer formulated by the Carothers team during the mid-1930s. PCL came to be commercially accessible, necessitating a number of researches to discover artificial polymers which are easily degraded by microbes [50]. PCL could be made *via* free-radical ring-opening polymerization of 2-methylene-1-3-dioxepane or by ring-opening polymerization using cationic, anionic, or co-ordination catalysts [51]. PCL is semi-crystalline and hydrophobic, and crystallinity decreases as the molecular weight increases. PCL's solubility, low melting point (59–64°C), and excellent blending compatibility have driven extensive research into its diverse biological applications [52]. Figure 16.6 shows the biomedical applications of PCL.

PCL and related copolymers were widely utilized in a range of pharmaceutical delivery systems for the duration of the absorbable polymers thriving in the 1970–1980. PCL has quite a lot of benefits, comprising of the capacity to tailor its degradation kinetics and mechanical properties, the

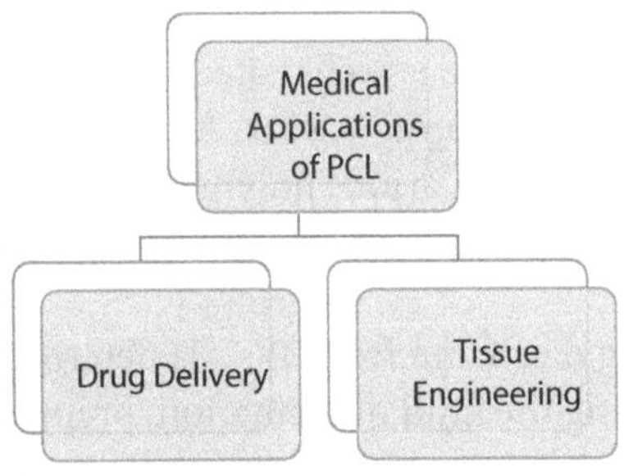

Figure 16.6 Biomedical applications of polycaprolactone.

ease with which it could be shaped and manufactured, the ability to create a porous surface which is suitable for developing tissue, and the ability to deliver drugs encapsulated in the interior of the matrix. Functional groups are added to the polymer to increase its hydrophilicity, adhesiveness, or biocompatibility, allowing for a more favorable cell response. Owing to the fact that PCL decomposes at a far sluggish pace than polyglycolide (PGA), poly D, L-lactide (PDLA), and related copolymers, it was formerly utilized in drug delivery procedures which lasted over a year and suture supplies (Maxon TM) that degraded slowly [53, 54].

Although PCL has insufficient mechanical qualities to be utilized for heavy load-bearing utilities, the medical expedient business was keen to swap metal maneuvers (such as plates, screws, nails, and so on) with eco-friendly materials to produce grafts. Moreover, the medical device and drug delivery communities mutually agreed that quicker resorbable polymers had less apparent drawbacks related to abiding deterioration (PCL has a degradation time of 3 to 4 years) and intracellular resorption routes.

Through the emergence of an innovative discipline, *viz.*, tissue engineering, PCL has made a comeback into the realm of biomaterials. PCL's resurgence in the 1990s and 2000s was fueled by the realization that it had superior rheological and viscoelastic qualities than countless resorbable-polymer rivals, making this simple to construct and operate for a wide variety of scaffolds [55, 56]. In actuality, PCL may be employed in an extensive array of scaffolding expertise, and its cost-effective production pathways, when related to other aliphatic polyesters, are quite beneficial.

16.8.1 PCL in Drug Delivery Systems

PCL possesses several advantageous attributes for precise drug delivery, including enhanced permeability for various medications, robust biocompatibility, and the ability to be completely eliminated from the body upon bio-resorption. Due to its comparatively slower biodegradation rate compared to other polymers, PCL is well suited for long-lasting drug delivery systems that can endure for up to a year. Additionally, PCL's capability to form compatible blends with various polymers allows for the manipulation of degradation kinetics, facilitating the attainment of desired drug release profiles [57].

The degree of drug discharge from PCL is determined by a selection of parameters, comprising the kind of formation, preparation processes, PCL content, proportion, and dimension of the medication contained inside the microcapsules. Since PCL has greater penetrability, it has been combined with additional polymers to improve stress resistance, fracture resistance, and medication discharge rate control. PCL has become a prominent field of study in the last several years to produce regulated drug delivery systems, mostly for proteins and peptides [58]. The release of medicines, for instance, isometamidium chloride and ethidium bromide, from shafts constructed of poly caprolactone-poly D-lactic acid (PCL–DLLA) and poly caprolactone-poly L-lactic acid (PCL–PLLA) has been examined *in vivo* and *in vitro* by Lemmouchi and his colleagues [58].

16.8.2 PCL in Tissue Engineering

Tissue engineering is a multidisciplinary area of science which combines life science and engineering principles to create biological replacements that aid in the replacement, preservation, or improvement of the utilities of an entire body part or muscles (comprising bone, cartilage, and blood vessels) [59]. Tissues engaged in the mending process of tissues require certain structural and mechanical qualities in order to operate properly. Tissue engineering also refers to the use of cells inside an artificially built support system to accomplish certain biochemical processes (including an artificial liver or pancreas) [60].

Some significant advancements in tissue engineering have resulted in a distinct set of deployment methodologies for tissue substitutes. By constructing scientific progressions in stem cells, development and distinction elements, biomaterials, and biomimetic surroundings, a unique opportunity for building tissues *in vivo* using combinations of tailored extracellular milieus or scaffolds, biologically active fragments, and cells has been created [61].

PCL has gained considerable recognition as a biomaterial in cardiovascular as well as bone tissue engineering due to its advantageous qualities, such as a reduced melting point and remarkable rheological and mechanical properties. PCL is an ideal biomaterial for scaffold fabrication. With its exceptional rheological properties, PCL serves as a multipurpose bioresorbable polymer suitable for various polymer processing techniques, enabling the creation of diverse scaffold designs.

The scaffolds are proficient in sustaining cell linkage, propagation, and differentiation *in vitro*, along with transplantation *in vivo*. There are a variety of approaches to fabricate scaffolds for tissue engineering, but it is important to pay attention to the scaffolds' requirements and to comprehend the interplay of elements that influence the material's configuration and conformation standards. The greatest desired property of any polymeric scaffold implant could be the coordination of decomposed polymer through the addition of usual tissues created by cells [62].

The scaffold's resorption and decomposition kinetics are intended to permit embedded cells to multiply quickly and discharge their specific scaffold during the dynamic and static cell implantation stages (between 1 and 12 weeks), while during the gradual resorption of the scaffold, it provides ample space for cellular multiplication and tissue development to occur. The scaffold slowly resorbs, leaving enough space for cell multiplication and tissue development. The three-dimensional matrices were utilized to keep the physical sustenance going until the modified tissues were strong enough to sustain it. Using the electrospinning technique [63], Chen *et al.* (2015) created 3D nanofibrous scaffolds made of PCL. The following characteristics are beneficial to scaffold candidates [64]: cell development and the movement of metabolic waste and nutrients necessitate three-dimensional, extremely permeable configurations with a unified pore system. Bioresorbable and biocompatible products, such as biopolymers, have a regulated rate of resorption to match cell/tissue development *in vivo* and/or *in vitro* [65]. The bond, proliferation, and differentiation of cells all require the right surface chemistry of biomaterials. The biomaterial's mechanical qualities should be compatible with the tissues at the implantation site.

16.9 Conclusion

Plastic is a global concern and a threat to all living beings. Accumulation of plastic in the environment imposes many health hazards; therefore, the need for a replacement of plastic is inevitable. Bioplastics are biologically degradable plastics that can be broken down by microorganisms, thereby lessening the global burden. Bioplastics are of various types and have diverse uses, including the construction of roads, medicine, and medical applications.

References

1. Bano, K., Pandey, R., Jamal-e-Fatima, Roohi, New advancements of bioplastics in medical applications. *Int. J. Pharm. Sci. Res.*, 9, 2, 402–416, 2018.
2. Derraik, J.G., The pollution of the marine environment by plastic debris: A review. *Mar. Pollut. Bull.*, 44, 9, 842–852, 2002.
3. Rosenboom, J.G., Langer, R., Traverso, G., Bioplastics for a circular economy. *Nat. Rev. Mater.*, 7, 2, 1–21, 2022.

4. Chuetor, S., Tantayotai, P., Cheenkachorn, K., Cheng, Y.S., Tawai, A., Sriariyanun, M., Cellulose-based blends and composites, in: *Biodegradable Polymers, Blends and Composites*, pp. 87–122, Woodhead Publishing, United Kingdom, 2022.
5. Li, S., *Scaffolding in Tissue Engineering*, P.X. Ma and J. Elisseeff (Eds.), pp. 335–352, Taylor and Francis Group, Boca Raton, FL, 2006.
6. Shrivastava, A., Dubey, S., Mishra, A., Bhattacharya, S., Mishra, S., Sustainable development of polyhydroxyalkanoates and their application as an effective tool in drug delivery and tissue engineering, in: *Handbook of Research on Nano-Drug Delivery and Tissue Engineering: Guide to Strengthening Healthcare Systems*, p. 181, 2022.
7. Leskinen, T., Salas, C., Stephen, S.K., Dimitris, S.A., Wood extractives promote cellulase activity on cellulosic substrates. *Biomacromolecules*, 16, 10, 3226–3234, 2015.
8. Jia, S., Zhao, L., Wang, X., Chen, Y., Pan, H., Han, L., Zhang, H., Poly (lactic acid) blends with excellent low-temperature toughness: A comparative study on poly (lactic acid) blends with different toughening agents. *Int. J. Biol. Macromol.*, 201, 662–675, 2022.
9. Holten, C.H., Mueller, A., Rehbinder, D., *Lactic Acid Properties and Chemistry of Lactic Acid and Derivatives*, pp. 566–567, Verlag Chemie, Germany, 1971.
10. Carothers, W.H., Dorough, G.L., Van Natta, F.J., Polymerization and ring formation reversible polymerization of six-membered cyclic esters. *J. Am. Chem. Soc.*, 54, 5, 761–72, 1932.
11. Kulkarni, R.K., Pani, K.C., Neuman, C., Leonard, F., Polylactic acid for surgical implants. *J. Healthc. Eng.*, 93, 5, 839–843, 1966.
12. Ilyas, R.A., Zuhri, M.Y.M., Aisyah, H.A., Asyraf, M.R.M., Hassan, S.A., Zainudin, E.S., Sari, N.H., Natural fiber-reinforced polylactic acid, polylactic acid blends and their composites for advanced applications. *Polymers*, 14, 1, 202, 2022.
13. Bano, K., Pandey, R., Jamal-e-Fatima, Roohi, New advancements of bioplastics in medical applications. *Int. J. Pharm. Sci. Res.*, 9, 2, 402–416, 2018.
14. Zhang, Q., Mochalin, V.N., Neitzel, I., Knoke, I.Y., Han, J., Klug, C.A., Zhou, J.G., Lelkes, P.I., Gogotsi, Y., Fluorescent PLLA-nano diamond composites for bone tissue engineering. *Biomaterials*, 32, 87–94, 2011.
15. Chang, P.C., Liu, B.Y., Liu, C.M., Chou, H.H., Ho, M.H., Liu, L.C., Wang, D.M., Hou, L.T., Bone tissue engineering with novel rhBMP2-PLLA composite scaffolds. *J. Biomed. Mater. Res. A.*, 81, 771–780, 2007.
16. Li, G., Wang, Z.X., Fu, W.J., Hong, B.F., Wang, X.X., Cao, L., Xu, F.Q., Song, Q., Cui, F.Z., Zhang, X., Introduction to biodegradable poly-lactic acid ureteral stent application for treatment of ureteral war injury. *BJU Int.*, 108, 901–906, 2011.

17. Qin, Y., Yuan, M., Li, L., Guo, S., Yuan, M., Li, W., Xue, J., Use of poly lactic acid/poly trimethylene carbonate blends membrane to prevent post-operative adhesions. *J. Biomed. Mater. Res. Part B: Appl. Biomater.*, 79, 312–319, 2006.
18. Brekke, J.H., Olson, R.A.J., Scully, J.R., Osbon, D.B., Influence of polylactic acid mesh on the incidence of localized osteitis. *Oral. Surg. Oral. Med. Oral. Pathol.*, 56, 240–245, 1983.
19. Kumari, A., Yadav, S.K., Yadav, S.C., Biodegradable polymeric nanoparticles based drug delivery systems. *Colloids Surf. Biointerfaces*, 75, 1–18, 2010.
20. Hamad, K., Kaseem, M., Yang, H.W., Deri, F., Ko, Y.G., Properties and medical applications of polylactic acid: A review. *Express Polym. Lett.*, 9, 5, 435–455, 2015.
21. Ling, Y. and Huang, Y., Preparation and release efficiency of poly (lactic-co-glycolic) acid nanoparticles for drug loaded paclitaxel. *IFMBE Proc.*, 19, 514–517, 2008.
22. Rancan, F., Papakostas, D., Hadam, S., Hackbarth, S., Delair, T., Primard, C., Verrier, B., Sterry, W., Blume-Peytavi, U., Vogt, A., Investigation of poly lactic acid (PLA) nano particles as drug delivery systems for local dermatotherapy. *Pharm. Res.*, 26, 2027–2036, 2009.
23. Esmaeili, F., Ghahremani, M.H., Ostad, S.N., Atyabi, F., Seyedabadi, M., Malekshahi, M.R., Amini, M., Dinarvand, R., Folate-receptor-targeted delivery of docetaxel nano particles prepared by PLGA–PEG–folate conjugate. *J. Drug Targeting*, 16, 415–423, 2008.
24. Haers, P.E., Suuronen, R., Lindqvist, C., Sailer, H., Biodegradable polylactide plates and screws in orthognathic surgery: Technical note. *J. Craniomaxillofac. Surg.*, 26, 87–91, 1998.
25. Böstman, O., Hirvensalo, E., Vainionpää, S., Mäkelä, A., Vihtonen, K., Törmälä, P., Rokkanen, P., Ankle fractures treated using biodegradable internal fixation. *Clin. Orthop. Relat. Res.*, 238, 195–203, 1989.
26. Danhier, F., Ansorena, E., Silva, J.M., Coco, R., Breton, A.L., Preat, V., PLGA-based nano particles: An overview of biomedical applications. *J. Control. Release*, 161, 2, 505–522, 2012.
27. Lanao, R.P.F., Jonker, A.M., Wolke, J.G.C., Jansen, J.A., Van Hest, J.C.M., Leeuwenburgh, S.C.G., Physico-chemical properties and applications of poly (lactic-co-glycolic acid) for use in bone regeneration. *Tissue Eng. Part B Rev.*, 19, 380–390, 2013.
28. Pan, Z. and Ding, J.D., Poly (lactide-co-glycolide) porous scaffolds for tissue engineering and regenerative medicine. *Interface Focus*, 2, 366–377, 2012.
29. Zhang, L.J. and Webster, T.J., Nanotechnology and nanomaterials: Promises for improved tissue regeneration. *Nano Today*, 4, 66–80, 2009.
30. Zhou, S.B., Deng, X.M., Li, X.H., Jia, W.X., Liu, L., Synthesis and characterization of biodegradable low molecular weight aliphatic polyesters and their use in proteindelivery systems. *J. Appl. Polym. Sci.*, 91, 1848–1856, 2004.

31. Kim, S.S., Park, M.S., Jeon, O., Choi, C.Y., Kim, B.S., Poly (lactide-co-glycolide)/ hydroxy apatite composite scaffolds for bone tissue engineering. *Biomaterials*, 27, 1399–1409, 2006.
32. Ebrahimian, H.M., Ashrafizadeh, F., Etemadifar, M., Venkatraman, S.S., Evaluating and modeling the mechanical properties of the prepared PLGA/ nano-BCP composite scaffolds for bone tissue engineering. *J. Mater. Sci. Technol.*, 27, 1105–1112, 2011.
33. Huang, Z.M., Zhang, Y.Z., Kotaki, M., Ramakrishna, S., A review on polymer nano-fibers by electro spinning and their applications in nano composites. *Compos. Sci. Technol.*, 63, 2223–2253, 2003.
34. Nie, H.M. and Wang, C.H., Fabrication and characterization of plga/hap scaffolds for delivery of BMP-2 plasmid composite DNA. *J. Control. Release*, 120, 111–121, 2007.
35. Morgan, S.M., Tilley, S., Perera, S., Ellis, M.J., Kanczler, J., Chaudhuri, J.B., Oreffo, R.O., Expansion of human bone marrow stromal cells on poly-(D,L-lactide-co-glycolide) (PDL LGA) hollow fibres designed for use in skeletal tissue engineering. *Biomaterials*, 28, 5332–5343, 2007.
36. Mouthuy, P.A., Ye, H., Triffitt, J., Oommen, G., Cui, Z., Physico-chemical characterization of functional electrospun scaffolds for bone and cartilage tissue engineering. *Proc. Inst. Mech. Eng. H.*, 224, 1401–1414, 2010.
37. Bergman, K., Engstr, T., Hilborn, J., Ossipov, D., Piskounova, S., Bowden, T., Injectable cell-free template for bonetissue formation. *J. Biomed. Mater. Res. A*, 91A, 1111–1118, 2009.
38. Dhillon, A., Schneider, P., Kuhn, G., Reinwald, Y., White, L.J., Levchuk, A., Rose, F.R.A.J., Muller, R., Shakesheff, K.M., Rahman, C.V., Analysis of sintered polymer scaffolds using concomitant synchrotron computed tomography and *in situ* mechanical testing. *J. Mater. Sci. Mater. Med.*, 22, 2599–2605, 2011.
39. Rahman, C.V., Ben-David, D., Dhillon, A., Kuhn, G., Gould, T.W.A., Muller, R., Rose, F.R.A.J., Shakesheff, K.M., Livne, E., Controlled release of bmp-2 from a sintered polymer scaffold enhances bone repair in a mouse calvarial defect model. *J. Tissue Eng. Regen. Med.*, 8, 59–66, 2014.
40. Eyrich, D., Brandl, F., Appel, B., Wiese, H., Maier, G., Wenzel, M., Staudenmaier, R., Goepferich, A., Blunk, T., Longterm stable fibrin gels for cartilage engineering. *Biomaterials*, 28, 55–65, 2007.
41. Habraken, W.J., Wolke, J.G., Mikos, A.G., Jansen, J.A., Injectable plga microsphere/calcium phosphate cements: Physical properties and degradation characteristics. *J. Biomater. Sci. Polym. Ed.*, 17, 1057–1074, 2006.
42. Wang, Q., Gu, Z., Jamal, S., Detamore, M.S., Berkland, C., Hybrid hydroxyl apatite nano particle colloidal gels are injectable fillers for bone tissue engineering. *Tissue Eng. Part A.*, 19, 2586–2593, 2013.
43. Ferretti, C., A prospective trial of poly-L-lactic/poly glycolic acid co-polymer plates and screws for internal fixation of mandibular fractures. *Int. J. Oral. Maxillofac. Surg.*, 37, 3, 242–248, 2008.

44. Park, S., Kim, J.H., Kim., H., Evaluation of poly (lactic-coglycolic acid) plate and screw system for bone fixation. *J. Craniofac. Surg.*, 24, 3, 1021–1025, 2013.
45. Stockmann, P., Bohm, H., Driemel, O., Muhling, J., Pistner, H., Resorbable versus titanium osteo synthesis devices in bilateral sagittal split ramus osteotomy of the mandible-the results of a two center and omised clinical study with an eight-year follow up. *J. Craniomaxillofac. Surg.*, 38, 7, 522–528, 2010.
46. Do, M.P., Neut, C., Delcourt, E., SeixasCerto, T., Siepmann, J., Siepmann, F., *In situ* forming implants for periodontitis treatment with improved adhesive properties. *Eur. J. Pharm. Biopharm.*, 88, 2, 342–350, 2014.
47. Selim, M., Bullock, A.J., Blackwood, K.A., Chapple, C.R., MacNeil, S., Developing biodegradable scaffolds for tissue engineering of the urethra. *BJU Int.*, 107, 2, 296–302, 2011.
48. Gala-Garcia, A., Carneiro, M.B.H., Silva, G.A.B., *In-vitro* and *in-vivo* evaluation of the biocompatibility of a calcium phosphate / poly (lactic-co-glycolic acid) composite. *J. Mater. Sci. Mater. Med.*, 23, 7, 1785–1796, 2012.
49. Tokiwa, Y., Buenaventurada, P.C., Charles, U.U., Seiichi, A., Biodegradability of plastics. *Int. J. Mol. Sci.*, 10, 9, 3722–3742, 2009.
50. Sivabalan, A., Harihara, R., Subramani, H., Meenarathi, B., Palanikumar, S., Anbarasan, R., Synthesis and characterization of poly (ε-caprolactone): A comparative study. *Int. J. Sci. Res. Eng. Technol.*, 1, 2278–0882, 2014.
51. Nair, L.S. and Laurencin, C.T., Biodegradable polymers as biomaterials. *Prog. Polym. Sci.*, 32, 762–798, 2007.
52. Luciani, A., Coccoli, V., Orsi, S., Ambrosio, L., Netti, P.A., PCL micro spheres based functional scaffolds by bottomup approach with pre-defined micro structural properties and release profiles. *Biomaterials*, 29, 4800–4807, 2008.
53. Pulkkinen, M., Malin, M., Bohm, J., Tarvainen, T., Wirth, T., Seppalal, J., *In vivo* implantation of 2, 2-bis (oxazoline) linked poly-epsilon-caprolactone: Proof for enzyme sensitive surface erosion and biocompatibility. *Eur. J. Pharm. Sci.*, 36, 310, 2009.
54. Lam, C.X.F., Hutmacher, D.W., Schantz, J.T., Woodruff, M.A., Teoh, S.H., Evaluation of poly caprolactone scaffold degradation for 6 months *in vitro* and *in vivo*. *J. Biomed. Mater. Res. A.*, 90, 906–919, 2008.
55. Persenaire, O., Alexandre, M., Degee, P., Dubois, P., Mechanisms and kinetics of thermal degradation of poly (epsilon-caprolactone). *Biomacromolecules*, 2, 288–294, 2001.
56. Sivalingam, G., Vijayalakshmi, S.P., Madras, G., Enzymatic and thermal degradation of poly (epsiloncaprolactone), poly (D, L-lactide), and their blends. *Ind. Eng. Chem. Res.*, 43, 7702–7709, 2004.
57. Merkli, A., Tabatabay, C., Gurny, R., Heller, J., Biodegradable polymers for the controlled release of ocular drugs. *Prog. Polym. Sci.*, 23, 563–580, 1998.
58. Freiberg, S. and Zhu, X., Polymer microspheres for controlled drug release. *Int. J. Pharm.*, 282, 1–18, 2004.

59. Sinha, V.R., Bansal, K., Kaushik, R., Kumria, R., Trehan, A., Poly-epsilon caprolactone microspheres and nanospheres: An overview. *Int. J. Pharm.*, 278, 1–23, 2004.
60. Lemmouchi, Y., Schacht, E., Kageruka, P., De Deken, R., Diarra, B., Diall, O., Biodegradable polyesters for controlled release of trypanocidal drugs: *In vitro* and *in vivo* studies. *Biomaterials*, 19, 1827–1837, 1998.
61. Dalton, P.D., Woodfield, T., Hutmacher, D.W., Snap shot: Polymer scaffolds for tissue engineering. *Biomaterials*, 30, 701–702, 2009.
62. Pitt, C.G., Gratzl, M.M., Kimmel, G.L., Surles, J., Schindler, A., Aliphatic polyesters: The degradation of poly (DL-lactide), poly (ε-caprolactone), and their copolymers *in vivo*. *Biomaterials*, 2, 4, 215–220, 1981.
63. Peter, X.M., Biomimetic materials for tissue engineering. *Adv. Drug Deliv. Rev.*, 60, 2, 184–198, 2008.
64. Chen, Y., Zeng, D., Ding, L., Li, X.-L., Liu, X.-T., Li, W.J., Wei, T., Yan, S., Jiang, H.X., Wei, L., Zheng, Q.S., Three-dimensional poly-(ε caprolactone) nano fibrous scaffolds directly promote the cardio myocyte differentiation of murine-induced pluripotent stem cells through Wnt/βcatenin signalling. *BMC Cell Biol.*, 16, 22–25, 2015.
65. Shekh, M.I., Annu, Ahmed, S., Chapter 1-Biopolymers: An overview, in: *Advanced Applications of Biobased Materials*, S. Ahmed, and Annu (Eds.), pp. 3–21, Elsevier, Netherlands, ISBN: ISBN 978-0-323-91677-6, 2023.

17

Biopolymers in Biosensors

Jincymol Kappen[1], Feba Anna John[2] and Ajith James Jose[2*]

[1]*Department of Physical Chemistry and Technology of Polymers, The Silesian University of Technology, Gliwice, Poland*
[2]*Department of Chemistry, St. Berchmans College, Changanasserry, Kerala, India*

Abstract
The advanced research in the development of economically feasible biosensors to detect biomolecules is crucial for this era. The advantages of biopolymers such as biodegradability, biocompatibility, inexpensive, and easy availability make it a promising area in the field of biosensing. The incorporation of biopolymers with other polymers and metal nanocomposites enhances the sensitivity towards the detection of an analyte. Biopolymers like chitin, chitosan, cellulose, lignin, alginate, starch, gelatin, pectin, etc., and their composites are reported for biosensing application. These biopolymers are used for both optical and electrochemical sensing applications of glucose, urea, uric acids, DNA, H_2O_2, etc. The immobilization of corresponding enzymes like glucose oxidases triggers the sensing mechanism. The strong cross-linking between such enzyme and biopolymer enhances the sensitivity, selectivity, and stability of the corresponding sensor. The development of metal nanocomposites with such polymers helps non-enzymatic detection of such analytes. Moreover, the introduction of biopolymers can assist the evolution of single use solid state sensors.

***Keywords*:** Biosensors, biopolymers, cellulose, chitosan, alginate, carrageenan, dextran, gelatin

17.1 What are Biosensors?

A biosensor is an analytical device which is used for the detection of a chemical substance related to the biological system. The chemical substance called analytes includes proteins, DNA, enzymes, antibodies, and

**Corresponding author*: ajithjamesjose@gmail.com

Annu (ed.) Biopolymers for Biomedical Applications, (439–458) © 2024 Scrivener Publishing LLC

biomolecules involved in the biological process, like glucose, uric acid, xanthane, ascorbic acid, cholesterol, etc. The sensing can follow an optical or electrochemical method based on the sensor and analyte [1]. Optical sensing includes both colorimetric and according to the color change. Organic molecules like methylene blue and metal nanoparticles have been extensively exploited for colorimetric sensing. Because of the surface plasmon resonance (SPR) properties of silver and gold nanoparticles, they are widely used for the optical sensing of analytes [2, 3]. The incorporation of polymers with such nanoparticles can open up the way to solid-state sensor.

17.2 Biopolymers

Biopolymers are polymers which exist in natural sources itself or are synthesized chemically from natural sources. It can be extracted from plants, bacteria, fungi, and animals. They are classified as polysaccharides, polyesters, hydrocarbons, polyamides, etc. The advantages of biopolymers include easy availability, biodegradability, biocompatibility, and inexpensive [4]. Chitin, chitosan, cellulose, starch, alginate, carrageenan, dextran, gelatin, gum, lignin, pectin, polycaprolactones, pullulan, and glycosaminoglycans are examples of biopolymers [69–73]. These biopolymers have been widely exploited in various domains in biomedical sectors, viz., drug delivery, encapsulation, tissue engineering, biosensing, and scaffolds in ligaments [4, 5]. In the present chapter, the biosensing applications of various biopolymers are discussed in detail.

17.2.1 Cellulose-Based Biosensors

The most abundant and widely spread biopolymer is cellulose, which is the chief constituent of the cell walls of plants. It can be easily modified for the fabrication of disposable sensors with paper. Its biodegradability, biocompatibility, and large abundance make it more suitable for sensor applications. Paper-based electrochemical sensors are widely developed in recent years as disposable sensors [6]. The quality of cellulose sensors can be enhanced through the functionalization and introduction of metal nanocomposites on it [7]. Metal-based composites help to upgrade the conductivity and firmness of the sensor. Mahadeva *et al.* have presented the development of cellulose–SnO_2 hybrid nanocomposite (immobilized with glucose oxidase, GOx) biosensor for glucose oxidation [8]. In 2014, S.N. Tan and his coworkers developed a cellulose paper-based glucose sensor.

They have used grade 1 filter paper immobilized with enzyme glucose oxidase. Then, it was placed on a screen-printed electrode as working electrode and used for electrochemical experiments. The authors claimed that the present green and low-cost electrode maintained its stability throughout the experiments [9]. Cellulose nanocrystals (CNCs) functionalized with poly-vinyl alcohols (PVA) were studied for the fluorescence sensing of biomolecules. A large surface area and a higher number of reactive groups on the nanocomposites were utilized for the intercalation of the analytes. The CNCs–PVA composite was coated on a glass substrate and used to detect the protease activity. Furthermore, the same substrate was also outlined to find trypsin by fluorimetry [10]. A magnetic nanoparticle-embedded cellulose film was developed to monitor the level of glucose directly from saliva, sweat, and non-bloody body fluids. The (N-CNC)–Fe_3O_4 as well as (S-CNC)–Fe_3O_4 hybrid systems possess a peroxidase-like activity and shows a color change with the glucose as well as 2,2′-azino-bis (3-ethylbenzothiazoline-6-sulphonic acid). The above-mentioned magnetic composite was used to sense the glucose concentration as low as 5 mM [11]. Since cellulose containing disposable sensors are cheap and biodegradable, Brumer and his co-workers introduced an enzyme-based chemo-sensor for esterase detection [12]. High-purity cellulose membrane-based strips were introduced by Luo *et al.* for the enzyme-based colorimetric detection of a low concentration of glucose [13]. After the oxidation of hydroxyl groups on cellulose by sodium perodate, glucose oxidase was immobilized on it *via* chemical cross-linking. During the analysis of glucose, a cellulose strip is immersed in glucose or glucose is dropped onto the strip. The surface characterization of the strip shows the porous network-like structure of cellulose which helps to improve the interaction with the analyte. After the oxidation of cellulose, the pore diameter was found to be reduced, and the immobilization of glucose oxidase enzyme reduced the porous nature, and network-like structures arose. A new fiber-optic fluorescence glucose biosensor using fluorescent carbon quantum dots and glucose oxidase enzymes was introduced by Yu *et al.* An exceptionally selective and sensitive fluorescence sensor for glucose was developed by immobilizing through dip coating in a cellulose acetate film [14]. Xiangling and his coworkers presented a new way to prepare an amperometric glucose biosensor with least detection limit, reproducibility, good sensitivity, and remarkable storage stability. The immobilization has been achieved *via* combining the glucose oxidase with gold nano-rods (AuNRs) and subsequently cross-linking with a cellulose acetate medium [15]. A potentiometric biodegradable sensor using cholesterol oxidase enzymes has been developed by Pandey *et al.* to detect cholesterol. The team immobilized the cholesterol oxidase enzymes with SWNTs loaded on a cellulose acetate membrane. The prepared electrode exhibits good responses towards cholesterol oxidation with exceptional sensitivity, having a detection limit of 10^{-8} M [16]. Gilmartin *et al.* presented a uric acid sensor by making a composite of cobalt phthalocyanine and cellulose acetate loaded with uricase enzyme [17]. Therefore, cellulose is a promising material in the field of disposable biosensors, whether it is optical or electrochemical. Its low cost, biodegradability, availability, and easy functionalization make it more attractive in disposable sensors [7, 18].

17.2.2 Chitin/Chitosan-Based Biosensors

Chitosan (CS) is a well-known naturally occurring linear polysaccharide and is a deacetylated product of chitin. Both chitin and chitosan and its various derivatives such as beads, films, fibers, gels, spheres, suspensions, and microscopic threads are studied in various applications [19]. Chitosan, an important natural polymer, has physico-chemical advantages such as hydrophilicity, fair mechanical strength, processability, biocompatibility, environmental stability, and ability of chemical alteration. Chitosan and its polymer and metal nanocomposites are developed as immunosensor, enzyme sensor, and biomolecule sensor [20]. Shrestha *et al.* followed a one-step electrochemical polymerization technique to synthesize a polypyrrole/Nafion/functionalized multiwalled carbon nanotube (MWCNT) immobilized with CS/glucose oxidase. The obtained biodegradable-nanohybrid thin sheets were used to detect glucose in real serum samples [21]. The resultant sensor was characterized by electrochemical techniques, and it was sensitive towards the detection of glucose. An enzymatic glucose sensor was established *via* developing a composite of chitosan and nitrogen-doped graphene. The obtained electrode exhibits a sensitivity of 9.7 $\mu A\ mM^{-1}\ cm^{-2}$, within a broad linear range of 12 mM, and a fair detection limit of 9.5 μM. Even after a month the sensor was found to be stable [22]. A film of the nanocomposite iron oxide nanoparticles/CS-grafted polyaniline has been synthesized by Deviand Pundir in order to detect uric acid [23]. An amperometric hydrogen peroxide sensor was reported by using a chitosan-based sol–gel matrix [24]. In 2009, M.E. Ghica *et al.* delineate about an amperometric biosensor for glucose grounded on enzyme-immobilized chitosan film containing carbon nanotubes (CNT) [25]. For the application of chitosan-modified electrodes in biosensing, the impedance study of such electrodes was also analyzed [26]. Therefore, chitosan and its composites are promising materials towards the sensing of biomolecules, enzyme, and DNA [19].

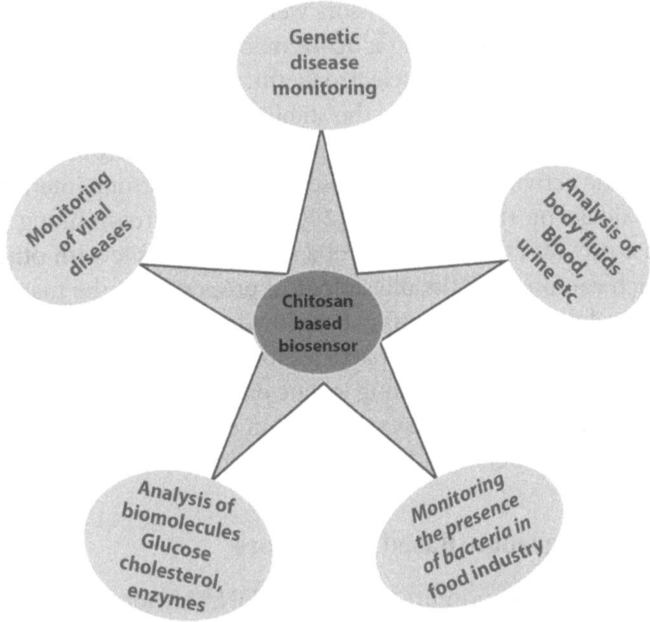

Figure 17.1 Schematic representation of the biological applications of chitosan-based biosensors.

The applications of chitosan-based biosensors in various biological fields are illustrated in Figure 17.1.

17.2.3 Alginate-Based Biosensors

Alginate is a biopolymer observed in the cell wall of brown algae. Alginates can be converted into viscous gum during hydration. S. Cosnier and his co-workers outlined the effectiveness of the biosensors grounded on the entrapment of cells of *Chlorella vulgaris* into a regular alginate gel as well as a freshly prepared pyrrole–alginate matrix. They have conducted a comparative study upon the amperometric current measurements to p-nitrophenyl phosphate when used as substrate for the detection of an algal alkaline phosphatase activity [27]. An efficient, new, and practical colorimetric test strip based on alginate microcapsules has been invented for the recognition of urea in serum. Here tricyanofuran hydrazone molecules and urease enzyme-coated calcium alginate microcapsules were synthesized and applied on cotton fibers successfully. The limit of detection attained for urea was ~250 ppm at ambient conditions [28]. A color change of the present sensor on cotton after the addition of various concentrations was observed.

The amperometric glucose biosensors were designed and reported by R.E. Ionescu *et al.* by incorporating glucose oxidase into polypyrrole–alginate hydrogel coatings. Here the electrochemical polymerization of pyrrole on the alginate matrix enhances the stability of alginate and improves the enzyme retention on the matrix. The polypyrrole–alginate electrode shows good catalytic activity towards glucose through the strong intercalation of glucose oxidase on the gel matrix [29]. The rapid cross-linking property of alginate provides the formation of a stable composite with other materials, which can improve the physiological properties of the material [30]. Ismael *et al.* developed an alginate-based hydrogel with poly(3,4-ethylenedioxythiophene) and gold nanoparticles for the electrochemical detection of L-lactate. The sensor containing lactate oxidase enzyme converts lactate to hydrogen peroxide, and the concentration of lactate is assigned from the electrochemical reduction of hydrogen peroxide [31].

17.2.4 Carrageenan-Based Biosensors

A highly sensitive (at optimum conditions) DNA sensor rooted on the biocomposite of kappa-carrageenan polypyrrole–gold nanoparticles has been prepared, characterized, and its employment demonstrated for gender classification of Arowana fish. The DNA was reworked by introducing a thiol group, and then it was covalently attached to the Au surface [32]. Recently, an enzymatic glucose sensor grounded on chitosan/kappa-carrageenan (C/KC) doped with gold nanoparticles was reported by I. Rasas *et al.* The as-prepared electrode shows good reproducibility, selectivity, and sensitivity, along with a detection limit 5 μM [33]. With the incorporation of urease, Nile blue chromoionophore-doped KC was employed as an optical urea sensor. Here the urease enzyme catalyzed the hydrolytic decomposition of urea into ammonium hydroxide. On changing the pH to alkaline, the color of the sensor membrane was changed from blue to violet. Then, it was quantified by a fiber optic reflectance spectrophotometer. The limit of detection was found to be 0.0001 mM, and it required 10 min as response time [34]. In order to identify the low-level urea, a non-enzymatic electrochemical biosensor routed on the biocomposite of KC, PVA, and nano-eggshell was synthesized *via* ultrasonic sonochemistry. They achieved the limit of detection of 60 μM at room temperature [35]. κ-carrageenan functionalized with gadolinium tin oxide nanoparticles is recently reported for the electrochemical detection of the drug dimetridazole with a lower detection limit of 3.1 nM [36]. Carrageenan composites were also used for the detection of SARS-CoV-2 and pesticides [37, 38].

17.2.5 Dextran-Based Biosensors

Dextran, a polysaccharide derived from the condensation of glucose, has been reported for biosensing applications by optical and electrochemical methods. A photopolymerized poly(ethylene glycol) hydrogel incorporating fluorescein isothiocyanate dextran-based fluorescence sensor for glucose was reported by Michael V. Pishko and his co-workers in 1999. Glucose enhanced the fluorescence of the hydrogel by fluorescence resonance energy transfer [39]. The synthesis of a very sensitive biosensor for prostate-specific antigen was demonstrated in 2009. The reported sensor was based on surface plasmon-enhanced fluorescence by dextran-based hydrogel at a detection limit of low femtomolar range [40]. Amino dextran-protected gold and silver nanoparticles were studied by Y. Ma *et al.* for the optical sensing of lectin protein concanavalin A (Con A). The introduction of Con A into amino dextran–AuNPs leads to the aggregation of AuNPs because of the formation of a Au–Con A complex. This leads to the attenuation of the SPR band of AuNPs at 525 nm. This was dissociated by the disaccharide isomaltose which breaks down Au–Con A complex thereby, which, in turn, regenerates the optical properties of AuNPs [41]. Both functionalized dextran and its various composites were reported for the electrochemical determination of hydrogen peroxide and Con A [42, 43] and colorimetric detection [44].

17.2.6 Gelatin-Based Biosensors

Studies on the biosensing applications of biopolymers gelatin, gum, lignin, pectin, polycaprolactones, pullulan, and starch are scarce in literature. Some of the reports are discussed here. Glucose oxidase-immobilized gelatin was reported for the sensing of glucose, and the limit of detection of 0.25 mM was recorded with stability of the electrode even after 2 months [45]. An aptasensor based on a collagen–gelatin–gold nanoparticle composite was developed, with a detection limit of 1.95 nM [46]. A gelatin-based wearable sensor was developed by Li *et al.* to monitor the physical and physiochemical signals from the human body as illustrated in Figure 17.2 [47]. Gelatin-based composites are also reported for colorimetric and fluorimetric biosensing applications [48, 49].

17.2.7 Gum-Based Biosensors

The preparation of functionalized gums and their composites with various metal nanoparticles is exploited for the electrochemical and optical sensing

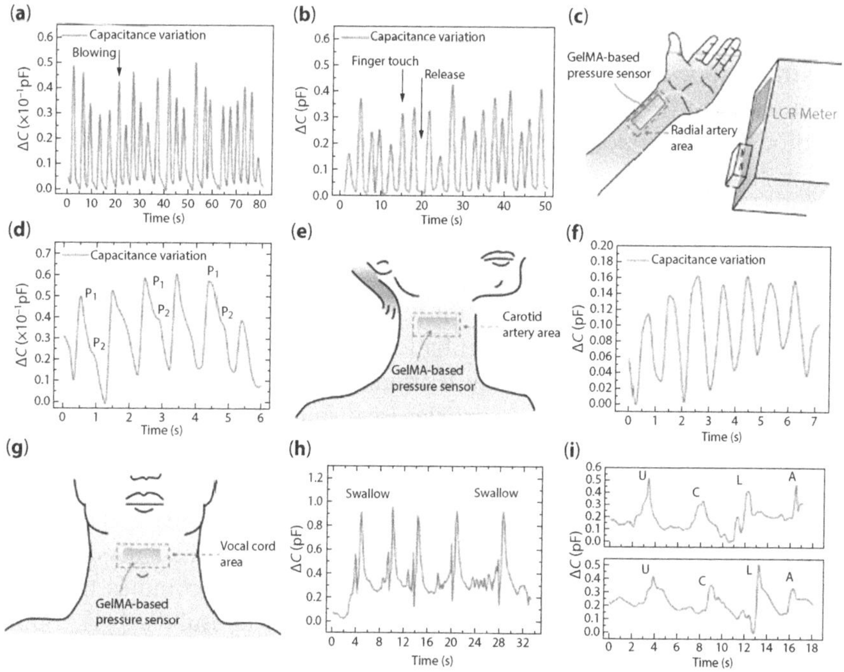

Figure 17.2 Applications of gelatin-based pressure sensors to detect physical or physiological signals from the human body. (a) Monitoring of air blowing from a nitrogen gun. (b) Sensing of finger touch force. (c) Human wrist pulse monitoring. (d) Results from wrist pulse detection technique. (e) Schematic representation of carotid artery pulse measurement. (f) Capacitance change according to carotid artery pulse. (g) Schematic diagram for vocal cord vibration detection. (h) Capacitance change during swallowing. (i) Capacitance variation while speaking the letters "U", "C", "L", and "A" [47].

applications of pollutants and biomolecules [50, 51]. S. Quasemy *et al.* used CdTe quantum dots-immobilized gum tragacanth nanoparticles along with the incorporation of glucose oxidase. The prepared biosensor was highly fluorescent, and quenching of fluorescence was observed as a result of enzyme-catalyzed oxidation of glucose, which results in the formation of H_2O_2. The emission quenching was in accordance with the glucose oxidation. An excellent detection limit of 1.5 mM was noticed with reproducibility and accuracy even in the presence of blood serum samples without pretreatment [52].

17.2.8 Lignin-Based Biosensors

Lignin, the biopolymer next to cellulose and its composites, has already been explored for various applications. The hydrogel of lignin was synthesized through various polymerization methods like cross-linking copolymerization and utilized in biosensing applications. The mechanical and porous structure of lignin hydrogel improves its applications in various fields, including biosensing. The hydrogel can be used as a solid-state sensor. The composites of lignin with graphene and carbon nanotubes are also reported as a good catalyst with an improved conductivity [53]. An enzyme biosensor was invented by Jedrzak *et al.* by developing silica–lignin composite and immobilizing glucose oxidase. The sensitivity of the electrode towards glucose was obtained to be 0.78 μA mM^{-1}. The detection limit was observed to be 145 μM from the linear range of 0.5–9 mM [54]. A lignin–silver nanoparticle–ionic liquid composite is reported for the non-enzymatic colorimetric detection of hydrogen peroxide [55].

17.2.9 Pectin-Based Biosensors

Pectin is a polysaccharide-based polymer consisting mainly of galacturonic acid and rhamnose. A chitosan–pectin composite-coated glassy carbon electrode is reported for the simultaneous electrochemical determination of an antibiotic drug metronidazole and herbicide metribuzin. Under optimized conditions, the detection of both analytes in the nanomolar range was obtained. The composite electrode was stable, selective, and reproducible towards the targeted analytes. The real sample analysis was also carried out in diluted blood serum and soil samples [56]. The simultaneous determination of hydrogen peroxide and glucose was attained by Mani *et al.* using graphene/pectin–copper nanoparticles (CuNPs). The CuNPs were electrodeposited on graphene/pectin, where pectin acted as a scaffold and graphene as support. The obtained electrode was utilized as an amperometric sensor to detect glucose and hydrogen peroxide with a detection limit of 2.1 and 0.35 μM, respectively [57]. Pectin-coated polyaniline nanoparticles are also available in the literature for the detection of glucose enzymatically [58]. Pectin-polymer/metal nanoparticle composites are studied for the electrochemical and colorimetric detection of biomolecules, pesticides, and pathogens [59, 60].

17.2.10 Polycaprolactone-Based Biosensors

The biocompatibility and flexibility of polycaprolactone contribute to its being used in various industrial applications including biosensing. An excellent amperometric sensor for hydrogen peroxide was developed by Zheng and his coworkers by immobilizing hemoglobin in polycaproloctone film. They have achieved a detection limit of 6.07×10^{-6} M. Furthermore, the electrode was used to determine the presence of hydrogen peroxide in eye drops [61]. Copper oxide nanoparticle-modified polycaprolactone@ polypyrrole fibers on indium–tin oxide were studied for the non-enzymatic detection of glucose. The obtained unique selectivity, reproducibility, stability, and repeatability are associated with the synergistic effect between CuO nanoparticles and polymer composite [62].

Gold-covered polycaprolactone is recently used for the electrochemical detection of superoxide in cell culture media, with the highest value of sensitivity of 40.1 μA mM^{-1} cm^{-2} [63]. Uricase-immobilized polycaprolactone–polyethylene imine fibers are utilized to monitor the concentration of urea in blood serum samples [64]. Gürel-Gökmen and coworkers developed polycaprolactone-silk fibroin nanofiber by electrospinning method to fabricate the lateral flow strip to monitor the presence of bisphenol A in breast milk samples as depicted in Figure 17.3 [65].

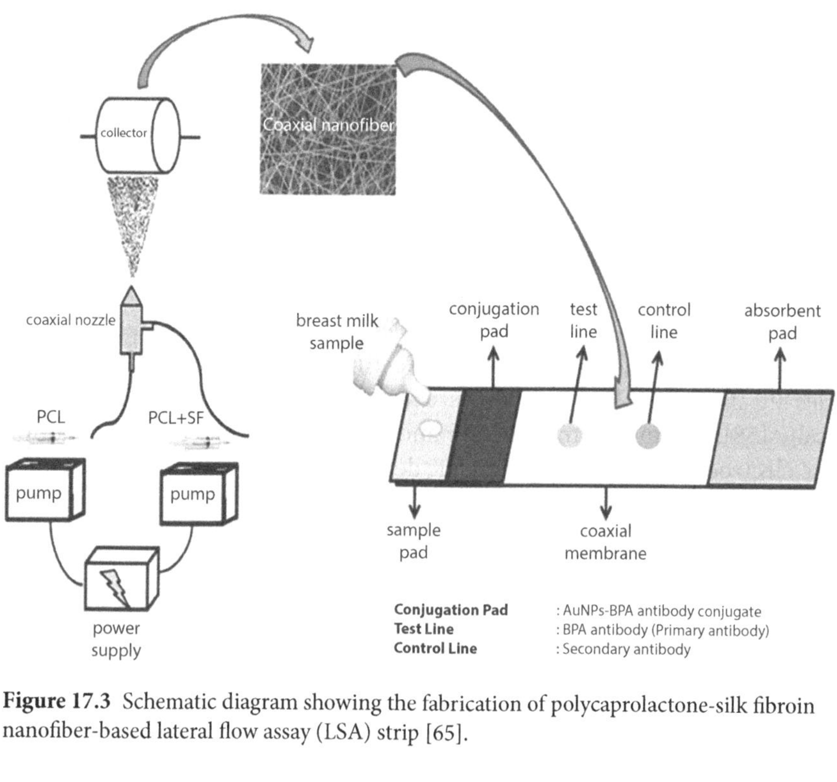

Figure 17.3 Schematic diagram showing the fabrication of polycaprolactone-silk fibroin nanofiber-based lateral flow assay (LSA) strip [65].

17.2.11 Pullulan-Based Biosensors

Pullulan is a polysaccharide-based polymer mainly consisting of maltotriose unit. A pullulan acetate/polyethylene glycol/heparin membrane was reported by Sung *et al.* for the amperometric detection of glucose. The resultant electrode was sensitive towards glucose even with possible interferences, viz., uric acid and ascorbic acid [66]. A paper-based colorimetric sensor for lactate dehydrogenase was reported by Kannan *et al.* in 2015. Here pullulan ink was printed onto paper surfaces and serum containing lactate dehydrogenase directly spotted on the paper to record the colorimetric data [67].

17.2.12 Starch-Based Biosensors

Starch, one of the most abundant, biodegradable, and low-cost biopolymers, has numerous glucose units. Gold nanoparticle (AuNP)-immobilized starch hydrogel was reported as a sensor for DNA by using colorimetry. The detection limit of 25 ng/mL was obtained from instrumental results, whereas it was 75 ng/mL from naked eye detection [68]. A summarized data regarding biopolymers in biosensing applications is tabulated in Table 17.1.

Table 17.1 Biosensing applications of biopolymers.

Biosensor	Method	Analyte	Ref.
Cellulose–SnO_2	Cyclic voltammetry	Glucose	[8]
Cellulose acetate film	Fluorescence	Glucose	[14]
Cellulose acetate-SWCNT	Potentiometry	Cholesterol	[16]
Chitosan	Amperometry	Hydrogen peroxide	[24]
Chitosan/CNT	Amperometry	Glucose	[25]
Fe_3O_4-NPs/chitosan/polyaniline	Cyclic voltammetry	Uric acid	[23]
Alginate	Colorimetry	Urea	[28]
polypyrrole-alginate hydrogel	Amperometry	Glucose	[29]

(*Continued*)

Table 17.1 Biosensing applications of biopolymers. (*Continued*)

Biosensor	Method	Analyte	Ref.
Chitosan/kappa-carrageenan/AuNPs	Cyclic voltammetry	Glucose	[33]
Kappa-carrageenan	Colorimetry	Urea	[35]
Polyethylene glycol/dextran	Fluorescence	Glucose	[39]
Dextran	Fluorescence	Prostate antigen	[40]
Dextran/Au and AgNPs	Colorimetry	Concanavalin A	[41]
Gelatin	Cyclic voltammetry	Glucose	[45]
Collagen–gelatin–AuNPs	Aptasensor		[46]
Gum/CdTe quantum dots	Fluorescence	Glucose	[52]
Silica/lignin	Cyclic voltammetry	Glucose	[54]
Graphene/pectin-CuNPs	Amperometry	Glucose and hydrogen peroxide	[57]
Polycaproloctone/hemoglobin	Amperometry	Hydrogen peroxide	[61]
Pullulan	Colorimetry	Lactate dehydrogenase	[67]
Starch/AuNPs	Colorimetry	DNA	[68]

17.3 Conclusion

The present chapter summarizes the biosensing applications of biopolymers. Biocompatibility, biodegradability, mechanical strength, easy availability, and inexpensive sources make biopolymers promising materials in the field of biosensing. Since the immobilization of enzyme leads to a strong binding with biopolymers, it can be used as a sensitive sensor for the enzymatic detection of biomolecules. In addition, the reactivity and differentiating capacity of the biosensors can be enhanced by the incorporation of other materials like polymer composites and metal nanocomposites with the biopolymers. The development of solid-state sensors like film and paper-based sensors is the major advantage of using biopolymers. This opens the way to disposable sensors with a visual detection of the analytes' interaction either by color change or by electrochemical sensing. Therefore, the fabrication of such devices definitely will help to make the diagnosis easy and with affordable prices.

References

1. Mehrotra, P., Biosensors and their applications-A review. *J. Oral. Biol. Craniofac. Res.*, 6, 153, 2016.
2. Aldewachi, H., Chalati, T., Woodroofe, M.N., Bricklebank, N., Sharrack, B., Gardiner, P., Gold nanoparticle-based colorimetric biosensors. *Nanoscale*, 10, 18, 2018.
3. Sabela, M., Balme, S., Bechelany, M., Janot, J.-M., Bisetty, K., A review of gold and silver nanoparticle-based colorimetric sensing assays. *Adv. Eng. Mater.*, 19, 1700270, 2017.
4. Pattanashetti, N.A., Heggannavar, G.B., Kariduraganavar, M.Y., Smart biopolymers and their biomedical applications. *Proc. Manuf.*, 12, 263, 2017.
5. George, A., Sanjay, M.R., Srisuk, R., Parameswaranpillai, J., Siengchin, S., A comprehensive review on chemical properties and applications of biopolymers and their composites. *Int. J. Biol. Macromol.*, 154, 329, 2020.
6. Mahato, K., Purohit, B., Kumar, A., Chandra, P., Chapter six-Paper-based biosensors for clinical and biomedical applications: Emerging engineering concepts and challenges, in: *Comprehensive Analytical Chemistry*, vol. 89, p. 163, 2020.
7. Kim, J.-H., Mun, S., Ko, H.-U., Yun, J.-Y., Kim, J., Disposable chemical sensors and biosensors made on cellulose paper. *Nanotechnology*, 25, 092001, 2014.
8. Mahadeva, S.K. and Kim, J., Conductometric glucose biosensor made with cellulose and tin oxide hybrid nanocomposite. *Sens. Actuators B: Chem.*, 157, 177, 2011.
9. Lawrencea, C.S.K., Tana, S.N., Floresca, C.Z., A "green" cellulose paper based glucose amperometric biosensor, Sens. *Actuators B: Chem.*, 193, 536, 2014.
10. Schyrr, B., Pasche, S., Voirin, G., Weder, C., Simon, Y.C., Foster, E.J., Biosensors based on porous cellulose nanocrystal–poly(vinylalcohol) scaffolds. *ACS Appl. Mater. Interfaces*, 6, 12674, 2014.
11. Tracey, C.T., Torlopov, M.A., Martakov, I.S., Vdovichenko, E.A., Zhukov, M., Krivoshapkin, P.V., Mikhaylov, V., II, Krivoshapkina, E.F., Hybrid cellulose nanocrystal/magnetite glucose biosensors. *Carbohydr. Polym*, 247, 116704, 2020.

12. Derikvand, F., Yin, D.T., Barrett, R., Brumer, H., Cellulose-based biosensors for esterase detection. *Anal. Chem.*, 88, 2989, 2016.
13. Luo, X., Xia, J., Jiang, X., Yang, M., Liu, S., Cellulose-based strips designed based on a sensitive enzyme colorimetric assay for the low concentration of glucose detection. *Anal. Chem.*, 91, 15461, 2019.
14. Yu, S., Ding, L., Lin, H., Wu, W., Huang, J., A novel optical fiber glucose biosensor based on carbon quantum dots-glucose oxidase/cellulose acetate complex sensitive film. *Biosens. Bioelectron.*, 146, 111760, 2019.
15. Ren, X., Chen, D., Meng, X., Tang, F., Du, A., Zhang, L., Amperometric glucose biosensor based on a gold nanorods/cellulose acetate composite film as immobilization matrix. *Colloids Surf. B Biointerfaces*, 72, 188, 2009.
16. Pandey, A., Pandey, P., Pandey, O.P., Shukla, N.K., Fabrication of potentiometric cholesterol biosensor by crosslinking of cholesterol oxidase and carbon nanotubes modified cellulose acetate membrane. *Sens. Lett.*, 14, 102, 2016.
17. Gilmartin, M.A.T. and Hart, J.P., Novel, reagentless, amperometric biosensor for uric acid based on a chemically modified screen-printed carbon electrode coated with cellulose acetate and uricase. *Analyst*, 119, 833, 1994.
18. Kamel, S. and Khattab, T.A., Recent advances in cellulose-based biosensors for medical diagnosis. *Biosensors*, 10, 67, 2020.
19. Suginta, W., Khunkaewla, P., Schulte, A., Electrochemical biosensor applications of polysaccharides chitin and chitosan. *Chem. Rev.*, 113, 5458, 2013.
20. Jiang, Y. and Wu, J., Recent development in chitosan nanocomposites for surface-based biosensor applications. *Electrophoresis*, 40, 2084, 2019.
21. Shrestha, B.K., Ahmad, R., Mousa, H.M., Kim, I.-G., Kim, J.I., Neupane, M.P., Park, C.H., Kim, C.S., High-performance glucose biosensor based on chitosan–glucose oxidase immobilized polypyrrole/Nafion/functionalized multi-walled carbon nanotubes bio-nanohybrid film. *J. Colloid Interface Sci.*, 482, 39, 2016.
22. Zou, R., Shan, S., Huang, L., Chen, Z., Lawson, T., Lin, M., Yan, L., Liu, Y., High-performance intraocular biosensors from chitosan-functionalized nitrogen-containing graphene for the detection of glucose. *ACS Biomater. Sci. Eng.*, 6, 673, 2020.
23. Devi, R. and Pundir, C.S., Construction and application of an amperometric uric acid biosensor based on covalent immobilization of uricase on iron oxide nanoparticles/chitosan-*g*-polyaniline composite film electrodeposited on Pt electrode. *Sens. Actuators B Chem.*, 193, 608, 2014.
24. Wang, G., Xu, J.-J., Chen, H.-Y., Lu, Z.-H., Amperometric hydrogen peroxide biosensor with sol–gel/chitosan network-like film as immobilization matrix. *Biosens. Bioelectron.*, 18, 335, 2003.
25. Ghica, M.E., Pauliukaite, R., Fatibello-Filho, O., Brett, C.M.A., Application of functionalised carbon nanotubes immobilised into chitosan films in amperometric enzyme biosensors. *Sens. Actuators B*, 142, 308, 2009.

26. Pauliukaite, R., Ghica, M.E., Fatibello-Filho, O., Brett, C.M.A., Electrochemical impedance studies of chitosan-modified electrodes for application in electrochemical sensors and biosensors. *Electrochim. Acta*, 55, 6239, 2010.
27. Ionescu, R.E., Abu-Rabeah, K., Cosnier, S., Durrieu, C., Chovelon, J.-M., Marks, R.S., Amperometric Algal Chlorella vulgaris Cell Biosensors Based on Alginate and Polypyrrole-Alginate Gels. *Electroanalysis*, 18, 1041, 2006.
28. Khattab, T.A., Fouda, M.M.G., Abdelrahman, M.S., Othman, S.I., Bin-Jumah, M., Alqaraawi, M.A., Fassam, H.A., Allam, A.A., Co-encapsulation of enzyme and tricyanofuran hydrazone into alginate microcapsules incorporated onto cotton fabric as a biosensor for colorimetric recognition of urea. *React. Funct. Polym.*, 142, 199, 2019.
29. Ionescu, R.E., Abu-Rabeah, K., Cosnier, S., Marks, R.S., Improved enzyme retention from an electropolymerized polypyrrole-alginate matrix in the development of biosensors. *Electrochem. Commun.*, 7, 1277, 2005.
30. Abraham, A.P., Victor, M., Kathelina, K., Robert, S.M., Alginate-based applications in biotechnology with a special mention to biosensors, in: *Alginate - Applications and Future Perspectives*, 2023.
31. Ismael, B., Anna, P.-J., Joan, J.R., Maria-Pau, G., Jose, G.T., Carlos, A., Hybrid conducting alginate-based hydrogel for hydrogen peroxide detection from enzymatic oxidation of lactate. *Int. J. Biol. Macromol.*, 193, 1237–1248, 2021.
32. Esmaeili, C., Heng, L.Y., Chiang, C.P., Rahman, Z.A., Safitri, E., Priya, R.S., Marugan, M., A DNA biosensor based on kappa-carrageenan-polypyrrole-gold nanoparticles composite for gender determination of arowana fish (*scleropages formosus*). *Sens. Actuators B: Chem.*, 242, 616, 2017.
33. Rassas, I., Braiek, M., Bonhomme, A., Bessueille, F., Raffin, G., Majdoub, H., Jaffrezic-Renault, N., Highly sensitive voltammetric glucose biosensor based on glucose oxidase encapsulated in a chitosan/kappa carrageenan/gold nanoparticle bionanocomposite. *Sensors*, 19, 154, 2019.
34. Esmaeili, C., Heng, L.Y., Ling, T.L., Nile Blue chromoionophore-doped kappa-carrageenan for a novelreflectometric urea biosensor. *Sens. Actuators B: Chem.*, 221, 969, 2015.
35. Taşaltın, N., Aydın, E., Karakuş, S., Kilislioğlu, A., Kcarrageenan/PVA/nanoeggshell biocompositebased nonenzymatic electrochemical biosensor for lowlevel urea detection. *Appl. Phys. A.*, 126, 827, 2020.
36. Subash, V.S., Ravichandran, J., Lincy, V., Adhimoorthy, P., Hong, P., Electrochemical detection of the antiprotozoal drug dimetridazole using κ-carrageenan functionalized gadolinium tin oxide nanoparticles. *J. Environ. Chem. Eng.*, 10, 107970, 2022.
37. Adilah, M.N., Ahmad, T.J., Farah, B.A., Immobilization of alpha-naphthyl acetate esterase (ANAE) on k-carrageenan for potential biosensor application in pesticide detection. *Mater. Today*, 57, 1296, 2022.
38. Patrícia, D.Z., Milena, M.C., Victor, B.S., Vitor, P.K., Thobias, T.S., Danielle, T.B., Tais, P., Amanda, F.L., Janice, L.G., Neftali, L.V., Andrés, M., Maria, S.A.-E., Claudio Martin, P., Rodrigo, A.V., Carrageenan from Gigartina

skottsbergii: A novel molecular probe to detect SARS-CoV-2. *Biosensors*, 13, 378, 2023.

39. Russell, R.J., Pishko, M.V., Gefrides, C.C., McShane, M.J., Cote, G.L., A fluorescence-based glucose biosensor using concanavalin a and dextran encapsulated in a poly (ethylene glycol) hydrogel. *Anal. Chem.*, 71, 3126, 1999.
40. Wang, Y., Brunsen, A., Jonas, U., Dostalek, J., Knoll, W., Prostate specific antigen biosensor based on long range surface plasmon-enhanced fluorescence spectroscopy and dextran hydrogel binding matrix. *Anal. Chem.*, 81, 9625, 2009.
41. Ma, Y., Li, N., Yang, C., Yang, X., One-step synthesis of amino-dextran-protected gold and silver nanoparticles and its application in biosensors. *Anal. Bioanal. Chem.*, 382, 1044, 2005.
42. Ou, X., Tan, X., Liu, X., Lu, Q., Chen, S., Wei, S., A signal-on electrochemiluminescence biosensor for detecting Con A using phenoxy dextran-graphite-like carbon nitride as signal probe. *Biosens. Bioelectron.*, 70, 89, 2015.
43. Zhang, H.-L., Lai, G.-S., Han, D.-Y., Yu, A.-M., An amperometric hydrogen peroxide biosensor based on immobilization of horseradish peroxidase on an electrode modified with magnetic dextran microspheres. *Anal. Bioanal. Chem.*, 390, 971, 2008.
44. Chang-Yue, C., Chien-Hsing, C., Chin-Wei, W., Fiber optic localized surface plasmon resonance sensor based on carboxymethylated dextran modified gold nanoparticles surface for high mobility group box 1 (HMGB1) analysis. *Biosensors*, 13, 522, 2023.
45. Sungur, S., Emregul, E., Günendi, G., Numanoğlu, Y., New glucose biosensor based on glucose oxidase-immobilized gelatin film coated electrodes. *J. Biomater. Appl.*, 18, 265, 2004.
46. Yilmaz, M.S., Derkuş, B., Emregül, E., Investigation of the use of collagen-gelatin-gold nanoparticle nanocomposite system as an aptasensor matrix. *J. Biol. Chem.*, 46, 523, 2018.
47. Zhikang, L., Shiming, Z., Yihang, C., Haonan, L., Libo, Z., Guoxi, L., Xiaochen, W., Martin, C.H., Hao, L., Yumeng, X., Reihaneh, H., KangJu, L., Wujin, S., HanJun, K., Junmin, L., Yichao, Z., Yepin, Z., Sam, E., Samad, A., Nureddin, A., Mehmet, R.D., Zhuangde, J., Ali, K., Gelatin methacryloyl-based tactile sensors for medical wearables. *Adv. Funct. Mater.*, 30, 2003601, 2020.
48. Yujie, G., Yu, C., Miaomiao, L., Lizhen, J., Lianbin, Z., Jintao, Z., Gelatin-based photonic hydrogels for visual detection of pathogenic Pseudomonas aeruginosa. *Sens. Actuators B Chem.*, 329, 129137, 2021.
49. Sanaz, M.F., Mehdi, D., Mohammad, A.K., Behnaz, K., Mohammad, A.K., Majid, R., Fluorometric detection of phytase enzyme activity and phosphate ion based on gelatin supported silver nanoclusters. *Food Chem.*, 396, 133711, 2022.
50. Melo, A.M.A., Oliveira, M.R.F., Furtado, R.F., Borges, M.F., Biswas, A., Cheng, H.N., Alves, C.R., Preparation and characterization of carboxymethyl cashew

gum grafted with immobilized antibody for potential biosensor application. *Carbohydr. Polym.*, 228, 115408, 2020.

51. Qasemi, S. and Ghaemy, M., Novel superabsorbent biosensor nanohydrogel based on gum tragacanth polysaccharide for optical detection of glucose. *Int. J. Biol. Macromol.*, 151, 901, 2020.
52. Qasemi, S. and Ghaemy, M., Highly sensitive and strongly fluorescent gum tragacanth based superabsorbent hydrogel as a new biosensor for glucose optical detection. *J. Mater. Chem. C*, 8, 4148, 2020.
53. Meng, Y., Lu, J., Cheng, Y., Li, Q., Wang, H., Lignin-based hydrogels: A review of preparation, properties, and application. *Int. J. Biol. Macromol.*, 135, 1006, 2019.
54. Jędrzak, A., Rębiś, T., Klapiszewski, L., Zdarta, J., Milczarek, G., Jesionowski, T., Carbon paste electrode based on functional GOx/silica-lignin system to prepare an amperometric glucose biosensor. *Sens. Actuators B: Chem.*, 256, 176, 2018.
55. Umar, N., Aatif, N., Nawshad, M., Muhammad, A., Azhar-ul-Haq, A.S., Naeem, K., Mansoor, K., Shaukat, S., Abdur, R., Non-enzymatic colorimetric biosensor for hydrogen peroxide using lignin-based silver nanoparticles tuned with ionic liquid as a peroxidase mimic. *Arab. J. Chem.*, 14, 103164, 2021.
56. Ranganathan, P., Mutharani, B., Chen, S.-M., Sireesha, P., Biocompatible chitosan-pectin polyelectrolyte complex for simultaneous electrochemical determination of metronidazole and metribuzin. *Carbohydr. Polym.*, 214, 317, 2019.
57. Mani, V., Devasenathipathy, R., Chen, S.-M., Wang, S.-F., Devi, P., Tai, Y., Electrodeposition of copper nanoparticles using pectin scaffold at graphene nanosheets for electrochemical sensing of glucose and hydrogen peroxide. *Electrochim. Acta*, 176, 804, 2015.
58. Thakur, B., Amarnath, C.A., Sawant, S.N., Pectin coated polyaniline nanoparticles for an amperometric glucose biosensor. *RSC Adv.*, 4, 40917, 2014.
59. Anjali, M.K., Bharath, G., Rashmi, H.M., Jaswal, A., Kumar, N., Raju, P.N., Raghu, H.V., Polyaniline-pectin nanoparticles immobilized paper based colorimetric sensor for detection of Escherichia coli in milk and milk products. *Curr. Res. Food Sci.*, 5, 823, 2022.
60. Meena, N., Tian, C.Z., Dinesh, K., Pectin modified metal nanoparticles and their application in property modification of biosensors. *Carbohydr. Polym. Technol. Appl.*, 2, 100164, 2021.
61. Zheng, W., Li, J., Zheng, Y.F., An amperometric biosensor based on hemoglobin immobilized in poly (ε-caprolactone) film and its application. *Biosens. Bioelectron.*, 23, 1562, 2008.
62. Xu, T., Jin, W., Wang, Z., Cheng, H., Huang, X., Guo, X., Ying, Y., Wu, Y., Wang, F., Wen, Y., Yang, H., Electrospun CuO-nanoparticles-modified

polycaprolactone @polypyrrole fibers: An application to sensing glucose in saliva. *Nanomaterials*, 8, 133, 2018.
63. Caroline, G.S., Melania, O., Anca, A., Madalina, M.B., Disposable superoxide dismutase biosensors based on gold covered polycaprolactone fibers for the detection of superoxide in cell culture media. *Talanta*, 241, 123255, 2022.
64. Fakhriy, M., Gamze, D., Seda, K., Kübra Karadaş, G., Büşra, B., Ahmet, U., Burhan, A., Design of highly selective, and sensitive screen-printed electrochemical sensor for detection of uric acid with uricase immobilized polycaprolactone/polyethylene imine electrospun nanofiber. *Electrochim. Acta*, 439, 141675, 2023.
65. Begüm, G.-G., Taslak, T.D., Ozan, Ö., Necla, I., Tughba, T.-A., Polycaprolactone/silk fibroin electrospun nanofibers-based lateral flow test strip for quick and facile determination of bisphenol A in breast milk. *J. Biomed. Mater. Res.*, 109, 1455, 2021.
66. Sung, W.J., Na, K., Bae, Y.H., Biocompatibility and interference eliminating property of pullulan acetate/polyethylene glycol/heparin membrane for the outer layer of an amperometric glucose sensor. *Sens. Actuators B: Chem.*, 99, 393, 2004.
67. Kannan, B., Jahanshahi-Anbuhi, S., Pelton, R.H., Li, Y., Filipe, C.D.M., Brennan, J.D., Printed paper sensors for serum lactate dehydrogenase using pullulan-based inks to immobilize reagents. *Anal. Chem.*, 87, 9288, 2015.
68. Heredia, F.L., Resto, P.J., Parés-Matos, E.I., Fast adhesion of gold nanoparticles (AuNPs) to a surface using starch hydrogels for characterization of biomolecules in biosensor applications. *Biosensors*, 10, 99, 2020.
69. Annu, and Ahmed, S., 1-Bionanocomposites: An overview, in: *Bionanocomposites in Tissue Engineering and Regenerative Medicine*, S. Ahmed, and Annu (Eds.), pp. 1–6, Woodhead Publishing, Cambridge, United Kingdom, 2021.
70. Aisverya, S., Ali, A., Sudha, P.N., 28-Pullulan-based bionanocomposites in tissue engineering and regenerative medicine, in: *Bionanocomposites in Tissue Engineering and Regenerative Medicine*, S. Ahmed, and Annu (Eds.), pp. 533–547, Woodhead Publishing, Cambridge, United Kingdom, 2021.
71. Annu, Ahmed, S., Kaur, G., Sharma, P., Singh, S., Ikram, S., Evaluation of the antioxidant, antibacterial and anticancer (lung cancer cell line A549) activity of: Punica granatum mediated silver nanoparticles. *Toxicol. Res. (Camb)*, 7, 5, 923–930, 2018.

72. Annu, Ahmed, S., Nirala, R.K., Kumar, R., Ikram, S., Green synthesis of chitosan/nanosilver hybrid bionanocomposites with promising antimicrobial, antioxidant and anticervical cancer activity. *Polym. Polym. Compos.*, 29, 9_suppl, S199–S210, 2021.
73. Annu, Ahmed, S., Ikram, S., Perspectives of chitosan and alginate membranes for biomedical applications, in: *Natural Polymers: Derivatives, Blends and Composites*, vol. 2, pp. 157–182, Nova Science Publishers, Incorporated, New York, USA, 2017.

18

Biopolymers in Dentistry: A Risk Assessment on Human Health

Stanley C. Onwubu[1]*, Chibuzor Stellamaria Okonkwo[2], Mokgadi Ursula Makgobole[1] and Phumlane Selby Mdluli[1]

[1]*Durban University of Technology, Durban, South Africa*
[2]*Shehu Idris College of Health Sciences and Technology, Makarfi, Nigeria*

Abstract
Biopolymers such as hydroxyapatite, collagen, and chitosan have long been recognized as favorable dental materials with a broad area of applications in dentistry. Moreover, these materials' unique characteristics such as toxicity, biodegradability, and biocompatibility, i.e., biodegradation, make them prime candidates and materials of choice in conservative and preventive periodontology and implant dentistry. Given that biobased polymers are commonly extracted from the biological system, it is of vital importance to assess any potential risk factors of the materials. This chapter assesses the characteristics, applications, and any possible toxicological risks to human health.

Keywords: Biopolymers, dentistry, risk assessment

18.1 Introduction

Biopolymer-based composites have been developed as substantial alternatives for synthetic polymers used in medical and drug applications [1]. Biopolymers, in the treatment of medical and dental lesions, can be traced back centuries before the advent of civilization as we know it [2]. Documented literature has shown that dentistry has long benefited from biomaterials, as they have applications in periodontology and oral surgery, primarily for restoring injured root systems, dentin, and dental pulp

**Corresponding author*: stanleyo@dut.ac.za; profstan4christ@yahoo.com

Annu (ed.) Biopolymers for Biomedical Applications, (459–478) © 2024 Scrivener Publishing LLC

[3] and wound repairs and even promote the survival/life expectancy of patients [2].

The use of biodegradable polymers in dental and maxillofacial surgery has supplanted the use of traditional non-degradable materials over the last two decades, notably in bone regeneration and periodontal therapy [4]. This is attributed to the unique properties of biodegradable materials such as biocompatibility, high bioactivity, ability to break down, and the capacity to dissolve and be absorbed by the body without releasing toxic products, i.e., biodegradation. Other characteristics of biopolymers that had endeared it to dentistry include its low antigenicity, ability to accelerate cell growth and proliferation, and extensive mechanical properties while maintaining its mechanical strength with a high potential for controlled drug administration, wound treatment [5], tissue engineering, and dental repair [1, 6].

Onwubu and colleagues, in their research, reviewed several biopolymers such as collagen, alginate, and cellulose and discussed their extensive applications and use in dentistry [7, 8]. Additionally, several documented research have been carried out recently on tissue engineering, which has employed many novel materials from biomaterials/biopolymers such as chitosan, collagen, gelatin, alginates, hyaluronic acids, hydroxyapatite, etc. [66–71], that can be employed in dental applications in the repair of tissues and wound healing in periodontology treatment, preventive and conservative dentistry, and implantology as well as the incorporation of bioactive substances to fabricate antimicrobial coatings for stents due to their low molecular weight and easy absorption into the human body [7, 9–11].

Despite all the laudable qualities of biopolymers, concerns about their impact on humans still arise. This is particularly so when one considers that the human body is made up of so many cells which function harmoniously to carry out different functions. Injury to a particular cell, tissue, or organ can result in a chain of reactions. It is, therefore, imperative to evaluate cell interactions with biopolymers in order to identify appropriate polymers for biomedical applications since this assists in establishing how polymeric materials impact cell survival, growth, anchorage dependency, proliferation, and cell–cell contact [3]. This chapter assesses the risks associated with biopolymers used in dentistry. The chapter focuses only on four biopolymers, namely, hydroxyapatite, collagen, chitosan, and hyaluronic acid. The chapter is divided into two sections: section one will highlight the basic characteristics, properties, and application of biopolymers in dentistry, while section two assesses the risk of these materials.

18.2 Biopolymers in Healthcare

Biopolymers or biomaterials are any materials that have been designed to interact with biological structures in order to perform a medical function, either healing or investigative [12]. When compared to synthetic polymers, biopolymers offer several advantages, including their well-defined, complex, and degradable structures as well as their renewable properties [1, 13]. The use of various biologically derived materials in amplifying or repairing the body stretches back to the dawn of time when natural materials such as wood were used in rejuvenating the tissues lost to disease or trauma [14].

According to several documented reviews, the functional efficiency of the biopolymer molecules can be majorly affected by the deflection in their composition, physicochemical properties, and structural features [5]. This clarifies that their structure is simpler and more random. Biopolymers, for example, are sophisticated molecular assemblages with well-specified 3D forms and structures. Their distinct shape and structure are crucial to their function. Bibire *et al.* [15] reveal that the well-defined shape and structure of biopolymers have favored its application in so many areas, namely, artificial skin, drug delivery systems, surgical sutures, vascular grafts, etc.

Moreover, the quest to find solutions to ailments affecting man as a result of exposure to synthetic biomaterial has led to research into natural polymers which can be obtained from natural sources [15, 16]. Natural polymers are biodegradable and obtained from renewable sources, making them sustainable and easily available abundantly to the public at a large scale [16]. The criteria for the selection of these biopolymers are based on the sustainable inflammatory response, appropriate permeability, availability, biocompatibility, low molecular weight with excellent mechanical stability, good biodegradability and processibility in relation to its function, and most importantly, the non-toxicity or low toxicity to human tissues has made them grow in popularity over the last decade [3, 9, 17].

18.3 Classification of Biopolymers

Biopolymers can be divided on the basis of degradability (bio-degradable and non-bio-degradable) [8, 17, 18], polymer backbone (polyesters, polysaccharides, polycarbonates, polyamides, and polycarbonates) [3, 19, 20], nature of their repeating units (polysaccharides, proteins, and nucleic

Table 18.1 Classification of biopolymers.

Classification	Origin	Biopolymers
Polysaccharides	Plant/algal Animal Bacterial Fungal Lipids/surfactants	Starch (amylose/amylopectin), cellulose, agar, alginate, carrageenan, pectin, konjac, various gums (e.g., guar) Chitin/chitosan, hyaluronic acid xanthan, dextran, gelian, levan, curdian, polygalactosamine cellulose (bacterial) Pullulan, elsinan, yeast glucans Acetoglycerides, waxes, surfactants, emulsan
Proteins		Silks, collagen/gelatin, elastin, resilin, adhesives, polyamino acids, soy, zein, wheat gluten, casein, serum albumin
Polyesters		Polyhydroxyalkanoates, polylactic acid
Specialty polymers		Shellac, poly-gamma-glutamic acid, natural rubber, synthetic polymers from natural fats and oils, nylon from castor oil

Source: Rebelo, Fernandes [21].

acids), and their application (bioplastics, biosurfacts, bioadhesives, etc.) [3]. The classification of biopolymers is illustrated in Table 18.1.

18.4 Biopolymers Used in Dentistry

Dentistry, like other medical professions, has long profited from the use of biopolymers in periodontology and oral surgery, notably for the regeneration of damaged root systems, dental pulp, and dentin [21]. Most of the biopolymers used in dentistry are polysaccharides sourced from plants, insects, and animal origins, including hydroxyapatite, chitosan, collagen, gelatin, bioactive glass, alginate, polylactic acid, and hyaluronic acid (Figure 18.1). Studies have shown that these materials have good

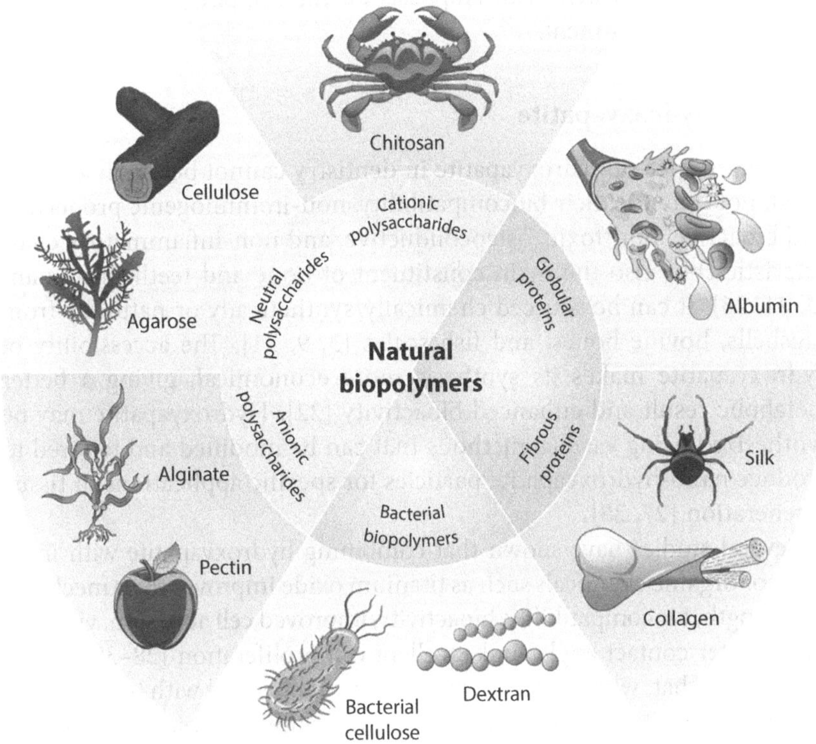

Figure 18.1 Biopolymers from natural sources [25].

mechanical properties, biocompatibility with cell tissues, regenerative ability, bioactivity, and biodegradability into non-toxic substances [3, 16]. They have been reported to be appealing for dental scaffold fabrication [3].

With the focus of medicine shifting from curative to preventive oral healthcare delivery, several biopolymer matrices have been developed to assist in the delivery of healthcare services. Reabsorbable regeneration membranes are currently used in periodontal surgeries to facilitate recovery without disruption, which is usually the case with the use of non-absorbable membranes [22, 23]—for example, biopolymer-derived membranes allow the passage of food and nutrients from the blood to the wound site while preventing bacteria from passing through and creating space for the regrowth of periodontal tissues [24]. Owing to this, there is a steady and noteworthy growth in studies concerning the use of biodegradable materials in dentistry. The next part of the chapter discusses some prominent biopolymers with emphasis on their application in dentistry and their risk assessment.

18.4.1 Hydroxyapatite

The importance of hydroxyapatite in dentistry cannot be overly emphasized, not only for their biocompatibility, non-immunogenic properties, and bioactive, non-toxic, osteoconductive, and non-inflammatory characteristics but also the main constituent of bone and teeth in humans [3, 21, 26]. It can be sourced chemically/synthetically or naturally from eggshells, bovine bones, and fish scales [3, 9, 21]. The accessibility of hydroxyapatite makes its synthesis more economical, giving a better metabolic result and enhanced bioactivity [22]. Hydroxyapatite may be synthesized using various methods that can be modified and tailored to produce nano-hydroxyapatite particles for specific applications in tissue regeneration [27, 28].

Several studies have shown that combining hydroxyapatite with inorganic or organic chemicals such as titanium oxide improves their mechanical strength, biocompatibility, bioactivity, improved cell adhesion, viability, larger water contact angle, and excellent cell proliferation [28–30]. Hung *et al.* report that, when tissue membranes are embedded with hydroxyapatite, the hydroxyapatite exhibits strong osteoconductivity and biocompatibility, hence establishing the ideal conditions for rapid bone regeneration [31]. This has been effectively utilized for the development of both porous and composite bone scaffolds, toothpaste additives, and excellent drug release carrier [31, 32]. This can also be applied to dental tissue regeneration as seen in dental implants [33].

18.4.2 Collagen

It is a major constituent of animal protein and essential for the extracellular matrix [33]. In humans, it consists of three types; I, II, and III, but type I which consists of three Alpha chains are commonly used for tissue engineering [34]. Collagen can be synthesized from several sources and tissues depending on their application [35]. Traditionally, it is sourced from bovine and has antihypertensive, antioxidant, anti-inflammatory, and antimicrobial properties but has been reported to be a source of swine flu among other health-related problems [36–38]. Thus, it has led to seeking alternative sources of collagen from marine sources such as jellyfish or sponges, Tilapia scales, and many other marine animals which all possess lower molecular weight and good antioxidant, bioactivity, and antimicrobial properties [39–42].

Human bone cells seeded onto collagen and collagen/calcium phosphate-coated substrates produced well-developed filopodia and lamellipodia as well as much greater cell proliferation than the other specimens, according to [35, 36]. This substantiated their claims of its biocompatibility, hemostasis enhancement, wound healing, and tissue restoration which can be attributed to its use for surgical wound treatment and commercial production of collagen films used in periodontal regeneration treatment [43].

Furthermore, collagen has been discovered to possess a high possibility for advancement as a material which usually is biocompatible as well as readily absorbed by tissues, rendering it a lot more acceptable than synthetic composites [44]. It has a record of effectiveness in dentistry as a root conditioning agent in orthopedic and maxillofacial procedures, a hemostatic agent, and a wound dressing agent [45, 46].

18.4.3 Chitosan

Chitosan is derived from alkaline deacetylation of chitin, a natural biomaterial obtained from the exoskeleton of marine shells and also found abundantly in plants such as mushrooms and fungi [47, 48]. However, due to its low solubility which limits its use, chitosan is preferred as it dissolves in dilute acids to produce cationic polymers that can be synthesized/tailored into different bioactive substances for different applications as required [48]. Consequently, it has been an interesting polymer of interest in so many fields such as biomedical sciences, pharmacology, and tissue engineering sciences [48].

Studies by Mato showed that the mechanical strength and flexibility of chitosan were increased when combined with bioglass in a humid environment, giving the material the capacity to stimulate bone regeneration and promote the proliferation of periodontal ligaments. Chitosan can be transformed into micro/nanoparticles, fibers, film sponges, gels, and injectable devices to treat periodontitis and dental caries and perform root canal procedures locally (endodontics). It keeps medicines in periodontal pockets with a maintained therapeutic concentration for long periods of time throughout the therapy [49].

18.4.4 Hyaluronic Acid

Hyaluronic acid (HA) is a naturally occurring linear hygroscopic polymer present in the extracellular matrix of connective tissue, synovial fluid, embryonic mesenchyme, vitreous humor, skin, and various parts of the body [50–52]. The gingiva and periodontal ligaments, as well as alveolar bone and cementum, are the main sources of native hyaluronic acid in oral tissues [34, 51]. It is made up of non-sulphated glycosaminoglycan with a high molecular weight of 4,000–20,000,000 Da and polyanionic disaccharide units of glucuronic acid and N-acetyl glucosamine linked by alternating bl-3 and bl-4 bonds [51, 52]. HA serves a diverse role in scar-free wound healing as well as a role in physiology in the oral cavity and in the area of dentistry [53].

HA plays a variety of physiological and structural tasks, including cellular and extracellular interactions with growth factors, osmotic pressure, and tissue lubrication, all of which contribute to the structural and homeostatic integrity of tissues. An extensive study into the chemical and physicochemical properties of HA, as well as its physiological role in humans, has shown that it is a great biomaterial for cosmetic, therapeutic, and pharmaceutical applications [48].

Moreover, a report by Kida *et al.* showed that HA regulates inflammatory responses by the suppression of immune response and undergoing extensive degradation to reduce molecular weights in chronically inflamed tissues [54]. This is attributed to the possible result of the reaction of the bacterial enzymes [55].

According to some documented research (*in vivo* and *in vitro*), HA has been a polymer of interest in medical interventions in wound healing stages as it can be seen to support the structural and homeostatic integrity of tissues controlling the osmotic pressure and tissue lubrication, thus preventing the penetration of viruses into the tissues due to its viscoelasticity [55, 56]. To improve HA's resistance to deterioration, its viscosity, elasticity, and hydrophilicity are often altered [57].

There have been multiple confirmed reports that HA is utilized to treat gingivitis, recessions, and periodontal pockets and can serve as a component of grafts and implants due to its anti-inflammatory and antibacterial qualities. HA also has a bacteriostatic effect and acts as a stable barrier against viral and bacterial penetration, which are important features in preventing the foreign pathogen contamination of surgical wounds and reducing the risk of postoperative infections, hence assisting in more predictable regeneration, including diseases of the oral cavity [53].

Several *in vivo* and *in vitro* reports on research conducted on the use of HA injectable gels for the reconstruction and treatment of periodontitis showed that HA is effective for minimally invasive therapy in enhancing papillary aesthetics and increases the healing prognosis for patients with impaired healing conditions [57, 58]. The local application of metronidazole and 0.2% HA gels had a good effect on the conventional management of periodontitis within 7 days of its application on patients with chronic periodontitis [59].

Furthermore, HA has been employed in dentistry for the regeneration of papilla and the promotion of osseointegration as well as the topical application of HA to treat oral ulcers, stomatitis, and discomforts due to permanent or removable dentures [55].

Using freeze-drying processes, hyaluronic acid–gelatin hydrogel polymers, tricalcium phosphate, and biphasic calcium phosphate ceramics may be designed to provide a dental bone replacement for bone reformation [60]. Such biopolymer and bioceramic mixing scaffold represented good bone formation *via* osteocytes while avoiding tissue collapse as a matrix to encapsulate stem cells for tissue regeneration and a scaffold for the promotion of osseointegration without collapse, thus ensuring that the remodeling of TMJ, salivary glands, dental pulp, enamel, root canal, and mucosa can be achieved [61].

18.5 Risk Assessment of Biopolymers in Dentistry

During the last several decades, there has been an increase in the need for safety evaluation and management of dental materials since it is used in contact with the human body. Biopolymers, although synthesized from biological resources such as microbes and plants or otherwise produced chemically from basic biological systems, could also pose a risk to human systems—for example, collagen extracted from bovine has been reported to be the source of infectious diseases such as spongiform encephalopathy [62]. Moreover, dental materials, in general, could pose adverse reactions to the human body system. These reactions could be related to allergic and or non-allergic reactions (Figure 18.2).

From a health and safety perspective, an allergic response is caused by substances (allergens) found in pollen, medicines, microorganisms, and food which provoke an allergic reaction. Unlike allergic reactions, non-allergic reactions are caused by the exposure to chemicals of low concentrations as emitted by dental materials and are thus dose-dependent. Given that bio-based polymers are commonly extracted from the biological

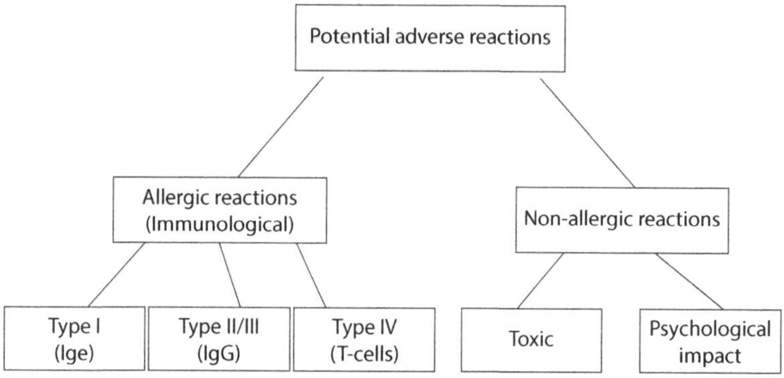

Figure 18.2 Potential adverse reactions to dental materials.

system, it is of vital importance to assess any potential risk factors of these materials. Moreover, according to Anju *et al.* [3], the evaluation of potential toxicological hazards is a crucial criterion to consider before confirming any biomaterial as safe for biomedical purposes, including dentistry.

Anju *et al.* [3] proposed that the analysis of test material breakdown products may aid in forecasting the probable toxicity. Another area of interest before selecting a biopolymer, particularly in dentistry, is material biocompatibility. It is therefore of critical importance for dental clinicians to carry out appropriate bench testing of biopolymer toxicity. This, according to Anju *et al.* [3], will reduce the biocompatibility risk.

18.5.1 Assessing the Toxicity of Biopolymers

The biopolymers in use in dentistry today are due to their properties which are non-toxic and biocompatible with living tissues. A study was carried out by Brum *et al.* [48] using an experimental bovine collagen membrane to test the biocompatibility using *in vitro* and *in vivo* tests. The proposed membrane is non-cytotoxic and does not induce inflammatory responses [48, 61]. Chitosan has also been found to be biocompatible, bioactive, and biodegradable with the human tissues due to its neutrality after degradation. It has become a material of interest to researchers as some novel materials have been produced from *in vitro* studies [48].

With reference to hydroxyapatite, a study by Coelho *et al.* [63] evaluated the cytotoxicity of commercially available hydroxyapatite nanoparticles (HA-NP) and an *in vitro* assessment using HET-CAM assay to evaluate other irritations in an attempt to imitate tooth brushing procedure for length of exposure, particle size, cell type, and quantity incorporated in oral care products [63]. The *in vitro* HET-CAM assay demonstrated that HA-NP had no irritation potential. In general, the cytotoxicity studies were carried out in the exposure scenario. Moreover, the usage of these HA-NP in oral care products has no safety concerns [53]. The results showed that 3.1% HA-NP suspension and its leachate are devoid of any irritation potential towards human gingival fibroblasts. Similarly, in another study carried out by Onwubu *et al.* [9], when the cell viability of nHAp obtained from eggshells was tested against NIH 3T3 mouse fibroblasts using the MTS endpoint assay, the researchers found that, even at very high concentrations, the cells had no effect on the nucleus after 4 days of incubation.

18.5.2 Assessing the Biocompatibility of Biopolymers

Biopolymers are selected in many dental restorations because of their biocompatibility with living tissues. A study has shown that these materials do not produce a carcinogenic response or cause harm to the oral tissues or their associated structures [64]. This can also be determined by the degradability or absorbability of the polymer [64]. Forecasting the concentration of byproducts present in the tissue affects the pace of breakdown and the resulting host reaction [64]. Biopolymers are degraded from large polymeric chains into smaller oligomeric or monomeric forms for easy absorption [3]. The factors—mainly heat, electromagnetic radiation, moisture, mechanical strength, and biological aspects—altogether contribute to the degradation of biopolymers, which results in the alteration in their characteristics, i.e., polymeric composition, molecular weight, and structural and chemical characteristics [65]. While the degradability properties of biopolymers give it some advantage over the synthetic counterpart, this is, however, risky and usually produce toxic leachable products. Another concern is that the degradation of biopolymers may occur in an unanticipated manner, rendering them useless for practical purposes [3]. This concern has led to the selection of biopolymers based on their degradation routes and the advantages they can offer without sustained injury to the cells [3].

18.6 Discussion

The evaluation of toxicity and biocompatibility is crucial for ensuring the safe use of biopolymers in dentistry. Biopolymers, derived from biological resources or chemically synthesized from basic biological systems, may pose risks to human systems despite their biocompatible and non-toxic properties. Therefore, it is important to assess the potential risk factors associated with these materials.

Toxicity assessment plays a vital role in determining the safety of biopolymers. Anju *et al.* [3] suggest that analyzing the breakdown products of test materials can help predict their probable toxicity. Various studies have been conducted to evaluate the cytotoxicity and inflammatory responses induced by biopolymers used in dentistry—for instance, a study by Brum *et al.* [48] examined the biocompatibility of an experimental bovine collagen membrane using *in vitro* and *in vivo* tests. The results indicated that the proposed membrane was non-cytotoxic and did not induce inflammatory responses. Chitosan, another biopolymer, has been found to be biocompatible, bioactive, and biodegradable with human tissues [48]. These findings highlight the non-toxic and biocompatible nature of certain biopolymers, making them suitable for dental applications.

In addition to toxicity assessment, evaluating the biocompatibility of biopolymers is crucial. Biocompatibility refers to the material's ability to interact with living tissues without causing harm. The biopolymers used in dentistry are selected based on their biocompatibility with living tissues. It is important to consider factors such as degradation, absorbability, and the production of harmful byproducts.

Degradability is a critical aspect of biopolymer biocompatibility. Biopolymers should degrade into smaller forms for easy absorption without producing toxic leachable products [3]. However, the degradation process must be controlled to avoid adverse effects. Factors such as heat, radiation, moisture, mechanical strength, and biological aspects can influence the degradation of biopolymers [65]. Therefore, it is crucial to select biopolymers that degrade predictably and safely, preserving their desirable characteristics without causing harm to surrounding cells.

The assessment of toxicity and biocompatibility of biopolymers in dentistry has several implications. Firstly, by evaluating the potential toxicological hazards associated with biopolymers, researchers and dental clinicians can ensure the safety of these materials for biomedical purposes, including dentistry [3]. This assessment provides valuable information for selecting biopolymers that are non-toxic and pose minimal risk to human health.

Furthermore, the biocompatibility assessment of biopolymers enables dental clinicians to choose materials that interact favorably with living tissues. Biopolymers with high biocompatibility offer advantages over synthetic counterparts as they are more compatible with the human body [64].

By considering factors such as degradation, absorbability, and the avoidance of harmful byproducts, dental clinicians can select biopolymers that are more likely to integrate well with oral tissues and minimize adverse reactions.

The research studies mentioned provide evidence of the biocompatibility and non-toxicity of certain biopolymers—for instance, Brum *et al.* [48] demonstrated the non-cytotoxic and non-inflammatory properties of an experimental bovine collagen membrane, while Onwubu *et al.* [9] showed that nHAp obtained from eggshells had no effect on cell viability. These findings support the safety and potential benefits of using biopolymers in dentistry.

In summary, assessing the toxicity and biocompatibility of biopolymers is essential for ensuring their safe use in dentistry. By conducting thorough evaluations, dental clinicians can select biopolymers that are non-toxic, biocompatible, and capable of degrading in a controlled and safe manner. This assessment process enhances patient safety and contributes to the development of advanced and safer dental materials.

18.7 Conclusion

The need to provide better and safe medical alternatives to the treatment of debilitating diseases in humans has led to research into alternatives to synthetic polymers which are more compatible with the cell tissues. These materials elicit little or no toxic response upon biodegradation into smaller units. From a dentistry perspective, this decision has benefited many disciplines such as conservative, preventative, implantology, etc., which has allowed them to explore their horizons extensively and provide diagnostic, therapeutic, preventive, and curative services as the needs arise comprehensively. However, like any other dental materials in contact with soft tissues, dental biopolymers are not without their risk-associated factors considering that they are used directly on human tissues. The selection of the best biomaterial to use generally depends on the fact that it should not be injurious to the oral tissues. It is therefore imperative that a critical analysis of their byproducts should be done extensively at different host levels to determine their interaction at the cell, tissue, organ, and system levels so as to be able to proffer solutions as the needs arise.

References

1. Al-Ghraibah, A.M., Al-Qudah, M., AL-Oqla, F.M., Medical implementations of biopolymers, in: *Advanced Processing, Properties, And Applications of Starch and Other Bio-Based Polymers*, F.M. AL-Oqla and S.M. Sapuan (Eds.), pp. 157–171, Elsevier, 2020.
2. Marin, E., Boschetto, F., Pezzotti, G., Biomaterials and biocompatibility: An historical overview. *J. Biomed. Mater. Res. A*, 108, 8, 1617–1633, 2020.
3. Anju, S., Prajitha, N., Sukanya, V.S., Mohanan, P.V., Complicity of degradable polymers in health-care applications. *Mater Today Chem.*, 100236, 16, 2020.
4. Pepla, E., Besharat, L.K., Palaia, G., Tenore, G., Migliau, G., Nano-hydroxyapatite and its applications in preventive, restorative and regenerative dentistry: A review of literature. *Ann. Stomatol.*, 5, 3, 108–14, 2014.
5. Wahab, M.A., Islam, N., Hoque, M.E., Young, D.J., Recent advances in silver nanoparticle containing biopolymer nanocomposites for infectious disease control–A mini review. *Curr. Anal. Chem.*, 14, 3, 198–202, 2018.
6. Reis, C.H.B., Buchaim, D.V., Ortiz, A.D.C., Fideles, S.O.M., Dias, J.A., Miglino, M.A., Teixeira, D.D.B., De Souza Bastos Mazuqueli Pereira, E., da Cunha, M.R., Buchaim, R.L., Application of fibrin associated with photobiomodulation as a promising strategy to improve regeneration in tissue engineering: A systematic review. *Polymers*, 14, 15, 3150, 2022.
7. Onwubu, S.C., Mdluli, P.S., Singh, S., Thakur, S., Biodegradable natural materials in dentistry: Fiction or real?, in: *Applications of Advanced Green Materials*, S. Ahmed (Ed.), pp. 77–88, Elsevier, New Delhi, 2021.
8. Conte, R., Di Salle, A., Riccitiello, F., Petillo, O., Peluso, G., Calarco, A., Biodegradable polymers in dental tissue engineering and regeneration. *AIMS Mater. Sci.*, 5, 6, 1073–1101, 2018.
9. Onwubu, S.C., Naidoo, D., Mkhize, S.C., Mabaso, N.L.N., Mdluli, P.S., Thakur, S., An investigation in the remineralization and acid resistant characteristics of nanohydroxyapatite produced from eggshell waste via mechanochemistry. *J. Appl. Biomater. Funct. Mater.*, 18, 1–8, 2020.
10. Negut, I., Floroian, L., Ristoscu, C., Mihailescu, C.N., Mirza Rosca, J.C., Tozar, T., Badea, M., Grumezescu, V., Hapenciuc, C., Mihailescu, I.N., Functional bioglass—Biopolymer double nanostructure for natural antimicrobial drug extracts delivery. *Nanomaterials*, 10, 2, 2020.
11. Sowmya, S., Bumgardener, J.D., Chennazhi, K.P., Nair, S.V., Jayakumar, R., Role of nanostructured biopolymers and bioceramics in enamel, dentin and periodontal tissue regeneration. *Prog. Polym. Sci.*, 38, 10–11, 1748–1772, 2013.
12. Biswas, J. and Datta, B., Biomaterials: An introduction to materials for biomedical applications, in: *Nanostructured Materials and their Applications*, B.P. Swain (Ed.), pp. 43–53, Springer, Singapore, 2021.
13. Aziz, T., Ullah, A., Ali, A., Shabeer, M., Shah, M.N., Haq, F., Iqbal, M., Ullah, R., Khan, F.U., Manufactures of bio-degradable and bio-based polymers for

bio-materials in the pharmaceutical field. *J. Appl. Polym. Sci.*, 139, 29, 1–21, 2022.
14. Huebsch, N. and Mooney, D.J., Inspiration and application in the evolution of biomaterials. *Nature*, 462, 7272, 426–432, 2009.
15. Bibire, T., Yilmaz, O., Ghiciuc, C.M., Bibire, N., Dănilă, R., Biopolymers for surgical applications. *Coatings*, 12, 2, 211, 2022.
16. Wróblewska-Krepsztul, J., Rydzkowski, T., Michalska-Pożoga, I., Thakur, V.K., Biopolymers for biomedical and pharmaceutical applications: Recent advances and overview of alginate electrospinning. *Nanomaterials*, 9, 3, 404, 2019.
17. Mohan, S., Oluwafemi, O.S., Kalarikkal, N., Thomas, S., Songca, S.P., Biopolymers–Application in nanoscience and nanotechnology, in: *Recent Advances in Biopolymers*, vol. 1, p. 1, 2016.
18. Pires, J.R.A., Souza, V.G.L., Fuciños, P., Pastrana, L., Fernando, A.L., Methodologies to assess the biodegradability of bio-based polymers—Current knowledge and existing gaps. *Polymers*, 14, 7, 1359, 2022.
19. Mazuki, N.F., Saadiah, M.A., Fuzlin, A.F., Khan, N.M., Samsudin, A.S., 1 Basic aspects and properties of biopolymers, in: *Biopolymers in Nutraceuticals and Functional Foods*, S. Gopi, P. Balakrishnan, M. Bračič (Eds.), pp. 1–35, The Royal Society of Chemistry, London, 2022.
20. Carafa, R.N., Foucher, D.A., Sacripante, G.G., Biobased polymers from lignocellulosic sources. *Green Chem. Lett. Rev.*, 16, 1, 2023.
21. Rebelo, R., Fernandes, M., Fangueiro, R., Biopolymers in medical implants: A brief review. *Proc. Eng.*, 200, 236–243, 2017.
22. Bordea, I.R., Candrea, S., Alexescu, G.T., Bran, S., Băciuţ, M., Băciuţ, G., Lucaciu, O., Dinu, C.M., Todea, D.A., Nano-hydroxyapatite use in dentistry: A systematic review. *Drug Metab. Rev.*, 52, 2, 319–332, 2020.
23. Lee, W.H., Loo, C.Y., Rohanizadeh, R., Functionalizing the surface of hydroxyapatite drug carrier with carboxylic acid groups to modulate the loading and release of curcumin nanoparticles. *Mater. Sci. Eng. C*, 99, 929–939, 2019.
24. Kaushal, S., Kumar, A., Khan, M.A., Lal, N., Comparative study of nonabsorbable and absorbable barrier membranes in periodontal osseous defects by guided tissue regeneration. *J. Oral. Biol. Craniofac. Res.*, 6, 2, 111–117, 2016.
25. Jo, Y.K. and Lee, D., Biopolymer microparticles prepared by microfluidics for biomedical applications. *Small*, 16, 9, 1903736, 2020.
26. Oni, O.P., Hu, Y., Tang, S., Yan, H., Zeng, H., Wang, H., Ma, L., Yang, C., Ran, J., Syntheses and applications of mesoporous hydroxyapatite: A review. *Mater. Chem. Front.*, 7, 9–43, 2013.
27. Derkus, B., Arslan, Y.E., Emregul, K.C., Emregul, E., Enhancement of aptamer immobilization using egg shell-derived nano-sized spherical hydroxyapatite for thrombin detection in neuroclinic. *Talanta*, 158, 100–109, 2016.

28. Maidaniuc, A., Miculescu, F., Voicu, S.I., Andronescu, C., Miculescu, M., Matei, E., Mocanu, A.C., Pencea, I., Csaki, I., Machedon-Pisu, T., Ciocan, L.T., Induced wettability and surface-volume correlation of composition for bovine bone derived hydroxyapatite particles. *Appl. Surf. Sci.*, 438, 158–166, 2018.
29. Wijedasa, N.P., Broas, S.M., Daso, R.E., Banerjee, I.A., Varying fish scale derived hydroxyapatite bound hybrid peptide nanofiber scaffolds for potential applications in periodontal tissue regeneration. *Mater. Sci. Eng. C*, 109, 110540, 2020.
30. Sathiyavimal, S., Vasantharaj, S., LewisOscar, F., Selvaraj, R., Brindhadevi, K., Pugazhendhi, A., Natural organic and inorganic–hydroxyapatite biopolymer composite for biomedical applications. *Prog. Org. Coat.*, 147, 105858, 2020.
31. Chen, K.C., Lee, T.M., Kuo, N.W., Liu, C., Huang, C.L., Nano/micro hierarchical bioceramic coatings for bone implant surface treatments. *Materials*, 13, 7, 1548, 2020.
32. Ramesh, N., Ratnayake, J.T., Moratti, S.C., Dias, G.J., Effect of chitosan infiltration on hydroxyapatite scaffolds derived from New Zealand bovine cancellous bones for bone regeneration. *Int. J. Biol. Macromol.*, 160, 1009–1020, 2020.
33. Huang, Y., Wu, C., Zhang, X., Chang, J., Dai, K., Regulation of immune response by bioactive ions released from silicate bioceramics for bone regeneration. *Acta Biomater.*, 66, 81–92, 2018.
34. Chai, Y., Okuda, M., Otsuka, Y., Ohnuma, K., Tagaya, M., Comparison of two fabrication processes for biomimetic collagen/hydroxyapatite hybrids. *Adv. Powder Technol.*, 30, 1419–1423, 2019.
35. Gelli, R., Del Buffa, S., Tempesti, P., Bonini, M., Ridi, F., Baglioni, P., Enhanced formation of hydroxyapatites in gelatin/imogolite macroporous hydrogels. *J. Colloid Interface Sci.*, 511, 145–154, 2018.
36. Wang, R.C.C., Hsieh, M.C., Yang, S.P., Chuang, P.K., Lin, J.C., Yang, C.Y., Lee, T.M., Characteristics and cyto-compatibility of collagen/Ca–P coatings on Ti6Al4V substrate. *Surf. Coat. Technol.*, 205, 19, 4683–4689, 2011.
37. Siaili, M., Chatzopoulou, D., Gillam, D., An overview of periodontal regenerative procedures for the general dental practitioner. *Saudi Dent. J.*, 30, 1, 26–37, 2018.
38. Lee, S.B., Kwon, J.S., Lee, Y.K., Kim, K.M., and Kim, K.N., Bioactivity and mechanical properties of collagen composite membranes reinforced by chitosan and β-tricalcium phosphate. *J. Biomed. Mater. Res.*, 100, 7, 1935–1942, 2012.
39. Chang, B., Ahuja, N., Ma, C., Liu, X., Injectable scaffolds: Preparation and application in dental and craniofacial regeneration. *Mater. Sci. Eng. R.*, 111, 1–26, 2017.
40. Abdollahi, M., Rezaei, M., Jafarpour, A., Undeland, I., Sequential extraction of gel-forming proteins, collagen and collagen hydrolysate from gutted silver

carp (Hypophthalmichthys molitrix), a biorefinery approach. *Food Chem.*, 242, 568–578, 2018.
41. O'Sullivan, S.M., Lafarga, T., Hayes, M., O'Brien, N.M., Bioactivity of bovine lung hydrolysates prepared using papain, pepsin, and alcalase. *J. Food Biochem.*, 41, 6, 1–10, 2017.
42. Lima, C.A., Campos, J.F., Lima Filho, J.L., Converti, A., da Cunha, M.G.C., Porto, A.L., Antimicrobial and radical scavenging properties of bovine collagen hydrolysates produced by Penicillium aurantiogriseum URM 4622 collagenase. *J. Food Sci. Technol.*, 52, 7, 4459–4466, 2015.
43. Rogina, A., Ressler, A., Matić, I., Ferrer, G.G., Marijanović, I., Ivanković, M., Ivanković, H., Cellular hydrogels based on pH-responsive chitosan-hydroxyapatite system. *Carbohydr. Polym.*, 166, 173–182, 2017.
44. Villamil, O., Váquiro, H., Solanilla, J.F., Fish viscera protein hydrolysates: Production, potential applications and functional and bioactive properties. *Food Chem.*, 224, 160–171, 2017.
45. Felician, F.F., Xia, C., Qi, W., Xu, H., Collagen from marine biological sources and medical applications. *Chem. Biodivers.*, 15, 5, 1–18, 2018.
46. Chen, J., Li, L., Yi, R., Gao, R., He, J., Release kinetics of Tilapia scale collagen I peptides during tryptic hydrolysis. *Food Hydrocoll.*, 77, 931–936, 2018.
47. Kwon, K.-J. and Seok, H., Silk protein-based membrane for guided bone regeneration. *Appl. Sci.*, 8, 8, 1214, 2018.
48. Brum, I.S., Elias, C.N., de Carvalho, J.J., Pires, J.L., Pereira, M.J., de Biasi, R.S., Properties of a bovine collagen type I membrane for guided bone regeneration applications. *e-Polymers*, 21, 1, 210–221, 2021.
49. Qasim, S.B., Delaine-Smith, R.M., Fey, T., Rawlinson, A., Rehman, I.U., Freeze gelated porous membranes for periodontal tissue regeneration. *Acta Biomater.*, 23, 317–328, 2015.
50. Muxika, A., Etxabide, A., Uranga, J., Guerrero, P., De La Caba, K., Chitosan as a bioactive polymer: Processing, properties and applications. *Int. J. Biol. Macromol.*, 105, 1358–1368, 2017.
51. Moghadam, E.T., Yazdanian, M., Alam, M., Tebyanian, H., Tafazoli, A., Tahmasebi, E., Ranjbar, R., Yazdanian, A., Seifalian, A., Current natural bioactive materials in bone and tooth regeneration in dentistry: A comprehensive overview. *J. Mater. Res. Technol.*, 13, 2078–2114, 2021.
52. Mota, J., Yu, N., Caridade, S.G., Luz, G.M., Gomes, M.E., Reis, R.L., Jansen, J.A., Walboomers, X.F., Mano, J.F., Chitosan/bioactive glass nanoparticle composite membranes for periodontal regeneration. *Acta Biomater.*, 8, 11, 4173–4180, 2012.
53. Casale, M., Moffa, A., Vella, P., Sabatino, L., Capuano, F., Salvinelli, B., Lopez, M.A., Carinci, F., Salvinelli, F., Hyaluronic acid: Perspectives in dentistry. A systematic review. *Int. J. Immunopathol. Pharmacol.*, 29, 4, 572–582, 2016.
54. Kida, D., Zakrzewska, A., Zborowski, J., Szulc, M., Karolewicz, B., Polymer-based carriers in dental local healing—Review and future challenges. *Materials*, 14, 3948, 2021.

55. Dahiya, P. and Kamal, R., Hyaluronic acid: A boon in periodontal therapy. *N. Am. J. Med. Sci.*, 5, 5, 309, 2013.
56. Sánchez, D.C., Ocampo, B.R.Y., Chirino, C.A.E., Use of hyaluronic acid as an alternative for reconstruction of interdental papilla. *Rev. Odontológica Mexicana*, 21, 3, 205–213, 2017.
57. Iviglia, G., Kargozar, S., Baino, F., Biomaterials, current strategies, and novel nano-technological approaches for periodontal regeneration. *J. Funct. Biomater.*, 10, 1, 3, 2019.
58. Radojkova-Nikolovska, V., Popovska, M., Minovska, A., Belazelkovska, Z., Influence of hyaluronic acid in periodontal tissue regeneration. *Balkan. J. Stomatol.*, 17, 2, 61–64, 2013.
59. Pitale, U., Pal, P.C., Thakare, G., Verma, M., Dhakad, S., Pandey, R., Minimally invasive therapy for reconstruction of lost interdental papilla by using injectable hyaluronic acid filler. *J. Indian Soc Periodont.*, 25, 1, 22, 2021.
60. Mahmood, A.A., Abdul-Wahab, G.A., Al-Karawi, S.I., Effect of hyaluronan and metronidazole gels in management of chronic periodontitis. *J. Int. Oral. Health*, 11, 3, 158, 2019.
61. Canciani, E., Sirello, R., Pellegrini, G., Henin, D., Perrotta, M., Toma, M., Khomchyna, N., Dellavia, C., Effects of vitamin and amino acid-enriched hyaluronic acid gel on the healing of oral mucosa: *In vivo* and *in vitro* study. *Medicina*, 57, 3, 285, 2021.
62. Ahmed, R. and Chun, B.-S., Subcritical water hydrolysis for the production of bioactive peptides from tuna skin collagen. *J. Supercrit. Fluids*, 141, 88–96, 2018.
63. Coelho, C.C., Grenho, L., Gomes, P.S., Quadros, P.A., and Fernandes, M. H., Nano-hydroxyapatite in oral care cosmetics: Characterization and cytotoxicity assessment. *Sci. Rep.*, 9, 1, 11050, 2019.
64. Anderson, J.M., Rodriguez, A., Chang, D.T., Foreign body reaction to biomaterials. *Semin. Immunol.*, 20, 2, 86–100, 2008.
65. Glaskova-Kuzmina, T., Starkova, O., Gaidukovs, S., Platnieks, O., Gaidukova, G., Durability of biodegradable polymer nanocomposites. *Polymers*, 13, 3375, 2021.
66. Annu, Ahmed, S., Ikram, S., Perspectives of chitosan and alginate membranes for biomedical applications, in: *Natural Polymers: Derivatives, Blends and Composites*, vol. 2, pp. 157–182, Nova Science Publishers, Incorporated, New Delhi, 2017.
67. Annu, Ahmed, S., Nirala, R.K., Kumar, R., Ikram, S., Green synthesis of chitosan/nanosilver hybrid bionanocomposites with promising antimicrobial, antioxidant and anticervical cancer activity. *Polym. Polym. Compos.*, 29, 9_suppl, S199–S210, 2021.
68. Annu, Pandit, P., Maity, S., Bhattacharya, T., Shekh, M.I., Ahmed, S., Chapter 30-Chitosan biobased materials in textile industry, in: *Advanced Applications of Biobased Materials*, S. Ahmed, and Annu (Eds.), pp. 717–735, Elsevier, New Delhi, 2023.

69. Annu, Manzoor, K., Ahmad, S., Soundarajan, A., Ikram, S., Ahmed, S., Chapter 30-Chitosan based nanomaterials for biomedical applications, in: *Handbook of Nanomaterials for Industrial Applications*, C. Mustansar Hussain, (Ed.), pp. 543–562, Elsevier, New Delhi, 2018.
70. Annu, Ahmed, S., Ahmed, S., Ikram, S., Chitin and chitosan: history, composition and properties, in: *Chitosan*, pp. 1–24, John Wiley & Sons, Ltd, 2017.
71. Shekh, M.I. and Ahmed, S., Chapter 1-Biopolymers: An overview, in: *Advanced Applications of Biobased Materials*, S. Ahmed, and Annu (Eds.), pp. 3–21, Elsevier, 2023.

19

Progress of Biopolymers in 3D and 4D Printing for Biomedical Applications

Randeep Singh* **and Young-Ho Ahn**†

Department of Civil Engineering, Yeungnam University, Gyeongsan, Republic of Korea

Abstract

This chapter takes an in-depth look at the diverse 3D- and 4D-printed biopolymers applications in the realm of biomedical research. It commences with a succinct outline of 3D and 4D printing technologies. Subsequently, it delves into the captivating world of 4D bioprinting, showcasing the dynamic characteristics of these printed materials. The chapter extensively examines a variety of biopolymers suitable for 3D and 4D bioprinting, particularly focusing on the stimuli responsible for inducing their transformative behavior. A crucial aspect of this study involves the exploration of techniques for processing 3D and 4D materials. We have elaborated on diverse methods, including extrusion, powder bed fusion, and material jetting printing, and vat polymerization printing, like stereolithography, digital light processing, and 2-photon polymerization printing particularly for biopolymeric aspects. Furthermore, the chapter offers a thorough summary of the biopolymers and bioinks utilized in 3D and 4D printing, providing insights into their rheological properties and potential applications. The subsequent sections present an extensive array of the medical uses of 3D- and 4D-printed biopolymers, encompassing tissue engineering, drug delivery, biomedical devices, and even food engineering. In conclusion, future perspectives on the technological advancements in the development of biopolymers and bioinks in 3D-and 4D-printed bioconstructs is highlighting for their immense potential in revolutionizing biomedical applications. This chapter serves as an invaluable resource for researchers and professionals seeking to harness the power of these state-of-the-art technologies for medical and bioengineering purposes.

Keywords: Additive manufacturing techniques, extrusion, digital light processing, stereolithography, inkjet, bioconstructs, stimuli-responsiveness, crosslinking

**Corresponding author*: randeep@yu.ac.kr
†*Corresponding author*: yhahn@ynu.ac.kr

Annu (ed.) Biopolymers for Biomedical Applications, (479–508) © 2024 Scrivener Publishing LLC

19.1 Introduction

3D printing is making significant progress in the recent past, leading to the development of advanced functional parts [1]. Additionally, the recent developments in 4D printing further brought critical changes in additive manufacturing technology, resulting in the printing of highly functional parts. 4D is the higher version of 3D printing and involves stimuli-responsive materials printing. The 4D-printed materials show responsiveness under an external stimulus, namely, pH, temperature, electrical, magnetic, moisture, etc., to a change in shape, function, or property [2–6]. Presently, 3D and 4D printing techniques are employed in diverse applications, for example, automotive, space, therapeutic, biotechnology, and medicine [7–13].

In general, polymers are utilized for 3D and 4D printing. Polymers help to fabricate commercial products in high quantities with great precision [14]. Importantly, the use of biopolymers is recommended for sustainable growth and keep the Earth green. Biopolymers are biodegradable, hence easily degraded by enzymes and microorganisms an producing by-products, namely, H_2O and CO_2 [15]. Biopolymers can be categorized into two groups, specifically, natural and synthetic. Natural biopolymers are commonly obtained from biomass. Synthetic biopolymers are synthesized from fermentation or microbial production, and biomass or petroleum products are used for their chemical synthesis. Biopolymers showed immense potential in the biomedical field, especially target-based drug delivery, tissue engineering, and anticancer activity. Biopolymers in biomedical applications are employed due to their better biocompatibility, accommodating biodegradation, and corresponding mechanical attributes [16].

This chapter discusses the various 3D and 4D printing techniques along with the mechanisms involved in 4D printing for a better understanding of the technique. Subsequently, important 3D and 4D printing biomedical applications were discussed.

19.2 Overview on 3D and 4D Printing Technologies

3D and 4D printing are additive manufacturing techniques that are capable of providing significant growth to biomedical industries. These technologies along with the biocompatible inks and 3D and 4D products help to rapidly evolve medical treatment from diagnosis to therapy. These techniques help to synthesize the needed structures obtained through magnetic resonance imaging (MRI) or computer-aided design (CAD). Presently, numerous types of products are developed using these prominent technologies, such as precise replicas of body parts, adaptable to anatomy, required for assisting a planned surgery, orthopedics implants, and progressive medications (target-based drug delivery and tissue regeneration).

The terms, such as additive (AM) or layered (LM) manufacturing, rapid prototyping (RP), and solid freeform fabrication are alternatives for 3D printing technology. In the 1980s, the invention of stereolithography (SLA) by Hull resulted in the development of 3D printing technology [17]. In 1986, Hull's patent was granted, marking the first patent for a 3D printer. To commercialize the SLA technique, Hull created 3D Systems, which became the pioneer in commercial 3D printers. The SLA involves using an ultraviolet (UV) laser to focus on a liquid monomer bath that can be cured by UV light used to create patterns that represent cross-sectional slices. Monomers are scanned and cross-linked with light, creating hardened 2D sections, while uncured monomers were left in the bath. Hull's groundbreaking contribution also involved developing a method for CAD files to commune with the RP system for constructing computer-modeled parts. Figure 19.1 shows a 3D printing setup along with the corresponding CAD model and different scaffolds obtained for various parameters [18].

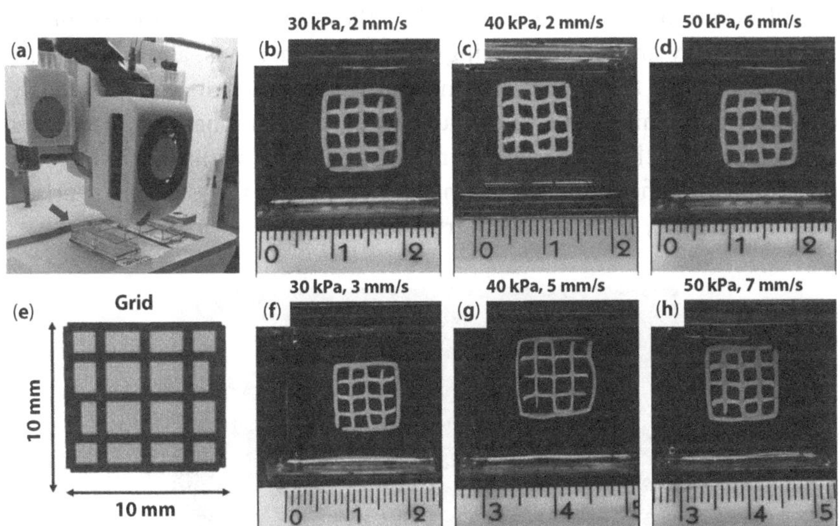

Figure 19.1 Images of (a) 3D printing set-up, (b, c, d, f, g, and h) scaffolds obtained for different printing parameters, and (e) corresponding CAD model [18].

Primitive 3D technologies were advanced during the development of SLA. Selective laser sintering (SLS) was invented by Deckerd in Beaman's group at the University of Texas at Austin as a graduate student [19]. In the SLS process, first, a CAD file with the required printing output is developed. This CAD file helps to selectively sinter the material provided in powder form on a build plate in the designated style and format [20]. Later, the same technology was used by the Sachs group to develop 3D printing at the Massachusetts Institute of Technology (MIT). Furthermore, inkjet printing, wherein the powder is spread over plates and a binder was printed on these plates based on the CAD file information, was developed. These techniques made it possible to print and fabricate complex 3D structures by using polymers, ceramics, and metals [21]. Fused deposition modeling (FDM) that uses a thermoplastic filament in its semi-liquid state was developed by Scott and Lisa Crump. This semi-liquid thermoplastic is then extruded via a nozzle and deposited over the substrate, producing different fragments in a layer-by-layer style and exercising the information delivered by the CAD file [22, 23]. The first 3D printer was launched by Sanders aligned with the inkjet printing of thermoplastic polymers [24].

In 2014, NASA introduced the first 3D printing system (Figure 19.2) to the International Space Station, enabling the construction of parts in zero gravity [20]. That same year, Oak Ridge National Laboratory researchers utilized additive manufacturing to create an entire car body, which was successfully driven at the International Manufacturing Technology Show in collaboration with Local Motors [25]. These advancements in 3D printing technology are revolutionizing production methods, allowing for the creation of both traditional and innovative designs that were previously challenging or impossible to produce using conventional manufacturing approaches. Furthermore, 3D printing is making significant strides in the biomedical field, encompassing device patterns, tissue engineering, implant printing, and controlled drug delivery.

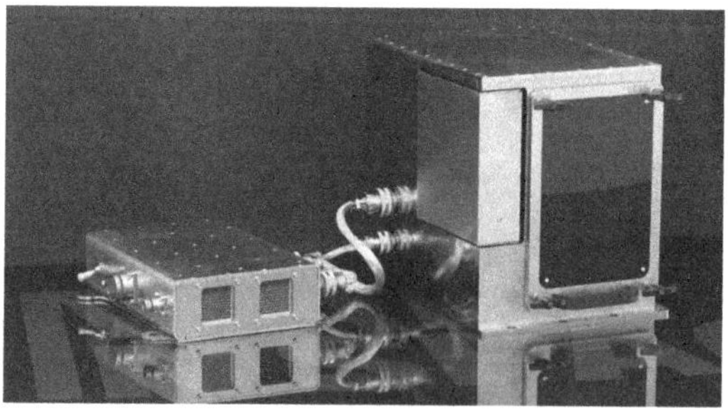

Figure 19.2 3D printer test bed installed on the International Space Station [26, 27].

After closely monitoring the advancements in technology, market trends, and published research for several years, it is clear that we have now reached a crucial stage. Key stakeholders in industry, media, research, and finance are increasingly acknowledging the extensive reach of 3D printing, extending beyond prototyping and manufacturing. This transition is anticipated to have a profound and far-reaching impact on multiple facets of our lives, with the potential to drive remarkable progress in the 3D printing industry [28].

19.3 4D Printing

4D printing is an innovative technology that enhances the capabilities of 3D printing by introducing a time dimension [29]. This advancement allows objects to be created using materials that can autonomously change their shape or properties in response to external stimuli, such as temperature, water, light, or pressure. As a result, these printed objects can dynamically transform, self-assemble, or adapt to different conditions without external intervention.

The advancement of 3D printing in the 1980s laid the groundwork for the emergence of 4D printing, which introduced time as a dimension to the creation of objects. The 4D printing concept got prominence in the 2010s, when Skylar Tibbits, a researcher from MIT, coined the term [3]. Tibbits described 4D printing as an evolution of 3D printing, incorporating the time element to generate objects capable of transforming and adapting over time. This visionary perspective inspired extensive research and development, leading to the exploration of novel materials and printing techniques that could achieve shape-changing properties.

One significant breakthrough in 4D printing arose via the integration of smart materials. These materials, including shape memory polymers and hydrogels, own the ability to alter their physicochemical properties or functionality in response to external stimuli [30, 31]. Researchers embarked on experiments utilizing these materials, harnessing their inherent characteristics to create objects capable of self-assembly, shape-shifting, or responsiveness to specific environmental triggers. Figure 19.3 shows the 4D printing of cellulose-based composites. The novel configuration arises from a significant strain differentially achieved through increased dehydration at higher temperatures. To return to the original flat configuration, immersion in a water bath transforms the shape at room temperature (Figure 19.3(f)). Figure 19.3(g) shows that the petal form, on drying, seamlessly returns to its 3D configuration. These cyclic changes amid the flat

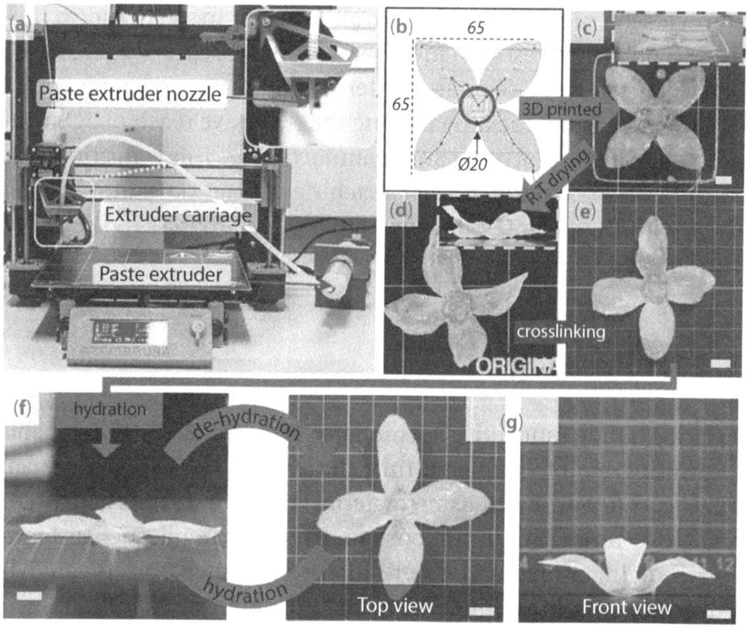

Figure 19.3 (a) Setup of the 3D printer and paste extruder, with an insert displaying the extruder housing and nozzle. (b–g) Process of fabricating the petal form, (b) print path generation from the CAD model, (c) 3D-printed configuration of the petal, (d) drying, which initiates the transformation process, (e) 3D shape crosslinking, (f) petal flat conformation on hydration, and (g) original 3D shape recovery of the petal on drying [32].

and 3D configurations are possible to repeat multiple times (>5), demonstrating remarkable shape memory behavior [32].

The applications of 4D printing expanded across diverse fields from the 2010s to the present day. Researchers explored its potential in architecture, aerospace, biomedical engineering, robotics, and consumer products [34]. In the architectural domain, innovative concepts emerged, envisioning structures that could adapt to changing conditions, optimize energy usage, and exhibit dynamic responses [34].

4D printing demonstrated promise for developing self-folding medical implants, scaffolds for tissue engineering, and controlled drug delivery approaches. Ongoing research and technological advancements continue to refine 4D printing techniques, materials, and applications.

The utilization of smart materials, often known as shape memory or programmable materials, is significant for 4D printing [35]. These materials possess the remarkable ability to reversibly alter their physical properties, namely, shape, size, rigidity, or color, when triggered by external factors. By integrating these materials into the 3D printing process, it becomes feasible to fabricate objects capable of reshaping, folding, unfolding, or even self-repairing upon exposure to specific stimuli.

4D printing shows significant potential in various fields, including architecture, space, bioengineering, robotics, and consumer products [30]—for instance, in architecture, 4D printing could enable the creation of structures that can adapt to varying weather conditions or optimize energy efficiency by transforming their shape. In healthcare, 4D-printed implants or devices could dynamically adjust to suit individual patient requirements or assist in targeted drug delivery [36].

While 4D printing is yet in its nascent stage, it possesses the potential to revolutionize manufacturing and give rise to a new generation of objects that are dynamic, adaptable, and intelligent. Ongoing research and technological advancements in materials science, design, and fabrication techniques are expected to unlock further possibilities in this exciting field.

19.4 Overview on Biopolymers for 3D and 4D Printing

Numerous biopolymers, including chitosan, alginate, and collagen, have undergone extensive investigation as potential materials for 3D-printed devices in the realm of medical applications [37, 38]. Moreover, synthetic polymers such as poly(ethylene glycol) (PEG), poly(ethylene oxide) (PEO), polycaprolactone (PCL), and poly-L-lactide acid (PLLA) have gained significant utilization in the biomedical field [39]. It is crucial to bear in mind, however, that not every polymer or polymer-based material is universally suitable for every manufacturing method. Take hydrogels, for instance, which are unsuitable for molding techniques due to the inherent challenges associated with their removal without compromising the structural integrity, potentially leading to implant fractures [40].

The degradation rate of biomaterials is essential to ensure their natural breakdown once their intended purpose is fulfilled, thereby eliminating the need for subsequent surgical procedures for implant removal–for instance, PLLA has a biodegradation rate on implantation that does not align adequately with the tissue regeneration process, and the resulting degradation byproducts, for example, lactic acid, can induce inflammation in the surrounding tissue [40]. In contrast, PCL possesses hydrolytic aliphatic-ester bonds that are inherently unstable, enabling a degradation period of up to

24 months until complete breakdown. Consequently, other materials are frequently added to PCL to attain appropriate degradation properties. This characteristic makes PCL highly suitable for implants with longer lifespans and controlled drug delivery techniques. Remarkably, a study by Honda *et al.* utilized poly(L-lactide-ε-caprolactone)-made biodegradable sponge to develop cartilage-like structures in a 4-week experiment involving nude mice [39]. Furthermore, polymers can undergo chemical modifications to align with the rate of tissue regeneration and degradation, thereby offering a plethora of properties suitable for various biomedical applications, including hydrophilicity, density, diffusivity, and other mechanical characteristics. These modifications also optimize precise interactions between the cells and polymer material, promoting comfortable interactions between cells and proteins binding to the material surface, particularly in the case of biopolymers [40].

Through the combination of diverse biomaterials, composite polymeric materials emerge, showcasing enhanced qualities when compared to using a single biomaterial. These remarkable composites find application in 3D printing, elevating the biocompatibility between implants and tissues and effectively evading rejection upon implantation by the body's defenses [37]. An intriguing example by Inzana *et al.* involves blending collagen, Tween 80 polymers, and calcium phosphate as a natural filler, concocting osseous implants that strive for amplified osteointegration and bolstered mechanical properties [41]. Yet the path to achieving this innovation faces a hurdle in the form of high-temperature manufacturing processes, impeding the inclusion of bioactive molecules and therapeutic agents during 3D printing and thereby limiting their potential in stimulating bone regeneration and combating infections [42]. Conversely, a breakthrough has been made in 3D printing composite biopolymers at lower temperatures, employing low-melting-point matrices [38, 41]. This pioneering approach allows for the integration of drugs and cells and the printing of synthetic or biological polymers like collagen, ushering in the development of bioactive implants [38]. The incorporation of collagen into mineralized bone cements significantly enhances their biomechanical prowess while imparting osteoconductive and osteoinductive characteristics. As an illustration, the addition of collagen to hand-mixed calcium phosphate cement not only fosters improved cell attachment, proliferation, activity, and viability but also enhances their mechanical properties [41]. A distinct advantage of 3D printing over traditional methods lies in its ability to create bespoke geometries tailored to individual patients, accompanied by precise delivery of drugs or polymers in the scaffold, unlocking a realm of possibilities.

4D printing involves the utilization of smart materials with controlled modifications in terms of physical attributes on exposure to external stimuli. The stimuli can be any physical parameter like temperature, moisture, electricity, and magnetism and chemical factors, for example, pH and ions [42–44]. These materials exhibit reversible transformations in physical or functional properties of bioinks sustained for numerous cycles that allow them to show diverse applications [45]. Under the influence of an external stimulus, the printed parts progressively change their properties to pre-programmed alterations alterations as observed in digital light processing (DLP) technique in case of fabrication of silk fibrion hydrogels [46]. By incorporating stimuli-responsive biopolymers into 3D printing, exciting possibilities emerge in the field of biomedicine, including the development of implants, skin bandages, splints, and scaffolds [47, 48].

19.5 Stimuli for 4D Transformation

4D transfiguration takes place with any of the physical, chemical, or biological external stimuli. The different types of stimuli are explained in the succeeding section.

19.5.1 Physical Stimuli

19.5.1.1 Temperature Responsiveness

Temperature-responsive materials possess a remarkable ability to endure reversible metamorphosis in physical or functional properties upon exposure to temperature stimuli [49–52]. This dynamic characteristic finds practical application in inducing transformative changes within printed structures [53]. When confronted with temperature fluctuations, a distinctive temperature-responsive material can gracefully change its physical or functional attributes as shown in Figure 19.4. Among the prominent temperature-responsive biopolymers are PNIPAM (poly(N-isopropylacrylamide))-based polymers and the versatile poly(ethylene oxide)/poly(propylene oxide)/poly(ethylene oxide) (PEO-PPO-PEO) triblock copolymers [54, 55]. Recently, an intriguing study showcased the marvel of 4D-printed temperature-responsive polymers capable of autonomously transitioning from normal cell culture supplements to histological cartridges. These extraordinary cartridges help to assess the drug responses in tissue models, paving the way for rapid assessment in the realm of pharmaceutical research [56].

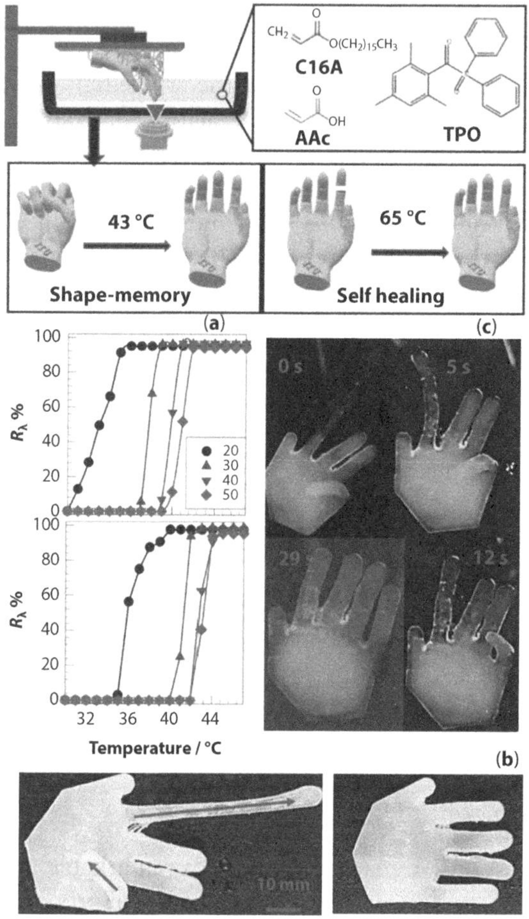

Figure 19.4 Schematics of the temperature-dependent shape recovery of printed hydrogels. These hydrogels are manufactured using a solvent-free copolymerization technique. (a) The shape recovery ratios of the printed hydrogel samples after bending (Rθ) and stretching (Rλ) are dependent on temperature. (b) 4D-printed robotic hand in both temporary and permanent shapes. (c) The shape recovery process shown in a water bath [57].

19.5.1.2 Moisture Responsiveness

Moisture fluctuations hold the potential to instigate captivating transformations within moisture-responsive materials like hydrogels [58, 59]. These materials gracefully embrace varied degrees of swelling and deformation across time and space. Immerse 3D-printed structures crafted from such materials in water and witness their orchestrated metamorphosis as they absorb moisture. In the realm of 4D printing, PEG emerges as a prominent player in moisture-responsive materials [60]. Nonetheless, further exploration is necessary to master the art of controlling the degree of swelling and contracting, thereby preserving the structural integrity of the printed components [61]. In a captivating recent investigation, bi-layered PEG, adorned with different molecular weights, unveiled its wondrous capability to self-fold through the magic of differential swelling on exposure to a solution [62].

19.5.1.3 Electro-Responsiveness

Electro-responsive materials offer a fascinating category of stimuli-responsive materials that transform their properties under the influence of an electrical stimulus [63, 64]. Extensive research was carried out to explore these electro-responsive materials to assess their notable characteristics [65]. Within this realm, various biopolymers exhibit a stimuli-responsive behavior, such as polypyrrole [66], polythiophene [67], and polyaniline-based hydrogels [68].

19.5.1.4 Magnetic Responsiveness

Unveiling a captivating realm of possibilities, magnet-responsive materials awaken under the influence of magnetic fields, granting them the power to reshape their physical properties [69, 70]. Typically, these materials encompass polymeric networks adorned with encapsulated magnetic nanoparticles, forming a harmonious union [71]. Carefully contemplating the size, uniformity, and concentration of the magnetic particles in the hydrogel matrix, alongside the unique characteristics of the hydrogels themselves, becomes paramount in this domain [72]. An exceptional study illuminated the remarkable potential of magnet-responsive materials in matrix configuration, harnessed through the enchantment of a magnetic field [73]. By infusing iron nanoparticles coated with streptavidin blended with type I collagen and agarose, the printing process orchestrated the alignment of collagen fibers, resulting in a matrix adorned with harmoniously arranged threads. Astonishingly, this magnificently aligned matrix demonstrated higher compression in comparison to its arbitrarily oriented counterpart, a testament to the power of alignment in shaping material properties.

19.5.2 Chemical Stimuli

Unveiling their remarkable adaptability, chemical stimuli-responsive materials gracefully transform their physical and functional properties when triggered by specific stimuli, including alterations in pH, light, sonication, or ion concentration [74–78]. The polymers with pH-responsiveness fall into two distinctive categories: those bearing acidic groups and those embracing basic groups [79–81]. Within the realm of 4D printing, noteworthy pH-responsive materials encompass PEO-based hydrogels, such as PEO-chitosan, PEO-gelatin, and chitosan-alginate [82–84]. On a different note, ion-sensitive materials unveil a captivating behavior as their properties metamorphose in response to fluctuations in the ion concentration. A captivating illustration emerges with alginate-based hydrogels, embarking on an ionic crosslinking journey when immersed in a solution rich in divalent cations [85]. The captivating utilization of Zn^{2+} ions further exemplifies the potential of ion-sensitive materials—for instance, the poly(acrylonitrile) chains' dipole–dipole interactions result in the fabrication of hydrogel constructs with reversible shape memory and extreme strength, while imidazole-zinc ion coordination gives rise to a hydrogel with improved properties [86, 87]. The addition of chelating agents helps to retain the shape of the printed hydrogels. The chelating agents remove zinc ions from their shape memory structures, solidifying their eternal form.

19.6 Biopolymers Used for 3D and 4D Printing

Various biopolymers, including chitosan, alginate, and collagen, were considered for 3D printing in medical applications, alongside broadly utilized synthetic polymers like PEG, PEO, PCL, and PLLA [37–39]. However, not every polymer is suitable for every manufacturing technique—for instance, molding is not appropriate for hydrogel-based medical implants due to the risk of damaging the implant during removal [39].

Another important consideration is the biomaterial's degradation rate. Ideally, the material should degrade once its purpose is fulfilled, avoiding the necessity to operate for a second time to remove the implant. PLLA's biodegradation rate is too slow for tissue regeneration, and its acidic degradation products may cause inflammation [39]. On the contrary, PCL has hydrolytic aliphatic-ester bonds, leading to a longer degradation duration of up to 24 months. Therefore, other materials are often added with PCL to achieve the preferred degradation properties, making it suitable for its long-term and precise applications [40].

Chemical modifications of polymers allow the tailoring of tissue regeneration and degradation rates as well as obtaining various characteristics for biomedical applications, including physical and functional properties. Moreover, these modifications enable particular interactions between the cells and polymer material, particularly biopolymers [39].

Combining multiple biomaterials leads to the development of composite polymeric materials, improving implant–tissue biocompatibility to prevent rejection upon implantation [39]—for example, osseous implants have been developed using polymer blends of collagen and Tween 80 with calcium phosphate as a natural filler to enhance osteointegration and the mechanical properties [42]. However, the development of these materials at high temperatures prevents the amalgamation of degradable biomaterials and drugs, limiting their potential for biomedical applications [42].

Conversely, the feasibility of 3D printing composite biopolymers at low temperatures has been demonstrated using low-melting-point matrices [37, 42]. This allows the incorporation of drugs, cells, and 3D printing of synthetic or biological polymers, like collagen, enabling the development of bioactive implants [37]—for instance, adding collagen to mineralized bone cement enhances its biomechanical characteristics that promote better cellular attachment, viability, proliferation, and mechanical properties. 3D printing outperforms conventional processes and other conventional techniques due to their potential to fabricate specific geometries and controlled spatial patterning of polymers and drugs [42].

19.7 Biomedical Applications of Biopolymers Based 3D- and 4D-Printed Bioconstructs

19.7.1 Tissue Engineering

In tissue engineering, some significant challenges are lack of vascularization and biomimetic microenvironments in engineered constructs. However, 3D printing was developed to create anatomically relevant structures with improved biomimicry and vascularization. The cellular arrangement and composition, including extracellular matrix (ECM), are vital for the design and production. 3D bioprinting techniques were employed for producing various tissue and artificial organs.

The human skin is very important for the human body as it serves crucial functions like protecting against harmful UV rays and harmful microbes. The human skin consists of a multifaceted microenvironment with varying physicochemical properties, such as porosity, density, and cell types. 3D bioprinting enables the precise placement of diverse cells at predetermined sites, facilitating enhanced cellular and biomaterial interactions. Type I collagen is a usual bio-ink used for skin bioprinting owing to its significance as a major structural protein in native human skin. Skin bioprinting has rapidly evolved from a simple layer-by-layer construct to added intricate bioengineered skin.

Recently, researchers printed diverse human skin cells, including fibroblasts, melanocytes, and keratinocytes, with porous graded constructs in a biomimetic 3D microenvironment. The engineered constructs mimicked the native human skin, with small and larger pores in the epidermal and dermal regions, respectively. This approach led to improved interactions between cells and materials [88].

An additional study utilized 3D bioprinting to fabricate a complete skin model with pigmentation [89]. Multiple layers of fibroblast were printed by adding collagen, and sodium bicarbonate was used for crosslinking. Then, the dermal layer was coated by printing melanocytes and keratinocytes over it, which results in induced skin pigmentation.

Moreover, researchers confirmed the possibility of *in situ* bioprinting for hair follicle regeneration in mice [90]. They mixed Matrigel with epidermal stem cells and skin-derived precursor cells taken from neonatal mice.

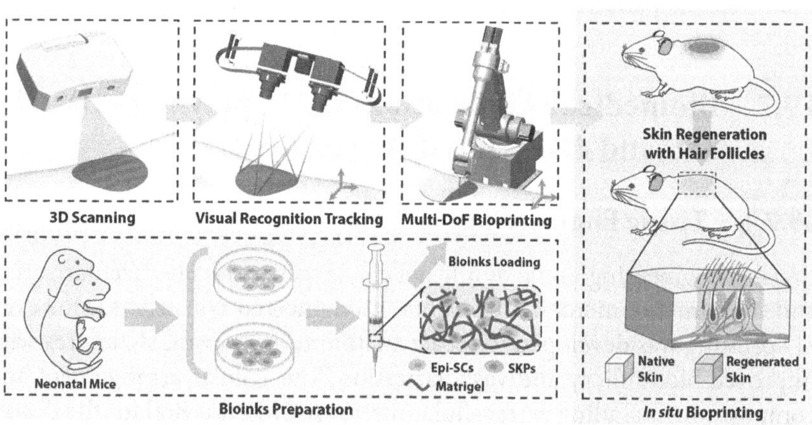

Figure 19.5 An illustrated diagram depicting the comprehensive workflow of an *in situ* bioprinting robot with adaptive multi-degree-of-freedom, utilized for achieving skin restoration in mice [90].

The mixture was directly printed on the injured site. This approach led to comprehensive healing and regeneration of the wound and functional skin tissue, respectively (Figure 19.5). This method shows promise for potential translational clinical applications.

Additionally, another bioprinting study induced hair follicle-like structure formation utilizing gelatin-alginate hydrogel and different cell types, namely, human fibroblasts, human umbilical vein endothelial, dermal papilla cells, and epidermal cells [91]. Furthermore, in a unique approach, a bioprinter was used to deposit collagen droplets comprising mouse embryonic mesenchymal and epithelial cells onto surgical guides, producing highly aligned guide-inserted hair microgels. The use of this innovative technique amended graft alignment in the skin and considerably improved hair extension. So far, at the utmost, every bio-printed skin construct showed alike skin biomarkers with a morphological look resembling human skin [92]. This success demonstrates the promising potential of skin bioprinting for achieving functional skin constructs.

The heart is a vital organ responsible for blood circulation, transporting oxygen and nutrients and removing metabolic wastes. It consists of specialized cardiac tissues with four layers: pericardium, epicardium, myocardium, and endocardium. A recent study fabricated a human heart using 3D bioprinting [93]. Thermo-responsive decellularized bio-ink from human omental tissues, along with induced pluripotent stem cell-derived cardiomyocytes (CMs) and endothelial cells, was used for the first time in personalized materials. The study also introduced a novel bioprinting approach using alginate microparticles for printing whole organs or tissues. However, creating complex vascular networks in the heart is still in the early stages, requiring further research. In another study, researchers achieved 3D printing of heart patches in a single step [94]. The resulting bio-electronics construct enables real-time monitoring and control of engineered tissue function.

For the human lung, prior bioprinting efforts focused on the air–blood barrier, critical for gaseous exchange [95]. Air–blood barrier models with two layers were successfully fabricated with drop-on-demand bioprinting. An alveolar lung model with triple layers was also developed with AECs, endothelial cells, and lung fibroblasts using microvalve printing.

Bone tissue engineering saw a significant progress in designing biomimetic implants [96, 97]. A tissue-engineered bone with flexible mechanical features was fabricated by employing an integrated tissue–organ printer.

In recent years, novel 4D printing techniques were used to produce biomimetic blood vessels. Recently, vascular cells stimulated by growth factors expedited signs of vascularization [98]. Hydrogel substrates patterned with different muscle cells likewise facilitated cell migration and aggregation over 2 weeks, resulting in the formation of vascular-like networks [99]. Another approach involved fabricating the bioprinting of agarose rods loaded with cells, which transformed into cylindrical tubes with matured tissue and fused cells [100, 101].

A different strategy relied on cells generating tangential tension to induce self-folding, resulting in vascularization. Cylindrical tubes resembling blood vessels were created, employing tangential tension to the human umbilical vein endothelial cells along diagonal lines, thus allowing for adjustable tube diameters [102].

Importantly, most researchers focused on simple structures. However, the vascular network in the human body consists of three different layers, namely, fibroblasts, smooth muscle cells, and endothelial cells. Additionally, thorough mechanical testing must be conducted to analyze the properties of the fabricated blood vessels [103].

These pioneering studies validated the vast capability of 3D and 4D bioprinting to fabricate organs and tissues, opening new avenues for medical applications and therapeutic advancements.

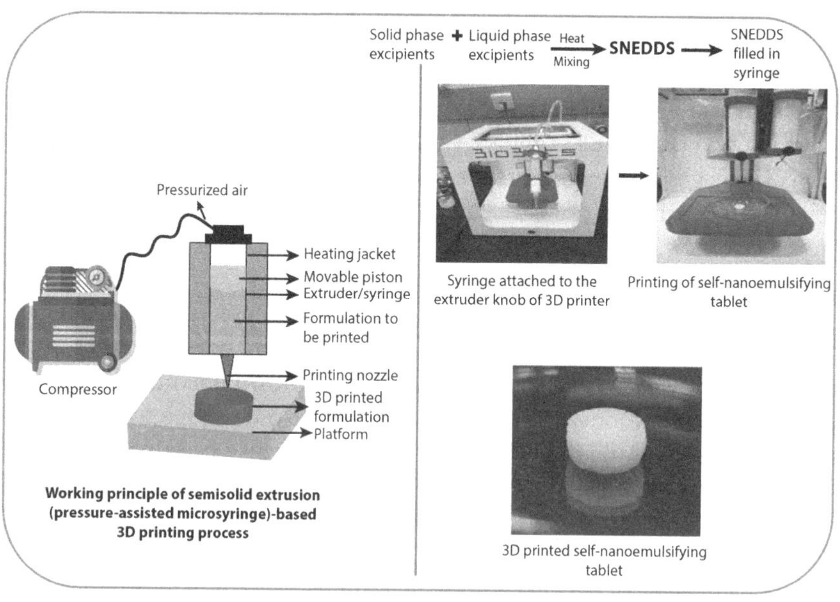

Figure 19.6 Schematics of the 3D-printing-based synthesis of dapagliflozin containing self-nano-emulsifying tablets [104].

19.7.2 Drug Delivery

3D and 4D printing finds an intriguing application in the realm of controlled drug release, where drugs can be released in a predefined way on trigger by an external stimulus. Reversible folding and delayed drug release are the two approaches used for controlled drug release. Figure 19.6 shows the schematics of the 3D printing-based synthesis of self-nano-emulsifying tablets loaded with dapagliflozin.

One approach involves printing aqueous droplets of specific compositions into an oil emulsion, leading to the formation of interface bilayers known as multisomes. The printed multisomes are utilized to deliver drugs in a controlled fashion under the influence of an external stimulus, such as temperature or pH. Additionally, osmolarity gradients can be programmed to carry out reversible folding and unfolding of intricate assemblies, enabling the encapsulation and on-demand release of drugs [105]. In another study, poly(vinyl alcohol) was 3D-printed with shape memory properties by employing the FDM printing technique [106].

19.7.3 Biomedical Devices

The relevance of 3D and 4D printing in the biomedical domain is fascinating. These techniques helped to create stimuli-responsive biomedical devices, such as stents. Stents are miniature tube-like biomedical devices used to keep blocked arteries open during angioplasty. They are inserted alongside a balloon catheter, which inflates at the site of a blocked artery to enlarge the stent. 4D printing allows for the production of stents with programmable variations in their dimensions when stimulated, reducing the invasiveness of angioplasty surgery [107–109]. Figure 19.7 shows an interesting example of a gripper fabricated using 4D printing.

The stents are also fabricated using temperature-responsive biopolymers using 4D printing [110, 111]. Recently, vascular stents with shape memory were 4D-printed using FDM with PLA fibers. The stent working mechanism relies on PLA's glass transition temperature (T_g), wherein it is heated above T_g (around 66°C). At this temperature, it becomes a soft and rubbery material representing its temporary form. Cooling then fixes this temporary form. Reheating it above PLA's T_g results in the original stent shape within 5 s [112]. Previous research demonstrated that brief balloon dilatation at 80°C does not result in major blood vessel damage [113]. Similarly, PLA stents were fabricated using 4D printing at 60°C using FDM [111, 112]. Furthermore, a polymer consisting of polycaprolactone (PCL) arms and a β-cyclodextrin (β-CD) core was fabricated using 4D printing

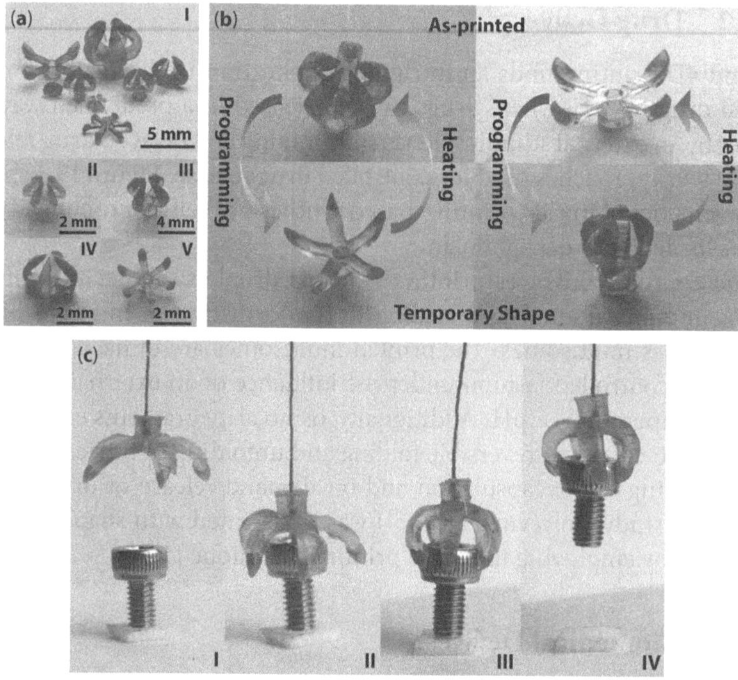

Figure 19.7 Development of 3D-printed grippers with multiple materials: (a) various designs of the multimaterial grippers, (b) displaying the shape transformation, and (c) photographs illustrating the gripping process [35].

to produce vascular stents with shape memory. The transition temperature (T_r) used was 40°C, which enables the shape memory effect within 5 s, showing a shape recovery ratio of ~98% [107].

Magnet-responsive materials have also been explored for stent fabrication. The iron oxide (Fe_3O_4) particles were mixed with PLA to synthesize a UV photo-curable ink to be employed to fabricate smart stents using 4D printing, activated under the influence of a magnetic field [108]. FDM printing was likewise used to fabricate a PLA/Fe_3O_4 composite tracheal stent with a shape recovery rate of >99% within 40 s [109].

19.8 Other Applications

19.8.1 Food Engineering

Cultured meat (CM) is an innovative approach to cellular agriculture where specific muscle-based cells are obtained from desired animals through biopsies [114]. The isolated cells are then multiplied *in vitro* using bioreactors. The major challenge associated with the CM is the scale-up

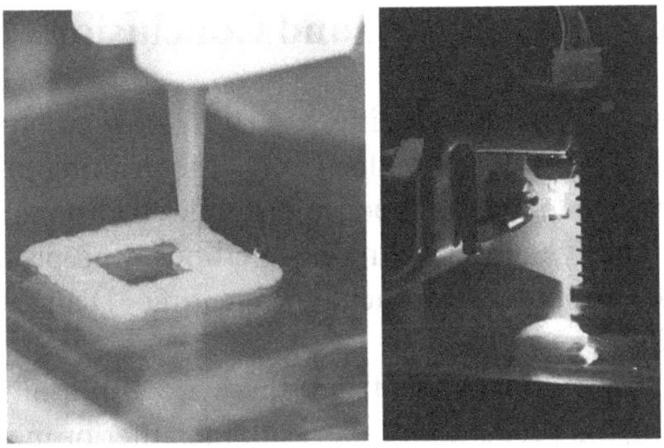

Figure 19.8 3D printing of raw chicken using a food printer. (a) An up-close view of raw chicken precisely printed using a food printer. (b) Raw chicken targeted by a blue laser beam [116].

of industrial production. This is due to the huge number of muscle cells (approximately 10^{13} cells) that are required to produce 1 kg of CM, necessitating bulky bioreactors (capacity ~5,000 L) [115]. Figure 19.8 demonstrates the 3D printing of raw chicken using a food printer.

One crucial aspect of cell multiplication is the growth media due to the use of fetal bovine serum, which is a concern in terms of animal welfare and ethics. Furthermore, the use of antibiotics may result in the development of antibiotic resistance.

In CM development, a 3D porous scaffold serves as the base for cell attachment, proliferation, and differentiation into muscles, fats, and connective tissues that ultimately resemble real meat in terms of texture and structure [117]. Many fabrication techniques were employed to fabricate 3D porous scaffolds for CM applications, including microbial transglutaminase-crosslinked electrospun gelatin fibers, decellularized cellulose scaffolds from apple, micropatterned film scaffolds fabricated by laser cutting, 3D micro-carriers, 3D bio-printed textured soy protein scaffolds, and freeze-dried scaffolds [118]. Remarkably, "meat paste" that consists of plant-based proteins or conventional meat with water is generally utilized to 3D print CM [119].

An ideal CM scaffold should possess biocompatibility, biodegradability, design, physicochemical properties, and effective manufacturing methods [120]. 3D food printing results in the precise deposition of desired cells at predetermined sites on the 3D-printed CM, which is an advantage compared to the conventional 3D cell culture technique. This enhancement in the interactions among cells and matrix is crucial to regulate tissue maturation processes.

19.9 Future Perspectives and Conclusion

Inks made of natural polysaccharides are commonly employed in 3D and 4D printing-based additive manufacturing. The printing process relies on various factors like polymer type, hydrophilic/hydrophobic properties, viscosity, printability, and crosslinking methods. The presence of better physicochemical properties is crucial for creating 3D and 4D objects and ensuring quality.

While there are evident advantages, there are also limitations, such as time-consuming ink preparation, short ink life, inferior mechanical properties, and storage challenges. Efforts were made to address these issues through formulation improvements and advancements in 3D and 4D printer technology.

Continuously striving for improvement and innovation in materials used for 3D and 4D printing of medical devices, the goal is to enhance stability, functional activity, cell attachment, proliferation, and viability by closely imitating the body's environment. Combining polysaccharides with other substances and employing multi-crosslinking approaches can improve their mechanical properties.

Polysaccharide-based composite materials with customizable features are desired to expand the application possibilities of 3D and 4D printing. Particularly, bioprinting natural polysaccharides are valuable for its ability to imitate the extracellular matrix composition, which eases the fabrication of tissues, organs, sensors, and various diagnostic tools.

In the healthcare field, 3D and 4D printing represents a revolutionary promise, allowing the production of incredibly adapted point-of-care medical devices. It offers the advantage of rapid production, even during emergencies like the early stages of the COVID-19 pandemic, where supply chain issues can be mitigated.

Despite the unquestionable advantages of 3D and 4D printing for the fabrication of medical devices, several challenges remain. The medical uses of 3D and 4D printing include various products, from complete devices to individual components like skin patches, biosensors, disposable devices, and scaffolds. Orthopedic applications, using biomaterial-based porous structures, are widely prevalent, as they facilitate tissue growth and regeneration.

In the near future, there are three potential scenarios for using 3D and 4D printing for the medical field: hospitals adopting 3D and 4D printing processes, healthcare providers using on-site 3D and 4D printers for production, and outsourcing the service to companies that are manufacturing and delivering medical tools. Each of these perspectives presents benefits and contests, particularly related to risk evaluation and accommodating changes in 3D and 4D designs.

New regulations related to 3D and 4D printing in healthcare emphasize patient safety and product suitability. With the widespread popularity of 3D and 4D printing, clear guidelines and manufacturing parameters are essential to minimize potential risks and ensure that medical devices meet standardized specifications and safety requirements.

Looking ahead, the next frontier beyond 3D printing is 4D printing, where printed objects show the ability to structural transformations in response to external stimuli. This opens possibilities for advanced implantable devices with more natural functionality in the body.

Biopolymers suitable for 3D and 4D printing are actively researched, emphasizing the use of renewable sources to reduce environmental impact. As 3D technologies improve and material providers better understand the properties and challenges of polymers, we anticipate substantial growth in the medical industry due to the advantages offered by 3D printing.

Thus far, challenges are there to prevail in integrating 3D and 4D printing into the medical field, whereas ongoing research and investments are driving innovation and progress in this area, paving the way for future advancements.

References

1. Lee, J.-Y., An, J., Chua, C.K., Fundamentals and applications of 3D printing for novel materials. *Appl. Mater. Today*, 7, 120–133, 2017.
2. An, J., Chua, C., Mironov, V., A perspective on 4D bioprinting. *Int. J. Bioprinting*, 2, 3–5, 2015.
3. Tibbits, S., 4D printing: Multi-material shape change. *Archit. Des.*, 84, 1, 116–121, 2014.
4. Serjouei, A., Yousefi, A., Jenaki, A., Bodaghi, M., Mehrpouya, M., 4D printed shape memory sandwich structures: Experimental analysis and numerical modeling. *Smart Mater. Struct.*, 31, 5, 55014, 2022.
5. Barletta, M., Gisario, A., Mehrpouya, M., 4D printing of shape memory polylactic acid (PLA) components: Investigating the role of the operational parameters in fused deposition modelling (FDM). *J. Manuf. Process.*, 61, 473–480, 2021.
6. Amukarimi, S. and Mozafari, M., 4D bioprinting of tissues and organs. *Bioprinting*, 23, e00161, 2021.

7. Najmon, J.C., Raeisi, S., Tovar, A., Review of additive manufacturing technologies and applications in the aerospace industry, in: *Additive Manufacturing for the Aerospace Industry*, 2019.
8. Böckin, D. and Tillman, A.-M., Environmental assessment of additive manufacturing in the automotive industry. *J. Clean. Prod.*, 226, 977–987, 2019.
9. Yan, Q., Dong, H., Su, J., Han, J., Song, B., Wei, Q., Shi, Y., A review of 3D printing technology for medical applications. *Engineering*, 4, 5, 729–742, 2018.
10. Jamróz, W., Szafraniec, J., Kurek, M., Jachowicz, R., 3D Printing in pharmaceutical and medical applications-recent achievements and challenges. *Pharm. Res.*, 35, 9, 176, 2018.
11. Ng, W.L., Chua, C.K., Shen, Y.-F., Print me an organ! Why we are not there yet. *Prog. Polym. Sci.*, 97, 101145, 2019.
12. Lee, J.M., Ng, W.L., Yeong, W.Y., Resolution and shape in bioprinting: Strategizing towards complex tissue and organ printing. *Appl. Phys. Rev.*, 6, 1, 11307, 2019.
13. Ng, W.L., Chan, A., Ong, Y.S., Chua, C.K., Deep learning for fabrication and maturation of 3D bioprinted tissues and organs. *Virtual Phys. Prototyp.*, 15, 3, 340–358, 2020.
14. Goh, G.D., Yap, Y.L., Tan, H.K.J., Sing, S.L., Goh, G.L., Yeong, W.Y., Process–structure–Properties in polymer additive manufacturing via material extrusion: A review. *Crit. Rev. Solid State Mater. Sci.*, 45, 2, 113–133, 2020.
15. Shankar, S. and Rhim, J.-W., Bionanocomposite films for food packaging applications, in: *Reference Module in Food Science*, vol. 1, pp. 1–10, 2018.
16. Reddy, N., Reddy, R., Jiang, Q., Crosslinking biopolymers for biomedical applications. *Trends Biotechnol.*, 33, 6, 362–369, 2015.
17. Hull, C.W., Apparatus for production of three-dimensional objects by stereolithography. 3D Systems INC (US), US-6027324-A Patent. https://patents.google.com/patent/US4575330A/en
18. Montalbano, G., Calore, A.R., Vitale-Brovarone, C., Extrusion 3D printing of a multiphase collagen-based material: An optimized strategy to obtain biomimetic scaffolds with high shape fidelity. *J. Appl. Polym. Sci.*, 140, 10, e53593, 2023.
19. Deckard, C.R., Method and apparatus for producing parts by selective sintering. US4863538A (US Patent). https://patents.google.com/patent/US4863538A/en
20. The Story of the World's First Zero Gravity Printer. Made in space, Webpage, https://madeinspace.us/.
21. M. Cima, E. Sachs, T. Fan., J.F. Bredt, S.P. Michaels, A. S. Khanuja, S.J.J. Lee, D. Brancazio, A. Curodeau, H. Tuerck, Three-dimensional printing techniques. EP 1099534 A2.
22. C.S. Scott, Apparatus and method for creating three-dimensional objects. US 5121329 A.

23. Stratasys Advanced 3D Printing Technologies, Stratasys Ltd. Webpage. https://www.stratasys.com/en/3d-printing-technologies/.
24. R.C. Sanders Jr., J.L. Forsyth, K.F. Philbrook, 3-D model making. US 5740051 A.
25. ORNL shares its know-how, Manufacturing Demonstration Facility, Oak Ridge National Laboratory. Webpage. https://www.ornl.gov/blog/ornl-review/ornl-shares-its-know-how.
26. Researching 3D printing technology on the space station. Webpage. https://www.nasa.gov/image-article/researching-3d-printing-technology-space-station/
27. 3D printer activated aboard the international space station. Webpage. https://www.nasa.gov/image-article/international-space-stations-3-d-printer-2/
28. Hofmann, M., 3D printing gets a boost and opportunities with polymer materials. *ACS Macro Lett.*, 3, 4, 382–386, 2014.
29. Bajpai, A., Baigent, A., Raghav, S., Brádaigh, C.Ó., Koutsos, V., Radacsi, N., 4D printing: Materials, technologies, and future applications in the biomedical field. *Sustainability*, 12, 24, 10628, 2020.
30. Sydney Gladman, A., Matsumoto, E.A., Nuzzo, R.G., Mahadevan, L., Lewis, J.A., Biomimetic 4D printing. *Nat. Mater.*, 15, 4, 413–418, 2016.
31. Ge, Q., Qi, H.J., Dunn, M.L., Active materials by four-dimension printing. *Appl. Phys. Lett.*, 103, 13, 131901, 2013.
32. Mulakkal, M.C., Trask, R.S., Ting, V.P., Seddon, A.M., Responsive cellulose-hydrogel composite ink for 4D printing. *Mater. Des.*, 160, 108–118, 2018.
33. Momeni, F., M.Mehdi Hassani.N, S., Liu, X., Ni, J., A review of 4D printing. *Mater. Des.*, 122, 42–79, 2017.
34. Choi, J., Kwon, O.-C., Jo, W., Lee, H.J., Moon, M.-W., 4D printing technology: A review. *3D Print. Addit. Manuf.*, 2, 4, 159–167, 2015.
35. Ge, Q., Sakhaei, A.H., Lee, H., Dunn, C.K., Fang, N.X., Dunn, M.L., Multimaterial 4D printing with tailorable shape memory polymers. *Sci. Rep.*, 6, 1, 31110, 2016.
36. Subash, A. and Kandasubramanian, B., 4D printing of shape memory polymers. *Eur. Polym. J.*, 134, 109771, 2020.
37. Colosi, C., Costantini, M., Latini, R., Ciccarelli, S., Stampella, A., Barbetta, A., Massimi, M., Conti Devirgiliis, L., Dentini, M., Rapid prototyping of chitosan-coated alginate scaffolds through the use of a 3D fiber deposition technique. *J. Mater. Chem. B*, 2, 39, 6779–6791, 2014.
38. Hoque, M.E., Chuan, Y.L., Pashby, I., Extrusion based rapid prototyping technique: An advanced platform for tissue engineering scaffold fabrication. *Biopolymers*, 97, 2, 83–93, 2012.
39. Honda, M., Yada, T., Ueda, M., Kimata, K., Cartilage formation by cultured chondrocytes in a new scaffold made of poly(L-lactide-ϵ-caprolactone) sponge. *J. Oral. Maxillofac. Surg.*, 58, 7, 767–775, 2000.
40. Cheung, H.-Y., Lau, K.-T., Lu, T.-P., Hui, D., A critical review on polymer-based bio-engineered materials for scaffold development. *Compos. Part B Eng.*, 38, 3, 291–300, 2007.

41. Inzana, J.A., Olvera, D., Fuller, S.M., Kelly, J.P., Graeve, O.A., Schwarz, E.M., Kates, S.L., Awad, H.A., 3D printing of composite calcium phosphate and collagen scaffolds for bone regeneration. *Biomaterials*, 35, 13, 4026–4034, 2014.
42. Shahbazi, M. and Jäger, H., Current status in the utilization of biobased polymers for 3D printing process: A systematic review of the materials, processes, and challenges. *ACS Appl. Bio Mater.*, 4, 1, 325–369, 2021.
43. Ding, R., Du, Y., Goncalves, R.B., Francis, L.F., Reineke, T.M., Sustainable near UV-curable acrylates based on natural phenolics for stereolithography 3D printing. *Polym. Chem.*, 10, 9, 1067–1077, 2019.
44. Basara, G., Yue, X., Zorlutuna, P., Dual crosslinked gelatin methacryloyl hydrogels for photolithography and 3D printing. *Gels*, 5, 3, 34, 2019.
45. Mazzocchi, A., Devarasetty, M., Huntwork, R., Soker, S., Skardal, A., Optimization of collagen type I-hyaluronan hybrid bioink for 3D bioprinted liver microenvironments. *Biofabrication*, 11, 1, 15003, 2018.
46. Hong, H., Seo, Y.B., Kim, D.Y., Lee, J.S., Lee, Y.J., Lee, H., Ajiteru, O., Sultan, M.T., Lee, O.J., Kim, S.H., Park, C.H., Digital light processing 3D printed silk fibroin hydrogel for cartilage tissue engineering. *Biomaterials*, 232, 119679, 2020.
47. BeMiller, J.N., Carbohydrate chemistry preface, in: *Carbohydrate Chemistry for Food Scientists*, Third Edition, J.N. BeMiller, (Ed.), pp. xi–xii, AACC International Press, Sawston, Cambridge, 2019.
48. Poshina, D.N., Tyshkunova, I.V., Petrova, V.A., Skorik, Y.A., Electrospinning of polysaccharides for tissue engineering applications. *Rev. Adv. Chem.*, 11, 1, 112–133, 2021.
49. Purkait, M.K., Sinha, M.K., Mondal, P., Singh, R., *Stimuli Responsive Polymeric Membranes Smart Polymeric Membranes*, Academic Press, Elsevier, Cambridge, Massachusetts, 2018.
50. Singh, R., Sinha, M.K., Purkait, M.K., Stimuli responsive mixed matrix polysulfone ultrafiltration membrane for humic acid and photocatalytic dye removal applications. *Sep. Purif. Technol.*, 250, 117247, 2020.
51. Purkait, M.K., Sinha, M.K., Mondal, P., Singh, R., Chapter 3-Temperature-responsive membranes, in: *Stimuli Responsive Polymeric Membranes*, vol. 25, pp. 67–113, Elsevier, Cambridge, Massachusetts, 2018.
52. Purkait, M.K. and Singh, R., *Membrane Technology in Separation Science*, CRC Press, Boca Raton, 2018.
53. Wang, X., Sun, Y., Peng, C., Luo, H., Wang, R., Zhang, D., Transitional suspensions containing thermosensitive dispersant for three-dimensional printing. *ACS Appl. Mater. Interfaces*, 7, 47, 26131–26136, 2015.
54. Champeau, M., Heinze, D.A., Viana, T.N., de Souza, E.R., Chinellato, A.C., Titotto, S., 4D printing of hydrogels: A review. *Adv. Funct. Mater.*, 30, 31, 1910606, 2020.
55. Yang, Q., Gao, B., Xu, F., Recent advances in 4D bioprinting. *Biotechnol. J.*, 15, 1, 1900086, 2020.

56. Chadwick, M., Yang, C., Liu, L., Gamboa, C.M., Jara, K., Lee, H., Sabaawy, H.E., Rapid processing and drug evaluation in glioblastoma patient-derived organoid models with 4D bioprinted arrays. *iScience*, 23, 8, 101365, 2020.
57. Abdullah, T. and Okay, O., 4D printing of body temperature-responsive hydrogels based on poly(acrylic acid) with shape-memory and self-healing abilities. *ACS Appl. Bio Mater.*, 6, 2, 703–711, 2023.
58. Park, Y. and Chen, X., Water-responsive materials for sustainable energy applications. *J. Mater. Chem. A*, 8, 31, 15227–15244, 2020.
59. Verpaalen, R.C.P., Engels, T., Schenning, A.P.H.J., Debije, M.G., Stimuli-responsive shape changing commodity polymer composites and bilayers. *ACS Appl. Mater. Interfaces*, 12, 35, 38829–38844, 2020.
60. Morouço, P., Azimi, B., Milazzo, M., Mokhtari, F., Fernandes, C., Reis, D., Danti, S., Four-dimensional (bio-)printing: A review on stimuli-responsive mechanisms and their biomedical suitability. *Appl. Sci.*, 10, 24, 9143, 2020.
61. Ashammakhi, N., Ahadian, S., Zengjie, F., Suthiwanich, K., Lorestani, F., Orive, G., Ostrovidov, S., Khademhosseini, A., Advances and future perspectives in 4D bioprinting. *Biotechnol. J.*, 13, 12, 1800148, 2018.
62. Jamal, M., Kadam, S.S., Xiao, R., Jivan, F., Onn, T.-M., Fernandes, R., Nguyen, T.D., Gracias, D.H., Bio-origami hydrogel scaffolds composed of photocross-linked peg bilayers. *Adv. Healthc. Mater.*, 2, 8, 1142–1150, 2013.
63. Purkait, M.K., Sinha, M.K., Mondal, P., Singh, R., Chapter 6-Electric field-responsive membranes, in: *Stimuli Responsive Polymeric Membranes*, vol. 25, pp. 173–191, Elsevier, Cambridge, Massachusetts, 2018.
64. Manouras, T. and Vamvakaki, M., Field responsive materials: Photo-{,} electro-{,} magnetic- and ultrasound-sensitive polymers. *Polym. Chem.*, 8, 1, 74–96, 2017.
65. Lee, S., Kim, Y.K., Hong, J.-Y., Jang, J., Electro-response of MoS2 nanosheets-based smart fluid with tailorable electrical conductivity. *ACS Appl. Mater. Interfaces*, 8, 36, 24221–24229, 2016.
66. Fantino, E., Roppolo, I., Zhang, D., Xiao, J., Chiappone, A., Castellino, M., Guo, Q., Pirri, C.F., Yang, J., 3D printing/interfacial polymerization coupling for the fabrication of conductive hydrogel. *Macromol. Mater. Eng.*, 303, 4, 1700356, 2018.
67. Pairatwachapun, S., Paradee, N., Sirivat, A., Controlled release of acetylsalicylic acid from polythiophene/carrageenan hydrogel via electrical stimulation. *Carbohydr. Polym.*, 137, 214–221, 2016.
68. Pyarasani, R.D., Jayaramudu, T., John, A., Polyaniline-based conducting hydrogels. *J. Mater. Sci.*, 54, 2, 974–996, 2019.
69. Purkait, M.K., Sinha, M.K., Mondal, P., Singh, R., Chapter 7-Magnetic-responsive membranes, in: *Stimuli Responsive Polymeric Membranes*, vol. 25, pp. 193–219, Elsevier, Cambridge, Massachusetts, 2018.
70. Filipcsei, G., Csetneki, I., Szilágyi, A., Zrínyi, M., Magnetic field-responsive smart polymer composites, in: *Oligomers-Polymer Composites - Molecular*

Imprinting, pp. 137–189, Springer Berlin Heidelberg, Berlin, Heidelberg, 2007.
71. Frachini, E.C.G. and Petri, D.F.S., Magneto-responsive hydrogels: Preparation, characterization, biotechnological and environmental applications. *J. Braz. Chem. Soc.*, 30, 10, 2010–2028, 2019.
72. Liu, Z., Liu, J., Cui, X., Wang, X., Zhang, L., Tang, P., Recent advances on magnetic sensitive hydrogels in tissue engineering. *Front. Chem.*, 8, 124, 2020.
73. Betsch, M., Cristian, C., Lin, Y.-Y., Blaeser, A., Schöneberg, J., Vogt, M., Buhl, E.M., Fischer, H., Duarte Campos, D.F., Incorporating 4D into bioprinting: Real-time magnetically directed collagen fiber alignment for generating complex multilayered tissues. *Adv. Healthc. Mater.*, 7, 21, 1800894, 2018.
74. Singh, R., Mondal, P., Purkait, M.K., *pH Responsive Membranes Biomedical Applications*, CRC Press, Boca Raton, 2022.
75. Purkait, M.K., Sinha, M.K., Mondal, P., Singh, R., Chapter 2-pH-responsive membranes, in: *Stimuli Responsive Polymeric Membranes*, vol. 25, pp. 39–66, Elsevier, Cambridge, Massachusetts, 2018.
76. Purkait, M.K., Sinha, M.K., Mondal, P., Singh, R., Chapter 4-Photoresponsive membranes, in: *Stimuli Responsive Polymeric Membranes*, vol. 25, pp. 115–144, Elsevier, Cambridge, Massachusetts, 2018.
77. Purkait, M.K., Sinha, M.K., Mondal, P., Singh, R., Chapter 5-Biologically responsive membranes, in: *Stimuli Responsive Polymeric Membranes*, vol. 25, pp. 145–171, Elsevier, Cambridge, Massachusetts, 2018.
78. Purkait, M.K., Sinha, M.K., Mondal, P., Singh, R., Chapter 8-Ultrasound-responsive membranes, in: *Stimuli Responsive Polymeric Membranes*, vol. 25, pp. 221–237, Elsevier, Cambridge, Massachusetts, 2018.
79. Yang, K.-N., Zhang, C.-Q., Wang, W., Wang, P.C., Zhou, J.-P., Liang, X.-J., pH-responsive mesoporous silica nanoparticles employed in controlled drug delivery systems for cancer treatment. *Cancer Biol. Med.*, 11, 1, 34–43, 2014.
80. Kocak, G., Tuncer, C., Bütün, V., pH-responsive polymers. *Polym. Chem.*, 8, 1, 144–176, 2017.
81. Tang, H., Zhao, W., Yu, J., Li, Y., Zhao, C., Recent development of ph-responsive polymers for cancer nanomedicine. *Molecules*, 24, 1, 4, 2019.
82. Patel, V.R. and Amiji, M.M., Preparation and characterization of freeze-dried chitosan-poly(ethylene oxide) hydrogels for site-specific antibiotic delivery in the stomach. *Pharm. Res.*, 13, 4, 588–593, 1996.
83. Amiji, M., Tailor, R., Ly, M.-K., Goreham, J., Gelatin-poly(ethylene oxide) semi-interpenetrating polymer network with ph-sensitive swelling and enzyme-degradable properties for oral drug delivery. *Drug Dev. Ind. Pharm.*, 23, 6, 575–582, 1997.
84. Feng, W., Nie, W., He, C., Zhou, X., Chen, L., Qiu, K., Wang, W., Yin, Z., Effect of pH-responsive alginate/chitosan multilayers coating on delivery

efficiency, cellular uptake and biodistribution of mesoporous silica nanoparticles based nanocarriers. *ACS Appl. Mater. Interfaces*, 6, 11, 8447–8460, 2014.
85. Rudko, M., Urbaniak, T., Musiał, W., Recent developments in ion-sensitive systems for pharmaceutical applications. *Polym. (Basel)*, 13, 10, 1641, 2021.
86. Bai, T., Han, Y., Zhang, P., Wang, W., Liu, W., Zinc ion-triggered two-way macro-/microscopic shape changing and memory effects in high strength hydrogels with pre-programmed unilateral patterned surfaces. *Soft Matter*, 8, 25, 6846–6852, 2012.
87. Nan, W., Wang, W., Gao, H., Liu, W., Fabrication of a shape memory hydrogel based on imidazole–zinc ion coordination for potential cell-encapsulating tubular scaffold application. *Soft Matter*, 9, 1, 132–137, 2013.
88. Ng, W.L., Qi, J.T.Z., Yeong, W.Y., Naing, M.W., Proof-of-concept: 3D bioprinting of pigmented human skin constructs. *Biofabrication*, 10, 2, 25005, 2018.
89. Min, D., Lee, W., Bae, I.-H., Lee, T.R., Croce, P., Yoo, S.-S., Bioprinting of biomimetic skin containing melanocytes. *Exp. Dermatol.*, 27, 5, 453–459, 2018.
90. Zhao, W., Chen, H., Zhang, Y., Zhou, D., Liang, L., Liu, B., Xu, T., Adaptive multi-degree-of-freedom *in situ* bioprinting robot for hair-follicle-inclusive skin repair: A preliminary study conducted in mice. *Bioeng. Transl. Med.*, 7, 3, e10303, 2022.
91. Kang, D., Liu, Z., Qian, C., Huang, J., Zhou, Y., Mao, X., Qu, Q., Liu, B., Wang, J., Hu, Z., Miao, Y., 3D bioprinting of a gelatin-alginate hydrogel for tissue-engineered hair follicle regeneration. *Acta Biomater.*, 165, 19–30, 2023.
92. Pourchet, L.J., Thepot, A., Albouy, M., Courtial, E.J., Boher, A., Blum, L.J., Marquette, C.A., Human skin 3D bioprinting using scaffold-free approach. *Adv. Healthc. Mater.*, 6, 4, 1601101, 2017.
93. Noor, N., Shapira, A., Edri, R., Gal, I., Wertheim, L., Dvir, T., 3D printing of personalized thick and perfusable cardiac patches and hearts. *Adv. Sci.*, 6, 11, 1900344, 2019.
94. Asulin, M., Michael, I., Shapira, A., Dvir, T., One-step 3D printing of heart patches with built-in electronics for performance regulation. *Adv. Sci.*, 8, 9, 2004205, 2021.
95. Horváth, L., Umehara, Y., Jud, C., Blank, F., Petri-Fink, A., Rothen-Rutishauser, B., Engineering an *in vitro* air-blood barrier by 3D bioprinting. *Sci. Rep.*, 5, 1, 7974, 2015.
96. Shuai, C., Yang, W., Peng, S., Gao, C., Guo, W., Lai, Y., Feng, P., Physical stimulations and their osteogenesis-inducing mechanisms. *Int. J. Bioprinting*, 4, 2, 138, 2018.
97. Bose, S., Roy, M., Bandyopadhyay, A., Recent advances in bone tissue engineering scaffolds. *Trends Biotechnol.*, 30, 10, 546–554, 2012.
98. Norotte, C., Marga, F.S., Niklason, L.E., Forgacs, G., Scaffold-free vascular tissue engineering using bioprinting. *Biomaterials*, 30, 30, 5910–5917, 2009.

99. Hong, S., Song, S.-J., Lee, J.Y., Jang, H., Choi, J., Sun, K., Park, Y., Cellular behavior in micropatterned hydrogels by bioprinting system depended on the cell types and cellular interaction. *J. Biosci. Bioeng.*, 116, 2, 224–230, 2013.
100. Huang, G.Y., Zhou, L.H., Zhang, Q.C., Chen, Y.M., Sun, W., Xu, F., Lu, T.J., Microfluidic hydrogels for tissue engineering. *Biofabrication*, 3, 1, 12001, 2011.
101. Yan, Y., Wang, X., Pan, Y., Liu, H., Cheng, J., Xiong, Z., Lin, F., Wu, R., Zhang, R., Lu, Q., Fabrication of viable tissue-engineered constructs with 3D cell-assembly technique. *Biomaterials*, 26, 29, 5864–5871, 2005.
102. Kuribayashi-Shigetomi, K., Onoe, H., Takeuchi, S., Cell origami: Self-folding of three-dimensional cell-laden microstructures driven by cell traction force. *PloS One*, 7, 12, e51085, 2012.
103. Camasão, D.B. and Mantovani, D., The mechanical characterization of blood vessels and their substitutes in the continuous quest for physiological-relevant performances. A critical review. *Mater. Today Bio.*, 10, 100106, 2021.
104. Algahtani, M.S., Mohammed, A.A., Ahmad, J., Abdullah, M.M., Saleh, E., 3D printing of dapagliflozin containing self-nanoemulsifying tablets: Formulation design and *in vitro* characterization. *Pharmaceutics*, 13, 7, 993, 2021.
105. Villar, G., Graham, A.D., Bayley, H., A tissue-like printed material. *Science*, 340, 6128, 48–52, 2013.
106. Melocchi, A., Uboldi, M., Inverardi, N., Briatico-Vangosa, F., Baldi, F., Pandini, S., Scalet, G., Auricchio, F., Cerea, M., Foppoli, A., Maroni, A., Zema, L., Gazzaniga, A., Expandable drug delivery system for gastric retention based on shape memory polymers: Development via 4D printing and extrusion. *Int. J. Pharm.*, 571, 118700, 2019.
107. Zhou, Y., Zhou, D., Cao, P., Zhang, X., Wang, Q., Wang, T., Li, Z., He, W., Ju, J., Zhang, Y., 4D printing of shape memory vascular stent based on βCD-g-polycaprolactone. *Macromol. Rapid Commun.*, 42, 14, 2100176, 2021.
108. Wei, H., Zhang, Q., Yao, Y., Liu, L., Liu, Y., Leng, J., Direct-write fabrication of 4d active shape-changing structures based on a shape memory polymer and its nanocomposite. *ACS Appl. Mater. Interfaces*, 9, 1, 876–883, 2017.
109. Zhang, F., Wen, N., Wang, L., Bai, Y., Leng, J., Design of 4D printed shape-changing tracheal stent and remote controlling actuation. *Int. J. Smart Nano Mater.*, 12, 4, 375–389, 2021.
110. Zhao, W., Li, N., Liu, L., Leng, J., Liu, Y., Origami derived self-assembly stents fabricated via 4D printing. *Compos. Struct.*, 293, 115669, 2022.
111. Wang, X., Zhang, Y., Shen, P., Cheng, Z., Chu, C., Xue, F., Bai, J., Preparation of 4D printed peripheral vascular stent and its degradation behavior under fluid shear stress after deployment. *Biomater. Sci.*, 10, 9, 2302–2314, 2022.
112. Lin, C., Zhang, L., Liu, Y., Liu, L., Leng, J., 4D printing of personalized shape memory polymer vascular stents with negative Poisson's ratio structure: A preliminary study. *Sci. China Technol. Sci.*, 63, 4, 578–588, 2020.

113. Tamai, H., Igaki, K., Kyo, E., Kosuga, K., Kawashima, A., Matsui, S., Komori, H., Tsuji, T., Motohara, S., Uehata, H., Initial and 6-month results of biodegradable poly-*l*-lactic acid coronary stents in humans. *Circulation*, 102, 4, 399–404, 2000.
114. Handral, H.K., Tay, S.H., Chan, W.W., Choudhury, D., 3D printing of cultured meat products. *Crit. Rev. Food Sci. Nutr.*, 62, 1, 272–281, 2022.
115. Stephens, N., Di Silvio, L., Dunsford, I., Ellis, M., Glencross, A., Sexton, A., Bringing cultured meat to market: Technical, socio-political, and regulatory challenges in cellular agriculture. *Trends Food Sci. Technol.*, 78, 155–166, 2018.
116. Blutinger, J.D., Tsai, A., Storvick, E., Seymour, G., Liu, E., Samarelli, N., Karthik, S., Meijers, Y., Lipson, H., Precision cooking for printed foods via multiwavelength lasers. *NPJ Sci. Food*, 5, 1, 24, 2021.
117. Specht, E.A., Welch, D.R., Rees Clayton, E.M., Lagally, C.D., Opportunities for applying biomedical production and manufacturing methods to the development of the clean meat industry. *Biochem. Eng. J.*, 132, 161–168, 2018.
118. Seah, J.S.H., Singh, S., Tan, L.P., Choudhury, D., Scaffolds for the manufacture of cultured meat. *Crit. Rev. Biotechnol.*, 42, 2, 311–323, 2022.
119. Wilson, A., Anukiruthika, T., Moses, J.A., Anandharamakrishnan, C., Customized shapes for chicken meat–based products: Feasibility study on 3D-printed nuggets. *Food Bioprocess Technol.*, 13, 11, 1968–1983, 2020.
120. O'Brien, F.J., Biomaterials & scaffolds for tissue engineering. *Mater. Today*, 14, 3, 88–95, 2011.

20

Future Perspectives of Biopolymers for Biomedical Applications

Hitesh Chopra[1], Shivani Sharma[2], Mohd Shabbir[3], Annu[4*] and Dong-Kil Shin[4†]

[1]*Department of Biosciences, Saveetha School of Engineering, Saveetha Institute of Medical and Technical Sciences, Chennai, Tamil Nadu, India*
[2]*Institute of Pharmaceutical Sciences, Bhaddal, Punjab, India*
[3]*Department of Applied Sciences and Humanities (Chemistry), GL Bajaj Institute of Technology and Management, Greater Noida, India*
[4]*Thin-Film and Materials Laboratory, School of Mechanical Engineering, Yeungnam University, Gyeongsan, Republic of Korea*

Abstract

Materials of bio-origin are in great demand for their associated characteristic advantages over synthetic or petroleum-based materials. Biopolymers are produced by living cells in animals, plants, and microorganisms. They are similar to other polymers in their chemical structure such as made up of monomeric units connected in long chains, but added degradation properties. Cellulose, chitosan, starch, polylactic acid etc. are few examples of biopolymers utilized for almost every aspect of life. Better degradability and biocompatibility of these biopolymers make them more advantageous in biomedical fields. Biopolymers are heavily used in 3D printing, drug delivery, tissue engineering and transplants. This chapter highlights the modern applications of biopolymers in biomedical field and explores the future perspectives of biopolymers for biomedical applications. Current innovations in this area are highlighted with the support of recent patents and research literature.

Keywords: 3D-4D printing, hydrogel, stimulus-responsive, hyphenated drug, bioprinting, self-healing, patent, cancer model

**Corresponding author*: annuchem92@gmail.com
†*Corresponding author*: dkshin@yu.ac.kr

Annu (ed.) Biopolymers for Biomedical Applications, (509–530) © 2024 Scrivener Publishing LLC

20.1 Introduction

Biopolymer is derived from the Greek word for nature or living things. Living organisms generate polymeric biomolecules in long chains. A covalent bond is formed between the monomer units of these polymers, which results in the formation of biopolymers [1]. Gelatin, keratin, and cellulose are some examples of biopolymers. Numerous biopolymers (DNA and RNA) are capable of transmitting genetic information from generation to generation. Plants rely on cellulose and starch as building blocks. Unlike manufactured plastics, these polymers have existed for billions of years. A biopolymer has a well-defined structure, while synthetic polymers are unpredictable in their arrangement [2]. Thus, biological forces break big chains into smaller chains to form biopolymers.

Because they possess less C–C bond energy, enzymes, moisture, heat, and other elements readily break them down. This is due to the presence of a hydroxyl group. Numerous biopolymers are hydrophilic or water-absorbent. There is no universally accepted method for assessing degradation. They, like carbohydrate and protein, are synthesized through hydrolysis and condensation. However, hydrolysis results in the formation of a variety of synthetic biopolymers (polylactic acid, PVA, PCL, etc.). Despite their similarities, their fundamental mechanisms are very distinct. They include chitosan derived from crustaceans and plant derivatives (xanthol, gums, starch, etc.) [3, 4]. The overuse of fossil fuels and the ensuing oil crises need the creation of biopolymers. Several polymer companies are focusing on biopolymers as a synthetic raw material alternative. Imperial Chemical Industries launched the "BIOPOL" initiative to produce thermoplastic biopolymers [5]. Biopolymers need more energy than synthetic polymers in recent years. However, modern science is moving towards producing low-energy and economically viable biopolymers [6, 7].

Due to the fact that biodegradable biopolymers are less damaging to the environment and its inhabitants than fossil-based polymers, they are widely used. Since 1850, latex and cellulose have been utilized to manufacture rubber and plastic [8]. PHA is a polymer that stores carbon and energy, while polylactic acid (PLA) is derived from lactic acid fermentation. PHA is extensively used in a variety of industries, from agriculture to medicine. It is produced *via* the fermentation of renewable feedstocks such as glucose, sucrose, and vegetable oil. To enhance heat resistance or durability, PLA may be combined with synthetic or natural petroleum-based polymers or fibers. PLA-based polymers are biodegradable and compostable, which provides additional disposal options. Thermoplastic starch (TPS) has been shown to be a suitable alternative for synthetic polymers used in packaging. TPS is usually mixed with synthetic polymers to produce marketable products. Their components may be biodegradable or compostable, allowing them to be utilized in films—stiff materials such as plates and cutlery, packaging, and foams. Chitin, the second most common natural biopolymer, is a supporting and protective substance in arthropod exoskeletons and fungus and yeast cell walls [9]. Chitin molecules create 3–50-nm-wide fibrils encased in a protein and $CaCO_3$ matrix, similar to cellulose. Chitosan is made by deacetylating chitin, a marine protein. Water-insoluble chitosan is first dissolved in acidic solutions before being integrated into polymer films [10]. For its unmatched availability, cheap cost, light weight, strength, and toughness, biopolymers have been explored to strengthen polymer composites. Their uses have recently expanded into new areas including biology, optoelectronics, energy, and the environment [11–13].

20.2 Role of Biopolymers in 3D Printing

Biopolymer 3D printing has significant academic, environmental, and social interest and future application in aerospace, textile, food, medical, and industrial sectors [14–17]. 3D printing is a set of techniques that transform a CAD design into a metallic, polymeric, ceramic, or composite layer-by-layer structure. 3D printing provides a variety of benefits over conventional manufacturing techniques, such as cost-effectiveness, fast manufacture of complex designs, modifying the model, cutting the production time, and multi-materials production [14, 18]. How 3D things are produced depends heavily on the specification of the material. While some materials, like starch, may directly be printed from an extruder, other materials, such as meat, vegetables, and grain, must first be powdered or pasted before being 3D-printed [19, 20]—for example, a wide variety of printability, bio-compatibility, degradability, and mechanical characteristics must apply to biomedical materials. On the other hand, textile goods must be flexible, durable, and tensile enough to the comfort and mobility of users. In addition, edible 3D-printed food items should be easy to manufacture and retain their shape during post-processing.

20.2.1 Role of Cellulose in 3D Printing

Research on bio-based alternatives has been driven by environmental and social concerns. Scientists have lately studied the usage in 3D printing of cellulose, one of the most common biopolymers in nature [21].

The intrinsic rigidity of intra-/intermolecular hydrogen networks and their high crystallinity prevent their use in additive manufacturing (AM). The cellulose surface may be chemically modified/functionalized *via* esterification, etherification, selective oxidation, graft copolymerization, or intermolecular cross-linking [22–25].

Cellulose fibers are excessive, cheap, and flexible. Cellulose fibers contain a high number of functional groups (especially OH groups) on its surface as a naturally hierarchical and porous substratum that may firmly link other materials [26]. One example is the combination of the macroporous structure and high strength of cellulose paper with the electrical conductivity and the electroactivity of graphic nanosheets [27]. For decades, the two main derivatives of cellulose, cellulose ethers and esters, were marketed. Ethyl cellulose, HEC, HPC, and HPMC are only a few of water-soluble cellulose esters widely used in industry (HPMC).

Cellulose may be transformed through esterification into liquids, fibers, and three-dimensional structures. In 3D printing, the characteristics of the ink are important [28]. 3D printing inks must be viscoelastic (high viscosity and shear-shinning). To achieve this, the shear dilution characteristics of polymer solutions are frequently employed. Cellulose ethers have been used for many years to change the viscosity of industrial goods. Hydrogen bonds may develop, limiting water movement and increasing the viscosity between the surrounding cellulose strands [29]. Due to shear-thinning or pseudoplastic activity, as a result of the alignment, the cellulose chains become less viscous. The viscosity of cellulose ether solutions is low because of their rheological characteristics, with high shear rates and high with low shear rates. Since cellulose ethers are thixotropic, they provide good 3D inks for printing. The hydration and expansion of cellulose ether by aqueous solutions on tablets lead to intertwining chains and physical hydrogel formation [30]. The medication spreads throughout the polymer network after dissolution in water, with the cellulose ether extending from the surface of the tablet into its core [31]. Because of the physical characteristics of the hydrogel network, chain breaking and swelling may occur depending on the kind of cellulose ether employed [32–35]. Therefore, a combination of swelling, diffusion, and erosion leads to drug release. Erodible delivery methods often utilize cellulose ethers like HPMC, HPC, EC, and HEC.

Nanocelluloses may be synthesized by physical, chemical, and enzyme methods in addition to cellulose nanofibrils and nanocrystals [36, 37]. Microfibrillated cellulose (MFC) is a flexible material over 1 m in length and nanoscale diameter [37, 38]. The cellulose nanofibrils (CNC) consist of stiff rod-like particles, with 3–10 nm in diameter and 100–200 nm in length [39, 40]. Nanocelluloses are an interesting new class of natural nanomaterials with an unusually strong combination, with the surface and surface chemistry being adjustable. Nanocelluloses, including medicines, energy, and construction, may be used in a number of applications. Modern 3D printing technology allows sophisticated 3D functional living tissues to be produced using biocompatible materials, cells, and supporting components.

More effort is nevertheless necessary to produce bioinks suitable for printing. Nanocellulose may be used as a substratum for printing or a component of bioinks in 3D printing [41–43]. It has been demonstrated that nanocelluloses are not cytotoxic in a variety of cell lines [44, 45]. Thus, 3D nanocellulose printing is an appealing way of developing sustainable structures.

20.2.2 Role of Chitosan in 3D Printing

As with glycosaminoglycans, the mechanical properties of chitosan are stable under physiological conditions [46]. Additionally, the breakdown products of chitosan are often involved in the production of cartilage [47]. Chitosan's advantages have a broad range of applications in bone and cartilage tissue engineering [48–51]. In PCL scaffolds, the chitosan coating enhanced cell adherence and penetration. It is suggested to combine chitosan with other materials such as bioglass or collagen to improve its biological, osteoconductive, and mechanical properties [52–57]. The technique involves manufacturing chitosan/hydroxyapatite scaffolds through robotic dispensing [58]. Researchers supplemented the scaffold with HA chitosan to enhance the osteoconductivity and mechanical strength. They produced chitosan and chitosan–HA hydrogels in acetic acid and then extruded them into a bath of sodium hydroxide and ethanol to precipitate gels. Layer attachments were low or nonexistent as a consequence of fast precipitation and restricted mobility of chitosan molecules at structure contact edges. Low NaOH concentrations resulted in a decrease in precipitation rate and form retention. The researchers built the structure layer by layer using a Teflon-lined nuzzle to extrude a viscous solution of chitosan and chitosan–hydroxyapatite into a dispensing medium. The chitosan matrix exhibited a significant degree of layer adhesion [59–64]. As a result, the final design featured connected pores. There were acceptable results in cell culture experiments. As a result, rapid prototyping robotic dispensing may be utilized to manufacture chitosan-based materials. Ang *et al.* reported a difficulty in determining the starting dispensing parameters such as dispensing pressure, speed, and initial dispensing height. These parameters changed according to the solution's viscosity for each composition and concentration. Macroscopical analysis was utilized to overcome this barrier and produce homogeneous strands.

Choosing the best 3D printer binder may be challenging. Chavanne *et al.* evaluated a variety of binders for chitosan 3D printing and found that lactic acid produced virtually similar results [58]. They retested printing using chitosan/HA and lactic, citric, and acetic acids as binder materials. This was accomplished *via* medium viscosity, good wettability, and rapid solidification time. The construct was more stable when a binder mixture of 20% chitosan and 40% lactic acid was used as well as a post-treatment with 40% lactic acid. Due to the high concentration of lactic acid, the open pores closed, resulting in poor porosity. They believe that the chemical structure of chitosan offers a pleasurable biostimulation. Their results demonstrate that chitosan/HA-printed scaffolds are a feasible alternative for bone regeneration.

20.2.3 Role of Starch in 3D Printing

3D printing has recently been used in the food sector. Starch is frequently used in 3D food printing as a thickener/gelling agent or as a rheological modifier [65]. As with cellulose, starch has numerous hydroxyl groups that operate as an active site. Lam *et al.* utilized quick prototyping to produce porous scaffolds in three dimensions using a unique mix of maize starch, dextran, and gelatin starch polymers [65]. A solid cylindrical grinder has the greatest rigidity and the first compressive output strength. In order to improve the water resistance and mechanical strength of the scaffolds, they were dried for 1 h at 100°C.

Feng *et al.* also investigated pea protein's impact on 3D potato starch printing materials [66]. The added strength of the materials was increased by 45% potato starch, 1% pea protein, 9% butter, and 45% water. As the quantity of pea protein increased, the degree of interconnection decreased, which affected the form and resolution of the printed 3D items.

In addition, Yang *et al.* examined the rheological and mechanical characteristics of potato starch lemon juice gel (10–20 wt.%) [67]. With a shear speed of 0.1 s^{-1}, an ideal concentration of 15 wt.%, and an apparent viscosity of 8,079.3 Pas, a storage module of 4,924.2 Pa and a loss module of 760.8 Pa were determined. A high amount of starch has demonstrated that fluidity and extrusion have fallen, whereas a low content of starch causes loss of shape following extrusion. Chen *et al.* also stressed the connection between rheological and impressive characteristics [68]. They assessed three starches for 3D hot extrusion printing (potato, rice, and maize). The expansion of starch granules led to the development of a three-dimensional form for the molecules of amylose. This increased the storage module, which decreased the intermolecular connections (primarily hydrogen bonds) and the lowering of the chain interstick owing to the rupture of the storm granule with further heating.

20.3 Biopolymers in Recent Intelligent Biomedical Systems

20.3.1 Stimulus-Responsive Systems

Sol–gel transitions have recently been a hot topic in the field of functional polymers that are triggered by external stimuli such as temperature, pH, and light [69–74]. Physical modifications to stimulating polymers are also reversible. The development of stimulating polymer-based drug delivery systems with prolonged and regulated release of medications has recently accelerated [75]. Stimulus-responsive polymers' critical solution temperature (CST) is defined as the temperature at which the polymer solution divides from the isotropic to the anisotropic phase [76]. Less critical polymers show a solution phase below the CST and become insoluble or form a hydrogel when heated through CST. The LSCT temperature is the point at which a polymer solution phase transitions from isotropic to anisotropic. LCST polymers are employed mainly in the delivery of pharmaceutical drugs [77]. Activated gelling materials derived from natural and manmade polymers are being utilized to create regenerative medical products and drug delivery systems [78]. Stimulating polymers' switchable physical properties enable a wide variety of common biological applications [79, 80]. Thermoresponsive polymers exhibit sol–gel transitions with ambient temperature changes [81–83], and some thermoresponsive polymers may produce body temperature hydrogels [84–87]. Thermoresponsive polymers such as poly(N-isopropylacrylamide) and pluronic F127 are extensively utilized in pharmaceutical delivery because they can form a gel at 37°C [88–91]. When the hydrogel cools under LCST, the polymers in the solvent combination degrade, releasing the medication. The swelling–shrinking property of pH-responsive polymers makes them ideal for medication control. Chitosan, PDMAEMA, and PDEAEMA are cationic polymers that expand at acidic pH and contract at basic pH, while albumin and PAA are fundamental polymers that expand at acidic pH and contract at basic pH [92, 93].

Hydrogels and microgels, which are crosslinked polymer networks, are critical pacemakers for controlled pharmaceutical delivery. Hydrogels are helpful in a wide variety of biological applications due to their porous nature and swell ability [94]. Their porosity allows drugs to be put into the gel matrix and subsequently released according to the diffusion coefficient of small molecules or macromolecules in the gel network. As previously stated, many hydrogels exhibit variable swelling properties in response to environmental changes. Due to the structure and characteristics of hydrogels, they are often utilized to deliver transdermal or injectable medication. Microneedles significantly improve the efficacy of transdermal medications by providing a clear route for medicinal vaccinations (such as small molecules, protein, and DNA) [95–97]. To provide regulated medication release, hydrogel-forming microneedle arrays may quickly absorb interstitial skin fluid, producing discrete *in situ* hydrogel bulbs. For microneedle arrays that generate hydrogel, the swelling-induced release mechanism is critical [98]. Stimulatory microneedles allow on-demand medication administration by altering the environmental stimuli [99]. The McCoy Group has created new on-demand transdermal medicine delivery systems utilizing hydrogel-forming microneedles and light-sensitive pharmacological conjugates [100]. This device is held together by poly(2-hydroxyethyl methacrylate) and dimethacrylate ethylene glycol. Microneedles manufactured by micro-molding have superior mechanical characteristics. For up to 160 h, the gadgets produced ibuprofen in reaction to light. Additionally, the light may be utilized to switch on and off the release. The researchers created a biphasic microneedle range that improves adhesiveness by physically interlocking the tissue through swelling microneedle points. Photothermal treatment and chemotherapy patches developed from the microneedles' light-sensitive components. The photosensitive heating and release action may be precisely regulated and cycled on and off.

Recently, Chen *et al.* created smart insulin microneedle array patches and a method for regulated antibody delivery [101]. The device is composed of a microneedle array that contains insulin and glucose oxidase vesicles that are hypoxia-sensitive hyaluronic acid (HS-HA)-based (GOx). They take advantage of local hypoxia induced by hyperglycemia to promote insulin secretion. In a hypoxic environment, 2-nitroimidazoles transform into hydrophilic 2-aminoimidazoles. Researchers synthesized this hyaluronic acid and utilized it to self-assemble insulin and GOx vesicles. When blood glucose levels are elevated, dissolved oxygen is quickly depleted. This results in a hypoxic environment that decreases 2-nitroimidazole, causing the vesicle to dissociate and release insulin. Additionally, the vesicle has been incorporated into the microneedle array, enabling painless and continuous insulin delivery. The smart insulin patch's novel triggering mechanism may allow closed-loop insulin administration. Additionally, it will aid in the development of a drug delivery platform based on artificial vesicles for additional illnesses. Moreover, they investigated the combination of pancreatic cells with synthetic glucose signal enhancers to create smart insulin. To begin with, a synthetic amplifier was used to increase the physiological signal, "glucose level", to ensure its efficient transmission and adequate activation of beta cell insulin production.

Recently, numerous small molecules were released consecutively and under control from a microgel-based reservoir device [102]. The release of tiny molecules and their doses may be regulated and changed by combining microgels with various chemistries that enable them to be ionized in response to changes in solution pH. The apparatus was constructed using pNIPAm-CoAAc microgels (AAc-MG) and pNIPAm-Co-APBA microgels (APBA-MG) (MB). Both models have a positive impact, which enables them to interact substantially with negatively charged microgels. The electrostatic interactions of positively charged CV and MB with negatively charged pH-10 AAc- and APBA-modified microgels were substantial. In this instance, MB was obtained exclusively from APBA-MG. Once again, lowering the pH to 3.0 neutralizes the AAc groups and liberates CV from AAc-MG (as shown in Figure 20.1). These devices enable the successive addition of tiny compounds to a system to have a controlled release kinetics. Modifying the microgel chemistry creates ionizing reservoir devices at various pH values to alter their release behavior. This would enable the delivery of tiny compounds to a pH-activated system. The adjustable release kinetics and the possibility to show many devices on a single substrate distinguish and strengthen this delivery method [102].

Cao *et al.* have created phase-changeable polymers that release DOX and perfluoropentane using low-intensity ultrasound (LIFU) to enhance the delivery of medications to tumors as shown in Figure 20.2 [103].

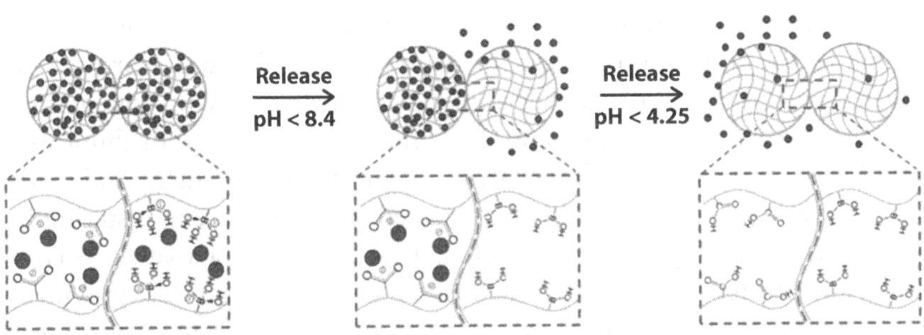

Figure 20.1 Release of medication as a pH-sensitive behavior (reproduced with permission from [102]).

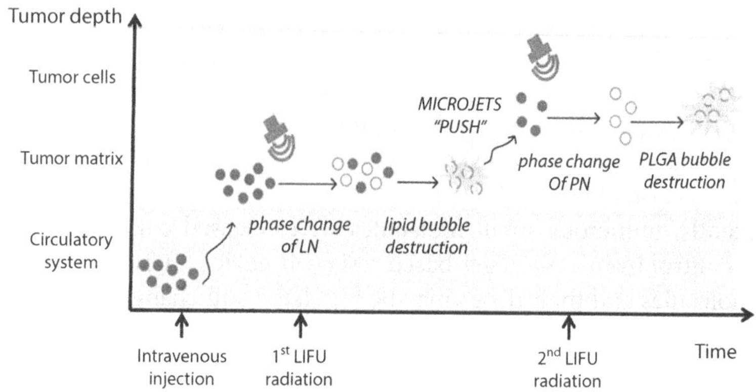

Figure 20.2 Release of drug from the PLGA bubble as a function of ultrasound waves and time (reproduced with permission from [103]).

Nanodroplets, more effectively than untreated materials, inhibited the formation of tumors, thus increasing animal survival. There was increased animal longevity as compared to a single LIFU burst and increased vascular permeability of tumors.

Raising the pulse count, Salgarella *et al.* synthesized and evaluated five distinct micelles of poly(2-oxazoline) as a potential supply for the ultrasonic delivery system [104]. The hydrophilic–hydrophobic block copolymer ratio allowed for controlled dexamethasone release. Gai *et al.* created free-standing PLA nano- and micro-chamber arrays that are biocompatible [105]. This research demonstrated the long-term encapsulation of minute hydrophilic compounds in microchambers with and without HIFU. HIFU microchambers emit about four times as much as non-HIFU microchambers.

Pearson *et al.* created light-responsive glycopolymer micelles with azo and galactose units for cancer therapy [106]. The galactose units were designed to bind to the melanoma cells' galectin-3 receptors. This is due to the control group azo for copolymer breakdown. Hardy *et al.* created an "on-demand" transdermal drug delivery array by conjugating 2-hydroxyethyl methacrylate and 3,5-dimethoxybenzene with ibuprofen [107]. The microneedle hydrogel array containing the conjugate and medication was applied to the skin and exposed to UV radiation, resulting in the release of conjugate and ibuprofen. This device was activated *in vitro* using an optical trigger, and several doses of the medication were provided.

Numerous clinical studies have shown that the SPION's magnetic boost is well tolerated *in vivo* and can be continuously sustained at temperatures between 40°C and 45°C in glioblastoma and prostate cancer [108–110]. SPION hyperthermia may be utilized in small doses to increase the medication release without substantially increasing the tissue temperature.

20.3.2 Hyphenated Techniques in Drug Formulations

With the advancement in drug formulation techniques, hyphenated techniques recently play a vital role in the analytical development and quality control of drug substances and dosage form. Since the dosage form is mainly dependent on the medical condition as well as on the response to the treatment, the hyphenated techniques used have a major importance. Hyphenated techniques have been developed by combining two techniques together: one is to carry out the separation or isolation of a component—for instance, chromatography—and another is for the identification of the separated component by spectral techniques such as NMR and mass spectroscopy. The most commonly utilized hyphenated techniques are GC–MS, LC–MS, LC–NMR-MSLC–PDA-MS, LC–MS–MS, and LC–NMR-MS. These techniques provide a tool to the analytical chemists to study the contamination or impurities associated with the drug (impurity profile) and their degradation (force degradation) behavior [111]. As far as medical condition is concerned, the impurities lie in the medicines of diabetic, heart, and high blood pressure patients. Due to lifelong medication, it may cause different adverse effects in their body such as cytological and gene toxicity. Therefore, it is required to be separated and identified even if present in minimal quantity [112]. Regarding polymeric materials, a hyphenated technique, viz., gel permeation chromatography (GPC)-IR, can be used. It involves the combination of separation of polymeric molecules of hydrodynamic size by GPC and identification of the functional group by FT-IR to understand the structural composition of the copolymers—for instance, copovidone, a copolymer of vinyl pyrrolidone and vinyl acetate, which has been popularly used as excipient in controlled drug release and comparative drug release profile studies of different drug suppliers, can be studied by these techniques as approved by FDA [113].

20.3.3 Patent

A 4D chitosan-based thermosensitive hydrogel preparation technique and a medical material preparation method are disclosed in US patent US20210079170A1. Firstly, chitosan is dissolved in acetic acid solution and lyophilized after solvent extraction to obtain lyophilized chitosan. Subsequently, aqueous sodium glycerophosphate solution is prepared with ultrapure water and sodium glycerophosphate, and then aqueous carboxymethyl chitosan solution is prepared with ultrapure water and aqueous sodium glycerophosphate solution which are charged into and mixed. Using scientific and trustworthy principles, this innovation tackles an issue that existing thermosensitive hydrogels have unequal pore diameters and increase the entrapment efficiency and ability of limbal stem cells [114].

In another patent, US20180251649A1, the investigators described an aqueous suspension of anisotropic filler particles, clay particles, one or more monomers including an acrylamide and/or an acrylate, a polymerization initiator, and an oxygen-scavenging enzyme to form a hydrogel composite ink formulation for 4D printing [115]. A hydrogel matrix is formed by polymerizing one or more monomers, which expands rapidly when exposed to water. When included as a rheological assistance in the ink formulation, clay particles may help to enhance the viscoelastic properties essential for 3D printing [115].

Self-healing hydrogel is an essential 4D-printed material that can be exposed to many environmental stimuli, including light and electrical impulses. Under situations of thermostimulation, redox stimulation, and healing, colloidal gel and colloidal sol are transformed. As a result of this invention, the usage of hydrogels has been broadened, for example, by the invention of persons who employ volume change induced by self-healing hydrogel. Cyclodextrin and its derivatives have been grafted to create a self-healing hydrogel that responds to light stimuli. When azo and its derivatives are grafted to another major polymer chain on rigid backbones like glycan, two types of polymers are combined [116]. Colloidal sol or gel is produced *via* clathration of cyclodextrin and azo in the cis–trans of distinct wave lengths, i.e., in the UV of 430 nm. In order to create the self-healing effect, colloidal sol is irradiated with light, and the gel is then irradiated with UV light of 365 nm. This approach, however, is difficult to prepare because of the strict ratio requirements of the two types of polymers. The environmental response hydrogels currently available for printing can only be prepared by two-dimensional photoetching technology; the supporting force of the active response hydrogel material based on the three-dimensional laser direct writing technology is not sufficient, and a suspended structure cannot be printed, so a complex three-dimensional structure designed randomly in space cannot be realized. A 4D micro–nano-printed object and a printing test technique are all disclosed in the invention. DPEPA serves as a crosslinking agent, while the photosensitizer triethanolamine and the photoinitiator EMK serve as reactants in the copolymer gel. The copolymer gel also contains functional monomers of

N-isopropylacrylamide and acrylic acid. Adding monomers of N-isopropyl acrylamide and acrylic acid causes the structure to shrink under the influence of temperature, pH, and organic solvent [116].

20.4 Critical Aspects of Biopolymers in Biomedical Applications

Along with the mammalian sources of polymers such as collagen, gelatin, and different types of glycosaminoglycan, non-mammalian sources of polymers such as alginate, carrageenan, chitin, and chitosan have been extensively used for developing 3D polymeric matrix environment. Cells can be either seeded on microporous, electro-spinned fibrous scaffold, microdroplets, or microfilaments, and hydrogel cell encapsulation has been developed for a 3D cancer model. Since the synthetic materials require some chemical modification to enhance their biological properties, biopolymers have their inherent biocompatibility. However, they often evoke the activation of immune cells due to uncontrolled degradation. Synthetic polymers exhibit high mechanical and controlled degradation properties. Therefore, linear synthetic polymers such as polyesters, polyglycolic acid, polylactic acid, polyvinylidene fluoride, and copolymers polycaprolactone are most commonly employed for 3D cancer models [117] in addition to the biopolymers based nanocomposites [3].

20.5 Conclusion and Future Perspective

Biopolymers have been extensively used in the fabrication of different biomaterials due to their biodegradability, biocompatibility, mucoadhesiveness, and low immunogenic properties. From seaweed to collagen, a large number of biopolymers have been prepared either by blending along with synthetic polymeric materials or as a nanocomposite for their application in tissue engineering and regenerative medicine. Recently, 3D bioprinting has gained tremendous attention in biomedical applications. Novel bioinks and advanced additive manufacturing technologies have been developed by scientists to fabricate improved biomaterials. The cell encapsulation in 3D polymeric scaffolds is widely used for tissue engineering applications and development of 3D cancer models. Stimuli-responsive polymers, thermosensitive polymers, and activated gelling materials are fabricated for advanced drug delivery and regenerative medical products. Such materials can open a passage to design hyphenated drugs and pharmaceutical products. Furthermore, 3D and 4D bioprinting techniques are extensively used in developing hydrogels for tissue engineering and regenerative medicine.

References

1. Yadav, P., Yadav, H., Shah, V.G., Shah, G., Dhaka, G., Biomedical biopolymers, their origin and evolution in biomedical sciences: A systematic review. *J. Clin. Diagn. Res.*, 9, ZE21-ZE25, 2015.
2. Chen, F.M. and Liu, X., Advancing biomaterials of human origin for tissue engineering. *Prog. Polym. Sci.*, 53, 86–168, 2016.
3. Annu, Bhat, Z.I., Imtiyaz, K., Rizvi, M.M.A., Ikram, S., Shin, D.K., Comparative study of ZnO-and-TiO2-nanoparticles-functionalized polyvinyl alcohol/chitosan bionanocomposites for multifunctional biomedical applications. *Polym. (Basel)*, 15, 3477, 2023, doi: 10.3390/polym15163477.
4. Elieh-Ali-Komi, D. and Hamblin, M.R., Chitin and chitosan: Production and Application of versatile biomedical nanomaterials. *Int. J. Adv. Res.*, 4, 411–427, 2016.
5. Zinn, M., Development of biopolyesters (PHA) as part of the swiss priority program in biotechnology. *Chimia (Aarau)*, 74, 398–401, 2020, doi: 10.2533/CHIMIA.2020.398.
6. Annu, and Ahmed, S., 1-Bionanocomposites: An overview, in: *Bionanocomposites in Tissue Engineering and Regenerative Medicine*, S. Ahmed, and Annu (Eds.), pp. 1–6, Woodhead Publishing Series in Biomaterials, Woodhead Publishing, Cambridge, United Kingdom, 2021, ISBN: 978-0-12-821280-6.
7. Shekh, M.I., Annu, Ahmed, S., Chapter 1-Biopolymers: An overview, in: *Advanced Applications of Biobased Materials*, S. Ahmed, and Annu (Eds.), pp. 3–21, Elsevier, Amsterdam, Netherlands, 2023, ISBN: 978-0-323-91677-6.
8. Morton, M., Introduction to polymer science, in: *Rubber Technology*, pp. 1–19, 1999.
9. Rathore, A.S. and Gupta, R.D., Chitinases from bacteria to human: Properties, applications, and future perspectives. *Enzyme Res.*, 2015, 791907, 2015.
10. Szymańska, E. and Winnicka, K., Stability of chitosan-a challenge for pharmaceutical and biomedical applications. *Mar. Drugs*, 13, 1819–1846, 2015.
11. Koide, S.S., Chitin-chitosan: Properties, benefits and risks. *Nutr. Res.*, 18, 1091–1101, 1998.
12. Morin-Crini, N., Lichtfouse, E., Torri, G., Crini, G., Applications of chitosan in food, pharmaceuticals, medicine, cosmetics, agriculture, textiles, pulp and paper, biotechnology, and environmental chemistry. *Environ. Chem. Lett.*, 17, 1667–1692, 2019.
13. Zhao, D., Yu, S., Sun, B., Gao, S., Guo, S., Zhao, K., Biomedical applications of chitosan and its derivative nanoparticles. *Polym. (Basel)*, 10, 462, 2018.

14. Ngo, T.D., Kashani, A., Imbalzano, G., Nguyen, K.T.Q., Hui, D., Additive manufacturing (3D Printing): A review of materials, methods, applications and challenges. *Compos. Part B Eng.*, 143, 172–196, 2018.
15. Saravanakumar, K., Ali, D.M., Kathiresan, K., Wang, M.-H., Antimicrobial, anticancer drug carrying properties of biopolymers-based nanocomposites-A mini review. *Curr. Pharm. Des.*, 24, 3859–3866, 2018, doi: 10.2174/1381612825666181120161300.
16. Shahbazi, M. and Jäger, H., Current status in the utilization of biobased polymers for 3D printing process: A systematic review of the materials, processes, and challenges. *ACS Appl. Bio Mater.*, 4, 3859–3866, 2021.
17. Wasti, S. and Adhikari, S., Use of biomaterials for 3D printing by fused deposition modeling technique: A review. *Front. Chem.*, 8, 2020.
18. Jiménez, M., Romero, L., Domínguez, I.A., Espinosa, M.D.M., Domínguez, M., Additive manufacturing technologies: An overview about 3D printing methods and future prospects. *Complexity*, 2019, 9656938, 2019, doi: 10.1155/2019/9656938.
19. Pereira, T., Barroso, S., Gil, M.M., Food texture design by 3D printing: A review. *Foods*, 10, 320, 2021, https://doi.org/10.3390/foods10020320.
20. Pulatsu, E. and Lin, M., A review on customizing edible food materials into 3D printable inks: Approaches and strategies. *Trends Food Sci. Technol.*, 107, 68–77, 2021.
21. Bahçegül, E.G., Bahçegül, E., Özkan, N., 3D printing of hemicellulosic biopolymers extracted from lignocellulosic agricultural wastes. *ACS Appl. Polym. Mater.*, 2, 2622–2632, 2020, doi: 10.1021/acsapm.0c00256.
22. Wohlhauser, S., Delepierre, G., Labet, M., Morandi, G., Thielemans, W., Weder, C., Zoppe, J.O., Grafting polymers from cellulose nanocrystals: Synthesis, properties, and applications. *Macromolecules*, 51, 6157–6189, 2018.
23. George, J. and Sabapathi, S.N., Cellulose nanocrystals: Synthesis, functional properties, and applications. *Nanotechnol. Sci. Appl.*, 8, 45–54, 2015, doi: 10.2147/NSA.S64386.
24. Leszczyńska, A., Radzik, P., Szefer, E., Mičušík, M., Omastová, M., Pielichowski, K., Surface modification of cellulose nanocrystals with succinic anhydride. *Polym. (Basel)*, 11, 866, 2019, https://doi.org/10.3390/polym11050866.
25. Wang, Y., Wang, X., Xie, Y., Zhang, K., Functional nanomaterials through esterification of cellulose: A review of chemistry and application. *Cellulose*, 25, 3703–3731, 2018.
26. Sharma, A., Thakur, M., Bhattacharya, M., Mandal, T., Goswami, S., Commercial application of cellulose nano-composites–A review. *Biotechnol. Rep.*, 21, e00316, 2019.
27. Larraza, I., Vadillo, J., Calvo-Correas, T., Tejado, A., Olza, S., Peña-Rodríguez, C., Arbelaiz, A., Eceiza, A., Cellulose and graphene based polyurethane

nanocomposites for fdm 3D printing: Filament properties and printability. *Polym. (Basel)*, 13, Article ID 839, 2021, doi: 10.3390/polym13050839.

28. Chopra, H., Kumar, S., Singh, I., Bioinks for 3D printing of artificial extracellular matrices, in: *Advanced 3D-Printed Systems and Nanosystems for Drug Delivery and Tissue Engineering*, pp. 1–37, Elsevier, Amsterdam, Netherlands, 2020.
29. Chami Khazraji, A. and Robert, S., Interaction effects between cellulose and water in nanocrystalline and amorphous regions: A novel approach using molecular modeling. *J. Nanomater.*, 2013, 2013, doi: 10.1155/2013/409676.
30. Parhi, R., Cross-linked hydrogel for pharmaceutical applications: A review. *Adv. Pharm. Bull.*, 7, 515–530, 2017.
31. Shahzad, Y., Ibrar, N., Hussain, T., Yousaf, A.M., Khan, I.U., Rizvi, S.A.A., Relevancy of nizatidine's release from floating tablets with viscosity of various cellulose ethers. *Sci.*, 3, 2021, doi: 10.3390/sci3020022.
32. Asri, L.A.T.W., Rahmatika, A., Fahreza, M.Z., Insanu, M., Purwasasmita, B.S., Preparation and release behavior of carboxylated cellulose nanocrystals-alginate nanocomposite loaded with rutin. *Mater. Res. Express*, 5, 095303, 2018, doi: 10.1088/2053-1591/aad5d3.
33. Shah, K.U. and Khan, G.M., Regulating drug release behavior and kinetics from matrix tablets based on fine particle-sized ethyl cellulose ether derivatives: An *in vitro* and *in vivo* evaluation. *Sci. World J.*, 2012, 842348, 2012, doi: 10.1100/2012/842348.
34. Jelvehgari, M., Atapour, F., Nokhodchi, A., Micromeritics and release behaviours of cellulose acetate butyrate microspheres containing theophylline prepared by emulsion solvent evaporation and emulsion non-solvent addition method. *Arch. Pharm. Res.*, 32, 1019–1028, 2009, doi: 10.1007/s12272-009-1707-y.
35. Rastogi, V.K., Dattatreya, J.G., Chattopadhyay, S., Ray, A.K., Drug loading in pulp cellulose fiber and its release behavior, in: *Proceedings of the AIChE Annual Meeting, Conference Proceedings*, 2012.
36. Yu, S., Sun, J., Shi, Y., Wang, Q., Wu, J., Liu, J., Nanocellulose from various biomass wastes: Its preparation and potential usages towards the high value-added products. *Environ. Sci. Ecotechnology*, 5, 100077, 2021, doi: 10.1016/j.ese.2020.100077.
37. Trache, D., Tarchoun, A.F., Derradji, M., Hamidon, T.S., Masruchin, N., Brosse, N., Hussin, M.H., Nanocellulose: From fundamentals to advanced applications. *Front. Chem.*, 8, 392, 2020.
38. Chinga-Carrasco, G., Cellulose fibres, nanofibrils and microfibrils: The morphological sequence of MFC components from a plant physiology and fibre technology point of view. *Nanoscale Res. Lett.*, 6, 417, 2011, doi: 10.1186/1556-276X-6-417.
39. Henschen, J., Li, D., Ek, M., Preparation of cellulose nanomaterials via cellulose oxalates. *Carbohydr. Polym.*, 213, 208–216, 2019, doi: 10.1016/j.carbpol.2019.02.056.

40. Dhali, K., Ghasemlou, M., Daver, F., Cass, P., Adhikari, B.A., Review of nanocellulose as a new material towards environmental sustainability. *Sci. Total Environ.*, 775, 145871, 2021.
41. Fourmann, O., Hausmann, M.K., Neels, A., Schubert, M., Nyström, G., Zimmermann, T., Siqueira, G., 3D printing of shape-morphing and antibacterial anisotropic nanocellulose hydrogels. *Carbohydr. Polym.*, 259, 117716, 2021, doi: 10.1016/j.carbpol.2021.117716.
42. Giubilini, A., Siqueira, G., Clemens, F.J., Sciancalepore, C., Messori, M., Nyström, G., Bondioli, F., 3D-printing nanocellulose-poly(3-hydroxybutyrate- Co-3-hydroxyhexanoate) biodegradable composites by fused deposition modeling. *ACS Sustain. Chem. Eng.*, 8, 10292–10302, 2020, doi: 10.1021/acssuschemeng.0c03385.
43. Ee, L.Y. and Yau Li, S.F., Recent advances in 3D printing of nanocellulose: Structure, preparation, and application prospects. *Nanoscale Adv.*, 3, 1167–1208, 2021.
44. Wang, X., Wang, Q., Xu, C., Nanocellulose-based inks for 3d bioprinting: Key aspects in research development and challenging perspectives in applications—A mini review. *Bioengineering*, 7, 40, 2020.
45. Bergonzi, C., Remaggi, G., Graiff, C., Bergamonti, L., Potenza, M., Ossiprandi, M.C., Zanotti, I., Bernini, F., Bettini, R., Elviri, L., Three-dimensional (3D) printed silver nanoparticles/alginate/nanocrystalline cellulose hydrogels: Study of the antimicrobial and cytotoxicity efficacy. *Nanomaterials*, 10, 844, 2020, doi: 10.3390/nano10050844.
46. Croisier, F. and Jérôme, C., Chitosan-based biomaterials for tissue engineering. *Eur. Polym. J.*, 49, 780–792, 2013.
47. Yang, R., Chen, F., Guo, J., Zhou, D., Luan, S., Recent advances in polymeric biomaterials-based gene delivery for cartilage repair. *Bioact. Mater.*, 5, 990–1003, 2020, doi: 10.1016/j.bioactmat.2020.06.004.
48. Pandey, A.R., Singh, U.S., Momin, M., Bhavsar, C., Chitosan: Application in tissue engineering and skin grafting. *J. Polym. Res.*, 24, 125, 2017.
49. Ojeda-Hernández, D.D., Canales-Aguirre, A.A., Matias-Guiu, J., Gomez-Pinedo, U., Mateos-Díaz, J.C., Potential of chitosan and its derivatives for biomedical applications in the central nervous system. *Front. Bioeng. Biotechnol.*, 8, 2020.
50. Mittal, H., Ray, S.S., Kaith, B.S., Bhatia, J.K., Sukriti, Sharma, J., Alhassan, S.M., Recent progress in the structural modification of chitosan for applications in diversified biomedical fields. *Eur. Polym. J.*, 109, 402–434, 2018.
51. Rodríguez-Vázquez, M., Vega-Ruiz, B., Ramos-Zúñiga, R., Saldaña-Koppel, D.A., Quiñones-Olvera, L.F., Chitosan and its potential use as a scaffold for tissue engineering in regenerative medicine. *BioMed. Res. Int.*, 2015, 2015.
52. Andonegi, M., Las Heras, K., Santos-Vizcaíno, E., Igartua, M., Hernandez, R.M., de la Caba, K., Guerrero, P., Structure-properties relationship of chitosan/collagen films with potential for biomedical applications. *Carbohydr. Polym.*, 237, 116159, 2020, doi: 10.1016/j.carbpol.2020.116159.

53. Fernandes, L.L., Resende, C.X., Tavares, D.S., Soares, G.A., Castro, L.O., Granjeiro, J.M., Cytocompatibility of chitosan and collagen-chitosan scaffolds for tissue engineering. *Polimeros*, 21, 2011, doi: 10.1590/S0104-14282011005000008.
54. Susanto, A., Susanah, S., Priosoeryanto, B.P., Satari, M.H., Komara, I., The effect of the chitosan-collagen membrane on wound healing process in rat mandibular defect. *J. Indian Soc. Periodontol.*, 23, 113–118, 2019, doi: 10.4103/jisp.jisp_232_18.
55. Pasaribu, K.M., Gea, S., Ilyas, S., Tamrin, T., Radecka, I., Characterization of bacterial cellulose-based wound dressing in different order impregnation of chitosan and collagen. *Biomolecules*, 10, 1511, 2020, doi: 10.3390/biom10111511.
56. Zhang, J., Lynch, R.J.M., Watson, T.F., Banerjee, A., Chitosan-bioglass complexes promote subsurface remineralisation of incipient human carious enamel lesions. *J. Dent.*, 84, 67–75, 2019, doi: 10.1016/j.jdent.2019.03.006.
57. Clavijo, S., Membrives, F., Quiroga, G., Boccaccini, A.R., Santillán, M.J., Electrophoretic deposition of chitosan/bioglass® and chitosan/bioglass®/Tio2 composite coatings for bioimplants. *Ceram. Int.*, 42, 14206–14213, 2016, doi: 10.1016/j.ceramint.2016.05.178.
58. Ang, T.H., Sultana, F.S.A., Hutmacher, D.W., Wong, Y.S., Fuh, J.Y.H., Mo, X.M., Loh, H.T., Burdet, E., Teoh, S.H., Fabrication of 3D chitosan-hydroxyapatite scaffolds using a robotic dispensing system. *Mater. Sci. Eng. C*, 20, 35–42, 2002, doi: 10.1016/S0928-4931(02)00010-3.
59. Annu, Ahmed, S., Ahmed, S., Ikram, S., Chitin and chitosan: History, Composition and properties, in: *Chitosan*, pp. 1–24, John Wiley & Sons, Ltd, ISBN: 9781119364849, New Jersey, United States, 2017.
60. Annu, Ahmed, S., Nirala, R.K., Kumar, R., Ikram, S., Green synthesis of chitosan/nanosilver hybrid bionanocomposites with promising antimicrobial, antioxidant and anticervical cancer activity. *Polym. Polym. Compos.*, 29, S199–S210, 2021, doi: 10.1177/0967391121993977.
61. Annu, Manzoor, K., Ahmad, S., Soundarajan, A., Ikram, S., Ahmed, S., Chapter 30-Chitosan based nanomaterials for biomedical applications, in: *Handbook of Nanomaterials for Industrial Applications, Micro and Nano Technologies*, C. Mustansar Hussain, (Ed.), pp. 543–562, Elsevier, Amsterdam, Netherlands, 2018, ISBN: 978-0-12-813351-4.
62. Annu, Ahmed, S., Ikram, S., Perspectives of chitosan and alginate membranes for biomedical applications, in: *Natural Polymers: Derivatives, Blends and Composites*, S. Ahmed, and S. Ikram, (Eds.), vol. 2, pp. 157–182, Nova Science Publishers, Incorporated, New York, United States, 2017, ISBN: 9781536104400.
63. Ali, A., Annu, Ahmed, S., 17-Physical and chemical modification of chitosan-based green materials, in: *Advanced Green Materials*, S. Ahmed, (Ed.), 2021, 379–397, Woodhead Publishing in Materials, Woodhead

Publishing, ISBN: ISBN 978-0-12-819988-6, Cambridge, United Kingdom, 2021.

64. Ahmed, S., Annu, Ali, A., Sheikh, J., A review on chitosan centred scaffolds and their applications in tissue engineering. *Int. J. Biol. Macromol.*, 116, 849–862, 2018, doi: 10.1016/j.ijbiomac.2018.04.176.
65. Liu, Z., Chen, H., Zheng, B., Xie, F., Chen, L., Understanding the structure and rheological properties of potato starch induced by hot-extrusion 3D printing. *Food Hydrocoll.*, 105, 105812, 2020, doi: 10.1016/j.foodhyd.2020.105812.
66. Chuanxing, F., Qi, W., Hui, L., Quancheng, Z., Wang, M., Effects of pea protein on the properties of potato starch-based 3D printing materials. *Int. J. Food Eng.*, 14, 2018, doi: 10.1515/ijfe-2017-0297.
67. Yang, F., Guo, C., Zhang, M., Bhandari, B., Liu, Y., Improving 3D printing process of lemon juice gel based on fluid flow numerical simulation. *LWT*, 102, 89–99, 2019, doi: 10.1016/j.lwt.2018.12.031.
68. Chen, H., Xie, F., Chen, L., Zheng, B., Effect of rheological properties of potato, rice and corn starches on their hot-extrusion 3D printing behaviors. *J. Food Eng.*, 244, 150–158, 2019, doi: 10.1016/j.jfoodeng.2018.09.011.
69. Norioka, C., Okita, K., Mukada, M., Kawamura, A., Miyata, T., Biomolecularly stimuli-responsive tetra-poly(ethylene glycol) that undergoes sol-gel transition in response to a target biomolecule. *Polym. Chem.*, 8, 6378–6385, 2017, doi: 10.1039/c7py01370a.
70. Onoda, M., Ueki, T., Tamate, R., Shibayama, M., Yoshida, R., Amoeba-like self-oscillating polymeric fluids with autonomous sol-gel transition. *Nat. Commun.*, 8, 15862, 2017, doi: 10.1038/ncomms15862.
71. Qin, C., Zhou, J., Zhang, Z., Chen, W., Hu, Q., Wang, Y., Convenient one-step approach based on stimuli-responsive sol-gel transition properties to directly build chitosan-alginate core-shell beads. *Food Hydrocoll.*, 87, 253–259, 2019, doi: 10.1016/j.foodhyd.2018.08.001.
72. He, X., Fan, J., Wooley, K.L., Stimuli-triggered sol-gel transitions of polypeptides derived from α-amino acid n-carboxyanhydride (nca) polymerizations. *Chem.-Asian J.*, 11, 437–447, 2016.
73. Wang, L., Zhou, W., Wang, Q., Xu, C., Tang, Q., Yang, H., An injectable, dual responsive, and self-healing hydrogel based on oxidized sodium alginate and hydrazide-modified poly (ethyleneglycol). *Molecules*, 23, 546, 2018, doi: 10.3390/molecules23030546.
74. Tsao, C.T., Hsiao, M.H., Zhang, M.Y., Levengood, S.L., Zhang, M., Chitosan-PEG hydrogel with sol-gel transition triggerable by multiple external stimuli. *Macromol. Rapid Commun.*, 36, 332–338, 2015, doi: 10.1002/marc.201400586.
75. Patra, J.K., Das, G., Fraceto, L.F., Campos, E.V.R., del Pilar Rodriguez-Torres, M., Acosta-Torres, L.S., Diaz-Torres, L.A., Grillo, R., Swamy, M.K., Sharma, S. *et al.*, Nano based drug delivery systems: Recent developments and future prospects. *J. Nanobiotechnology*, 16, 71, 2018, doi: 10.1186/s12951-018-0392-8.

76. Wei, M., Gao, Y., Li, X., Serpe, M.J., Stimuli-responsive polymers and their applications. *Polym. Chem.*, 8, 127–143, 2017.
77. Chopra, H., Singh, I., Kumar, S., Bhattacharya, T., Rahman, M.H., Akter, R., Kabir, M.T., Comprehensive review on hydrogels. *Curr. Drug Deliv.*, 18, 658–675, 2021, doi: 10.2174/1567201818666210601155558.
78. Catoira, M.C., Fusaro, L., Di Francesco, D., Ramella, M., Boccafoschi, F., Overview of natural hydrogels for regenerative medicine applications. *J. Mater. Sci. Mater. Med.*, 30, 115, 2019, doi: 10.1007/s10856-019-6318-7.
79. Roy, I. and Gupta, M.N., Smart polymeric materials: Emerging biochemical applications. *Chem. Biol.*, 10, 1161–1171, 2003.
80. Guo, B. and Ma, P.X., Synthetic biodegradable functional polymers for tissue engineering: A brief review. *Sci. China Chem.*, 57, 490–500, 2014, doi: 10.1007/s11426-014-5086-y.
81. Miyazaki, M., Maeda, T., Hirashima, K., Kurokawa, N., Nagahama, K., Hotta, A., PEG-based nanocomposite hydrogel: Thermoresponsive sol-gel transition controlled by PLGA-PEG-PLGA molecular weight and solute concentration. *Polym. (Guildf).*, 115, 246–254, 2017, doi: 10.1016/j.polymer.2017.03.016.
82. Kim, M.S., Hyun, H., Seo, K.S., Cho, Y.H., Lee, J.W., Lee, C.R., Khang, G., Lee, H.B., Preparation and characterization of MPEG – PCL diblock copolymers with thermo-responsive sol - gel - sol phase transition. *J. Polym. Sci. Part A Polym. Chem.*, 44, 5413–5423, 2006, doi: 10.1002/pola.21659.
83. Wang, C., Hashimoto, K., Tamate, R., Kokubo, H., Watanabe, M., Controlled sol–gel transitions of a thermoresponsive polymer in a photoswitchable azobenzene ionic liquid as a molecular trigger. *Angew. Chemie - Int. Ed.*, 57, 227, 2018, doi: 10.1002/anie.201710288.
84. Liu, M., Song, X., Wen, Y., Zhu, J.L., Li, J., Injectable thermoresponsive hydrogel formed by alginate-g-poly(N-isopropylacrylamide) that releases doxorubicin-encapsulated micelles as a smart drug delivery system. *ACS Appl. Mater. Interfaces*, 9, 35673–35682, 2017, doi: 10.1021/acsami.7b12849.
85. Xu, X., Liu, Y., Fu, W., Yao, M., Ding, Z., Xuan, J., Li, D., Wang, S., Xia, Y., Cao, M., Poly(N-isopropylacrylamide)-based thermoresponsive composite hydrogels for biomedical applications. *Polym. (Basel)*, 12, 580, 2020.
86. Chatterjee, S., Hui, P.C.L., Kan, C.W., Thermoresponsive hydrogels and their biomedical applications: Special insight into their applications in textile based transdermal therapy. *Polym. (Basel)*, 10, 480, 2018.
87. Cao, M., Wang, Y., Hu, X., Gong, H., Li, R., Cox, H., Zhang, J., Waigh, T.A., Xu, H., Lu, J.R., Reversible thermoresponsive peptide-PNIPAM hydrogels for controlled drug delivery. *Biomacromolecules*, 20, 3601–3610, 2019, doi: 10.1021/acs.biomac.9b01009.
88. Nguyen, D.D. and Lai, J.Y., Advancing the stimuli response of polymer-based drug delivery systems for ocular disease treatment. *Polym. Chem.*, 11, 6988–7008, 2020, doi: 10.1039/d0py00919a.
89. Kim, M.R. and Park, T.G., Temperature-responsive and degradable hyaluronic acid/pluronic composite hydrogels for controlled release of human

growth hormone. *J. Control. Release*, 80, 69–77, 2002, doi: 10.1016/S0168-3659(01)00557-0.
90. Sarwan, T., Kumar, P., Choonara, Y.E., Pillay, V., Hybrid thermo-responsive polymer systems and their biomedical applications. *Front. Mater.*, 7, 2020.
91. Chatterjee, S., Hui, P.C.L., Kan, C.W., Wang, W., Dual-responsive (PH/temperature) pluronic F-127 hydrogel drug delivery system for textile-based transdermal therapy. *Sci. Rep.*, 9, 11658, 2019, doi: 10.1038/s41598-019-48254-6.
92. Schmaljohann, D., Thermo- and PH-responsive polymers in drug delivery. *Adv. Drug Deliv. Rev.*, 58, 1655–1670, 2006.
93. Du, H., Liu, M., Yang, X., Zhai, G., The design of PH-sensitive chitosan-based formulations for gastrointestinal delivery. *Drug Discovery Today*, 20, 1004–1011, 2015.
94. Chopra, H., Kumar, S., Singh, I., Bioadhesive hydrogels and their applications, in: *Bioadhesives in Drug Delivery*, pp. 147–170, Wiley, New Jersey, United States, 2020.
95. Tuan-Mahmood, T.M., McCrudden, M.T.C., Torrisi, B.M., McAlister, E., Garland, M.J., Singh, T.R.R., Donnelly, R.F., Microneedles for intradermal and transdermal drug delivery. *Eur. J. Pharm. Sci.*, 50, 623–637, 2013.
96. Kim, Y.C., Park, J.H., Prausnitz, M.R., Microneedles for drug and vaccine delivery. *Adv. Drug Deliv. Rev.*, 64, 1547–1568, 2012.
97. Jung, J.H. and Jin, S.G., Microneedle for transdermal drug delivery: Current trends and fabrication. *J. Pharm. Investig.*, 51, 5, 503–517, 2021.
98. Donnelly, R.F., McCrudden, M.T.C., Alkilani, A.Z., Larrañeta, E., McAlister, E., Courtenay, A.J., Kearney, M.C., Raj Singh, T.R., McCarthy, H.O., Kett, V.L. *et al.*, Hydrogel-forming microneedles prepared from "super swelling" polymers combined with lyophilised wafers for transdermal drug delivery. *PloS One*, 9, e111547, 2014, doi: 10.1371/journal.pone.0111547.
99. Makvandi, P., Jamaledin, R., Chen, G., Baghbantaraghdari, Z., Zare, E.N., Di Natale, C., Onesto, V., Vecchione, R., Lee, J., Tay, F.R. *et al.*, Stimuli-responsive transdermal microneedle patches. *Mater. Today*, 47, 206–222, 2021.
100. McCoy, C.P., Rooney, C., Edwards, C.R., Jones, D.S., Gorman, S.P., Light-triggered molecule-scale drug dosing devices. *J. Am. Chem. Soc.*, 129, 9572–9573, 2007, doi: 10.1021/ja073053q.
101. Chen, G., Yu, J., Gu, Z., Glucose-responsive microneedle patches for diabetes treatment. *J. Diabetes Sci. Technol.*, 13, 41–48, 2019.
102. Gao, Y., Wong, K.Y., Ahiabu, A., Serpe, M.J., Sequential and controlled release of small molecules from poly(: N-isopropylacrylamide) microgel-based reservoir devices. *J. Mater. Chem. B*, 4, 5144–5150, 2016, doi: 10.1039/c6tb00864j.
103. Cao, Y., Chen, Y., Yu, T., Guo, Y., Liu, F., Yao, Y., Li, P., Wang, D., Wang, Z., Chen, Y. *et al.*, Drug release from phase-changeable nanodroplets triggered by low-intensity focused ultrasound. *Theranostics*, 8, 1327–1339, 2018, doi: 10.7150/thno.21492.
104. Salgarella, A.R., Zahoranová, A., Šrámková, P., Majerčíková, M., Pavlova, E., Luxenhofer, R., Kronek, J., Lacík, I., Ricotti, L., Investigation of drug release

modulation from poly(2-Oxazoline) micelles through ultrasound. *Sci. Rep.*, 8, 9893, 2018, doi: 10.1038/s41598-018-28140-3.
105. Gai, M., Frueh, J., Tao, T., Petrov, A.V., Petrov, V.V., Shesterikov, E.V., Tverdokhlebov, S.I., Sukhorukov, G.B., Polylactic acid nano- and micro-chamber arrays for encapsulation of small hydrophilic molecules featuring drug release via high intensity focused ultrasound. *Nanoscale*, 9, 7063–7070, 2017, doi: 10.1039/c7nr01841j.
106. Pearson, S., Vitucci, D., Khine, Y.Y., Dag, A., Lu, H., Save, M., Billon, L., Stenzel, M.H., Light-responsive azobenzene-based glycopolymer micelles for targeted drug delivery to melanoma cells. *Eur. Polym. J.*, *69*, 616–627, 2015, doi: 10.1016/J.EURPOLYMJ.2015.04.001.
107. Hardy, J.G., Larrañeta, E., Donnelly, R.F., McGoldrick, N., Migalska, K., McCrudden, M.T.C., Irwin, N.J., Donnelly, L., McCoy, C.P., Hydrogel-forming microneedle arrays made from light-responsive materials for on-demand transdermal drug delivery. *Mol. Pharm.*, 13, 907–914, 2016, doi: 10.1021/acs.molpharmaceut.5b00807.
108. *A Phase I Clinical Trial of Neoadjuvant Chemotherapy With/Without SPIONs/SMF for Patients With Osteosarcoma*, -Full Text View-ClinicalTrials.Gov.
109. *In Vivo Characterization of Inflammation with Ferumoxytol, an Ultrasmall Superparamagnetic Iron Oxide Nanoparticle, on 7 Tesla Magnetic Resonance Imaging*, Full Text View-ClinicalTrials.Gov.
110. *Radiotherapy with Iron Oxide Nanoparticles (SPION) on MR-Linac for Primary & Metastatic Hepatic Cancers*, -Full Text View-ClinicalTrials.Gov.
111. Supriya Bhatt, S., Thakur, G., Nune, M., Preparation and characterization of PVA/chitosan cross-linked 3D scaffolds for liver tissue engineering. *Mater. Today Proc.*, 2023, doi: https://doi.org/10.1016/j.matpr.2023.02.251.
112. Patel, K.N., Patel, J.K., Patel, M.P., Rajput, G.C., Patel, H.A., Introduction to hyphenated techniques and their applications in pharmacy. *Pharm. Methods*, 1, 2–13, 2010, doi: 10.4103/2229-4708.72222.
113. Carson, W.W., Kearney, T., Zhou, M., An LC–IR hyphenated approach to characterize polymeric excipients in pharmaceutical formulations. Available online: https://www.chromatographyonline.com/view/lcir-hyphenated-approach-characterize-polymeric-excipients-pharmaceutical-formulations (accessed on 22 March 2023).
114. Fan, Y., Patent Application Publication (10) Pub. No.: US 2021/0079170 A1, 2021.
115. Lewis, J.A., Gladman, A.S., Matsumoto, E., Mahadevan, L., Hydrogel composite ink formulation and method of 4D printing a hydrogel composite structure, Patent Application Publication US20180251649A1, 2018.
116. Li, Y., Zheng, W., Zhang, J., Xu, L., Li, B., Dong, J., Gao, G.-L., Jiang, Z., 3D printed thermo-responsive electroconductive hydrogel and its application for motion sensor. *Front. Mater.*, 10, 2023, DOI: 10.3389/fmats.2023.1096475.
117. Neufeld, L., Yeini, E., Pozzi, S., Satchi-Fainaro, R., 3D bioprinted cancer models: From basic biology to drug development. *Nat. Rev. Cancer*, 22, 679–692, 2022, doi: 10.1038/s41568-022-00514-w.

Index

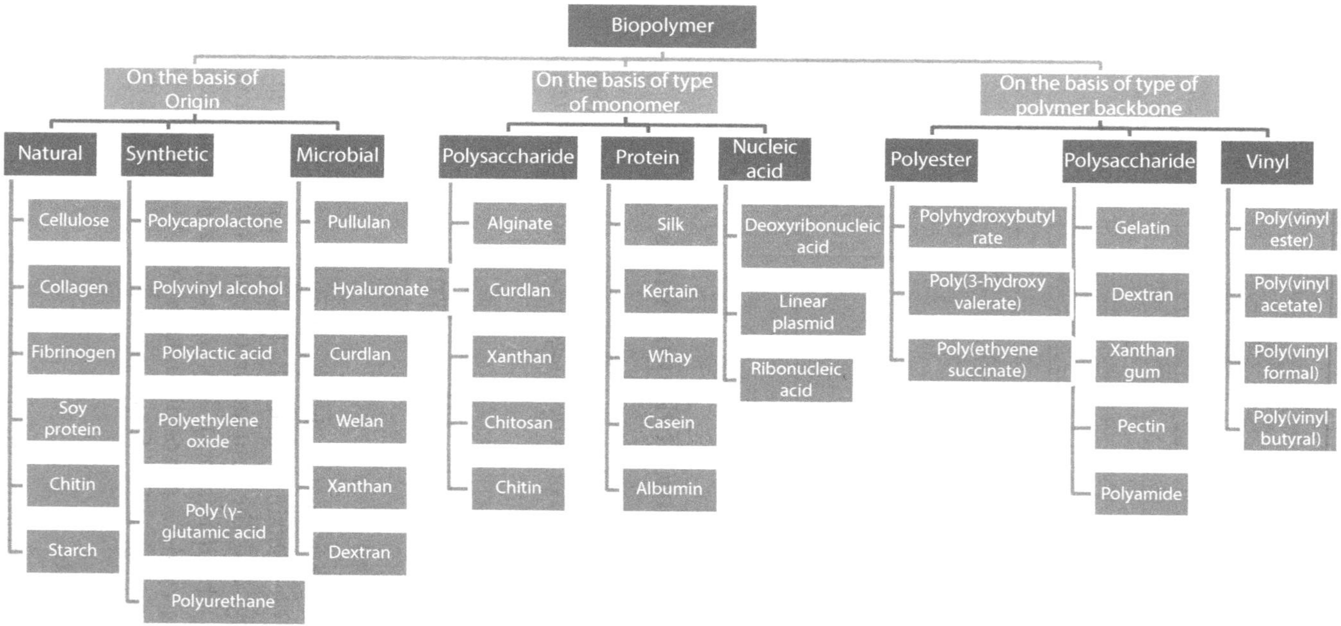

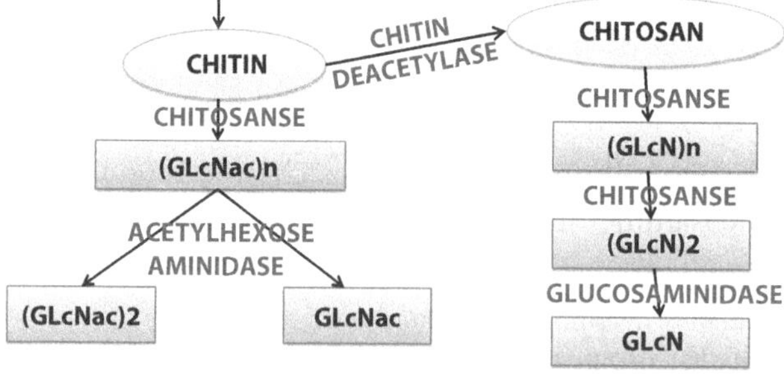

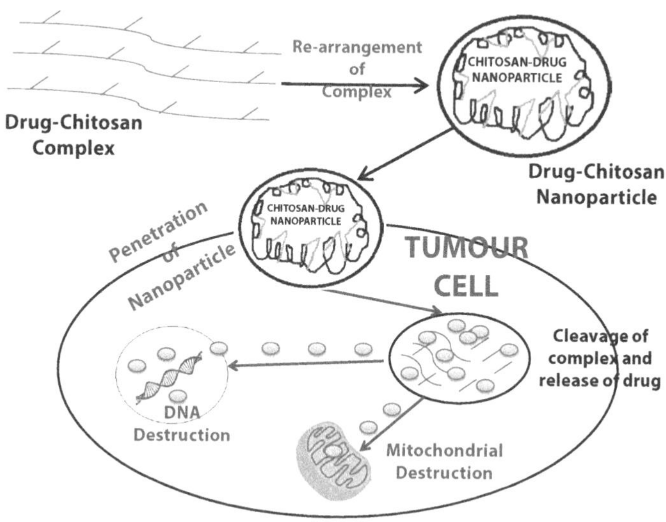

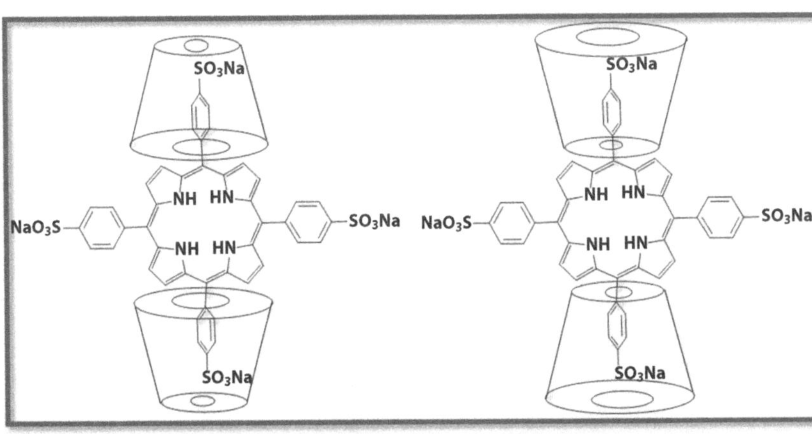

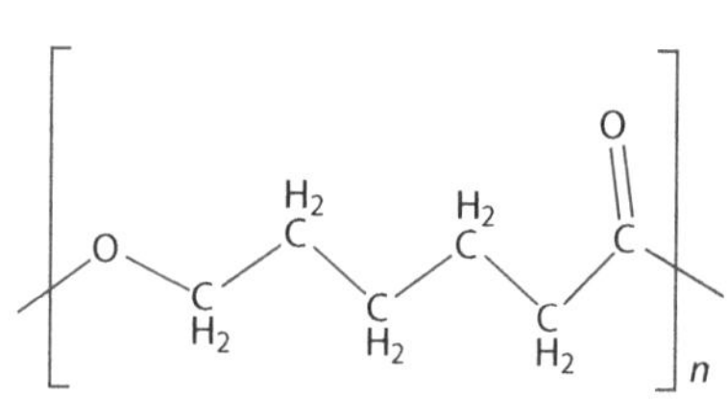

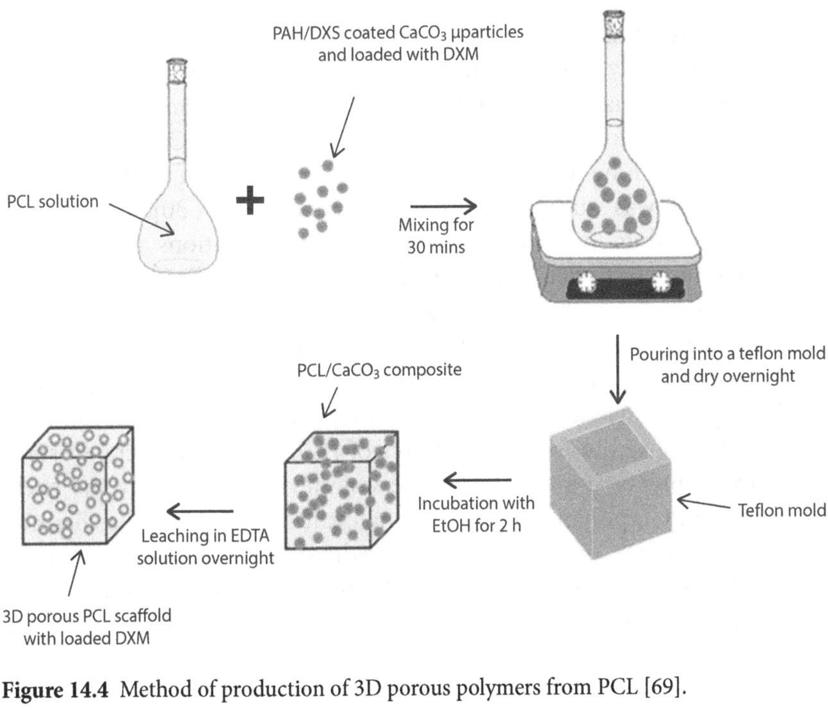

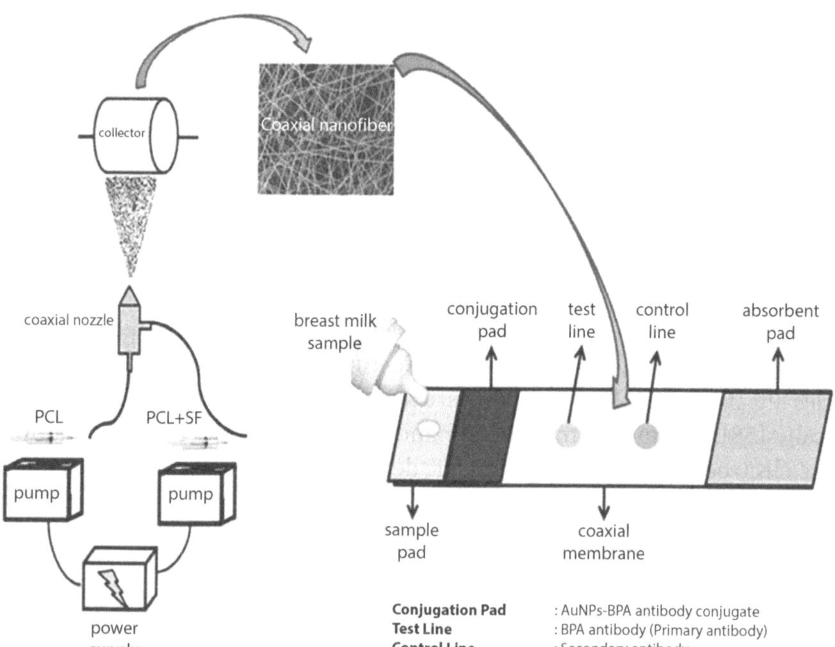

105. Gai, M., Frueh, J., Tao, T., Petrov, A.V., Petrov, V.V., Shesterikov, E.V., Tverdokhlebov, S.I., Sukhorukov, G.B., Polylactic acid nano- and micro-chamber arrays for encapsulation of small hydrophilic molecules featuring drug release via high intensity focused ultrasound. *Nanoscale*, 9, 7063–7070, 2017, doi: 10.1039/c7nr01841j.
106. Pearson, S., Vitucci, D., Khine, Y.Y., Dag, A., Lu, H., Save, M., Billon, L., Stenzel, M.H., Light-responsive azobenzene-based glycopolymer micelles for targeted drug delivery to melanoma cells. *Eur. Polym. J.*, *69*, 616–627, 2015, doi: 10.1016/J.EURPOLYMJ.2015.04.001.
107. Hardy, J.G., Larrañeta, E., Donnelly, R.F., McGoldrick, N., Migalska, K., McCrudden, M.T.C., Irwin, N.J., Donnelly, L., McCoy, C.P., Hydrogel-forming microneedle arrays made from light-responsive materials for on-demand transdermal drug delivery. *Mol. Pharm.*, 13, 907–914, 2016, doi: 10.1021/acs.molpharmaceut.5b00807.
108. *A Phase I Clinical Trial of Neoadjuvant Chemotherapy With/Without SPIONs/SMF for Patients With Osteosarcoma*, -Full Text View-ClinicalTrials.Gov.
109. *In Vivo Characterization of Inflammation with Ferumoxytol, an Ultrasmall Superparamagnetic Iron Oxide Nanoparticle, on 7 Tesla Magnetic Resonance Imaging*, Full Text View-ClinicalTrials.Gov.
110. *Radiotherapy with Iron Oxide Nanoparticles (SPION) on MR-Linac for Primary & Metastatic Hepatic Cancers*, -Full Text View-ClinicalTrials.Gov.
111. Supriya Bhatt, S., Thakur, G., Nune, M., Preparation and characterization of PVA/chitosan cross-linked 3D scaffolds for liver tissue engineering. *Mater. Today Proc.*, 2023, doi: https://doi.org/10.1016/j.matpr.2023.02.251.
112. Patel, K.N., Patel, J.K., Patel, M.P., Rajput, G.C., Patel, H.A., Introduction to hyphenated techniques and their applications in pharmacy. *Pharm. Methods*, 1, 2–13, 2010, doi: 10.4103/2229-4708.72222.
113. Carson, W.W., Kearney, T., Zhou, M., An LC–IR hyphenated approach to characterize polymeric excipients in pharmaceutical formulations. Available online: https://www.chromatographyonline.com/view/lcir-hyphenated-approach-characterize-polymeric-excipients-pharmaceutical-formulations (accessed on 22 March 2023).
114. Fan, Y., Patent Application Publication (10) Pub. No.: US 2021/0079170 A1, 2021.
115. Lewis, J.A., Gladman, A.S., Matsumoto, E., Mahadevan, L., Hydrogel composite ink formulation and method of 4D printing a hydrogel composite structure, Patent Application Publication US20180251649A1, 2018.
116. Li, Y., Zheng, W., Zhang, J., Xu, L., Li, B., Dong, J., Gao, G.-L., Jiang, Z., 3D printed thermo-responsive electroconductive hydrogel and its application for motion sensor. *Front. Mater.*, 10, 2023, DOI: 10.3389/fmats.2023.1096475.
117. Neufeld, L., Yeini, E., Pozzi, S., Satchi-Fainaro, R., 3D bioprinted cancer models: From basic biology to drug development. *Nat. Rev. Cancer*, 22, 679–692, 2022, doi: 10.1038/s41568-022-00514-w.

Index

Printed in the USA/Agawam, MA
September 16, 2026

910661.163